The Great
Contemporary
Issues Series
Set 1 Vol. 15

UPDATE
1980

The Series

The Great
Contemporary
Issues Series
Set 1 Vol. 15

UPDATE
1980

The New York Times

ARNO PRESS

NEW YORK/1980

Library of Congress Cataloging in Publication Data
Main entry under title:

Update 1980
    (The Great contemporary issues series; set 1,
v. 15)
    Bibliography: p.
    Includes index.
    1. History, Modern—1945-   —Addresses,
essays, lectures. I. Keylin, Arleen, II. Bowen,
Douglas John. III. Series: Great contemporary
issues; set 1, . 15.
D848.U62   909.82   79-27511
ISBN 0-405-13086-4

Manufactured in the United States of America.

The editors express special thanks to The Associated Press, United Press International, and Reuters for permission to include in this series of books a number of dispatches originally distributed by those news services.

Edited by Arleen Keylin and Douglas John Bowen.

Contents

Publisher's Note About the Series

It would take even an accomplished speed-reader, moving at full throttle, some three and a half solid hours a day to work his way through all the news The New York Times prints. The sad irony, of course, is that even such indefatigable devotion to life's carnival would scarcely assure a decent understanding of what it was really all about. For even the most dutiful reader might easily overlook an occasional long-range trend of importance, or perhaps some of the fragile, elusive relationships between events that sometimes turn out to be more significant than the events themselves.

This is why "The Great Contemporary Issues" was created—to help make sense out of some of the major forces and counterforces at large in today's world. The philosophical conviction behind the series is a simple one: that the past not only can illuminate the present but must. ("Continuity with the past," declared Oliver Wendell Holmes, "is a necessity, not a duty.") Each book in the series, therefore has as its subject some central issue of our time that needs to be viewed in the context of its antecedents if it is to be fully understood. By showing, through a substantial selection of contemporary accounts from The New York Times, the evolution of a subject and its significance, each book in the series offers a perspective that is available in no other way. For while most books on contemporary affairs specialize, for excellent reasons, in predigested facts and neatly drawn conclusions, the books in this series allow the reader to draw his own conclusions on the basis of the facts as they appeared at virtually the moment of their occurrence. This is not to argue that there is no place for events recollected in tranquility; it is simply to say that when fresh, raw truths are allowed to speak for themselves, some quite distinct values often emerge.

For this reason, most of the articles in "The Great Contemporary Issues" are reprinted in their entirety, even in those cases where portions are not central to a given book's theme. Editing has been done only rarely, and in all such cases it is clearly indicated. (Such an excision occasionally occurs, for example, in the case of a Presidential State of the Union Message, where only brief portions are germane to a particular volume, and in the case of some names, where for legal reasons or reasons of taste it is preferable not to republish specific identifications.) Similarly, typographical errors, where they occur, have been allowed to stand as originally printed.

"The Great Contemporary Issues" inevitably encompasses a substantial amount of history. In order to explore their subjects fully, some of the books go back a century or more. Yet their fundamental theme is not the past but the present. In this series the past is of significance insofar as it suggests how we got where we are today. These books, therefore, do not always treat a subject in a purely chronological way. Rather, their material is arranged to point up trends and interrelationships that the editors believe are more illuminating than a chronological listing would be.

"The Great Contemporary Issues" series will ultimately constitute an encyclopedic library of today's major issues. Long before editorial work on the first volume had even begun, some fifty specific titles had already been either scheduled for definite publication or listed as candidates. Since then, events have prompted the inclusion of a number of additional titles, and the editors are, moreover, alert not only for new issues as they emerge but also for issues whose development may call for the publication of sequel volumes. We will, of course, also welcome readers' suggestions for future topics.

Introduction

1979, the last turbulent year of a tempestuous decade, may also have been a significant and, alas, foreboding turning point. There were ominous developments, most particularly on the international scene, when accelerated in intensity and blossomed into major world problems during the autumn and winter months of 1979. At its outset, however, they were only vaguely definable, still in embryo.

Very special events with far-reaching consequences rendered the delicate balance between the superpowers and their respective blocs ever more tenuous and increasingly threatened by political, economic and religious factors in the third, heretofore, less committed bloc of neutral and non-aligned states. The harsh reality that the entire structure of "free society" (every aspect of it) was dependent upon oil from the Middle East manifested itself in crisis terms in the United States, Western Europe and Japan. Indications of such dependence were not, of course, entirely couched from astute public view earlier in the decade. Nonetheless the horrifying implications of vulnerability to OPEC manipulations — price increases, diminutions of production, and threats of oil cut-offs — became crystal clear as more and more Americans and their political and financial leaders became aware of and felt the tangible consequences of this meance. This was not the straight-forward albeit alarming prospect of military confrontation. This was the spectre of the gradual undermining of the economic, political and social fabric of Western industrial societies because of the diminution of their primary energy sources and, concurrently, the rapid inflationary effects of the alarming rise in oil prices.

Energy, in fact, became one of the most controversial issues in American politics, and energy costs began to nibble away at the purses and the lifestyle of Americans of every class. Gasoline shortages and heating oil prices were dark consequences of the weakened position of the United States, no longer self-sufficient enough to meet even its most basic energy need. Grim scenarios were conceived by policy-makers and strategists: the oil-producing countries could force the West to its knees merely by raising the price of oil to unbearable levels; OPEC could profoundly influence Western political and diplomatic decisions, using oil as blackmail; the USSR could decide to strike a mortal blow by cutting off the West's "lifeline" — but that would certainly result in the most unthinkable scenario of all — nuclear confrontation! America's dependence upon OPEC for oil most assuredly underlay the concern of U.S. diplomats politicians and strategists as they followed developments in the Middle East. Clearly, Western Europe and Japan shared this concern.

During the final weeks of 1978, Israel and Egypt seemed to be on the verge of an almost miraculous agreement, initiated and inspired by Sadat's dramatic overtures to Israel and his courageous visit to Jerusalem, yet forcefully nurtured and carefully guided by President Carter and his aides at a series of meetings. In March 1979, as a result of agreements reached at Camp David and subsequent negotiations, Egypt and Israel signed a formal treaty, terminating thirty years of bitter hostility and violence. The coming together of these two states, however, failed to resolve several basic problems at the very heart of the Arab-Israeli controversy. The Palestinian problem was in no way reconciled. When and where could the Palestinians, now in exile throughout the area and the world over, establish their own sovereign state? Was the P.L.O. a legitimate representative of that people or merely a well-organized (financed by the Soviet bloc) terrorist group? Even though a P.L.O. observer actively participated in sessions and discussions at the United Nations, U.S. Ambassador Andrew Young resigned his post as the result of a furor over talks with the representatives of the organization. Egypt, too, felt the sting of isolation. As a result of the Egyptian-Israeli accord, Sadat became the "outcast" of the Arab world, and Israel, having gained relative security on one of her frontiers, faced more determined foes on her other borders. The issue of oil was infinitely more clear. Although Israel could provide the United States with bases and a well-equipped military force, it had no oil. Egypt had potential oil resources which, however, would always be limited. Sadat's control over the destiny of his people might also prove to be limited. The Arab states, no matter what their political affiliation, remained hostile to Israel and overtly, at least, loyal to the Palestinian cause.

Not bordering on Israel, but decidely of importance to her security due to the Shah's policy of normal relations and trade between the two nations, was Iran. Because of the successful revolution of militant Iranian Moslems, the Shah's controversial regime ended abruptly, as did his policy vis à vis Israel. This international change proved to be of momentous import not only to Israel but to the United States, the stability of the Middle East, and relations between the superpowers. How ironic it is that when Iranians rushed the United States embassy in February, one hundred Americans were released at Khomeini's insistence, or that, almost at the same time, an American envoy was shot in an increasingly turbulent Afghanistan. In November, when Moslem students took American hostages after occupying the same embassy, the act was endorsed by Khomeini and release of the hostages was predicated on the return of the Shah to Iran together with his legendary fortune. What had previously been reported as Iran's cultural revolution rapidly exploded into an international crisis. By the end of 1979, Washington was planning contingency military, naval and economic moves in the area to release the hostages and guarantee the flow of oil to the West. Subsequent events in Afghanistan added new and dangerous dimensions to the volatile situation.

Dramatic trends already underway continued in 1979 and drew the United States and Communist China a few steps closer, at least in terms of economic and business co-interests. Because of the rapid modernization sought by his regime, China's Minister Deng Xiaoping paid a coast-to-coast visit to the United States for a friendly first-hand assessment of American technology suitable for Chinese needs. Foreign business with China increased markedly as major American firms flocked to Peking in search of new markets and new products for their consumers. China's altered economic policies and outlook brought domestic changes, and reports of small yet visible pockets of private entrepreneurship reached the outside world. On the political front, however, there was still disagreement with the West, particularly when Chinese troops and airplanes invaded Vietnam, now under the influence of the USSR. The Soviets delivered an ultimatum to Peking. The United States urged a prompt withdrawal of Chinese forces and Hanoi pleaded for help at the United Nations. China eventually pulled her forces back and seemed content, publicly at least, to have taught Hanoi a lesson. A conflict between the USSR and its populous Asiatic neighbor was, for the time being, happily avoided, but warfare still ravaged the hapless peoples of Cambodia as the Soviet-backed Vietnamese armies invaded their country and established a new regime. The Pol Pot government, supported by China, withdrew to the jungles and mountains and continued to struggle for control and the continuation of its harsh agriculturization program which was reputed to have cost the lives of millions of Cambodian men, women and children.

On the African continent, superpower competition in the guise of surrogate military confrontatons was overshadowed by events in Rhodesia and Uganda although a new East German Afrika Korps demonstrated that East-West struggles in the area were far from diminished.

On January 30th, Rhodesian whites voted to accept the limited rule of a black majority, but the terms of such a government and the great privileges still to be granted to the white minority under it were vigorously attacked by Nkomo and his nationalist followers. From bases in Zambia and Botswana, they continued their raids into Rhodesia and invited reprisals by the Rhodesian military. This was the volatile situation even after the Muzorewa Party won a bare majority in a reorganized Rhodesian parliament in April, and the name of the country was officially changed to Zimbabwe at the end of the summer. Matters in the area worsened dramatically as military confrontations, terrorist acts and raids initiated by both sides continued to augment. The entire area's stability seemed to be in jeapordy and the confrontation threatened to spread into neighboring countries. Representatives of the Muzorewa Party and Nkomo's Black Nationalists were invited to London for a series of negotiations under the aegis of Margaret Thatcher's newly formed government. Despite deeply-felt antagonisms and fears, particularly on the part of Zimbabwe's white minority, the guerillas accepted a charter proposed by the British. By mid-November an accord had been reached on the terms and mechanics of transition to Black rule. Only the details of a cease-fire remained an open issue.

Idi Amin's insane dictatorship of Uganda, marked by corruption and officially-sanctioned murder, came to a final and appropriately bloody end as Tanzanian forces occupied the capital. The Tanzanian troops uncovered horrifying evidence of Amin's crimes against his people and, using Ugandan exiles, attempted to establish a provisional government under the leadership of Yusufu K. Lule. At the end of the year, Uganda was still battling the bitter Amin legacy. Amin, who had been supported by the Soviet bloc, seemingly found refuge in Libya.

The political scene in the United States was greatly affected by international events, particularly so in 1979 inasmuch as it heralded the vague beginnings of the 1980 presidential campaign. One central issue seemed to be the cost of energy and its obvious relation to skyrocketing inflation as OPEC nations continued to raise the price of oil. There was an increasing awareness on the part of the American Government and people that the United States was dangerously dependent upon foreign energy sources. Americans looked to the President for a solution. Another important issue manifested itself in a debate over whether or not the American Congress ought to ratify the newly-signed SALT agreements. These agreements were an integral part of Carter's foreign policy. Questioning their advisability would lead, therefore, to a re-examination of US -USSR relations. The Administration attempted to win support for SALT on Capitol Hill and across the nation. By the spring of 1979, the greatest challenge within the Democratic Party to the Carter bid for re-election came from Edward Kennedy. Kennedy, aided by his family mystique, was ahead in the polls. Within the Republican Party, competitors for the Presidential candidacy were the seemingly cool and self-controlled Howard Baker, tough-talking millionaire John Connally, former C.I.A. director George Bush, and the perennial conservative Ronald Reagan. An avenue of attack for all of these men, Democrats and Republicans, was Carter's image as a strong leader. Connally officially opened the mass media aspect of the 1980 campaign with a television commercial on October 30th. When the American embassy and fifty hostages fell into the hands of Iranian students, a grave international crisis ensued which demanded Carter's full attention and precluded his participation in political debates especially in the realm of Middle Eastern policy. As he advocated stronger and stronger measures to counteract the taking of the hostages yet demonstrated patience in negotiating their release, Carter's public image as a resolute and capable leader rose from coast to coast. Yet, even as the crisis in Afghanistan marked the turn-of-the-year, it seemed inevitable that Carter would have to go on the campaign trail and debate his radically changing outlook on defense and foreign policy with his competitors.

New trends in city living became more and more apparent in 1979. An affluent, talented young elite was beginning to re-examine the attractions of urban life and to reassess the advisability of returning to core cities in order to fulfill personal ambitions in the company of peers. As racial strife continued to haunt Boston, and decay eroded New York's South Bronx, areas of other urban centers were being cleaned up, rebuilt, revital-ized. Long Beach, California, too long in the shadow of giant Los Angeles, attempted to assert its own image by creating a new civic center. Los Angeles, in turn, where the need to traverse great distances by car was legendary, revealed plans for a rapid transit "people mover." A Chicago bank offered easy loans to rebuild deteriorating neighborhoods and in Salt Lake City, officials were proudly anticipating a lavish arts center. In New York, the fight against pornography and drug traffic in the Times Square area seemed to be gathering momentum with each theater restoration and each peep-show shop closing.

Crime and its punishment continued to preoccupy law enforcement agencies and government officials as well as ordinary citizens. One of the major issues of the year was the reinstatement of capital punishment in several states which led to executions in Nevada and Florida. The debate over capital punishment has always been agitated in the United States yet, in 1979, many Americans seemed to favor extreme measures as the only answer to a rising national crime rate, particularly in smaller cities and rural areas. Drug-related crimes also increased as new sources for marijuana, opium and cocaine opened up to American dealers and their business boomed from coast to coast. In addition, health and law enforcement authorities were increasingly concerned about the over-use of prescription drugs, the fact that Americans, in greater numbers, were becoming dependent upon dangerous and often lethal drugs often prescribed by less-than-ethical physicians.

1979 was a year in which Jane Margaret Byrne became the Mayor of Chicago and immediately displayed her astute political know-how; when Margaret Thatcher was elected British Prime Minister and took her position as a world leader; and when the American women's movement demonstrated its strength by advocating the right of battered wives to self-defense and attacking the sexual abuse and humiliation of women by pornographers. At the same time, Pope John clearly stated his opposition to women as priests in the Catholic Church and Iranian women retreated in the face of Khomeini's religous revival. It was a year of tremendous strides in astronomy as scientists continued to investigate the size, the age and the rate of the expansion of the universe. Voyager reached Jupiter in March and humanity gazed upon the eddying gases of the planet's giant red spot and delved into the secrets of its four, astoundingly diverse moons. The *Star Trek* crew voyaged to movie screens nation-wide in a $40 million film version of the popular television series. The muppets, on the other hand, were far more conservative travelers. They went as far as Hollywood!.

Any introduction to this tumultuous year can only provide an outline to the wealth of information contained in this volume. The articles, reprinted exactly as they first appeared in *The New York Times* are an invaluable first-hand look at history as it unfolded in 1979.

Sanford Louis Chernoff

Drugs

A More Potent Marijuana Is Stirring Fresh Debates

Most marijuana now being sold throughout the United States is three to 10 times more potent than the marijuana that was sold two years ago, according to Federal drug enforcement officials and pro-marijuana groups.

Both sides agree this is because most of the marijuana being sold today is from Colombia. Colombian marijuana is stronger than the marijuana from Mexico, which once dominated the United States market. A.though some Mexican marijuana is still sold domestically, especially in California and the Southwest, that share of the market has dwindled from about 75 percent five years ago to as little as 10 percent today.

Mexican Supply Diminished

The Mexican supply diminished because of intensive policing efforts by customs officials at the Mexican-American border, because of the disclosure that Mexican marijuana was being sprayed with a dangerous herbicide, paraquat, and because of an increased demand for the stronger Colombian strain by users.

"Almost all of the pot sold here is now Colombian," a New York City marijuana dealer said recently, "and that's quickly becoming true everywhere. People won't accept that weak Mexican stuff anymore."

It is no more difficult to purchase marijuana now than it was when the Mexican variety was prevalent, according to many dealers and users. The price of an ounce or "lid" of marijuana has, however, increased from $20 to $30 two years ago to $40 to $50 today. Some exotic strains (such as the Hawaiian-grown Maui Wowie) sell for as much as $150 an ounce.

The potency of marijuana is determined by the amount of tetrahydrocannabinol (THC), the hallucinogenic agent, it contains. Mexican marijuana contains 1 to 2 percent of THC, whereas the Colombian strain has from 6 to 10 percent.

Both advocates and foes of marijuana announced major policy shifts in recent weeks as a result of spreading public knowledge about the strength of the current marijuana.

Federal drug enforcement officials have adopted a new aggressive stance against use of the drug, believing that recent negative reports about marijuana use have made the pro-marijuana movement more vulnerable than ever.

"We've definitely mounted a much more significant effort to interdict the flow of marijuana recently," said Peter Bensinger, administrator of the Federal Drug Enforcement Administration.

Meanwhile, the National Organization for the Reform of Marijuana Laws (NORML) decided last week to switch its strategy from working for decriminalization of marijuana use to supporting full legalization.

"It is more apparent than ever that full legalization and hence product control on the open marketplace is the only way to prevent misuse of the drug," said Frank Fioramonti, a member of the organization's board of directors.

School officials have expressed concern that teen-agers and pre-teen-agers may be underestimating the strength of the marijuana they are smoking.

"We don't like to talk about it," said a counsellor at a public school in Manhattan, "but we're getting a lot of kids flipped out and bummed out — sick and in serious trouble in the middle of the school day — and they say all they did was smoke some pot. We don't know if it's just the pot or if someone slipped something into it."

Spokesmen for the government and for NORML strongly disagree about how much the stronger marijuana should affect traditional medical and moral assumptions about the drug and about whether it is realistic to think that the marijuana industry can be suppressed.

Earlier Thinking Held Outdated

"Everyone's earlier thoughts that marijuana is harmless are now clearly incorrect," said Mr. Bensinger recently. " While we can't shut down the marijuana industry overnight, we can make a renewed effort against the drug by better arrangements with foreign governments and stronger sentences for major marijuana traffickers."

"This is a new scare tactic, just paranoid overreaction," said Keith Stroup, the outgoing director of NORML. "The dope is definitely stronger, and it's good that people realize it, but it's not the killer drug the government is making it out to be."

Mr. Stroup also denied a recent report in The New York Post that he was leaving his post because of chronic bronchitis caused by his use of marijuana. "I've never had bronchitis, chronic bronchitis or doubts about pot, " Mr. Stroup said. "In fact, I'm smoking a joint right now."

"It's insane to think that the marijuana industry can be suppressed," said Mr. Fioramonti, the NORML official. "It's so entrenched in the growing areas that it has raised many countries from below the poverty line to relative prosperity. What we will now try to persuade the general public of is that marijuana is here to stay, so let's get it off the black market and away from racketeers and organized crime."

The Government's own figures for the volume of the marijuana industry, which have ranged from $20 billion to $48 billion, are higher than NORML's estimate. The latter figure would put marijuana sales in third position on the "Fortune 500" list of corporations, just behind General Motors and Exxon.

Both sides seem to feel that the law on marijuana may be decided within the next two years.

"Legalization is a completely unrealistic proposal," said the drug agency's Mr. Bensinger. "We have already assigned more agents from our department to investigate marijuana traffickers. There's a misconception that we can't interdict the marijuana trade: But if someone had said that we could get Turkey to wipe out its poppy fields 10 years ago, who would have believed it?"

"We won the the first half of the war when President Carter advocated decriminalization in 1977," Mr. Fioramonti said, "but we have to be one step ahead of everyone else now. By the 1980 campaign, either we'll have made some serious inroads toward legalization with politicians and candidates, or the movement may be in serious trouble."

December 28, 1978

Fear of Increase In Use of Drugs By Police Cited

By LEONARD BUDER

Concern over a suspected increase in the use of marijuana and cocaine by police officers during offduty hours has been voiced in a report by the New York City Police Department.

"Although the problem does not appear to be widespread," the report declared, "this area has become one of increasing concern among some commanders." It said that increased drug usage by officers might be the result of "liberalized laws and current philosophies" on the use of narcotics.

The report, submitted to the Police Commissioner, Robert J. McGuire, by John Guido, the chief of inspectional services, was based on a two-month departmentwide assessment of major "corruption hazards" facing police in 1978 and on the measures now under way to combat them.

Chief Guido said yesterday that 10 police officers had been arrested last year for the use of and, in some instances, the sale among themselves of drugs, chiefly cocaine and marijuana. In both 1976 and

1977, four officers were arrested on these grounds.

The assessment also found that many precinct commanders regarded as a potentially serious hazard the sale or misuse by officers of confidential information obtained from police computerized-information systems. But the report added that there was no evidence that this was now occurring with "great frequency" and that new controls had been instituted to minimize the problem.

In one case under investigation by the state's special anticorruption prosecutor, a Brooklyn officer is believed to have sold names of luxury car owners obtained from the precinct computer to suspected thieves.

As did the reports for other recent years, the 1978 report said that "in general, the assessment indicates that there are no organized forms of corruption now nor is there any evidence of its reappearance in the immediate future."

But Chief Guido said that the number of complaints of corruption against police personnel rose substantially last year — to about 2,700 last year from 2,100 complaints in 1977. In 1973, in the aftermath of the Knapp Commission's finding of widespread and systematic police corruption here, the department received 3,395 complaints of alleged corruption.

Chief Guido said that while the department was not taking this increase lightly, the increase could also mean that officers and their supervisors were now more diligent in accepting and acting upon corruption complaints from the public.

Among the more traditional types of corruption hazards facing police officers cited in the new report were the theft of property in department custody, bribes offered by drug dealers to avoid arrest and gratuities and discounts given to the police by businesses.

January 23, 1979

Egypt Fights Opium Cultivation, On the Increase Among Villagers

By CHRISTOPHER S. WREN
Special to The New York Times

ASSIUT, Egypt — As the winter harvests get under way around this provincial capital, some of the fields watered by the Nile will blossom with vivid pink poppies cunningly concealed among the ripening cotton, wheat and other crops.

Before the illicit poppies can be slit and milked of their opium sap, a grim game of hide-and-seek takes place between khaki-clad policemen and heavily armed villagers in the outlying countryside, traditionally known for clannishness and resistance to authority.

Confrontations have sometimes escalated into battles lasting up to several days.

Opium growing in Egypt is as old as the pyramids, but it remains modest in comparison with operations in countries like Turkey, Burma or Mexico. Egyptian authorities are worried that opium planting is increasing, however, and they know that some international narcotics traffickers visited Egypt last year to check the potential.

Sami Farrag, a police general who heads Egypt's narcotics administration, said recently that the Government would propose legislation providing for confiscation of land where narcotic plants were found, in addition to the present fines and prison sentences of up to 25 years.

Unlikely to Deter Growers

The threat seems unlikely to deter the inhabitants of the poor and remote villages around Assiut, which are accustomed to fighting over land and water.

Feuds have given the region the highest crime rate in an otherwise largely violence-free Egypt. According to a top official in Cairo, 700 people were killed in one extended family vendetta in a village across the Nile from Assiut.

Police sources say that many farmers grow opium to pay for arms to protect

themselves. Last summer, a police search outside Assiut yielded 8,000 illegal weapons, mostly Soviet-designed automatic rifles bootlegged from Egyptian Army supplies.

National police units operating out of Assiut seized 200 acres of opium-producing poppies last year, according to Col. Mohammed Zahran, who is in charge of the regional crackdown on narcotics. An additional 50 acres were taken by the local police, he said.

A rise in seizures reflects what an Egyptian Government report said was an increase in opium cultivation. Colonel Zahran said his units seized 61 acres of poppies in 1977 and 55 acres in 1976.

Fired On by Growers

The tough police colonel, who has the build of a heavyweight wrestler, said that the raids were usually "very dangerous" since police were fired upon by growers.

He said that one policeman was killed last year in a clash near Tell el-Amarna, a small village near here, but he would not give other information on casualties.

Colonel Zahran said that 16 operations were carried out around Assiut last year. As for the suspects captured in each raid, he said: "We have never caught fewer than 20 and it has sometimes gone into the hundreds."

Cairo press reports suggest that the operations can be enormous. In one last year, 600 paramilitary policemen armed with automatic rifles and light mortars took on opium growers. In another, several thousand policemen swept through cave-pocked desert hills to flush out local gangs.

Such clashes are not new. In 1970, the police called upon artillery and armored vehicles in a four-day battle before overrunning a hashish plantation in the El Ghabayem area of Assiut Province.

Growers in the Nile Valley now seem to be switching from hashish, which is said

The New York Times

to be of mediocre quality, to opium, which locally has a morphine content as high as 12 percent. The usual content runs 8 to 10 percent.

Opium growing can yield up to $35,000 an acre at Cairo prices, according to a law enforcement source. This makes it attractive to poor villagers who are otherwise lucky if they earn a few dollars a week. What is not seized or sent to Cairo is consumed locally, either dissolved in tea or rolled up and chewed.

Western specialists believe that Egyptian opium is not yet being refined into heroin or exported. But authorities monitored the movements of some known American and European drug dealers who met local gangsters in Cairo last year. To help preclude such traffic, the United States Government has sent technical advisers to assist the Egyptians.

With an estimated 500,000 opium addicts and three million to five million hashish users, the country is an importer of narcotics. About three to six tons of opium and up to 200 tons of hashish are believed to be smuggled in annually.

January 28, 1979

They're Finding a Way Out of the Prescription Drug Trap

By LESLIE BENNETTS

She seems a more likely candidate for a country club tea than for drug addiction. A delicate 60-year-old who lives in Westchester, Margaret wears expensively understated clothes and speaks in a low, well-bred voice. Because she suffers chronic pain — from arthritis, severe migraine headaches and an agonizing pinched nerve, among other causes — she has long taken a regular dosage of analgesics. She hardly noticed when the amount began to creep insidiously upward.

"For over 40 years it didn't become a problem," she recalled. But last summer, deeply discouraged about her health, she found herself taking more drugs, more often: to mute the constant pain, to sleep, to cheer her up. "I started being unable to wait four hours for the next dose; I'd only wait three," confessed Margaret. "That's how I fell into the trap."

The trap of drug dependency is one most "respectable" people are loath to recognize, but one with which an increasing number are struggling. "Prescription drugs are the primary source of the drug abuse problem in this country," said Thomas Coffey, director of the Bureau of Prescription Analysis of the New York State Health Department. "You hear so much about illegal substances, but the real problem is with legal drugs. More people apparently die from Darvon than from heroin."

Report to Congress

According to a 1978 report to Congress by the Comptroller General of the United States, "The diversion and abuse of legal drugs may be involved in as many as seven of every 10 drugs reportedly being abused or resulting in deaths."

The National Institute on Drug Abuse says that in 1977, more than 120 million prescriptions were filled in this country for sleeping pills and tranquilizers alone. An estimated 51 million people have used tranquilizers prescribed for them by a doctor; about 28 million have used sedatives, and 17 million have resorted to stimulants of various kinds.

"Only recently has this problem come out of the closet," said Dr. Herbert Peyser, a psychiatrist and consultant to the Smithers Alcoholism Rehabilitation Center in Manhattan, which also treats pill-dependent people. "It's a middle-class thing, quiet and secret, and there's a lot of shame attached to it. As a result, we really have no idea of the extent of the problem, except that it's a really serious one — and it's growing."

Until recently, pill dependents often felt there was nowhere to turn for specialized help. But in 1975 a group of dually addicted members of Alcoholics Anonymous — people dependent on both pills and alcohol — banded together to found a new organization called Pills Anonymous (P.A.). Though it is not affiliated with A.A., P.A. uses the same program and approach to aid those dependent on prescription drugs.

It was at P.A. that Margaret finally found help, after her doctor confronted her about her escalating drug abuse and recommended that she join the group. There Margaret not only found a network of supportive new friends, but she also learned about — and is now using — a variety of other pain treatments that don't involve drugs.

"I never could have done it alone," she maintained. "Initially it was difficult for me to identify with the group. Unlike some others, I had valid reasons for taking medicine. But no matter what the genesis of the problem, the end result is the same."

Although many people, like Margaret, turn to pills for legitimate medical reasons, more than 30 million Americans admit to having used sedatives, tranquilizers or stimulants "for recreational purposes," according to the National Institute on Drug Abuse. That was how Steven, a ruggedly handsome 30-year-old lawyer, got started eight years ago. He would take a Quaalude or a Seconal on a Saturday night, "just for fun" — a habit that eventually cost him two jobs and four wrecked cars, among other things.

Steven's mother finally became so alarmed by his condition that she flew to New York from Florida and prodded him to attend a P.A. meeting. That was in September, and since then, Steven said happily, he has given up pills with the help of the group and has picked up the threads of his broken life.

At a recent P.A. meeting, many members, including Steven, expressed anger at their doctors' role in facilitating the addiction process. "I started taking diet pills," said one young P.A. member, "but even after I lost the weight, my doctor was constantly prescribing amphetamines for me."

A Gradual Withdrawal

A 39-year-old social worker said she took tranquilizers, amphetamines and antidepressants prescribed by her doctor for 22 years. "I finally told him I really wanted to get off of them, and he said, nonsense!" she reported. "He said I needed to take what I was taking, and that there was no problem with my drug use." However, after joining P.A., this woman managed to withdraw gradually. "I kept cutting down the dosage, until at the end I was using a scissors to cut the pills into fine fragments," she said. "I did it because people in the group told me they had done it that way and survived."

Drug abuse professionals acknowledge that partial responsibility for the pill problem lies with the medical community. For one thing, many doctors are simply unaware of the pervasiveness and the symptoms of pill addiction. "Doctors are not well trained in detecting substance abuse," said Dr. Peyser, who has specialized in the field. "Emergency room lay people will often recognize the pill or alcohol abuser before a doctor will."

The Quaalude Problem

Mr. Coffey explained, "A doctor can be a party to this knowingly or unknowingly. There's a problem with people manipulating their physicians to get pills, and you have the patients who shop around from doctor to doctor. Some doctors are just careless in overprescribing. For example, an awful lot of the Valium problem happens this way; it's easier to write a prescription than to sit and talk to someone for half an hour about what's really bothering them," Mr. Coffey said. "And then there are the doctors prescribing for profit. Almost all the Quaalude problem is with doctors who are in it for profit, and Quaalude is a major problem now."

Pharmacists have a corresponding responsibility, he added. "Some of them don't question prescriptions when they should, and some dispense drugs illegally without a prescription." But even when they themselves are the perpetrators, doctors and pharmacists, said Mr. Coffey, can also fall victim to the problem, and may even be more susceptible because of their extensive access to drugs.

From Pills to Alcohol

Howard started taking pills when he was a medical student, to help him study. "I never got a prescription for a pill in my life; I got them all from the drug company representatives who give out samples," he said. That was the beginning of a harrowing 15-year nightmare. Despite intense paranoia, blackouts, trembling and vomiting, convulsions, car accidents and two suicide attempts, Howard insisted it never occurred to him he was addicted. "I always felt the only way I could deal with life was to take a pill," he explained. When he dropped out of medical school (after repeating the third year three times), his access to pills was drastically curtailed, and Howard did what many pill users do instead: He started drinking.

Many painful years later, he finally found help through Alcoholics Anonymous, where he learned that countless others are also dually addicted, and that the fact that he was able to substitute alcohol for pills made him no less chemically dependent.

Howard too is a member of P.A. now, and his wife goes once a week to PilAnon, an affiliated organization for the families of pill dependents.

Hearings End Monday

Both organizations, which can be reached by phone at 874-0700, were founded in New York, and at the moment have no chapters elsewhere, although the groups receive an increasing volume of mail requesting information on how to set up branches around the country. Trying to deal with the prescription drug problem from another angle, various state and Federal agencies are exploring different regulatory measures. The Senate Small Business Committee concludes hearings Monday on the use and abuse of Darvon and related drugs, including proposals to further restrict their sale.

But few experts in the field are hopeful such measures can solve the larger problem. "My personal opinion is that it will continue to grow," said Mr. Coffey. "Our society is becoming more dependent on drugs all the time, and people are less and less likely to tolerate any kind of pain or discomfort."

As Dr. Peyser said, "We have an increasing number of useful drugs which seem to alleviate this discomfort, but in many cases the cure is worse than the disease."

February 3, 1979

Cocaine: Big Business in Lawless Bolivian Lowlands

By JUAN de ONIS
Special to The New York Times

SANTA CRUZ, Bolivia, Feb. 7 — This long-isolated region of eastern Bolivia was opened up by an oil boom 20 years ago. Then came cotton and cattle ranches. Now it's cocaine.

Only 10 hours by truck from the coca-growing Chaparé region, and with airline connections to the rest of the world and little law enforcement, this prosperous city is a wide-open center for smuggling. It is a natural for cocaine traffic.

Under the palm trees along the streets, dealers sell cocaine-laced cigarettes — known as viboreros, or vipers — along with American cigarettes, Scotch whisky, Japanese radios, and all sorts of contraband from Panama.

Civic groups are alarmed at the increase in the local consumption of cocaine among young people, who picked up the practice from tourists. At the Cafe Pascana on the corner of the main square, long-haired youths in loose batik cotton shirts and beads bring a touch of San Francisco to this tropical city.

Small Fraction of Drug Trade

But, while chewing the coca leaf is common among the highland Indians of Bolivia, cocaine use is a small fraction of the drug trade, which is nearly all for export to the United States, Europe and, to a smaller extent, Brazil and Argentina. To make cocaine, coca leaves are dried and made into a paste, which is then purified in a laboratory.

The United States Government's efforts to control the narcotics flow from South America first identified Bolivia as a major source of the cocaine paste being shipped by processors of coca leaves here to laboratories in Colombia.

In 1976, then Secretary of State Henry A. Kissinger met with Gen. Hugo Banzer Suárez, then President of Bolivia, and proposed an effort, financed by $2 million in United States grants, to train and equip the Bolivian narcotics police for better enforcement.

A system to register coca producers was also proposed, and promises were made for a $45 million crop substitution program to replace coca bushes with citrus trees or coffee plants.

Two Main Growing Regions

About 13,000 coca producers have been

The New York Times / Feb. 14, 1979

registered in two growing regions: the Yungas region near La Paz, which mainly supplies local chewers of coca, and the Chaparé region near Cochabamba, the main source of the leaf for cocaine laboratories. Many growers are not registered.

But the new development in Bolivia is the production from paste of pure cocaine that can be transported in smaller quantities, and therefore more easily hidden, by American "tourists," by passengers on commercial airlines, or in hundreds of private planes flying over the uncontrolled Amazon region to Colombia.

The impact of this new export on Bolivia is enormous. The La Paz newspaper El Diario estimated that cocaine exports brought in $170 million last year. Some bankers say it was more like $200 million, which is equal to one-third of Bolivia's legal exports.

"There are so many dollars available that the rate of exchange doesn't go up despite internal inflation," said an American businessman here who asked not to be identified. "There are people who don't have any clear source of income who are suddenly buying up houses in Santa Cruz for $100,000 in cash. You can smell it everywhere."

Alerted by Large Seizures

Large seizures of pure cocaine carried by Bolivian travelers have alerted the United States Drug Enforcement Agency to the increase in Bolivian cocaine smuggling.

On Dec. 29, seven Americans flew from Santa Cruz to Rio de Janeiro, where a search turned up 39 pounds of Bolivian cocaine hidden in their chartered plane. On another occasion, 88 pounds of pure cocaine was found in the luggage of a family of five Bolivians when they arrived in Paris on a commerical airliner. They had flown from Santa Cruz to Rio de Janeiro, then to Paris, apparently without ever being searched.

A pound of pure cocaine sells in Santa Cruz for about $7,000, according to local narcotics agents, so these large shipments involve big investments in a market where payments are made in cash.

But there also seems to be a trickle of small shipments.

Hidden Under Vegetables

Julio Rivero, a former air force colonel who is now an air-taxi pilot at the El Trompillo terminal, flew five passengers to Puerto Suárez on the Brazilian border.

"One of the passengers was an old lady with a shopping bag filled with vegetables," Mr. Rivero said. "I was surprised that she would pay 1,500 pesos [about $70] for the flight, but she said she had s sickness in the family." Bolivian customs officials in Puerto Suárez found six pounds of cocaine under the vegetables.

The border with Brazil, and with Paraguay to the south, are wide open to smuggling. The vast expanses of jungle are streaked with rivers used to move cattle across the border. "It is one of the most open, uncontrolled borders in the world," said an American cattle dealer.

But for large-scale cocaine smuggling, the 200 private planes registered at Santa Cruz, as well as the Bolivian airline and military transport planes that fly regularly to Miami and Panama with stops in Colombia, are more suspect to narcotics agents.

Wrapped in Banana Leaves

In the Callatayud marketplace of Cochabamba on Mondays, a blocklong section is devoted to trading in 50-pound bales of coca wrapped in banana leaves. The bales sell for about 50 cents a pound and are trucked away by scores of buyers.

The shipments of coca leaves from the market, and their diversion to illegal laboratories, is controlled by a 10-man narcotics unit in the Cochabamba police department. In Santa Cruz, there are 20 men.

In Cochabamba, where cocaine-paste operations have been found in the city as well as in remote valleys, Capt. Edil Montellano, who heads the narcotics unit there, said the clandestine operators could offer a cooperative farmer "more money than he will earn in a year."

"When the peasants who work in a cocaine operation learn the art," Captain Montellano said, "they then make cocaine paste at home and supply the dealer."

Concern of Civic Group

The Pro-Santa Cruz Committee, a civic organization that manages the royalties paid to the region by the state oil company and rallies local efforts for development and urban planning, confronted Santa Cruz's new narcotics director, Col. Raúl Escobar, with the problem of enforcement.

Colonel Escobar told the committee that his men tried to jail known traffickers, but local judges were releasing 70 percent of those convicted on "medical permits." This allowed traffickers to disappear once out of jail, he said, or to continue their trade while reporting once a week to the jail.

Santa Cruz is believed to be a center of cocaine laboratories, and Colonel Escobar promised the committee he would try to determine why the town of Montero, about 15 miles to the north, was receiving weekly truckloads of coca leaves in amounts sufficient to supply a chewing population of 60,000 people. The town's population is only 12,000, and few lowlanders chew coca.

February 14, 1979

Colombia Cracks Down but the Marijuana Gets Through

By JUAN de ONIS
Special to The New York Times

BARRANQUILLA, Colombia, March 16 — The war between the armed forces and drug smugglers on the Atlantic coast of Colombia has produced huge seizures of marijuana and hundreds of arrests, but major shipments continue to move to the East Coast of the United States..

Since the armed forces began their campaign in October, 5,000 tons of marijuana has been seized, more than 750 people, including 30 Americans, have been arrested, and 16 planes have been impounded, according to the officer in charge, Gen. José María Villareal.

"The mafiosi are backed up against the wall like cats, but they have claws," said Col. Rafael Padilla Vergara, the Ministry of Defense information chief.

In this port city, a sprawl of shantytowns by the mangrove swamps and elegant homes on the higher ground, the underworld gossip is that the big operators, despite some losses, are still getting their shipments out.

Two Shipments Get Through

A source familiar with the fortunes of one marijuana smuggler said that since early January one shipment had been seized at sea but that two had reached the Florida coast, with a profit for the Colombian and his American partners of $10 million. In New York and Miami, primary markets for Colombian marijuana, street prices were not significantly higher this week than they were early this year. Drug Enforcement Administration sources said that the measures taken so far had not been effective in reducing supplies at the street level.

The campaign directed against the major marijauana smuggling region of Colombia, the wild Quajira peninsula east of here, at the base of the snowcapped Santa Marta mountain range, was ordered by President Julio César Turbay Ayala immediately after he took office in August as a sign of cooperation with the United States Government.

Diego Ascensio, the United States Ambassador, said that about $1.5 million in American aid had gone toward the campaign, mainly for equipment. Colombia has spent $3 million more.

The campaign was scheduled for the marijuana harvesting period in this region, before the start of the heavy rains. The Santa Marta range is now under heavy clouds and marijuana still in the fields will rot.

Crops Are Not Destroyed

But the army forces at Ríohacha and at an advance base at Buenavista near the mountains have not destroyed the crops in the field. The Colombian Government has refused to spray the marijuana plantations, some of which are extensive irrigated terraces, because of the threat of affecting food crops.

There may therefore be large amounts of marijuana in warehouses awaiting a relaxation of the military controls. Ambassador Ascensio has asked the United States to provide $1.3 million more to keep the campaign going for three or four months.

But influential voices are now being raised in Colombia against a continuation of the campaign. They say it hurts the peasants, who earn five times more from marijuana than from any other crop. It is estimated that 10,000 farmers grow marijuana and that the marketing provides a livelihood for 50,000 other people, from packers and truckers to the armed guards who protect the shipments.

In a television debate this week, Ernesto Samper Pizano, president of the National Association of Financial Institutions, suggested that the legalization of marijuana cultivation and export would provide tax revenues, eliminate the gangsters who now bribe judges and officials, or send gunmen to kill them, and reduce police expenses.

Legalization Is Called Immoral

Mr. Ascensio replied that any country that legalized marijuana exports would "become a pirate nation outside of international society." President Turbay and other administration officials have said that legalization would be immoral, but the issue has entered the area of public debate.

Army commanders are displeased with the drug enforcement task. "It is not something that we like because it exposes our officers and men to the danger of corruption," Colonel Padilla said. "A lieutenant who seized two trucks with several

The New York Times/March 22, 1979

Barranquilla is center of Colombian drive against marijuana traffic.

tons of marijuana was offered a cash bribe of $100,000 on the spot."

Colombia, in addition to providing more than half the marijuana entering the eastern United States, is the transit point for most of the cocaine that moves from the producing countries south of here, Peru, Bolivia and Ecuador. The drug traffic is known in local slang as "marimba," and those who are in the business are "marimberos".

The big operators are well known, but there have been no arrests or other measures taken against them.

In this city, where contraband is a way of life, with whisky, cigarettes, electronic gadgets and bolts of cloth on sale on the street, any official who interferes with corruption is exposed to gangland killing. Last week, Rodrigo Rodríguez Pacheco, the chief customs officer, was machinegunned to death in front of his home.

Big Dealers Are More Discreet

The guards for the drug operators have brought the violent habits of the Guajira badlands into this previously peaceful city. A woman who tooted her car horn at an intersection behind a new pickup truck, the vehicle favored by the Guajiro smugglers, was confronted by the driver with a revolver in hand.

The big dealers, who live in homes by the beach, are more discreet. They include Ivan Lafaurie, who jumped bail in Florida on a cocaine possession charge with his brother Paulo. As a result, they cannot enter the United States, but many other reputed "marimberos" have apartments in Florida and frequently travel on visas issued by the consulate here. These include César Molina, who passes as a rancher, Winces Velasco, who is a "export-import" merchant, and Julio Calderón, who owns a major share of the Colombian airline Aerocondor.

These are a few of the better known operators on the coast, but many more work out of Bogotá, Cali, Medellín and other interior cities.

One of the largest cocaine seizures by the special police antidrug unit took place at a large ranch near Turbo at the western end of the Atlantic coast on Feb. 24. A twin-engined Aero Commander carrying 500 pounds of refined cocaine and 15 tons of marijuana was seized. An American pilot arrested at the scene escaped from the local jail the same night.

Pilots Take Greater Risks

The Guajira campaign, backed up by air force Mirages and T-33 fighters, has forced pilots flying in from the United States to change tactics and assume greater risks by landing at night on airstrips lit by kerosene lamps.

Since October, 17 planes have been destroyed in accidents. In one a plane flew into the high-tension lines near Santa Marta on March 2, electrocuting the pilot and blacking out the coast.

But judging from the daily reports of seizures, sighting of aircraft without flight clearance, and the grapevine on the "marimba" circuit, the planes and the boats keep coming.

March 23, 1979

Washington Governor Signs Law Allowing Marijuana for Patients

OLYMPIA, Wash., March 28 (AP) — Washington has become the fifth state to permit the use of marijuana for medicinal purposes.

Gov. Dixy Lee Ray signed the legislation yesterday. It will provide marijuana free to cancer patients undergoing chemotherapy and radiation treatment and to glaucoma victims.

"I hesitated for a while, but I decided you are right," the Governor told Corleen Hapeman, a cancer patient from Seattle who had campaigned for the law.

"I decided that anything that could be used to relieve the pain and nausea is good," said Governor Ray. She said that the law was stringent enough to insure that it would not open the door to legalized recreational use of marijuana.

The other states permitting such use of marijuana are New Mexico, Florida, Illinois and Louisiana.

March 29, 1979

Grass Roots

Christopher Lehmann-Haupt

GRASS ROOTS. Marijuana in America Today. By Albert Goldman. 262 pages. Harper & Row. $12.95.

HOWEVER you feel about marijuana, you'll probably want to know where Albert Goldman stands on the subject before you consider reading his uneven but useful new book, "Grass Roots: Marijuana in America Today." Well, the best way to illustrate how he feels is to summarize a bit of first-person-singular reportage that appears at the end of Chapter Two, "The Threefold History of Hemp." It seems that a couple of years ago, Mr. Goldman, in a fit of hunger and curiosity, consumed three brownies cooked according to the recipe of Miss Alice B. Toklas. The initial rush he experienced was ecstatic enough, but during the downer that followed, Mr. Goldman became convinced that if he dared to swallow he would choke to death on his tongue.

Spreading his panic around, he persuaded a friend to call the local emergency number. Five minutes later there arrived a white fire department emergency vehicle, sirens screaming. Hearing Mr. Goldman's problem, a medical attendant suggested that if he was so frightened of swallowing his tongue, "Maybe you should hold it in your hand." "Right!" Mr. Goldman exclaimed. "Without a moment's hesitation, I reached up and took hold of the tip of my tongue with my left hand while I continued clasping [the attendant's] hand with my right. In this bizarre fashion, Professor Albert Goldman, A.M., Chicago; Ph.D., Columbia, Phi Beta Kappa; Who's Who in the East — the whole schmeer — rode to a drug clinic in an ambulance, tongue in hand."

In other words, Mr. Goldman has a sufficiently open mind about marijuana to have been willing to try it. Yet he is keenly aware of its hazards. Also, he has a sense of humor about it.

As far as how this book made me feel: All I had to learn was that smoking grass may be worse for you than smoking cigarettes, and the question was settled. But fortunately for my sense of reality, I didn't find that out until the last chapter, which evenhandedly sums up the pros and cons. By that point, I had learned that marijuana has to be judged in a lot broader context than whether it's good or bad for you.

Actually, what I had learned is not all that surprising. Most readers must surely have sniffed that "the age of pot is upon us," as Mr. Goldman puts it, although it may startle them somewhat to learn that there appear to be between 16 million and 20 million regular marijuana users in this country now, and that they will consume something like 50 million pounds of pot this year, at a street value of roughly $22.5 billion, or nearly four times the combined revenues of the two largest legitimate entertainment industries: movies and records. "Even the giant cigarette industry, averaging $16 billion annually, is far surpassed by dope."

Nor will it stun anyone to learn that all this grass is being smuggled into the country from somewhere by someone, at considerable risk and profit all along the line. Mr. Goldman makes it all seem more real and immediate by going out in the field, investigating the business at some risk to himself, and introducing us firsthand to everyone from the pot farmers in Santa Marta, Colombia, to the Drug Enforcement Administration investigators who try rather vainly to stamp out the smugglers. But there isn't much here that one couldn't construe for oneself with a small exercise of the imagination.

Indeed, the only historical chapter that is really must reading is Mr. Goldman's account of "Marijuana in the New World," wherein he struts his considerable knowledge of the Harlem jazz scene in the 1930's and reminds us that rock music is not the only legacy of black culture that this country was bequeathed in the 1960's.

'The Dreadful Example'

All the same, "Grass Roots" makes one reality of the marijuana underground powerfully clear. One may balance the pros and cons of pot down to the last gram, but the question of legalization has little to do with the facts on the negative side, that marijuana may be bad for people with heart disease, that smoking it is probably even more risky than inhaling cigarettes, that, contrary to cant, users often do combine it with alcohol and that, whether for physical or psychological reasons, a person who tries out pot is more likely than someone who does not to move on up to more powerful and dangerous drugs; or, on the positive side, that grass seems to be a specific for asthma, that it helps to relieve the nausea

7

associated with cancer chemotherapy, that for some reason it is useful in the treatment of glaucoma, or that it simply gives pleasure in a variety of forms.

The hard fact, according to Mr. Goldman, is that marijuana is here, and here to stay. Therefore: "Marijuana should be a source of pleasure, not of pain and shame. We should be free to cultivate and sell and buy this 'euphorant.' The only controls should be those imposed to protect the public from bogus or polluted merchandise. With the dreadful example of Prohibition before us, it seems nearly unthinkable that we should have done it again; taken some basic human craving and perverted it into a vast system of organized crime and social corruption. When will we learn that in a democracy it is for the people to tell the government, not for the government to tell the people, what makes them happy?"

June 25, 1979

Climbing Out of an Addiction

I'M DANCING AS FAST AS I CAN

By Barbara Gordon.
313 pp. New York: Harper & Row. $9.95.

HAVING BEEN THERE

Compiled and edited by Allan Luks.
Introduction by Ring Lardner Jr.
189 pp. New York: Charles Scribner's Sons. $8.95.

By JILL ROBINSON

David Schorr

BOOKS about the contemporary Fall, through alcohol and other drugs, and the movement toward recovery resemble fairy tales and other serious stories about the search for redemption. The emphasis is invariably on the more dramatic aspects of the Fall. The Devil is alluring; nightmares are more memorable than sweet dreams. Self-destruction is seductive. Survival seems mundane. Rimbaud, Baudelaire and certain modern alcoholic men of letters have inspired a new generation of plummeting rock balladeers and writers who have fallen in love with despair.

Writers such as Kate Braverman or Jonathan Schwartz, writing about the drug experience or being drunk, in their respective novels "Lithium for Medea" and "Distant Stations," enhance reality; and there is a tendency to hope one can do better writing through chemistry. But it is the art that makes the experience transcendent. Usage or pathology in the private life of the writer does not guarantee the art; and, sadly, recovery, an absence of pathology or an awareness of it do not guarantee art either. In "Asylum Piece and Other Stories" Anna Kavan, wild on heroin, is marvelous to read, and I'm not going to say she might have written a lot better straight: given her particular world, she might never have written without the goad of the drug. It is a miserable fact, though, that people who think they need alcohol or the other drugs the most are the ones who cannot control their use and are ultimately eaten alive.

Jill Robinson is the author of "Bed/Time/Story" and "Perdido."

This process of addiction is clearly shown in two new books. "I'm Dancing as Fast as I Can" by Barbara Gordon is a straightforward autobiography of a woman climbing out of an addiction to Valium, sometimes called a "soft" drug. There is nothing soft about its impact or danger. Miss Gordon, who is a producer of documentary films for television and the winner of four Emmy awards, has created here a kind of documentary of her own experience in the sense that the book is as much an indictment of psychiatry, ". . . a fragile science," and of some of its practitioners and institutions, as it is a terrifying personal story.

The subjugation, collapse and bitter fight for recovery of this competent, previously independent woman reminded me of "The Sleeping Beauty" — a story about a junkie if there ever was one — except the man she thinks of as her prince becomes her captor. She is paralyzed, a victim of infantilism because of her addiction; he beats her, imprisons her in her own apartment, ties her up and hides her from her friends. I thought, too, of "Gaslight," a touchstone of a title for many of us who, at our most vulnerable, attract "saviors" who are frequently more subtly, dangerously sick than we are — though some of them are licensed to treat us.

The observer, the journalist in Miss Gordon, was never asleep. She shows precisely what it is like to be in a mental institution: the way that life becomes one's only reality; the terror of moving out into the other world in new, tender skin. She quickly and astutely sizes up other patients, staff members and friends; her generosity when warranted is enormous, as is her justified anger. The woman who comes through here is not someone who speaks easily about herself, and one sees that the book is a genuine offering: it hides nothing, yet is not exploitive.

Her redemption comes partly through an understanding of the double bind Karen Horney writes about: the feverish, stimying love-work neurosis is believable and refreshingly free from easy, chic answers. We like to think that just getting off the drugs is enough; it is generally only the very beginning of the process, and Miss Gordon knows and explains this very well. None of us wants to grow the hard way; once an addict, you long to settle for the quick surcease, and it's easy to forget the bad parts; we fear certain revelations, long for postponement of pain. There are few women who will not settle for killing relationships, with destructive men or pills or liquor, rather than walk through loneliness. If you read this book, you will have met someone to guide you through that walk — there are no epiphanies, no miracles, only the promise of growth, and Miss Gordon makes us trust that promise.

"Having Been There," edited by Allan Luks, with a

fine introduction by Ring Lardner Jr., is a collection of stories by writers who seem to have achieved a perspective on the disease of alcoholism. The rationalizations, the denials and the nightmares will be familiar to any who know an alcoholic or have been practicing alcoholics. When I was drinking I would have thrown it against a wall. Recognizing the tricks, I would have nitpicked about the banality of a few of the lesser stories. I did not want to identify the problem, but these writers know all the games and show precisely how we play them with ourselves and, most hideously, with our families. I haven't read a much better story about a child with an alcoholic parent than "John Gardner and the Summer Garden" by Catherine Petroski. You will not readily forget the carving up of the dinner served this child.

It is all here: the man who longs to get home for dinner without one more drink, but cannot; the excuses of the talented woman who tries to control her husband's drinking by moving from place to place; the patient who tries to climb out of his own remorse to help a newer, sicker alcoholic; the flamboyant, cocksure attempt to help someone else, in which one falls on one's own face; the child's voice asking "Has Daddy called yet?" and the amazement when you do reach for coffee, rather than the booze you're longing for.

Story after story of the knowledge you have to have to stop (something I suspect nonalcoholics are not aware that we understand), the attempts, the failures and the tremulous, fear-filled successes. These writers in "Having Been There" know all the blind rages; the self-pity; the desolation — well shown in "A New Life" by Patricia Sieck — when the life you abandon turns and abandons you; and the catch phrases: "Just one more," "This is the last," "Tonight, but not tomorrow," "Maybe if I just get out of here," "I'll stop if I get this job" and everyone's favorite, "I'll stop if you just come back." Through all the stories there is the thread, the lure of denials, the bind of the disease with the flagrant symptom: "I drink. I'm different. I have a reason. But not a disease."

"Having Been There" is more valuable and precise than most textbooks. It is indispensable for anyone who wants to understand how the dynamic works, how it feels to try to stop and fail. The book deserves attention. ∎

July 1, 1979

F.D.A. Urges Banning Use of Amphetamines For Reducing Weight

By The Associated Press

WASHINGTON, July 16 — The Food and Drug Administration today proposed to crack down on the misuse of amphetamines as pep pills by banning their use as dieting aids.

If the agency is successful in having Government approval of amphetamines for weight reduction withdrawn, production of the pills could be reduced by 80 to 90 percent.

Notice of the proposed action was to be published in the Federal Register tomorrow. The proposal is the result of widespread abuse of amphetamines. The drug agency concluded that they are misused at a significantly higher rate than other drugs used in treating obesity.

Amphetamine is the generic name for pills sold as Delcobese by Delco Chemical Co., Inc. of Mount Vernon, N.Y.; Obetrol by Obetrol Pharmaceuticals, a division of the Rexar Pharmacal Corporation of Valley Stream, L.I.; Benzedrine sulfate by Smith Kline & French Laboratories of Philadelphia; and Biphetamine by the Pennwalt Corporation of Rochester, according to the Physicians' Desk Reference.

Amphetamines when used in treating obesity are called anorectics, although it has not been established that their action is through appetite suppression. Presumably, amphetamines have another type of action on the central nervous system to achieve the effect.

Short-term studies have shown that adult obese people who have been taught dietary management and treated with anorectic drugs lose more weight on the average than those treated with placebos, or dummy pills, and diet. But, the amount of increased weight loss of the anorectic group is just a fraction of a pound each week.

Even if the drug agency proposal is accepted, amphetamines could still be used for the treatment of narcolepsy, a rare ailment characterized by uncontrollable sleepiness, and for hyperactivity in children.

Some 3.3 million prescriptions for amphetamines were written in 1978.

The Pharmaceutical Manufacturers Association, a trade group for the industry, has already expressed its opposition to keeping amphetamines away from dieters. The industry argument is that the Government should crack down on amphetamine abusers instead, since the drugs have not been found to be unsafe or ineffective when taken as directed.

The Drug Enforcement Administration estimates that 50 million of the 500 million amphetamine pills produced in the United States each year wind up in the hands of drug abusers. A diversion to unlawful use occurs when physicians write too many prescriptions for the pills, when prescriptions are forged, through illegal sales or through dispensations at "fat clinics."

Some 90 percent of the amphetamines that are used illegally are diverted through these means, the D.E.A. said.

Dr. J. Richard Crout, director of the F.D.A.'s Bureau of Drugs, said the proposed limits on amphetamines "will not adversely affect patient care in any way."

Under the drug agency's proposal, manufacturers of amphetamines could request a hearing to challenge the new restrictions. If a hearing is justified, the agency would hold one before making a final decision on whether to withdraw approval of the drugs for weight loss.

July 17, 1979

THE DOCTOR'S WORLD

On Alert for The Potentially Fatal D.T.'s

By LAWRENCE K. ALTMAN, M.D.

AS an intern in San Francisco I treated a business executive, a member of a prominent West Coast family, who went to surgery for repair of a hip that he broke in a fall. Although he denied drinking more than an occasional cocktail, my suspicion — based on limited experience as a medical student at Boston City Hospital where alcoholism was a commonly treated problem — was that he was an alcoholic with impending delirium tremens, or the D.T.'s. It was something about his manner, perhaps the way he answered questions. My suggestion to the orthopedic surgeon in charge was that the man be treated for this potentially life-threatening condition.

The surgeon disagreed. He voiced surprise that someone would consider a man of such stature a silent alcoholic, let alone one who faced the D.T.'s.

But at 3 A.M., the nurse called, asking me to examine the patient, who, now in a cast, was agitated, restless and unable to sleep. When I arrived at his room, he was hallucinating. The diagnosis was clear: florid D.T.'s. He described spiders and other

The Bettmann Archive

Woodcut depicts effects of delirium tremens

bugs, often distorted by missing legs, crawling up the wall. He also imagined a panorama of animals in his room.

His arms were shaking as he picked at his sheets in the mistaken belief that he was counting dollar bills. At the same time, he was talking so rapidly and continuously that he might have aroused the envy of a politician in a fili-

buster. The content of his conversation was garbled. Yet he turned his head to listen to the people with whom he thought he was conversing. Clearly, he was misinterpreting the meaning of the sounds and shadows in his hospital room. In his disorientation, he thought he was on the street.

When he stopped talking and seemed lucid for a moment, I did a simple test.

My hands were held apart as if an imaginary string were being pulled taut. I asked him what color the string was. Blue, he responded. When I handed him an imaginary bottle of beer, he went through the motions of opening it. Then, holding it up, he "drank" it.

Variations on that scene are repeated elsewhere many times when patients deny their alcoholism and doctors do not suspect it. Sometimes when people of stature in the community, such as high-ranking military officials, are affected, doctors are asked to disguise the diagnosis.

Alcoholism is a cruel disease, causing damage in a variety of ways. The D.T.'s are the gravest of the complications of alcohol withdrawal. About 15 percent of those who experience the D.T.'s die, usually from the circulatory collapse, or shock, that follows the fevers that can reach 107 degrees. The executive's drinking was suddenly stopped by his accident, and his life was jeopardized by the delay in treating his D.T.'s.

Delirium is a common emergency due to altered brain function that can have many causes. Among them are overdoses of antidepressant drugs, fevers in children, brain tumors, injuries, and disorders of endocrine glands. Delirium also can be a complication of surgery as well as of heart attacks, presumably because of a diminished oxygen supply to the brain. However, the chief causes are alcohol and drug abuse.

The doctor's job is not only to distinguish among the causes of delirium, but also to guard against the D.T.'s in susceptible individuals during their care for other medical problems. The D.T.'s often follow an infection, or, as in the case of the business executive, an injury or surgery. Because delirium can be a symptom of meningitis one of the tests doctors do in such cases is a lumbar puncture, or spinal tap. The test is to determine if there are bacteria or other organisms in the spinal fluid due to meningitis, a potentially fatal infection affecting the lining of the brain and spinal cord.

•

The D.T.'s undoubtedly have occurred since antiquity but the problem presumably increased when technology made distilleries possible. Doctors first described the D.T.'s at the end of the 18th century.

Alcohol varies widely in its effects on people. The D.T.'s usually occur in an individual who has been an excessive and steady drinker for many years. But they can also occur after a drinking spree. Many alcoholics never get the D.T.'s. Others have repeated bouts. That fact is why doctors, when rou-

tinely taking medical histories, are careful to ask patients if they have had the D.T.'s.

Among those who died from the D.T.'s was Edgar Allan Poe, who was wildly delirious in a Baltimore hospital for four days before his death in 1849. According to his physician's description, Poe was drenched in sweat, his arms and legs trembled, and he answered questions incoherently. He talked constantly with imaginary objects on the walls.

Although physicians who treat alcoholism suspect that the incidence of the D.T.'s has dropped in recent years, the situation is not clear because accurate statistics are not kept about this condition. Doctors at Bellevue Hospital say they now treat fewer cases of D.T.'s. Among the suspected reasons: the population of alcoholic derelicts in the Bowery apparently has declined recently for unknown reasons.

Diagnosing the D.T.'s involves judgment calls because withdrawal from alcohol results in a continuing spectrum of symptoms. The hangover is the mildest. Many people have suffered slight muscle tremors, malaise and irritability on the morning after a few too many drinks. Others have seizures from "rum fits." The D.T.'s generally describes the severest of alcohol withdrawal symptoms, although some doctors use the term to cover milder reactions.

The precise cause of the D.T.'s is unknown. When brains of individuals who die from the D.T.'s are examined under a microscope, pathologists cannot identify a specific anatomical abnormality. Many doctors believe that the D.T.'s result from a biochemical disorder, but the specific type remains obscure.

•

In the early phases of alcohol withdrawal, the amount of magnesium in the blood drops and the critical balance between acids and bases in the body is upset. But these factors do not fully explain the D.T.'s.

The D.T.'s are a medical emergency and the treatment is complex. Because such patients usually have no appetite, sweat profusely and have fevers, they are fed intravenously large amounts of fluids — as many as seven quarts each day. Doctors also add sodium, potassium and magnesium to the intravenous bottles to replace the amount of these electrolytes that are lost by D.T. patients.

Equally important is the need to prescribe drugs to counteract the tremors and other effects of the D.T.'s. Many are used, among them: Librium, Valium, Thorazine and Compazine. Probably the most popular is paraldehyde, whose odor is recognizable in the emergency rooms of most hospitals.

Doctors must be alert to the special dangers that can result from including the sugar glucose in intravenous feedings to alcoholics whose diets may have been low in the B vitamins like thiamine. Intravenous administration of glucose can consume the last molecules of thiamine and produce Wernicke's disease, which is characterized by a disturbed mind, paralysis of the eye muscles and a staggered gait.

When patients with the D.T.'s develop fevers, they may be placed on cooling mattresses. Doctors treating patients with the D.T.'s must also consider their psychological needs. Lights may be kept on to reduce the patient's confusion. More than usual care is taken to explain even the most trivial thing that a doctor or nurse does in a D.T. patient's care to allay the fears and anxieties that accompany the disorder. When feasible, a family member is asked to stay with the patient.

Most D.T.'s end as abruptly as they began after about two days. Relapses are rare. Such a frightening, close call with death seemingly would scare alcoholics about the damage from excessive drinking. Yet many victims are oblivious to the events during the D.T.'s, a factor that complicates rehabilitation. When the San Francisco business executive recovered from the D.T.'s, he was exhausted, lucid and complaining of discomfort from the cast. But he said he had no recollection of the spiders and hallucinations, and denied that he had a drinking problem.

August 21, 1979

Marijuana Growing Becomes Big Business on Coast

By ROBERT LINDSEY
Special to The New York Times

RIVERSIDE, Calif., Oct. 29 — Commercial cultivation of marijuana has become a booming underground branch of agriculture in California and, according to investigators, is generating many millions of dollars yearly in illicit profits for a new version of the old-fashioned mountain moonshiner.

"These aren't a bunch of kids growing marijuana for their own use," said John Bitzer, an official of the Federal Drug Enforcement Administration. "These are sophisticated operations," he said, "a big business" that "has a major effect on the local economy" in some counties.

According to several recent public opinion polls, marijuana, which had long been regarded as an illegal and socially unacceptable drug in much of middle America, has gained increased acceptance among many people. More than 40 percent of adult Americans have tried it at least once and more than 15 percent use it with some regularity, according to pollsters' estimates.

Demand Expected to Rise

Although no one knows how much marijuana is consumed in the United States, some Federal drug officials put the amount at thousands of tons annually, and they say that the demand will probably continue to grow as the drug's social acceptance grows.

In the past, most marijuana sold in this country was smuggled in from Mexico, Colombia, Jamaica or some other foreign country. But in the last two years, drug investigators said, commercial production of the drug has increased significantly in California.

Officials say that there are two main reasons for the rapid growth in domestic cultivation:

¶Consumer concern exists about the quality and safety of marijuana smuggled into this country from Mexico. Under a program financed by the United States Government, Mexican plantations have been sprayed with the herbicide paraquat.

¶There is heavy demand for the type of marijuana most commonly produced in California, called sinsemilla, which, because of specialized farming techniques, is chemically more potent than most imported marijuana.

Harvest Time Arrives

It is now harvest time, and from the remote valleys of this rural county in southern California to clandestine farms in northern California, growers are picking and drying marijuana for sale in this state and in the East and Middle West.

Mr. Bitzer estimated that 77 tons of marijuana was grown commercially in California last year. Some law enforcement officials in rural counties have suggested that the figure may be twice that amount.

Investigators say that the illicit production has been concentrated in four counties in northern California: Mendocino, Humboldt, Del Norte and Lake, and two in the South, Riverside and San Diego. Some plantings have been found in six other counties.

Particularly in the four northern counties, which have experienced a sharp economic decline in recent years because of setbacks in the logging industry, the production of marijuana is an important part of the local economy, a source of income to merchants and other businessmen.

Two weeks ago, the Humboldt County Supervisors rejected a proposal by local law enforcement officials to seek a Federal grant to help defray overtime costs incurred in investigating marijuana growers. The rejection was interpreted by some local residents as an indication of the rising political influence of the growers.

Thomas Jondahl, the Mendocino County sheriff, said that it was all but impossible for small counties to police the rapidly expanding industry without outside financial help.

Initially, police officials said, the marijuana growers were mostly counterculture dropouts from the 1960's who had emigrated to the rural communities to find solitude and planted marijuana for their own use. Later, a few began producing the plant to earn a living, and then commercial production began to expand.

More recently, officials said, there have been indications that newcomers, including businessmen and several lawyers, have come to the area and financed establishment of marijuana farms.

The growers, Sheriff Jondahl said, operated farms ranging from "five or six plants to some in the hundreds or thousands."

"By no stretch of the imagination are these for home use," he said. "Some of these trees get 14, 16 feet tall; the stalks at the base are four or five inches thick; they are fertilized and irrigated just like on a regular farm."

Along with the growth in the illicit business, he added, has come increasing violence, usually in the form of confrontations between growers and people trying to steal their crop.

Here in Riverside County, which is 60 miles southeast of Los Angeles, deputy sheriffs so far this year have closed 51 commercial operations, arrested 72 persons and seized 26 tons of marijuana. Most of the arrests and seizures came in a 10-day period this month in which a Federal Drug Enforcement Administration aircraft was used, along with 50 policemen, to find growers in the county's remote mountainous reaches.

However, Capt. Jack Reid of the Riverside County Sheriff's Department said that the cultivation of marijuana had become so common, and the rural area that had to be checked so enormous, that undoubtedly many operations were continuing to thrive. "We really don't know how much there is in the county," he said.

The operations were so extensive, he added, that he believes growers have probably formed a cooperative, or several of them, to market their product. Most of the crop, he said, is being sold in the East.

"Some of these farms are very sophisticated, with miles of irrigation pipe and brass valves, and it's obvious somebody is making a lot of money from them," he said.

In California, Captain Reid said, the sinsemilla sells for up to $1,500 a pound; in Salt Lake City for $2,200, and in New York for $3,200.

October 31, 1979

CHAPTER 2

The Mass Media and Politics

John B. Connally bought the earliest Presidential campaign commercial ever to steal a march on his competitors.

Connally for President Committee

Keeping the Press, and the Judiciary, Independent

By Irving R. Kaufman

Increasingly, newspapers have been making news as well as reporting it. From a police search of The Stanford Daily to the citation of The New York Times for contempt of court, journalists have been in the public eye as never before.

The battle front over free expression has extended beyond the bounds of the Bill of Rights, reflecting an affirmative concern that, whatever the technical limits of his First Amendment freedoms, the journalist's ability to pursue his profession must not be hampered.

Thus, the crucial issue in the Myron Farber case was interpretation of New Jersey's "shield law," enacted to allow a reporter to preserve the confidentiality of his sources. And the United States Supreme Court's recent decision in Zurcher v. Stanford Daily, upholding the search of a newsroom by law-enforcement authorities, explicitly left the door open for legislative action.

This attention on insulating the press from official harassment brings sharply into view the importance of another fundamental American institution, the Federal judiciary. Different as the press and the Federal judiciary are, they share one distinctive characteristic: Both sustain democracy not because they are responsible to any branch of government but precisely because, except in the most extreme cases, they are not accountable at all. Thus they are able to check the irresponsibility of those in power.

For this reason, judicial independence, like free expression, is crucial — and vulnerable — in times of tension and intolerance. It is imperative, therefore, that the lines protecting Federal judges be inviolable if our society is to fulfill the promise emblazoned above the portals of the Supreme Court: "Equal Justice Under Law." And history teaches us that a precondition for judicial independence is secure tenure.

The heart of judicial independence is individualism. For the law to grow and respond, it must adopt views that were previously in disfavor, and often the groundwork is laid by dissenting judges. Justice Oliver Wendell Holmes stated it well when he declared that "time has upset many fighting faiths." The wisdom of today may not be that of tomorrow.

Recently, I wrote that "the lifeblood of the editorial process is human judgment." Clearly these words are equally true of judges. The judicial function must be performed by individuals, not cogs in a vast machine.

And yet, today there looms an ominous threat to the tenure, and thus the independence, of our Federal judges. The euphemistically named Judicial Tenure Bill was passed by the Senate last September and is certain to be reconsidered in the 96th Congress. Establishing a multi-tiered bureaucracy based in Washington, the bill would empower a special commission and court to remove Federal judges from office.

Our nation's Founders did all within their power to prevent the tenure of Federal judges from being aborted in this manner. Determined that abuses of the judiciary such as those perpetrated by George III would not be tolerated in the new Republic, they established in Article III of the Constitution a judiciary of unprecedented independence, bolstered by the assurance of secure tenure.

Of course, the constitutional draftsmen did not make the grant absolute. A Federal judge can be removed from office for "high Crimes and Misdemeanours," but only upon impeachment by the House and conviction by the Senate. For nearly two centuries this has been the only means by which disgraced judges have been ousted from the Federal bench. And recent historical research makes it patently clear that the Framers intended impeachment to be the exclusive methods of judicial removal.

Furthermore, the Judicial Tenure Bill appears both unnecessary and dangerous. The problem of the unfit or miscreant Federal judge is not nearly as great as is sometimes alleged. Federal judges have not suddenly fallen into collective dotage or indulged in a binge of bad behavior. The intensive selection and confirmation process to which candidates for the Federal bench are subjected — closer scrutiny than given to most state judges — provides substantial assurance that those chosen will be fit and competent. And if a judge overstays his period of useful service, his colleagues manage the problem efffectively in a personal and informal manner.

Any procedure short of impeachment that would guarantee the removal of the rare unfit judge would inevitably also threaten the dissenters who, although they may be unpopular with the public and with their colleagues, render useful, innovative service. Under the Judicial Tenure Bill, a judge could be removed, without resorting to impeachment procedures, for "conduct prejudicial to the administeration of justice that brings the judicial office into disrepute." It

requires little imagination to foresee the harassment courageous judges such as William O. Douglas or Frank Johnson of Alabama might suffer under so malleable a standard.

The protections of judges' tenure have not been erected for the sake of those who wear the robes but to safeguard the sacred responsibility of ensuring justice for the entire populace. It is not within the province of anyone, in or out of government, to undermine or dilute that independence. As with the journalist, then, so with the Federal judge. We must not tamper with the working processes of either. Both need "breathing space" to thrive. Fear of punishment must not chill their independence.

Irving R. Kaufman is Chief Judge of the United States Court of Appeals for the Second Circuit. This article is adapted from his recent Benjamin R. Cardozo lecture delivered to the Association of the Bar of New York City.

By Daniel Schorr

SAN FRANCISCO — Many Americans seems ready to believe — so soon after Watergate! — that reporters endanger national security by finding out about things the Government calls secret. This is despite the fact that the most outrageous security violations have nothing to do with the press.

Like Ambassador Graham Martin, bitter about Vietnam, roaming around North Carolina with a trunkful of secret documents that got lost with his stolen car. Like the C.I.A. letting a dozen copies of its hottest spy-satellite manual wander away, one to be sold to the Russians, unmissed for months until the C.I.A. heard from the F.B.I. Like a C.I.A. analyst passing top-secret documents to Senator Henry Jackson's staff to help fight an arms-control treaty.

Traveling around the country, I am rarely asked about these things, but I am frequently asked whether the press doesn't spill too many secrets and whether it hasn't become too powerful for the general good. "The bad-news media" is a game that any number in any position can play. It makes soul mates of President Carter and Mr. Nixon. Mr. Nixon's former speechwriter Ray Price talks of the "sustaining venom of the media," and Jody Powell, President Carter's press secretary, sneers at "the imperial press." In his recent interview with Bill Moyers on public television, President Carter talked about "the irresponsibility of the press." Bert Lance, now under grand-jury investigation, says he has been the victim of "careless, erroneous or biased reporting" of a kind that could lead to censorship in this country, and he is strumming a chord of popular sentiment.

We are getting to the point where a politician will be able to run against the news media as he used to run against Communism, crime or corruption — issues no longer available to some of them. The press may be a fair target but it is being hit with the wrong ammunition. There are dangers in a situation where the press, instead of being criticized, is turned into the enemy. If the press become totally discredited, its ability to fulfill its function of public watchdog will be destroyed.

It is time to take an open-minded look at this gulf of unease and suspicion that has opened between the people and their press. I have no final answer, but I have some tentative thoughts, and they have to do with the evolution of the press, both in terms of economic concentration of newspapers and the development of a television industry that subordinates the reporter to a large entertainment enterprise, an enterprise that is widely perceived as exploiting violence, sex and audience susceptibilities for bigger ratings and greater profits.

The press, once typically anti-Establishment, is perceived now as itself a huge Establishment. The picture of a great faceless corporation, manipulating tastes, brainwashing audiences, making and breaking public figures is hard to reconcile with a band of gallant gadflies, exposing the corrupt and powerful, goading the pompous, and deserving of special legal privilege so that they can wage our war for truth.

"The press," wrote Supreme Court Justice William O. Douglas, "has a preferred position in our constitutional scheme, not to enable it to make money, not to set newsmen apart as a favored class, but to bring fulfillment to the public's right to know."

If we could somehow convince the public that it is their right to know that is primarily involved in what we do, I think we would have better luck defending our privilege. But it is getting harder and harder to do.

It is only six years since Watergate, and I wonder what the situation would be today if a vast cover-up conspiracy were in progress. With confidential sources in danger, would there be a "Deep Throat" to blow the whistle? With reporters facing some future jail term for contempt, would there be a reporter to write the story? With the possibility of newsroom searches and ruinous fines for contempt, could you count on every paper, magazine and broadcaster to put out the story?

Consider the larger consequences of drying up confidential sources and giving reporters their comeuppance, making it easier for those in authority to manage the news. Our news media can be arrogant, affluent, sex-prone, violence-prone and sometimes downright smug. But it is the only press we have, and at crucial times has helped to save our free institutions. Fulfillment of our function, like fulfillment of your [lawyers'] function, depends on a fragile thing called privilege. Ours today is more fragile than yours, but yours is also being questioned. Privilege ultimately rests on society's recognition of its value. We live in times of passion when many values are being challenged. If the free press is eroded in the name of justice, then justice will surely be eroded next. There cannot, ultimately, be any victory for a civil liberty at the expense of another civil liberty.

Daniel Schorr, a columnist for The Register and Tribune Syndicate of Des Moines, resigned as CBS correspondent in 1976 after having been suspended for his role in the publication of a House Intelligence Committee classified report. This article is adapted from an address to the California State Bar Association.

January 16, 1979

The Media Event Comes of Age in British Politics

By R. W. APPLE Jr.
Special to The New York Times

COALVILLE, England, April 20 — In the last three days, Margaret Thatcher has reclined in an East Anglian pasture with a calf, operated a sewing machine in a Leicester garment factory and stuffed bonbons into boxes in a Birmingham chocolate plant.

She has also been photographed doing these things. The media event, a campaign stunt performed because it will look good in newspapers and on television, and not because it has the remotest connection with the job the candidate is seeking, has really arrived in British politics.

The media are devouring and clamoring for more, so much so that a group of people in a London retirement home were almost trampled today by the herd of cameramen and technicians who followed Mrs. Thatcher there. The Conservative leader's husband, Denis, had to escort a terrified 91-year-old out of the melee.

A few traditionalists resent the intrusion of the razzmatazz of American politics on the quieter, more substantive British style of campaigning still exemplified by Prime Minister James Callaghan and David Steel, the Liberal Party leader. Peter Jenkins, The Guardian's political columnist, is one of the objectors. After a morning in "the green grassroots of Suffolk," he reported caustically:

"First she was filmed looking at the Friesians and then at the Herefords. She seemed to like the Herefords best, with their mournful white faces; she let one of them suck her fingers for several minutes in a rather sensual fashion."

The most bizarre scene of the campaign for the May 3 general election unfolded yesterday in the Cadbury chocolate factory at Bournville, four miles southwest of Birmingham. Everyone in the Thatcher party, candidate, security men, aides, reporters and cameramen, had to put on long white coats and little paper butchers' hats. The workers were wearing identical outfits, so Mrs. Thatcher, unable to tell one group from the other, found herself shaking hands with an American journalist and asking him earnestly to support the Tories.

When she reached the assembly line where women spend the day fitting nougat crunches and mint cremes into their appointed positions in assortment boxes, the cameramen climbed onto the machinery in search of a better vantage point. They looked like demented Martian surgeons.

'Harder Than It Looks'

"It's harder than it looks," commented Mrs. Thatcher while trying her hand at the assembly line, "getting them into the correct places."

"You're hitting my mike," shouted a sound technician.

"Do you export much to Africa?" asked Denis Thatcher from the fringe of the crowd. "Really? But doesn't it melt?"

"You're a rather excitable lot, aren't you?" said one of the women farther down the line, glancing up from her stock of hazelnut clusters.

Then it was off to another part of the factory to watch a machine that wraps creme eggs in colored foil.

Half the time, because of the racket made by the machinery, she could not hear what the workers were saying to her. This, wrote Frank Johnson of The Daily Telegraph, "is the ideal arrangement for conversations between party leaders and voters at election time, since it cuts out a lot of the unnecessary detail."

Good at Small Talk

"You have more temptations than most of us," Mrs. Thatcher called to a woman stacking chocolate-laden trays. "We never eat them," the woman replied.

Mrs. Thatcher is good at this sort of thing. Somehow, in the midst of the shoving, cursing, rugby-scrum of journalists, she manages to establish a kind of intimacy with the person she is talking to. She is a toucher, a practitioner of the two-handed grip, the shoulder hold, the elbow clasp.

And she has a limitless fund of small talk. She sounds warmer and more natural than in her more studied appearances on television or on the stump. "How long have you worked here?" "Where do you live? Ah, my sister trained there as an orthopedic nurse." "How many boxes do you pack in a day?" "That's a pretty pair of earrings you're wearing."

Three hours later, it was all on television, and the next morning it was in the papers.

Bustling, enthusiastic, crisply turned out in a blue suit with red and white stitching on the collar, Mrs. Thatcher moved across the Midlands in a carefully calculated pattern.

In addition to generating pictures, the stop at Cadbury's was designed to help defeat Thomas Litterick, a Labor Member of Parliament who won the Selly Oak seat last time by only 326 votes. At her side was the Conservative candidate, Anthony Beaumont-Dark, a stockbroker who for campaign purposes has packed away his hyphen, shortened his name to Tony Dark and switched from his baby-blue Rolls-Royce to a more egalitarian Land-Rover.

There was a stop at a butcher's shop in Loughborough, another marginal constituency, where Mrs. Thatcher was photographed pointing with alarm at the list of prices. And a stop in Coalville, part of a constituency called Bosworth after the nearby battlefield where the Wars of the Roses ended in 1485 with the death of Richard III and the accession of Henry Tudor to the throne.

Adam Butler, Mrs. Thatcher's parliamentary private secretary, won Bosworth for the Tories in 1974 by 302 votes.

April 22, 1979

12 Texas State Senators, Claiming Political Victory, Come Out of Hiding

AUSTIN, Tex., May 22 — Twelve state senators dubbed the "Killer Bees" returned to the Texas Senate chamber today after a five-day absence, contending that the Presidential primary legislation they sought to block was "dead."

Their arrival, to a cheering gallery, produced a quorum in the Senate and tentative approval, by a 17 to 14 vote, of the primary bill.

But it appears that the two-thirds majority needed for final passage will not materialize and that the Killer Bees have destroyed, at least for the current legislative session, any prospects for a separate Presidential preference primary in Texas for 1980.

"We think the bill's dead," said Senator A. R. Schwartz, one of the Bees.

Benefit to Connally Argued

The senators went into hiding on Friday in an effort to kill a bill calling for a Presidential primary separate from the general primary, in which state and local candidates would be listed on the ballot. They believe a separate primary would favor the candidacy of John B. Connally, the former Texas Governor, who is seeking the Republican nomination for President in 1980.

The Killer Bees contend that the separate primary would allow conservative Democrats to vote in the Republican Presidential primary for Mr. Connally, a former Democrat, without having to desert local, conservative Democrats on the general primary ballot.

Senator Schwartz argued that there was "no legitimate reason" to give any political candidate an advantage in a Presidential primary at the cost of $5 million in tax money, the price of a separate primary.

Lieut. Gov. William P. Hobby Jr., who serves as President of the Senate, had ordered the arrest of the 12 absentees last Friday after placing a "call" on the Senate. Texas Rangers, Department of Public Safety Troopers and local police officers hunted the senators from border to border.

Supporters Clap and Buzz

The 12 had vowed to remain absent from the session, which ends next Monday at midnight, until Lieutenant Governor Hobby either agreed to remove the primary bill from the top of the Senate calendar, or until midnight tonight, when Senate rules would have removed it from consideration.

The senators returned to Senate galleries filled with supporters who applauded and buzzed — many were wearing yellow and black "Killer Bee" T-shirts.

Then the group held a press conference and reported that Mr. Hobby had agreed not to attempt any parliamentary maneuvers that would have brought the bill up for a final, simple-majority vote.

Texas has no Presidential preference primary law in effect for 1980. A bill providing for a one-time primary, concurrent with the regular party primaries, was enacted for the 1976 elections.

Named by Lieutenant Governor

All 12 of the Killer Bees, who received their name from Lieutenant Governor Hobby for the manner they attack legislation, are liberal and moderate Democrats.

Upon returning, the senators refused to reveal where, or with whom, they had been in hiding, other than to say that nine of them were together throughout the ordeal, in one room, with one toilet, and one telephone, apparently, near the Capitol.

Senator Carl Parker charged that those senators wanting a separate primary were interested in protecting their own legislative seats. "I don't know a Senator that's worth a million dollars, including myself," he added.

Senator Jones stressed that despite the call for their arrest, the 12 had not violated any civil or criminal law, or their constitutional oaths. An agreement with Mr. Hobby to remove the "call," and not seek the arrest of the members, resulted in their return before the Tuesday midnight deadline they had set.

Support by Officials of Parties

On Saturday the State Republican Party's executive committee reaffirmed the party's stand backing a same-day primary and commended the Killer Bees for their action. The state Democratic Party chairman also supported their "flight."

The concept of a separate Presidential primary also appears to have stiff opposition in the House, where "Killer Bee" supporters claim to have 80 of the 150 votes against such a plan.

On Sunday, officers searching for the senators arrested Clayton Jones, one of the seven brothers of Senator Gene Jones, at Senator Jones's Houston area residence. However, the legislator scrambled over an eight-foot fence, and then called the local radio station to announce he had not been arrested. His brother, meanwhile, was flown to Austin in a state helicopter before Senate officials discovered the wrong man had been arrested.

May 23, 1979

Killer Bees Head North

Let it not be thought that the adventures of the "killer bees" in the Texas State Senate represent an isolated instance of legislative irresponsibility. There's a strange buzz over Washington, too; the House Democratic caucus has voted, by a 2-to-1 margin, to oppose an energy measure it is powerless to stop.

The Texas hijinks at least carry the flavor of principle. Conservative Republicans usually cross over to vote against liberal candidates in the Texas Democratic primaries. The conservative Democrats who run the Legislature like that fine, but they have feared that next year, all the voting conservatives would be lured to the Republican side, to vote for John Connally, the former Democratic governor who seeks the Republican Presidential nomination. That could leave all the local Democratic nominations to the liberals. The conservative solution was a separate Presidential primary to boost the former governor in March, allowing con-servatives to become crossover Democrats in the usual May primary. By kidnapping themselves until the session was nearly over, the "killer bees" have now scotched that plan.

There may not be even a saving touch of principle in the vote of the House Democrats disapproving the decontrol of oil prices announced by the Democratic President. They all know that the idea was dictated by Congress in the first place. They all know that the Senate will insist on decontrol of some kind. So what are they up to? It certainly looks like pusillanimous politics. Oil prices must rise, but a majority of House Democrats want to be able to say "Don't blame me." The President has been stung; so has everyone else anxious that the country join in a responsible national policy on energy.

May 24, 1979

Wuxtry! The Story No One Dares to Tell!

By James Reston

EDGARTOWN, Mass., July 24 — Almost everybody seems to be wondering these days why President Carter lost his way recently, but the answer is fairly obvious.

He has been deprived, poor man, of the daily wisdom and guidance of Walter Cronkite, Art Buchwald, Mike Wallace, and other media philosophers who fled the continental United States to Martha's Vineyard at the first sign of a serious crisis.

It is not the President who abandoned The Press but The Press that abandoned the President.

Where was Cronkite when the Republic needed the calm authority of his voice? He was using up his energy on the Martha's Vineyard tennis courts, that's where!

And where was Buchwald, just when we needed his merry eye on the latest

James Reston, whose column appears in a major metropolitan newspaper, wrote this article, which tells all, for The Vineyard Gazette. He is otherwise on vacation.

bizarre antics in Washington? Like Cronkite, he was right here on the Vineyard complaining about the humidity and working on his backhand.

I will get to Mike Wallace later but Buchwald's activities recently are bound to arouse skepticism even in the most innocent of minds.

Where was he before he came to the Vineyard — when President Carter was in Vienna trying to strangle the nuclear arms race? He was in Peking defying the Soviets and peddling his column to the Chinese. And who was with him? None other than Joe Califano, at that time, but no longer, Secretary of Health, Education and Welfare.

The working press has explained that Califano was fired because he was too close to Teddy Kennedy, but *the idle press* here on the Vineyard suspects he was ditched because he was too close to Buchwald, whose spoofish character is not always appreciated in the White House.

There is a popular view, originally known as the Spiro Agnew Doctrine, that a President's serenity depends in large part on the absence of colum-nists and commentators, but the validity of this doctrine depends to some extent on the issues that are in, and the commentators who are out on vacation.

For example, no President of either party is likely to welcome the absence of Walter Cronkite during a political crisis. It is Cronkite's special genius that he could report the imminent destruction of the human race in such comforting tones as almost to make the event sound reasonable, if not actually desirable.

Mike Wallace of CBS and the "60 Minutes" program is also a loss to any President in trouble. For it is the studied technique of this program not to attack the people who are being attacked by the rest of the media, but to attack the people who are not being attacked.

Thus, one can imagine Mike, Harry Reasoner and Dan Rather grilling the Mafia on reports of its connection with the OPEC countries — thus diverting the fire from Carter — or demanding to know precisely what Ronald Reagan, John Connally, or Teddy Kennedy would do to end the energy crisis, reduce inflation and unemployment, and end the arms race.

This might not please the people who think Jimmy Carter personally invented all these disasters, but for the moment the nation is deprived of the gentle mercies of Mike Wallace's Sunday evening questions.

Perhaps more important from Jimmy Carter's point of view is the absence of Buchwald. Washington without Buchwald is like Rome without Nero. He is the master of the simple question in a topsy-turvy world, and if he'd been around, one would have suspected that this crisis was made, if not invented, by him.

Unlike the rest of the press, Buchwald would almost certainly have defended the President for canceling that energy speech — "if you have

nothing to say, why say it?" He might even have approved the shuffle of the Carter Cabinet on the ground that it was long overdue, though more than likely he would have been asking Jimmy: "Why Not The Best?"

There were, obviously, two clear ways to solve this dilemma. The first was for Cronkite, Buchwald, Wallace, and the other publicity saints to stay

on the job, and the other was for President Carter to extend his vacation by coming to the Vineyard. And of these two, the latter might have been the better. For it was only when the President stopped resting and thinking at Camp David and began acting in Washington that things began to fall apart.

July 25, 1979

News Organizations Balk at Curb On Photographers on Carter Trip

By MARTIN TOLCHIN
Special to The New York Times

WASHINGTON, Aug. 14 — Several major news organizations have told the White House that they would not send photographers to accompany President Carter on his Mississippi River boat trip, which begins Friday, because they consider the conditions set by the White House to be management of the news.

"The rules are so restrictive that we're going to decline the opportunity to accompany the President on the trip," said Fred William Lyon, vice president for news pictures of United Press International. "We will aggressively pursue other methods of covering the story."

Time and Newsweek magazines also have decided not to assign staff photographers to the trip because of the restrictions. Executives of The Associated Press were still trying to come to a decision late tonight.

The three major television networks

have not rejected the ground rules, although "we have a severe problem with them," said George Watson, ABC's Washington bureau chief, but they have offered a counterproposal.

Restriction on Photographers

The White House ground rules would restrict news photographers to certain areas of the Delta Queen, on which President Carter will journey from St. Paul to St. Louis with several stops along the way to campaign for his energy proposals.

The White House also is asking the right to approve the photographs taken and is insisting that news organizations agree not to purchase photographs from tourists on the ship and not to give tourists cameras or film with which to make photographs. In the event that a news service ignored the prohibitions and used a tourist's photograph, its own photographers would be ejected from the boat.

Another of the White House conditions, which were given orally to representatives of Time and Newsweek and the two news services by Jody Powell, the Presidential press secretary, is that they must agree not to distribute nationally any photographs obtained by local newspapers along the route.

Single Television Pool

The three television networks, meanwhile, would be limited to a single television pool under the White House rules. The networks have countered with a proposal that would allow all three to photograph the President at some time each day of the seven-day voyage.

The networks also have asked that they be allowed to purchase photographs from tourists.

News coverage of the President is often accomplished by the use of a pool, by which a small group of reporters and photographers represent their colleagues, because the entire press corps is too large and unwieldly a group to accompany the President always.

The conditions imposed for the Delta Queen trip, however, are far more restrictive than usual.

August 15, 1979

A Tale of Carter and the 'Killer Rabbit'

WASHINGTON, Aug. 29 (AP) — A "killer rabbit" penetrated Secret Service security and attacked President Carter on a recent trip to Plains, Ga., according to White House staff members who said that the President beat back the animal with a canoe paddle.

The rabbit, which the President later guessed was fleeing in panic from some predator, reportedly swam toward a canoe from which Mr. Carter was fishing in a pond. It was said to have been hissing menacingly, its teeth flashing and its nostrils flared, and making straight for the President.

Mr. Carter was not injured, and reports are unclear about what became of the rabbit. But a White House staff photographer made a picture of the attack and the President's successful self-defense. This was fortunate because some of the President's closest staff members reportedly refused to believe the story when Mr. Carter told them about the attack later.

"Everybody knows rabbits don't swim," one staff member said.

Mr. Carter, said to be stung by this skepticism in his

inner circle, ordered a print of the photograph to offer as proof. But this was not good enough.

"You could see him in the canoe with his paddle raised, and you could see something in the water," said the doubter. "But you couldn't tell what it was. It could have been anything."

So Mr. Carter ordered an enlargement. "It was a rabbit, all right," the staff member said after seeing the blown-up photo.

Another staff member who saw the picture agreed. "It was a killer rabbit," this staff member said. "The President was swinging for his life."

No news photographers were allowed within camera range of Mr. Carter on the fishing trip, which was made on April 20. And the White House declined to make public any photographs of the encounter with the rabbit. "There are just certain stories about the President that must forever remain shrouded in mystery," Rex Granum, the deputy White House press secretary, said today.

August 30, 1979

WASHINGTON

The Jimmy-Teddy Show

By James Reston

BOSTON, Sept. 25 — The first impression one has of political audiences in this Presidential campaign is that they are skeptical of all candidates, even here in Kennedy country, but are ready for serious discussion of the complicated issues that now confuse and divide the nation.

The immediate outlook for such a clarifying debate on the problems of the United States at home and abroad obviously is not bright. The horses are still milling around outside the starting gates.

But it is not too early to ask how the press, radio and television will report this campaign, for the questions they ask and the coverage they provide are likely to influence the character of the coming debates.

There is nothing that can be done in this election to change the torment of 35 state primaries, which almost all candidates condemn as a ridiculous and unnecessary bore. They divert the President from the public's business for over a year and make it difficult for whoever wins to prepare for the serious work of governing the country.

But one or two things might be done.

Couldn't we at least put an end to the nickname game? This "Jimmy" stuff has already gone on too long, and the thought of a "Jimmy-Teddy" show is almost, but not quite, enough to make a man vote for plain John Anderson or even fancy John Connally.

This is the tryout period in the campaign, when the candidates test their themes before carefully selected audiences. The response to these themes by their audiences and by the press usually determines whether the old arguments or tricks are repeated, magnified or discarded.

For example, Richard Reeves, one of the country's best political observers, tells what happened the other day when Senator Howard Baker of Tennessee was asked whether he had an obligation to the American people to raise the question of Chappaquiddick in relation to Senator Kennedy's "character and competence under pressure."

No, Senator Baker replied. There is a great danger, as a result of Watergate and other events, he said, that the political arena has become so tough and mean that Presidents, senators and candidates cannot function. He was so deeply concerned about the lack of civility, the "savagery" and the personalizing of politics, he concluded, that he simply would not get into the Chappaquiddick question.

Mr. Reeves reports that this got by far the most enthusiastic response from the 2,000 people at the meeting of the Detroit Economic Club than anything else the Senator said.

A few days later, President Carter met a group of editors at the White House and said that with the election coming up, it would be time for the American people to make an assessment of his record and his plans for the future.

"And then, of course," he added,

"your own character assessment — the reputation for being steady in an emergency . . . these things become much more important than the relatively transient popularity polls . . ."

Well, you can make your own assessment of what that means, but while the candidates are flying their kites, it might not be a bad idea to keep score on how high or how low the kites are flying.

We are clearly going to have a campaign about both the past and the future, about both character and personality, about the issues and about style — but it is important what *dominates* the campaign.

I have been hearing in the first preliminary soundings here what Reeves heard in Detroit — a longing for a more generous discussion of the nation's puzzles rather than a repetition of the regional bigotry against Mr. Carter in 1976 or the religious bigotry against John Kennedy in the campaign of 1960.

These ancient American political animosities were at least set aside in those two campaigns, but they still lie under the surface at the opening of the 1980 campaign, and could erupt again if the candidates and their supporters think they can evade the major issues of the day and get away with it.

How to cover so many candidates in so long a campaign in so vast a continental country — how to keep the critical issues for decision from being overwhelmed by secondary issues or even trivialities — these are questions that have baffled every editor I have known since the election of Franklin Roosevelt. And it is now — before the Jimmy-Teddy show goes too far — that the problem has to be analyzed all over again.

September 26, 1979

WASHINGTON

Hearing It From Henry

By James Reston

WASHINGTON, Oct. 26 — Henry Kissinger has written a four-pound book (1,521 pages) at a little over $5 a

pound, which is in keeping with the inflationary trend of these days, but the view here is that it is the most remarkable insider's political memoir of our time.

That is really a rather modest compliment, for most books written by or for our highest officials usually make more money than sense, and are almost as deficient in historical perspective as they are in literary distinction.

Kissinger's "White House Years" is unusual in several respects. He wrote it himself — an original idea in this town. It is written in clear and muscular English prose — maybe because he began, as a German immigrant boy, with respect for the English language and mastered its grammar, poetry and even its subtle wit.

Also, Kissinger almost avoided — but not entirely — the temptation of most high officials and their ghosts to try to prove that they never made a mistake, or if they did, it was somebody else's fault. His explanation of the bombing of Cambodia, which contributed to the unbearable tragedy of that silent and amiable people and their children, has not convinced even his most devoted friends.

But that is only one part of Kissinger's report on his White House years. He has been almost recklessly frank in reporting, with the utmost precision, the problems placed before our country while he was in the White House: the tangles and dilemmas and even the mistakes.

His story of the anatomy of this book

19

is interesting. When Kissinger left the State Department after the election of President Carter, he went to Acapulco with a footlocker full of books. "You have to remember," he told me, "that I hadn't read an unclassified serious book for eight years."

In the footlocker, he had most of the Presidential memoirs, beginning with General Grant's, which he still regards as the best. Also the memoirs of Churchill, de Gaulle, Adenauer, former Secretary of State Acheson and the books of the British Prime Ministers Eden and Macmillan.

"You understand," he said, "I'm not comparing myself to these people. I was merely looking for models. Macmillan impressed me very much, but the one who probably influenced me the most was de Gaulle. . . . He was a good model because he defines and because he had a conceptual framework, a larger purpose and a gift of explaining how the particular problem of the day related to his philosophy and his goals."

In 1977 Kissinger spent six weeks in Acapulco with his footlocker full of books. He lived in Puerto Marques in a house owned by former President Alemán of Mexico. He agrees that this was not an unbearable hardship.

He had a staff of four researchers.

They collected all the documents by subject and by year and these he read, "about 12-15 hours a day" from March until December of 1977. He then prepared a bibliography of the important documents, and by July began writing little essays on people — and these vignettes remain among the real gems of his book. In December of 1977, after a bit of a struggle to get the whole book in focus, he began writing chapters.

"I wrote about a chapter a week," he says, "and rewrote them four or five times. I would do one chapter in long-hand, then Peter Rodman, who was my chief researcher, would check it for accuracy and another researcher, who was an expert on that particular subject, would doublecheck it."

Then Kissinger would send a finished chapter to two of his former State Department aides, Winston Lord and Bill Rogers, the former Latin American expert at State, and he would rewrite the chapter again, and submit this version to outside objective scholars.

Whatever anybody thinks about the decisions made in Kissinger's White House years, this is probably the most serious attempt ever made to explain the decision-making process, good or bad, and to illuminate the interplay of politics, personalities, the press,

and even of accident on final decisions. In this sense, it is a remarkable achievement of reducing unbelievable diversity to some kind of entity that can be understood by a careful reader.

He has now finished six chapters of his second volume but is having a tussle with it. "You see, Volume One was sort of a linear thing in which we had a concept," he explains. "Volume Two deals really with catastrophe, the disintegration of executive authority. . . . There really is a sharp break at Watergate.

"The turmoil of the first term, bad as it was, was a debate about national policy and America's role in the world. The turmoil of the second term was the personal struggle of Richard Nixon and really had no relationship to the policy we were trying to pursue. . . ."

A popular complaint is that nobody really knows what's going on in Washington, or how officials here reach decisions about things that affect people's lives. Well, Kissinger has given about the best description of the process possible. If you can hold all 1,500 pages of it, it may break your back, and if you read it carefully, it may break your heart, but it is worth risking both

October 28, 1979

Phyllis George: Half of a Glamorous Campaign

By JUDY KLEMESRUD
Special to The New York Times

LOUISVILLE, Ky. — It's been called "the kissing campaign," and it's easy to see why. John Y. Brown Jr., Kentucky's Democratic candidate for Governor, and his wife of seven months, Phyllis George, constantly hug and kiss like newlyweds on the campaign trail. And everybody in Kentucky, it seems, wants to kiss the former Miss America turned television personality.

"Oh, Phyllis, Phyllis, please kiss my husband, *please!*" a plump young woman asked the candidate's wife during a recent campaign stop at the Harvest Festival in Crestwood, Ky. The 30-year-old Mrs. Brown smilingly obliged, then continued making her way through the crowd, signing autographs and kissing (and being kissed by) everyone from babies to senior citizens.

At times like this, her handsome, 45-year-old husband, who parlayed Colonel Sanders's Kentucky Fried Chicken recipe into an estimated $35 million personal fortune, is likely to say something self-deprecating to the crowd like, "I've been walking behind her ever since we met," or "Doesn't anyone want *my* autograph?"

Phyllis George Brown, as she began calling herself after a crusty mountaineer pointed out that Kentucky women use their husbands' last names, candidly admits that one of her main values to the campaign has been to draw crowds. "They used to call me Flypaper Phyllis," she said with her familiar, ever-present smile as her campaign car sped back to Louisville. "But what was John supposed to do, hide me in the closet? I am his wife, and I have the right to be beside him."

Taken State by Storm

The John and Phyllis Show, as some Kentuckians call it, seems to have taken the Bluegrass State by storm. Recent polls showed Mr. Brown well ahead of his Republican opponent in the Nov. 6 election, former Gov. Louie B. Nunn.

Commenting on Mrs. Brown's importance to the campaign, Robert Cobb, the Brown campaign coordinator, said he thought that "John Y.," as the candidate is known here, would not have won the crowded Democratic primary last May 29 without his wife's help on the campaign.

Almost every day since Mr. Brown entered the gubernatorial race shortly after the couple's St. Patrick's Day wedding, his wife has been at his

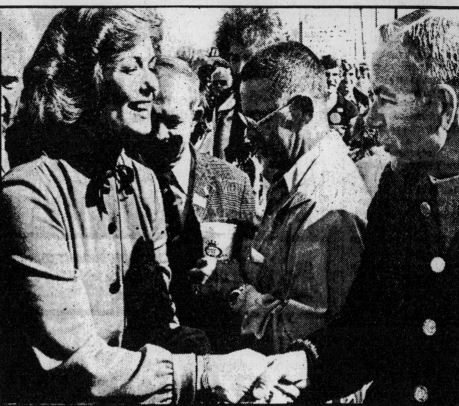

The New York Times / Keith Williams

John Y. Brown Jr., running for Kentucky Governor, and Phyllis George Brown, left. Mrs. Brown greets a voter, above.

side, sitting in on his strategy sessions and flying across the state with him, her huge pear-shaped diamond engagement ring glittering in the sun.

Invariably, she would introduce her husband at a campaign stop, adding an aside intended to blunt any suspicions that she was a "carpetbagger." In Crestwood, she told the crowd, "I'm from Texas, and lived there for 21 years, and never thought I could live anywhere else." Long pause. "But this is my old Kentucky home."

One of Mrs. Brown's biggest efforts for the campaign was a rally called "Kentucky's Salute to Women of the 80's," which drew almost 2,000 people to the Commonwealth Convention Center in Louisville on Sunday. The event, which Mrs. Brown insisted was "nonpolitical," drew such prominent female speakers as Representative Patricia Schroeder, Democrat of Colorado; Azie Taylor Morton, Treasurer of the United States; Representative Geraldine A. Ferraro, Democrat of New York; Dorothy Height, president of the National Council of Negro Women, and Mildred O'Neill, wife of the Speaker of the House.

But Mrs. Brown's repeated insistence on the nonpartisan nature of the rally, which was sponsored by the Kentucky State Democratic Party, angered some Kentuckians. "If she'd lie and say this wasn't a political rally, who knows what she'd say in the future about other things?" said one Louisville man who attended the rally.

Other Kentuckians have been criticial of the Browns' hugging and kissing, which seems to become more pronounced whenever cameras are

aimed at them. Beula Nunn, the 65-year-old wife of the Republican candidate, said in a recent inteview that she didn't plan to bite her husband's ear in public during the campaign.

When asked about the criticism that she and her husband were overly affectionate in public, Mrs. Brown said: "That's tough. I just hope maybe some of it will rub off on other people, and they will be as full of love as we are."

What was it like going from the glamorous world of television to campaigning for public office? "For the first two weeks, it was very difficult to adjust to the life," Mrs. Brown said. "I didn't have any friends here, and once I flew back to my apartment in New York and broke down and cried in my press agent's office. I made a big sacrifice in my career coming down here, and I remind John of that a lot. Once when I was feeling depressed, he told me he'd buy me a horse. I said, "How good a horse?" He said, "A *very* good horse.' "

Mrs. Brown added that she had been "disillusioned" by the "mudslinging tactics" used when the Republicans began to attack the Browns' "jet-set life style," specifically Mr. Brown's alleged fondness for high stakes gambling and the couple's attendance at a party at a New York discothèque (Xenon) that had been featured in Penthouse magazine.

"But basically, show business is politics," she added. "If you work for a network, as I did, it's who you know that counts. And politics, with its media blitz, is very much show biz. What matters is whether a candidate is handsome, and articulate, and has a good smile. The big difference is that politics is real, very real, and that show business is a fantasy world.''

The Browns first met two years ago at an professional football playoff game in Minneapolis, where they were introduced by the odds maker Jimmy (the Greek) Snyder. As the story goes, Mr. Brown had been wanting to meet Miss George ever since the night his divorce from his first wife, Ellie, who ran his unsuccessful Kentucky Colonels basketball team, became final, and he caught a glimpse of Miss George on television. He supposedly told a friend at the time, ''I think I'll go out with her.''

Miss George, however, was married soon after to Robert Evans, the Hollywood movie producer. Then last November, after her 11-month marriage had broken up, Miss George ran into Mr. Brown at a Hollywood party. He proposed two months later, and they were married last St. Patrick's Day in New York's Marble Collegiate Church by the Rev. Norman Vincent Peale, who urged them to ''go out and serve God together.''

Mrs. Brown said that if she becomes First Lady of Kentucky, her major projects will be working with retarded and handicapped children (she is honorary Kentucky chairman of the Special Olympics for the handicapped), and on women's issues. Both she and her husband support the equal rights amendment, which has been rescinded by the Kentucky legislature. And although both are personally opposed to abortion, they say it is up to the general assembly to decide the issue.

Wants to Start a Family

Does Mrs. Brown plan to continue her show business career? ''Yes,'' she replied, as she sat in her office in the Louisville campaign headquarters, ''but I'm so happily married, I don't want to be away from John that much. I also want to start a family. But I do have two years left to go on a contract with CBS, and I'm obligated to them.''

She said she also plans to do a motivational book for young women for Simon & Schuster, about how a girl from the small town of Denton, Tex., grew up to become Miss America, then the first successful female sportscaster, and, perhaps, the First Lady of Kentucky.

''John wants me to do all these things,'' she said, ''and you can't expect a person like me who's been an achiever to stay in the background. They used to say, 'Behind every strong man there's a strong woman.' Well, I like to say, '*Beside* every strong man there's a strong woman.' ''

October 30, 1979

Connally Is First to Air a 1980 Campaign Commercial

By ADAM CLYMER
Special to The New York Times

WASHINGTON, Oct. 30 — John B. Connally has bought the earliest Presidential campaign commercial ever, for broadcast tonight, using television to steal a march on his competitors and try to set himself up as the Republican Party's only realistic alternative to Ronald Reagan.

The commercial is a key element in a three-week blitz intended to establish the Republican Presidential contest as a two-man race. It advertises Mr. Connally's leadership and says he is the candidate of the ''forgottten American'' who works hard, pays his bills and taxes and ''believes in prayer in public schools.''

The five-minute program on the CBS network is expected to be followed tomorrow by an announcement that Mr. Connally has raised over $1 million in a two-day visit to Mr. Reagan's home state of California, by a major speech on taxes next week in New York, and, campaign aides hope, by a first-place or a close second-place showing against Mr. Reagan in a Florida state convention straw poll Nov. 17.

Julian Read, Mr. Connally's campaign director of communications, said that the former Texas Governor's success already in winning over Florida delegates ''proves that it's a two-man race'' and ''proves Reagan is not invincible.''

Mr. Connally and other challengers regularly describe Mr. Reagan as the Re-

United Press International
John B. Connally

publican front-runner, and national polls of Republicans consistently show him well ahead, even before his own nationally televised announcement of his candidacy, which is planned for Nov. 13.

But the Texan is not only the only chal-

lenger busy trying to chip away at the former California Governor's position. This week Senator Howard H. Baker Jr. of Tennessee, the Senate minority leader, will formally announce his candidacy. Then he will fly off to New England, ending up on Saturday at a Maine Republican convention where his alliance with Senator William S. Cohen of that state is expected to win him a straw poll victory.

Mr. Connally's commercial, following other recent attention-getting campaign moves, such as his controversial Middle East policy speech, emphasized one issue he has been stressing and was framed in terms to combat a major negative impression of the candidate.

The overt message is ''leadership,'' a word used five times in four and a half minutes. Mr. Connally is on camera saying, ''Leadership, first, is courage, leadership is vision, leadership is an understanding of our people.''

Integrity and Trust

But it goes on to address the problem of his negative image, one that ranks high in opinion polls, by having him say that the people ''have got to trust you, they have to believe in you'' and ''you have to have integrity in your dealings with them.''

Roger Ailes, the New York producer who made the commercial, said that while it was not his purpose to ''address some negatives,'' such as a ''wheeler-dealer image,'' ''I guess it does address them.''

Mr. Connally is described in the commercial, which CBS said it charged the campaign $28,500 to air and up to $2,500 to edit, as having been "gravely wounded in the same car in which President Kennedy died."

A picture of Mr. Kennedy, at the appointment of Mr. Connally as Secretary of the Navy, is shown, but although Mr. Connally is described as an adviser to four Presidents, pictures of the others (Presidents Johnson, Nixon and Ford) are not. Nor is Mr. Johnson's name mentioned when Mr. Connally is described as having worked for a Texas Congressman.

On camera, he complains about suggestions from some that Americans "lower our standard of living." He says, "Americans want to go forward, not backward."

He concludes by saying he is the "candidate of the forgotten American, who works hard and pays his bills and pays his taxes." That same forgotten American, he says, "goes to church on Sunday and believes in prayer in schools."

October 31, 1979

TV Political Season Under Way With Increased Confusion Likely

By LES BROWN

John B. Connally's paid telecast last night marked the opening of the political time-buying season, one that already promises to raise more than the usual amount of confusion and controversy because it has begun so early.

CBS is the only network that has agreed to sell time to the candidates this fall, and only in five-minute parcels. Nevertheless, it found itself faced with two political headaches yesterday: a news documentary now threatened because its subject, Senator Edward M. Kennedy, is becoming a candidate, and a formal complaint against it and the other networks filed with the Federal Communications Commission by the Carter-Mondale campaign committee.

Senator Kennedy's announcement on Monday that he would officially become a candidate for the Democratic Presidential nomination on Nov. 7 caused CBS-TV to alter its plan to televise a documentary on the Massachusetts Senator that had been scheduled for that very date. ·

The network was confronted with the problem of either canceling the program, "CBS Reports: Teddy," or finding an earlier air date for it, because to play it on the night Mr. Kennedy became a candidate, or any time thereafter, would touch off the application of the equal time rule. This rule requires television stations carrying a broadcast of a candidate to provide equivalent time to all other legally qualified candidates for the same office.

By officially declaring himself a candidate, a person running for public office effectively gives up opportunities to appear on talk shows and other forms of television entertainment programming, since these would invite equal time claims. What may be gained, however, is greater visibility in newscasts and news-interview programs, such as "Meet the Press," which are exempted from equal time considerations.

The one-hour Kennedy film report by Roger Mudd was described by a CBS News spokesman as "a fair, unbiased examination of Kennedy and not entirely favorable."

Officials of the CBS Broadcast Group, as of late yesterday afternoon, had not arrived at a decision on when to air the film or whether to air it at all.

Carter Campaign Announcement

The complaint against the networks at the F.C.C. concerns the refusal of all three to permit the purchase of a half-hour on Dec. 4, 5, 6 or 7 for the presentation of a film on the Carter Presidency to launch Mr. Carter's candidacy for re-election.

The complaint cites Section 315 of the Communications Act, which states that broadcast licenses may be revoked for "willful or repeated failure to allow reasonable access" to the airwaves for legally qualified candidates for Federal elective office.

The networks turned down the request on the ground that it was too early for national political campaigning and that since the primaries are held in individual states it was more appropriate for candidates to purchase the time locally. They argued, moreover, that under the equal time statute they would be obliged to sell numerous half hours to a multiplicity of candidates. ABC has stated flatly that it will not accept political advertising before 1980.

October 31, 1979

CHAPTER 3

China

China's Deputy Prime Minister Deng Xiaoping toured the U.S. in 1979 to look over American technology.
United Press International

CHINA AND AMERICA
A LOOK AT A LONG AND AMBIVALENT RELATIONSHIP

In the 1800's, Chinese products, like furniture in shop above, were exported to America; but Chinese laborers in the U.S. got rough treatment, as shown by Nast (inset).

Once staunchly anti-"Red China," Richard Nixon made two trips there in the 1970's. Here he is shown with an aide and Chinese officials on a cruise down the Li River.

By O. Edmund Clubb

Teng Hsiao - ping in 1978.

Tomorrow's historic visit to this country by Teng Hsiao-ping, China's powerful Deputy Prime Minister, marks an ending and a beginning. It ends what some Carter Administration officials describe as "a 30-year anomaly in international affairs," and it starts what the United States hopes will be a new and better era in world affairs.

The present euphoria over Mr. Teng's visit should not keep us from seeing some of the problems that may lie ahead. Granted the basic rationality of having regular diplomatic relations with a nation comprising one-quarter of the human race, there is no certainty that Chinese-American friendship is here to stay. Washington accepted Peking's three prime conditions for the normalization of relations: severance of American diplomatic ties with the Nationalist regime in Taipei, abrogation of the mutual-defense treaty of 1954 with that Government, and withdrawal of all remaining United States military forces from Taiwan. The establishment of ties was acknowledgment, said President Carter, of "simple reality."

The past history of Chinese-American relations has been far from simple. To many Americans, there has always been a special relationship between the two countries. But there have been frequent failures to understand each other and much ambivalence in policy. Indeed, Chinese-American relations have, since the 1940's, seesawed between the two poles of love and hate.

The American interest and good will of the 19th-century tea trade and missionary days were marred by passage of the Chinese Exclusion Acts of 1882 and 1902. President Wilson commended the Chinese for their revolution of 1911; yet he did not contest Japan's de-

Chinese trade group touring ironworks in Houston, 1975.

mand that it succeed to the German concession in Shantung in 1919, when Japan's good will was needed to facilitate the creation of the League of Nations. With the Japanese attack on Pearl Harbor, China became our ally, but after the Communist victory over the Nationalists in 1949 and the outbreak of the Korean War in 1950, China became an enemy to be "contained." For their part, the Chinese Communists during World War II viewed Americans as potentially valuable friends, but, after 1950, as outright "imperialists."

The two countries were now enemies, and the price of that enmity has been high — the Korean and Vietnam wars, and McCarthyism's distortion of American Asia policy. The man who began the *(Continued on Page 89)*

O. Edmund Clubb, last United States Consul General in Peking (1947-50), is author of "20th Century China" and "China and Russia: The 'Great Game.' "

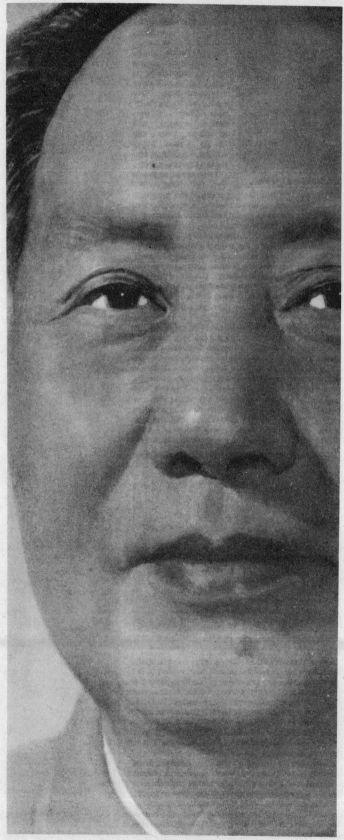

Mao Tse-tung held sway over the minds and lives of the Chinese for 30 years.

laborious process of extricating the United States from its boxed-in predicament was, paradoxically enough, Richard Milhous Nixon, who had first risen to national prominence by waging a crusade against Communism through the House Un-American Activities Committee. He had been Vice President when President Eisenhower entered into a mutual-defense treaty with the Nationalists in 1954. Mr. Nixon's trip to Peking in 1972, which marked the first visit to the People's Republic of China by an American President, paved the way for the visit tomorrow by Deputy Prime Minister Teng Hsiao-ping.

It is doubtful that Mr. Teng's visit will eradicate all political ambivalence toward that vast and populous land. The China question is raised once more as the 96th Congress deliberates on the constitutionality of President Carter's abrogation of the mutual-defense treaty with Taiwan. A recent Gallup poll indicated that while 58 percent of those polled favored the diplomatic recognition of the People's Republic, 47 percent felt that normalization should not be at the expense of Taiwan.

□

Since 1949, Taiwan has been the great issue dividing the United States and the People's Republic of China, and it threatens to be a major issue in this country for some time to come. As things stand, however, both Peking and Washington have agreed that the Government of the People's Republic of China is "the sole legal Government of China," and that Taiwan is part of China. Peking, as usual, has stated that the way in which Taiwan should be reunified with the mainland is "entirely China's internal affair."

But in ridding itself of a 30-year anomaly, the United States has created another one, albeit of shorter duration. Normalization of relations between China and the United States was in effect as of the first of this year and embassies will be established in Washington and Peking on March 1, but the treaty with the Nationalists will expire only on Dec. 31. Thus, for a full year, the United States will be in the position of standing committed to defend a government whose existence it has ceased to recognize.

Furthermore, the United States intends to maintain current commercial and other relations with Taiwan. But since the United States no longer recognizes the legitimacy of the Taiwan Govern-

ment, what is to take the place of the former diplomatic and consular offices? At present, there are some 60 formal U.S.-Taiwan accords and executive agreements (apart from the mutual-defense treaty), covering a host of matters ranging from atomic energy and investment guarantees to maritime matters and visas. American banks have extended some $2 billion in loans to Taiwan, the Export-Import Bank has lent, guaranteed or insured $1.7 billion more, and American business has invested another half-billion dollars on the island. U.S.-Taiwan two-way trade in 1978 is estimated at $7 billion.

To recognize the unrecognized, the Carter Administration is proposing to Congress the creation of a new, nominally private corporation called the American Institute in Taiwan, which will maintain informal diplomatic links to Taiwan as well as economic and military ties. The proposed legislation avoids all mention of the legally constituted government on the island, and instead refers to "the people on Taiwan."

Washington is evidently anticipating that Taipei will agree to the proposed new arrangements — if for no other reason than that the Nationalists have no practical alternative.

□

Just what does Washington stand to gain, in political and economic terms, from establishing full ties with Peking? Clearly, the American business community's hunger for commercial profit was one of the reasons for the Administration's decision. Peking has committed itself to achieving "four modernizations" — in industry, agriculture, defense,

and science and technology — in order to become a first-rank power by the year 2000. In another of China's periodic attempts to modernize — this time the effort is termed the "New Long March" — the country has mounted an economic plan emphasizing industrial development. According to Deputy Prime Minister Li Hsien-nien, this will involve a capital investment of $600 billion between now and 1985. There are speculations in the American business community that at least $200 billion of that will go toward the acquisition of plants, equipment, technology and services from abroad. Small wonder that foreign business interests have been galvanized.

Beginning with a Chinese-Japanese agreement of February 1978 for a $20-billion two-way exchange of goods in the upcoming eight years, the Chinese have signed a number of billion-dollar contracts. A Japanese consortium has agreed to build a $2-billion steel mill at Paoshan, near Shanghai. West Germany will construct a $14-billion steel complex in Hopei Province. A group of British banks late last year concluded an agreement with the Bank of China for a $1.2-billion line of credit to finance purchases of British industrial equipment. The United States is also in the picture. The Intercontinental Hotels Corporation, a Pan American World Airways subsidiary, has contracted to construct and operate a Chinese hotel chain; the cost will be $500 million. In addition, the United States Steel Corporation has contracted, for roughly $1 billion, to help China develop iron-ore mining facilities at Chitashan, in Manchuria.

At present, Peking's international credit rating is good,

for it has begun its huge acquisition program without foreign debts. Successful implementation of the overall import program, however, depends upon the buyer's capacity to pay. What is the extent of China's ability to generate foreign exchange? Obviously, Peking is counting heavily on profits from its budding petroleum industry. Current production is put at about 100 million tons per year, and output is expected to increase, according to a study prepared for the U.S. Department of Commerce, to a total of 335 million tons by 1985 — and to more than 400 million tons by 1988.

Yet, industry, agriculture and the military establishment will all be competing for China's limited foreign-exchange supply. And the country's industrial and transportation complex will be using more and more of the available petroleum, not to mention coal and natural gas.

There is another adverse factor in the equation: How can China, lacking technicians, a skilled labor force and managerial skills, hope to master and operate efficiently all that new equipment it is to acquire?

The Chinese, abandoning Mao Tse-tung's long-cherished principle of "self-reliance," are now ready to avail themselves of what is delicately termed "deferred payment finance." Further, Peking has indicated a willingness to permit private investments in China and to embark on joint ventures with foreign companies to develop such natural resources as petroleum, coal and iron ore. The foreign companies would be supplying capital, technology and managerial skills, and, upon completion of a project, they would be paid through a share in the market proceeds.

Foreign Trade Minister Li Chiang has revealed that China would need "several tens of billions of United States dollars" in foreign credits in order to finance its projects for modernizing the country by the year 2000. To that end, it is prepared to accept government-to-government loans.

Peking's modernization program is hampered by a series of complex factors. "Once [the Chinese] have amassed a debt load of $30 billion to $40 billion," said a U.S. Commerce Department official, "we'll start worrying."

□

Apart from economics, the political factor patently played an important role in sparking the Administration's December move. It is no secret that some high Adminis-

A TROUBLED COURSE

For almost 200 years the American vision of China and its people has derived as much from myths as from first-hand experience. In the late 18th century, attracted to that distant land primarily for the trade it had to offer, the newly independent United States saw China as a means of providing economic independence from Britain. In the 19th century, however, when China was on the verge of social, political and economic collapse, Americans came to view the Chinese as backward, heathen, ill-governed and weak. That notion in part fanned the intense prejudice — and subsequent legislation — against Chinese laborers in America. It was reinforced by the Boxer Rebellion of 1900, a popular uprising that

sought to expel all foreigners.

With the outbreak of civil war in China between the Nationalists and Communists in the late 1920's, America supported Nationalist leader Chiang Kai-shek, who had embraced both Christianity and capitalism. And when he lost the war in 1949, the U.S. considered China "lost." During the next two decades, the U.S. and China were on opposite sides of two major conflicts. But in the 1970's, they started to dismantle the Bamboo Curtain, through Ping-Pong diplomacy and the visits of two American Presidents. Tomorrow, at the beginning of the Chinese New Year, a third President will be welcoming China's de facto leader to the nation's capital.

At Versailles in 1919, the Chinese tried to reclaim Shantung province from Japan.   *During the Korean War in the 1950's, Americans fought against the Chinese.*

tration officials thought that full ties with China would help the United States "play the China card" against the Soviet Union.

There is, however, an initial price to pay. Over the years, the United States has professed concern for its "credibility," and time and again has assured Taipei that it intended to respect its ties to that "loyal ally." But Washington gave its ally only eight hours' advance notice before President Carter announced the dissolution of the alliance. The 1972 démarche toward Peking that came to be known as the "Nixon shock" has been duplicated by a "Carter shock," which will inevitably leave a telltale imprint on our cherished "credibility."

Over the last six years, the situation in the West Pacific has undergone a basic change, and the recognition of China has introduced new elements of instability into this strategic area The United States' relations with its remaining three West Pacific allies — Japan, the Philippines and South Korea — and with the Soviet Union are all affected by the break with Taiwan. All four can be expected to reassess their respective strategies in the light of current developments.

Japan in 1972 recognized the Peking regime (it now has an "Interchange Association" on Taiwan), and last August signed a treaty of peace and friendship with China. Manila and Peking bowed amicably toward each other last year. And Seoul, if still without diplomatic relations with China, is "talking" to Peking. Détente has indeed already taken place in these sectors. But now that the United States has recognized the People's Republic of China as the sole legitimate government of China *and* Taiwan, what will be the meaning and future functioning of the several American alliances originally created to "contain China"? The feeling of the several allies will inevitably be that the old ties have loosened fur-

ther — and the natural impulse will be to shift positions, to follow more independent courses in the changing conditions in Asia. Japan, in particular, the "pillar" of our West Pacific strategy, bids fair to become the independent-minded France of our West Pacific alliances.

What has been Moscow's reaction to the change in Chinese-American relations? In his letter to President Carter regarding the Dec. 15 communiqué, Soviet leader Leonid I. Brezhnev stated that Moscow deemed it proper for major sovereign nations to maintain full diplomatic relations. But in drafting his note, Mr. Brezhnev was doubtless keenly mindful that in this decade, under both Maoist and post-Mao leadership, Peking has moved closer to the United States and has stigmatized the American quest for détente with the Soviet Union as "appeasement."

Peking still adheres officially to Mao Tse-tung's "Three Worlds" strategy (first given public expression, ironically enough, by Teng Hsiao-ping before the United Nations Special Session on raw materials and development in April 1974) that forecasts an ultimate revolutionary confrontation between the third world of developing countries (including China) and the first world made up of the Soviet Union ("principal enemy") and the United States ("secondary enemy"). As these "worlds" move toward the final confrontation, Maoist tactics dictate the maneuvering of the secondary enemy into conflict with the principal enemy. It is not surprising that, according to fuller accounts of the contents of the Brezhnev note, the Soviet leader also communicated Moscow's concern that Washington not enter into the new Chinese-American relationship motivated primarily by the urge to manipulate the Chinese against the Soviet Union.

The Soviet Union has been given some assur-

ances in that regard. The United States and its NATO allies stand stoutly committed to the concept of détente. President Carter stated that "the normalization of relations between the United States and China has no other purpose than the advancement of peace." And his national security adviser, Zbigniew Brzezinski, was more categorical: The move was not directed, he said, at the Soviet Union or anyone else; "we designed to accomplish our objective of shaping a more open, pluralistic international system." The "antihegemony" clause contained in the joint communiqué ("hegemony" is a Chinese code word for Soviet expansionism) does not in fact have a minatory anti-Soviet thrust, for Washington seemingly succeeded in getting something like the formulation adopted in the Chinese-Japanese treaty, and the clause is qualified by a significant proviso: "Neither [party] is prepared to negotiate on behalf of any other third party or to enter into agreements or understandings with the other directed at other states." And it is further stated that "both believe that normalization of Chinese-American relations is not only in the interest of the Chinese and American peoples but also contributes to the cause of peace in Asia and in the world."

Most probably, while bound to watch developments closely, Moscow will remain relatively philosophical about the establishment of diplomatic ties between the United States and China. After all, the Soviet Union has an embassy in Peking, too.

From now on, the political and economic policies followed by the industrial-

ized countries will have a strong impact on Peking's strategic thinking. The hard reality is that economic development and global strife do not mix. Logic suggests that, even as Peking has substantially revised Maoist domestic policies, so it may well in due course discard or substantially de-Maoize the apocalyptic "Three Worlds" strategy as unworkable and unprofitable.

What, then, are the prospects for the future? Will China achieve the lofty aims fixed for it by its new leaders? Under the impact of major shocks, as from war or natural disaster, the country might again be thrown into turmoil. Actual achievements will almost certainly fall far short of the inflated goals. But assuming reasonable stability and continuity for the pragmatic Chinese leadership, China should by 1985 have made substantial progress toward the "four modernizations" it seeks.

I have expressed reservations about the manner and consequences of the Carter Administration's establishment of ties with China, but I nevertheless believe that there is good reason to be optimistic about the future. Even if the road ahead for China were to remain long and arduous, the country at least would be better integrated into the world community, and might be contributing in significant ways to the building of a global structure in which — again in the words of President Carter — "peace will be the goal and the responsibility of all nations."

And the troubled world could then proceed with an added margin of safety along the perilous road to the year 2000. ■

January 28, 1979

Teng's Visit Seeks to Tap the Rich Lode of America's Technology

China's Road to Progress Is Mostly Uphill

By FOX BUTTERFIELD

ATLANTA — When Deputy Prime Minister Teng Hsiao-ping inspected a Ford Motor Company assembly plant near Atlanta last week, he got a look at a fundamental difference between Chinese and American industry. One worker in the Ford factory turns out 50 cars a year. His Chinese counterpart produces only one. This statistic underscores one purpose of Mr. Teng's extraordinary mission to the United States: to look over United States technology, which Peking badly needs to carry out its ambitious plan to make China into a fully modern industrial state by the year 2000. "We in China are faced with the task of transforming our backwardness and catching up promptly with the advanced countries of the world," Mr. Teng, China's senior Deputy Prime Minister, told Southern political and business leaders. "We want to learn from you."

That is a popular message in America. The economy needs all the business it can get, and China looks like the new bonanza. But the obstacles to Chinese modernization by the end of the century are enormous, and Peking's orders for industrial plants and technology are not likely to be big enough to have a broad effect, much as they may benefit some individual firms. Among the obstacles are these:

• China's agricultural technology is so backward that roughly 75 percent of its 900 million people still must till the land, compared to less than 5 percent in the United States. Attempts to increase the arable land in China have been cancelled out by the takeover of land by new factories or growing cities.

• Given its poor management and often shoddy standards of work, China may not be able to absorb all the new technology which it has already begun to import. The Communist Party paper, Jenmin Jih Pao, recently reported that of the $4 billion worth of farm machinery stored in China, one-third was unsaleable because of low quality. In one district in Kirin Province, it said, an official ordered a commune to buy all the tractors produced in a local factory, but only three of 10 tractors sent actually arrived.

• Will Peking be able to rekindle the traditionally strong Chinese work ethic among an urban work force disenchanted by years without material incentives? American businessmen who offer free seminars on their products in Peking are often surprised to find that Chinese bureaucrats go to lunch promptly at 11:45 and do not come back until 2 P.M. "Just when we got to the crucial point, they took off," a shipping specialist recalled. "The next day they asked us to make our presentation shorter."

• How will China pay for the $70 billion or more in foreign commitments Peking is currently discussing? Last year China's foreign trade totalled only about $20 billion.

• Last but not least, will Mr. Teng, who is 74 years old, live long enough to insure that his pragmatic policy of turning outward for help will survive him?

"The problems are manifold, it is too early for answers," says Nicholas Lardy, a specialist in Chinese economics at Yale. "You can only ask the questions." Mr. Lardy believes that Mr. Teng's reforms have already overcome the dangerous economic stagnation of the mid-1970's, the result of political quarrels and worker unrest. He feels that China's economic performance last year shows there is nothing structurally wrong with the economy. A 12 percent increase in industrial output and a rise in grain production of 10 million tons resembled the remarkable growth from 1953 to 1973. But, according to a study released last month by the Central Intelligence Agency, even if China achieved an overall annual gain of 6 percent in its gross national product, by the year 2000 its G.N.P. would still be 15 percent below the 1975 United States G.N.P. And, with a possible population of 1.5 billion by then, its per capita G.N.P. would be less than $1,000, about that of Brazil today.

Much will depend on Peking's ability to push agricultural production, and Mr. Teng has ordered that it be given first priority. A party Central Committee meeting last month decided that the peasants will get 20 percent more for their grain, while prices of farm equipment will be cut. They are also to get a greater say over the crops they plant, with less interference by authorities. This means that, in its agricultural role at least, the commune may gradually wither away. Peking hopes that 13 huge new chemical fertilizer plants it purchased in 1973 and 1974, plus farm machinery it is now buying abroad, will also spur production. But the fate of the fertilizer factories illustrates one problem — they were all to be in production by now, but only half of them are and, reportedly, only two are operating at full capacity. This suggests the possibility of similar delays on the new factories, hotels, mines and oilfields for which Peking is now contracting with foreign companies.

Delays could be critical, for China evidently intends to use income from many of these projects to help pay for them — exporting oil, for example, from offshore oilfields, for the joint development of which they are now negotiating with companies such as Exxon and Philips Petroleum. Delay means an expensive gap between the time Peking borrows the money it needs, and the time when it pays it back. In Mr. Lardy's view, Peking is probably aware of many of these problems and will ultimately be more conservative in its purchases abroad than eager businessmen would like. Indeed, some analysts wonder if Peking may have deliberately overstated its interest in purchasing foreign technology to help create strong support in the American business community for the normalization of United States-Chinese relations. This may be too cynical a view but, in recent months, a steady succession of United States businessmen who have been to Peking to try to conclude deals have given China millions of dollars worth of free advice and information about new technology.

February 4, 1979

The China Trade: Companies Mob Peking

China—How Much Will The West Change It's Industrial Face?

Urumchi
Cement, Machinery

Lanchow
Oil Refining, Chemicals

Paotow
Iron, Steel

Taiyuan
Iron, Machinery, Chemicals

Taching
Oil Production

Harbin
Machinery

Changchun
Automobiles, Machinery

Shanyang
Iron, Chemicals, Electrical Equipment

Luta
Steel, Railroad Equipment

Peking-Tientsin
Textiles, Cement, Chemicals

Tsingtao
Textiles, Locomotives

Shanghai-Nanking
Machinery, Textiles, Shipbuilding, Paper

Wuhan
Iron, Steel, Chemicals

Foochow
Chemicals

Nanchang
Aircraft

Canton
Sugar Refining, Paper, Shipbuilding

Changsha
Nonferrous Metals, Steel, Cement

Chanking-Red Basin
Sugar Refining, Chemicals

Sian
Textiles, Cement

Urumchi
Taching
Harbin
Changchun
Shenyang
Paotow · Peking · Luta
Tientsin
Lanchow · Taiyuan
Tsingtao
Sian
Nanking
Wuhan · Shanghai
Chungking · Nanchang
Changsha · Foochow
Canton · Taipei
TAIWAN
Hong Kong
Lhasa
Yellow River
Yangtze River

By FOX BUTTERFIELD

PEKING

A VISITOR to the Peking Hotel might be excused for thinking that a convention of major American corporations is taking place. There's the man from United States Steel, or from Pan American World Airways or Union Carbide or Exxon or the Chase Manhattan Bank.

Representatives from dozens of Western businesses have taken up residence in Peking to negotiate what may be a bonanza of perhaps $70 billion worth of deals with China over the next few years. These deals will form a vital part of China's enormously ambitious program to turn itself into a modern industrial power by the end of the century.

Whether Peking will achieve its aim is a matter of debate. But in 1978, China's economy enjoyed its first real growth since 1975. And with Peking's pragmatic leaders committed to introducing a more market-oriented economy, improving the worker's lot and importing huge amounts of advanced technology, this year looks even better.

Industrial output, which had dropped in 1976 (the year Mao Tse-tung died) and then began to recover in 1977, soared by 12 percent in 1978, economists in Hong Kong estimate.

Significantly, that represents a return to China's remarkable average rate of industrial expansion of about 12 percent during the country's first two decades under the Communists. It suggests that troubles in China's factories in the mid-1970's were caused by political dislocations rather than by structural defects in the economy, in the view of Nicholas Lardy, professor of Chinese economy at Yale University.

At the same time, China's foreign trade, which stagnated for several years as China sought to reduce an unusually high trade deficit, suddenly jumped from $14.8 billion in 1977 to about $20 billion last year.

Foreign companies are now being allowed to invest directly in China and build factories, mines and other facilities — using their technology and China's cheap labor. Western technology and cheap labor, ironically, formed the happy combination that propelled the dramatic growth of Hong Kong, South Korea and Taiwan in recent years.

Japan appears likely to win a major share of this new business, given its close political ties and its geographic, linguistic and cultural proximity.

But some American companies have already made important breakthroughs. And with President Carter's decision finally to normalize relations with China in 1979, trade with the United States seems likely to surpass greatly the 1978 total of $1 billion, itself a record. Almost $900 million of that represented American exports to China.

Among the most important deals: Inter-Continental Hotels, a subsidiary of Pan-Am, reached agreement to build a chain of at least five hotels in China valued at $500 million. The Fluor Corporation of Los Angeles will design and possibly construct a copper smelter that could run to $800 million and the Bethlehem Steel Corporation will build an iron ore mine for $100 million.

Some of the largest deals may be signed early in 1979 — joint production contracts with one or more United States oil companies for the development of some of China's huge offshore oil fields. Other projects for coal and iron mines, heavy-construction equipment and railroads may also be forthcoming, according to Nicholas Ludlow of the National Council for U.S.-China Trade.

A number of United States banks, including Citibank, Chase Manhattan, the Bank of America and the Manufactuers Hanover Trust Company, have all sent senior executives to Peking to discuss financing this new business.

Despite these advances, China's economy still has its troubles. Most serious, there has been no substantial growth in agriculture since 1975, when grain production reached 285 million tons. The population has since increased by perhaps 40 million.

To some extent this stagnation has resulted from bad weather. But to help stimulate production, China is now offering peasants a greater say over what they plant on communal land and more freedom to work private plots.

Peking is also experimenting with industry. Factories are to sign contracts directly with each other and wages are to be tied to plant profits.

Some economists wonder whether China can adjust to all these changes at once. Does its work force, for instance, have enough skills to handle all the new technology? Will there be a backlash from Maoist conservatives?

"They are going to have some troubles," remarked a United States banker in Hong Kong. "But right now things look very good and anyone who doesn't sign a contract with China is going to have to explain why not to his board of directors." □

February 4, 1979

Party Leaders

Following are names of leading Chinese political figures spelled in the Pinyin system. In parentheses, for reference, are the spellings in a modified Wade-Giles system.

Chairman
of the Central Committee
of the Chinese Communist Party
Hua Guofeng (Hua Kuo-feng)

Deputy Chairmen
Chen Yun (Chen Yun)
Deng Xiaoping (Teng Hsiao-ping)
Li Xiannian (Li Hsien-nien)
Wang Dongxing (Wang Tung-hsing)
Ye Jianying (Yeh Chien-ying)

Members of the Politburo
Chen Xilian (Chen Hsi-lien)
Chen Yonggui (Chen Yung-kuei)
Chen Yun (Chen Yun)
Deng Xiaoping (Teng Hsiao-ping)
Deng Yingchao (Teng Ying-chao)
Fang Yi (Fang Yi)
Geng Biao (Keng Piao)
Hu Yaobang (Hu Yao-pang)
Hua Guofeng (Hua Kuo-feng)
Ji Dengkui (Chi Teng-kuei)
Li Desheng (Li Teh-sheng)
Li Xiannian (Li Hsien-nien)
Liu Bocheng (Liu Po-cheng)
Ni Zhifu (Ni Chih-fu)
Nie Rongzhen (Nieh Jung-chen)
Peng Chong (Peng Chung)
Su Zhenhua (Su Chen-hua)
Ulanhu (Ulanfu)
Wang Dongxing (Wang Tung-hsing)
Wang Zhen (Wang Chen)
Wei Guoqing (Wei Kuo-ching)
Wu De (Wu Teh)
Xu Shiyou (Hsu Shih-yu)
Xu Xiangqian (Hsu Hsiang-chien)
Ye Jianying (Yeh Chien-ying)
Yu Qiuli (Yu Chiu-li)
Zhang Tingfa (Chang Ting-fa)

Alternate Members
Chen Muhua (Chen Mu-hua)
Seypidin (Saifudin)
Zhao Ziyang (Chao Tzu-yang)

Pronunciation

Following is the Pinyin system alphabet showing the pronunciation with approximate English equivalents. In parentheses, for reference, are the corresponding characters that were used in a modified Wade-Giles system.

a (a) Vowel as in far

b (p) Consonant as in be

c (ts) Consonant as in its

ch (ch) Consonant as in church, strongly aspirated

d (t) Consonant as in do

e (e) Vowel as in her

f (f) Consonant as in foot

g (k) Consonant as in go

h (h) Consonant as in her, strongly aspirated

i (i) Vowel as in eat or as in sir (when in syllables beginning with c, ch, r, s, sh, z and zh)

j (ch) Consonant as in jeep

k (k) Consonant as in kind, strongly aspirated

l (l) Consonant as in land

m (m) Consonant as in me

n (n) Consonant as in no

o (o) Vowel as in law

p (p) Consonant as in par, strongly aspirated

q (ch) Consonant as in cheek

r (j) Consonant as in right (not rolled) or pronounced as z in azure

s (s, ss, sz) Consonant as in sister

sh (sh) Consonant as in shore

t (t) Consonant as in top, strongly aspirated

u (u) Vowel as in too, also as in the French tu or the German München

v (v) Consonant used only to produce foreign words, national minority words and local dialects

w (w) Semi-vowel in syllables beginning with u when not preceded by consonants, as in want

x (hs) Consonant as in she

y Semi-vowel in syllables beginning with i or u when not preceded by consonants, as in yet

z (ts, tz) Consonant as in zero

zh (ch) Consonant as in jump

Place Names

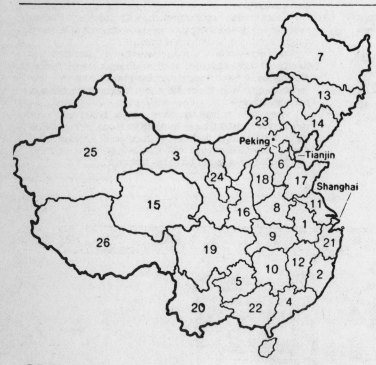

Following are names of municipalities, provinces, autonomous regions and well-known cities spelled in the Pinyin system. In parentheses, for reference, are the conventional forms.

Municipalities
Beijing (Peking) *
Shanghai (Shanghai)
Tianjin (Tientsin)

Provinces and selected cities
1 Anhui (Anhwei),
Hefei (Hofei)
Bengbu (Pengpu)
2 Fujian (Fukien)
Fuzhou (Foochow)
Xiamen (Amoy)
3 Gansu (Kansu)
Lanzhou (Lanchow)
4 Guangdong (Kwangtung)
Guangzhou (Canton) *

Shantou (Swatow)
5 Guizhou (Kweichow)
Guiyang (Kweiyang)
Zunyi (Tsunyi)
6 Hebei (Hopei)
Shijiazhuang (Shihkiachwang)
Tangshan (Tangshan)
7 Heilongjiang (Heilungkiang)
Harbin (Harbin)
Daqing Oilfield (Taching)
Qiqihar (Tsitsihar)
8 Henan (Honan)
Zhengzhou (Chengchow)
Luoyang (Loyang)
Kaifeng (Kaifeng)
9 Hubei (Hupeh)
Wuhan (Wuhan)
10 Hunan (Hunan)
Changsha (Changsha)
11 Jiangsu (Kiangsu)

Nanjing (Nanking)
Suzhou (Soochow)
Wuxi (Wusi)
12 Jiangxi (Kiangsi)
Nanchang (Nanchang)
Jiujiang (Kiukiang)
13 Jilin (Kirin)
Changchun (Changchun)
14 Liaoning (Liaoning)
Shenyang (Shenyang)
Anshan (Anshan)
Luda (Luta)
15 Qinghai (Chinghai)
Xining (Sining)
16 Shaanxi (Shensi)
Xian (Sian)
Yanan (Yenan)
17 Shandong (Shantung)
Jinan (Tsinan)
Qingdao (Tsingtao)
Yantai (Yentai)
18 Shanxi (Shansi)
Taiyuan (Taiyuan)
19 Sichuan (Szechwan)
Chengdu (Chengtu)
Chongqing (Chungking)
20 Yunnan (Yunnan)
Kunming (Kunming)
Dali (Tali)
21 Zhejiang (Chekiang)
Hangzhou (Hangchow)

Autonomous regions
22 Guangxi Zhuang (Kwangsi Chuang)
Nanning (Nanning)
Guilin (Kweilin)
23 Nei Monggol (Inner Mongolia) *
Hohhot (Huhehot)
Baotou (Paotow)
24 Ningxia Hui (Ningsia Hui)
Yinchuan (Yinchwan)
25 Xinjiang Uygur (Sinkiang Uighur)
Urumqi (Urumchi)
26 Xizang (Tibet), *
Lhasa (Lhasa)

*The four place names indicated by asterisks will continue to appear in The Times in their familiar forms.

Source: Adapted from Beijing Review; border from Map of the People's Republic of China

February 4, 1979

CHINESE TROOPS AND PLANES ATTACK VIETNAM; U.S. URGES WITHDRAWAL, HANOI IN PLEA TO U.N.

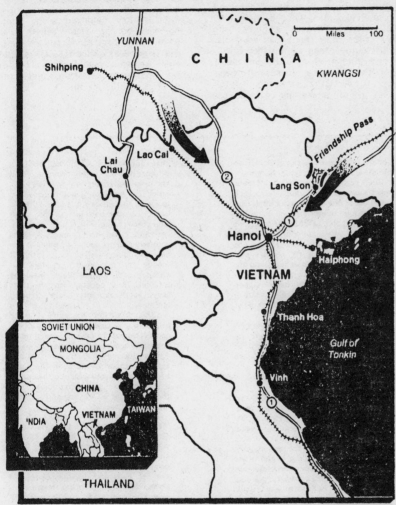

The New York Times / Feb. 18, 1979

Chinese forces reportedly attacked across most of the length of the Vietnam border. Arrows indicate movements thought to be the main thrusts.

A Classic Military Operation

China's Drive on a Wide Front in Vietnam Leaves Hanoi Guessing About Main Focus and Objective

By DREW MIDDLETON

Military Analysis

The Chinese invasion of Vietnam followed, at the outset, the classic pattern of attacks over a wide front — with the defenders left guessing about the principal axis of advance.

The operations of the Chinese Army developing in the Lao Cai area of Vietnam appeared the most dangerous to the Vietnamese at the close of the first day's fighting. If pushed successfully, the operations would give the Chinese control of the railroad to Hanoi from China and the north.

A second attack, in the Lang Son area, northeast of the capital, also appeared to be gaining momentum. But it was possible that one or both of these might prove to be holding operations and that the main thrust along the front of about 480 miles might come from elsewhere, possibly along the coast.

The most important missing element in the strategic equation is Chinese intentions about the duration of the action. Until very recently, it was thought that in any action against Vietnam, the Chinese Army would follow the pattern of the 1962-63 operation in India.

In the earlier operation, the Chinese made a series of successful attacks, for limited objectives, in India's northern frontier territory. The Chinese consolidated their positions and did not advance further.

Less Freedom of Action Today

The Chinese, experts emphasized, do not possess the same freedom of action today that they had during the invasion into northern India.

By attacking Vietnam, an ally of the Soviet Union, the Chinese may have jeopardized what they concede is a weak military position on the long border with the Soviet Union. The expectation is that the Chinese operations in Vietnam, although delivered with the maximum weight of all conventional weapons, will not be prolonged.

The opinion among experts was that the operation would prove to be a hit-and-run raid on a vast scale.

The terrain in the areas where the major Chinese advances have developed is rough and hilly, but it offers no serious obstacles to troops accustomed to moving without a great amount of truck transport. An advance along the coastal plain, if tried, might encounter difficulty as a result of the waterways and canals twisting across that region.

The Chinese apparently assembled about 30 regular divisions along the frontier facing the four Vietnamese provinces of Lang Son, Lai Chau, Huong Lien Son and Quang Minh. It is not known how many of China's 11 armored divisions were used in the operation, but tanks have been reported leading infantry in several local attacks along the front.

The normal strength of a regular infantry division is 12,000 and that of an armored division is 10,000 men.

The total Chinese force along the border may be as large as 360,000 men, although experts on the area regard 330,000 as probably more accurate.

The bulk of the divisions have been drawn from the Kunming and Canton military districts. The task of guarding communications is believed to have been given to the local force units in the two districts.

Main force units are regular troops under the direct command of central headquarters of the Chinese Army in Peking. They are generally the best-armed, trained and led divisions of an army that totals 135 divisions — 11 armored, 121 infantry and 3 airborne. In addition to these units there are also 40 artillery divisions.

Overall command in the Vietnam operation is believed to lie with Gen. Yang Teh-chih, who won recognition during the Korean War.

One of Vietnam's problems in the present situation is that many of its best troops — estimates range from 80,000 to 100,000 men — are in combat in Cambodia. They are fighting a guerrilla force armed and supplied to a great extent by the Chinese.

The Chinese may be counting on the absence of many of the best Vietnamese troops and commanders to insure a swift and wounding thrust into the northern heart of the country.

Drive on Hanoi Not Ruled Out

A drive on Hanoi cannot be discounted, according to experts.

The sources add that the Vietnamese raids across the Chinese frontier in the last week apparently pushed the Chinese command to the point where it asked for and received a mandate for much more than a limited punitive attack.

From the first battlefield reports, it appears that the Chinese Air Force, operating from a semicircle of bases, obtained early tactical control over the battlefield.

This may be a passing phase. Air power is the most flexible of weapons. The Vietnamese can switch their MIG-21's and other aircraft to the defense of Hanoi. The Chinese apparently are relying on their 80 MIG-21's and rebuilt MIG-19's for the air strike and cover operations.

A Major Calculated Risk

Senior American military officers, commenting anonymously on the operation, agreed on one point: the Chinese were taking a major calculated risk in attacking southward when the Soviet Union has a force of 44 divisions of relatively high quality deployed along China's northern border. This strength is supported by a large air force — some estimates put it at 1,300 aircraft — and a nuclear weapons arsenal whose accuracy is far superior to that of the Chinese medium-range ballistics missile

February 18, 1979

CHINA'S HIDDEN CAVE ART

For the first time, American tourists can visit China's three 'Treasure Houses, immense Buddhist temple-caves filled with murals and monumental sculptures. Text and photographs by Audrey Topping.

American devotees of the monumental Buddhist art of ancient China have long had to content themselves with museum sculptures and bas-reliefs taken from Chinese temple-caves by fortune hunters in the 1930's. But now, for the first time, some American visitors are being permitted to see the ancient sculptures and murals in the splendor of their original sites — three extensive temple-cave areas called the "Treasure Houses of China" that have been recently restored after a thousand years of neglect.

China's Department of Cultural Relics has spent 29 years slowly but systematically excavating and restoring the monumental art in these mysterious man-made grottoes: thousands of Buddhist stone sculptures, many more than 15 centuries old and ranging in size from the miniature to the gigantic, multitudes of deteriorating wooden statues of Buddhist deities and intricate religious frescoes from remote temples.

Even the most indefatigable tourist cannot expect to see all the caves in one trip. They are located in widely separate areas of China: The Tunhuang Grottoes are in Gansu in the west; the Yungang Caves in Shanxi to the north,

Audrey Topping is a writer and photographer specializing in Chinese culture.

and the Lungmen or Dragon Gate Temple caves in Henan, in central China.

Each comprises an array of enormous caverns and small caves hollowed out of steep sandstone cliffs that make whole mountainsides look like giant honeycombs. Within the curious black holes dwell hosts of stone Buddhas symbolizing the past, present and future. These serene figures are protected by fierce guardian kings and are surrounded by galaxies of compassionate bodhisattvas, dutiful lohans (disciples), flying fairies and a myriad of lesser deities covering every inch of the cave walls.

Wooden monasteries and golden tile-roofed temples, their curled eaves studded with mythical beasts to ward off evil spirits, cluster near the grottoes. They house vast numbers of exquisite wooden divinities and colorful frescoes depicting the story of Buddha.

The original excavations and art were executed by legions of devout Buddhist monks, who carried on the temple-cave tradition begun in India. The monks worked with skilled craftsmen and artists under the patronage of pious emperors. Building began in the fourth century and continued intermittently until the 14th century, spanning the period when Buddhism was China's primary religion.

Buddhism arrived at a crucial time in Chinese history. The glory of the Han dynasty (202 B.C.-A.D. 221) was eroded, and traditional Confucian standards had crumbled. After 200 years of

chaos and warfare, China was ripe for a religion that advocated peace on earth and promised a utopian life in the hereafter. The common people were eager to soothe their miserable lives with the balm of a nirvana that was open to all who practiced sufficient piety and self-restraint. And the emperors, who wanted to ensure their own place in this new paradise, poured vast sums of money and manpower into the construction of temple-caves honoring Buddha. Their heritage was the most wondrous monumental sculpture in all of China.

Although all the "Treasure Houses" are similar in concept and contruction, each retained its own unique style of sculpture and murals. The art often reflected the physical features of the reigning emperor and the clothing and life style of his court. One sixth-century emperor made such generous gifts to the caves that he was called the "Savior Bodhisattva."

Construction of the oldest of the "Treasure Houses," the Tunhuang Grottoes, was begun in A.D. 366 and continued for a thousand years. It includes the Makao Grottoes, the Western Cave of the Thousand Buddhas and the Ten Thousand Buddhas Gorge. Of these, the Makao are the most extensive; the statues and murals would fill a 15-mile-long gallery.

The history of the Tunhuang Grottoes is marked by neglect and thievery. From the late Ching dynasty (A.D. 1644-1912) through the rule of the Kuo-

Left: Visitors descend to the Temple of Celestial Guardians in Lower Hua Yan Monastery, in Shanxi. Right: In the temple, lacquered goddesses attend Buddha.

mintang, they were irreparably damaged by plundering warlords and foreign treasure hunters. Drifting sand buried the lower-level caves and walls crumbled. And there was flaking of the murals and statues caused by weathering, birds and insects. After the establishment of the People's Republic in 1949, restoration workers built five-story walkways to reach every cave along the 1,600-meter cliff face. Today, 492 of the caves in Tunhuang have been restored, together with 2,000 statues of divinities and 45,000 square meters of religious murals.

The second "Treasure House," a group of 53 temple-caves called the Yungang or "Cloud Hillock" caves, were begun in A.D. 460, 100 years after work began in Tunhuang. They form a honeycomb in a sandstone cliff in the Wuchow Mountain Pass on the edge of the Gobi desert near the Great Wall.

The form and style of the Yungang sculpture differ significantly from that of the Tunhuang, mainly because the caves were carved under the auspices of a foreign tribe of Mongol and Turkic descent called the Topa Tatars. The Topas, after a long period of warfare, founded the Northern Wei dynasty (A.D. 386-535) and promoted Buddhism as a state religion. The enormous sculptures of Buddha in the five major caves of Yungang were deliberately made to resemble the Northern Wei emperors, with distinctive foreign features and clothing. Soon, however, the Topa emperors and their imperial courts began losing their barbarian characteristics. In fact, when the emperor decreed in A.D. 486 that the court wear Chinese rather than nomadic dress and adopt Chinese customs, the Yungang statuary rapidly became more traditionally Chinese in style.

The Yungang caves were one of the first cultural sites the Communist Gov-

ernment decided to preserve. Even during the Cultural Revolution, when Red Guards rampaged through the countryside destroying traditional art, the Central Committee had a local militia stand a 24-hour guard by the caves. They were of special interest to the late Premier Zhou Enlai; on a 1973 visit, he left instructions that repairs be completed in three years, and they were.

At the end of the fifth century, the

This enormous Buddha, one of the earliest from Yungang, turns his palm outward in a gesture signifying "freedom from fear."

Northern Wei emperor decided to move to the traditional Chinese capital at-Loyang in Shanxi. The Yungang caves were forgotten and a new Buddhist treasure house, called Lungmen, was begun. Most of the Lungmen temple caves were carved between A.D. 640 and 720 in cliffs banking the Yi River. The most famous is called Fenghsien, where the 56-foot-high Buddha Vairocana sits majestically on a thousand-petaled lotus representing the universe, and gazes like an all-knowing colossus over the river and into the future.

Not long after the Fenghsien cave was finished, Buddhism went into decline. Under the Tang, the Buddhist Church grew so powerful and rich in land that even the emperor feared its influence. "Buddhist monks," he claimed, "are as thick as blades of grass." Hundreds of thousands of monks claimed exemption from taxes and military duties.

In A.D. 845 the emperor's anger resulted in a major Buddhist persecution. Thousands of monks were killed and great damage was done to the Lungmen caves. After the 10th century, Buddhism receded before the renascent Confucianism and the great Buddhist sculpture-caves were consigned to oblivion by the Confucian scholar class who considered Buddhism a foreign religion.

The temple-caves were virtually lost to civilization until the early 20th century, when

A restored fresco in Upper Hua Yan Monastery (A.D. 1140) in northern China shows an event in one of the past lives of Buddha.

foreign adventurers began to loot them systematically. A visitor today cannot fail to notice that large numbers of statues have lost their heads. In Yungang, 680 heads of Buddhas, bodhisattvas and flying fairies (Apsaras) are missing, 24 bas-reliefs have been torn away or mutilated and numerous statues have been removed completely.

Before 1949, foreigners who managed to reach the caves marked the statues or parts they wanted with chalk, then arranged — often with the Chinese caretakers — to chop the statue parts off and sell them. According to a 1930 Shanghai newspaper, the Yungang caves lost more than 90 pieces of sculpture this way. Nearly all, it stated, were sold to "foreign learned societies." (Most of them are now in European, Japanese or American museums.)

At the time, foreign buyers felt no guilt about buying the pieces because the sales were transacted with full knowledge of the Chinese Nationalist Government. In fact, many collectors felt they were rescuing neglected objects by sending them to places where they could be preserved, appreciated and marveled at by art lovers the world over.

Although the Chinese guides who take tourists through the caves today are obviously sensitive to this pillaging, they do not hold the buyers at fault but, rather, the Nationalist Government officials who ruled before the Communist takeover. Their tact is remarkable. When Philippe de Montebello, the director of the Metropolitan Museum of Art, was invited to visit the Yungang caves recently, no one mentioned the cave's sculptures on permanent exhibition at the Metropolitan.

In print, however, the Chinese are more direct. One 1977 pamphlet on the Yungang caves printed by the Cultural Relics Publishing House states: "From the start of the 20th century, foreign imperialists connived with reactionaries in China to plunder the caves, causing irreparable damage. Before Liberation, more than 1,400 sculptures or heads of images had been stolen or smashed. During the War of Resistance against Japan, some imperialists on the pretext of research hacked off certain images and damaged the caves. In 1934, three exquisite sculptures were stolen from the adjunct cave of Cave 15; they are now on exhibition in the Metropolitan Museum of Art in New York."

☐

When contemplating the huge amount of money and work involved in restoring the temple-caves, one wonders why an atheistic Communist Government attaches so much importance to ancient religious art. One explanation was given in a Committee of Cultural Relics pamphlet: "As Marx said, for a long period man explained history in terms of superstition, but now we are using history to explain superstition. Our task, therefore, is to study the sculptural art of the [Buddhist] caves from a class and historical standpoint and, by means of a Marxist analysis, to reverse the reversal of history."

But all of this ideological rhetoric crumbles when one is confronted by the awesome sight of these majestic stone sculptures rising gloriously from their desolate surroundings to give silent testimony to the endurance of the Chinese cultural heritage through universal beauty. ▨

June 17, 1979

China's New Market Policy Brings 'Sidewalk Capitalists'

By JAMES P. STERBA
Special to The New York Times

TAIYUAN, China, Aug. 18 — The plane dipped slightly over the western edge of the city and suddenly eight parachutes opened below it.

An airborne invasion?

"No, no," said Yang Yishi with a laugh. "They are up there all the time. It's just the Shanxi Sky Diving Club. They do it for sport."

The club, according to Mr. Yang, a provincial official, is made up of army veterans, office and factory workers, and even some peasants. Most provinces have similar ones, he said. They are subsidized by the Government, and people who are adventurous enough can take instruction.

The parachutists are nothing new for the 1.8 million people in this city of smokestacks 300 miles southwest of Peking. What is new is what is happening on street corners. Since the beginning of the year, sidewalk capitalists have flourished. After China's leaders in Peking gave their blessing to rural fairs and urban free markets last December, vendors have proliferated all over Taiyuan.

Peasants Peddle Produce

Individual peasants began by plopping themselves down on a sidewalk with a basketload of green peppers to sell. Others moved in to sell eggs, peanuts and a wide range of fruits and vegetables. Beside them, gradually, came hawkers of everything from cigarettes to sunglasses. Then came sellers of tea and other drinks, along with men who repair everything from shoes to watches.

"The free markets were around before, but they were more or less secret," said Mr. Yang. "Now they are out in the open and very convenient for both the buyer and the seller."

The free markets are, in many cases, in direct competition with state-run shops nearby, and in recent weeks newspaper articles have extolled the

The New York Times / Aug. 23, 1979

value of such competition. Mr. Yang said he thought it was a good idea, too.

"I sometimes buy fruits and vegetables there even though they may cost more," he said. "The reason is that they are fresher. Farmers bring them straight from their gardens."

The state-owned shops often have the lowest prices, selling eggs, for example, for about 15 percent less than farmers are paid for them. But again, the free market eggs are fresher.

Meanwhile, the ancillary free market services such as repairs are being encouraged because there has been a shortage of such operations in government-run shops. The Government is now encouraging people to set up collective businesses to provide such services. Here, the sidewalk merchants are called one-man collectives.

Taiyuan is the capital of Shanxi Province, one of China's great coal and iron ore mining regions. Its factories produce a wide range of industrial materials, ranging from silicon steel to locomotives, chemicals, farm equipment and textiles.

Shanxi is also the home province of China's leader, Chairman Hua Guofeng.

Modest Comeback for Buddhism

Buddhism is making a modest comeback in the city. The Chongshan Temple, which was built in 1381 and is the home of a famous thousand-hand image of Buddha along with hundreds of priceless Sung and Ming dynasty sutras, is now open daily for worship.

Wei Pujing, a 72-year-old former monk who now wears a Mao jacket, looks after the temple. He said that 30 or 40 Buddhists came to pray each day. Mr. Wei is deputy chairman of the Buddhist Association of Shanxi. He said that two temples were now open in Taiyuan compared to more than a dozen before the Cultural Revolution of the late 1960's.

A major change, designed to provide new incentives for crop production, is under way in several of the farm communes around Taiyuan.

Smaller Work Units Formed

In the past, commune and production brigade leaders assigned farm workers to fields on the basis of where they were needed. There was little incentive to work hard. With the start of this growing season, however, production brigades and teams were split into smaller units and given sole responsibility for a particular field, from planting to harvest.

"Instead of getting a day's pay for a day's work and the whole brigade or commune receiving equal benefits, each small group of people splits whatever they produce on their particular field," explained an official from the Huangling People's Commune, on the southern edge of Taiyuan.

If the small group can get the work done in their field in half a day, they are free to do something else, the official said. If they get a bumper crop, they split the rewards. If their crop fails, they share the responsibility and the losses, he said.

August 23, 1979

Some Vietnamese Ethnic Chinese Find Life Hard in China

By JAMES P. STERBA
Special to The New York Times

YUANJIANG, China, Sept. 5 — The Red River Valley, 300 miles upstream from Hanoi, is a dozen shades of green this time of year. Terraces of rice, sugar cane, and pineapple march up the foothills of the mountains to meet a thick fog creeping down from the peaks.

By midmorning, most of the 2,000 members of the Hong He state farm are busy in the fields. But Mong Li Cuong, a 26-year-old ethnic Chinese refugee from Vietnam, is not working. He cannot stand it here. It is too hot. The water and rice taste funny. He says that he coughs and gets nosebleeds a lot.

He says he was trained by the North Vietnamese Army to fire 122-millimeter rockets at Saigon's American-backed troops, not to be a farmer. Now his goal in life is to get out of China and his dream is to settle in the United States.

For the peasant farmers who make up the majority of the 27,000 Vietnamese refugees in Yunnan Province, life as new members of 15 state farms is a blessing, a fresh and peaceful beginning of a new and productive life.

Create Social Problem

But for a number of refugees who fled from Vietnamese cities, life here amounts to a Chinese version of the "new economic zones" that they fled Vietnam to escape. For Chinese officials, the refugees pose a perplexing social problem since most of them have no relatives abroad and thus cannot qualify to emigrate.

Unlike other Asian nations, China has set no conditions or limits on the numbers of Vietnam refugees it will accept. In two years, China has accepted 251,000, more than any other nation, and has offered permanent homes to all who want to stay.

Chinese officials say they will eventually sort out skilled workers and intellectuals from peasants and move them into jobs that fit their skills. Some of the refugees have already moved into cities. One professor from Hanoi is teaching English at Yunnan University in Kunming.

But many refugees interviewed by a group of American reporters on two farms here said that they wanted no part in China's system, not even in small collective businesses in which they can use their skills for profit. And they refuse to work in the fields.

Refuse to Do Farm Work

"These people want to do private business, but obviously this cannot be done in China," said Yao Bosheng, director of refugee resettlement for Yunnan Province. "They do not like farm work, so they just stay home and live off what we give them."

Unlike Vietnam's new economic zones, Yunnan's state farms do not force the refugees to survive on what they can produce. Instead, they are paid wages according to the amount of work they do, while old people and children are subsidized. The refugees also live and work among local residents, many of whom were once refugees themselves.

Yunnan resettled about 9,000 ethnic Chinese who fled Indonesia in 1960. Most were shopkeepers from Jakarta and other Indonesian cities and it took them two or three years to adjust, local officials said.

About 5,000 ethnic Chinese were resettled here from Burma and India, and a few hundred others came from Malaysia to help establish rubber plantations

Doubt Vietnamese Will Adjust

Chinese officials said that they had hoped the Vietnamese refugees would similarly adjust. But they have some doubts. Some of the Vietnamese refugees have already shown signs of deviousness, the Chinese officials said.

Mr. Yao, for example, cited the case of a refugee who had said he was a physician. But when he was assigned to a local hospital, it became obvious that he was not. Another refugee who said that he was a truck driver shook behind the wheel, obviously incapable of driving.

The southern part of Yunnan Province has been off limits to foreigners for years. Except for a United Nations geology team, which visited here two years ago, no foreigners have been allowed in Yuanjiang county since the Communists took control in 1949. Chinese officials, who recently asked the United Nations for financial help in resettling refugees here, granted a request by nine American reporters and two photographers to visit resettlement farms.

The countryside is largely undeveloped and living conditions are Spartan in the valleys and along the hillsides between rugged mountains. The Government has not been able to lure many young people here to help develop the region despite its enormous potential.

Only 6% of Land Cultivated

The Gan Zhuang state farm, which has 7,500 members, including 1,500 refugees from Vietnam, has 26,000 acres. But only 1,600 acres are under cultivation and much of that is on raw, newly plowed foothills.

Some 2,000 Indonesian settlers on the farm still speak their native language as well as Chinese. The new Vietnamese settlers have been organized into three of the farm's 50 production teams in isolated areas. They are further isolated because many of them do not speak Chinese.

When asked how she liked the farm, a young woman, Fong Lee Ha, burst into tears. Her family ran The World restaurant in Cholon, the Chinese section of Saigon, and another restaurant in Da Nang.

Miss Fong, 22, and her 18-year-old brother paid a Vietnamese official to drive them from Ho Chi Minh City to Hanoi 16 months ago. From there, she said they were expelled with other ethnic Chinese into China.

Brother in Virginia

Since Miss Fong and her brother have another brother who works as a cook in Arlington, Va., they expect to go to the United States some day. They have also applied to go to France, where an older sister lives. In the meantime, they are teaching other refugees Chinese.

The problems for those refugees with relatives abroad is that the Chinese Government has set up no organization to help them emigrate. Chinese officials in Yunnan said that, once the refugees are accepted by third countries, they will help arrange an exit. But all the preliminary paper work must be done by the refugees by mail.

Yunnan officials said Peking authorities had not yet decided whether to allow United Nations refugee officials or foreign embassy workers to visit the resettlement farms to interview those who might qualify for acceptance abroad.

Several refugees with relatives in the United States said that they had written to the American Embassy in Peking but received no reply. An embassy official said there was no one there who could read Vietnamese until two months ago when two Vietnamese wives of American

officers joined their husbands in Peking.

Chinese officials are more concerned with disgruntled refugees here who do not have relatives abroad. Already some have sneaked away from the farms and tried to enter Hong Kong illegally. Others, including many from the northern part of Vietnam, mope around the farms, refusing to work.

Mao Thuy Hoang, 23, wears his hair at shoulder length and, although he is not Roman Catholic, wears a wooden cross on a necklace. He says that he is an artist and cannot tolerate farm labor. So he does nothing. Although he has no relatives abroad, he appealed to visiting re-

porters to help him get out of the Red River Valley, which he said was too hot in the daytime and too cold at night. He said that he would go to any other country that would accept him.

Lam Thy San, 48, a former artilleryman in the French Army and cycle driver in Saigon, said he was an ethnic Chinese but described himself as an "international refugee." He said that he did not think he could ever adapt to the Chinese system regardless of how much it was liberalized.

September 11, 1979

A Tough New Drive On Births in China

By WALTER SULLIVAN

CANTON

USING a variety of contraceptives, including "vacation pills" for couples who are rarely united and "paper pills" prepared in perforated sheets like postage stamps, China is making an all-out effort to stem its population growth.

The initial goal is to reduce the annual growth rate to 1 percent by next year. In 1954, according to official figures, it reached a peak of 2.34 percent. From interviews with medical workers, pharmaceutical researchers and ordinary citizens it is evident that an extensive educational campaign is under way, using radio, television, newspapers, commune reading rooms and exhortative songs at concerts in the cities.

Economic incentives are being used, and next March they will become stronger when the salaries of those who have a third child will be reduced 10 percent until that child reaches the age of 14. The child will also be denied free education and medical care.

While intrauterine devices were the recommended and most widely used methods at a commune visited last month near Shanghai, virtually all techniques are used. In some communes and urban communities oral contraceptives appear to predominate.

It was not long ago that folk remedies against conception were prevalent. According to one formulation, "Fresh tadpoles coming out in the spring should be washed clean in cold boiled water and swallowed whole three or four days after menstruation."

"If a woman swallows 14 live tad-

poles on the first day and 10 more the following day, she will not conceive for five years," the prescription said.

As recently as 1964, as pointed out by Dr. Carl Djerassi of Stanford University in his forthcoming book, "The Politics of Contraception," reference was made to the tadpole method in an otherwise sophisticated Chinese review of contraceptive techniques.

Seven years earlier, however, the method had been tested on 65 women. Each swallowed 44 tadpoles and 43 percent became pregnant.

Dr. Djerassi was largely responsible for discovery of Norlutin, the first of the female oral contraceptives, which is widely used in China as "pill No. 1." While various contraceptive preparations are taken by millions of Chinese women, only 10,000 men are taking the birth-control substance gossypol experimentally. All the contraceptives appear to be made in China.

One difficulty in assessing the birth-control effort is learning how effective it is in remote rural areas where a large percentage of Chinese live. The newly enacted economic incentives, however, clearly indicate the Government's determination to halt population growth in order to improve living standards.

In interviews, birth-control and public health officials of Guangdong Province, of which this city is the capital, said nursery costs and charges for school supplies through senior middle school (comparable to high school in the United States) are waived for couples who have produced only one child and have pledged not to have another.

Women cannot register for marriage, they said, until age 23. The minimum age for men is 26 in urban areas and 25 in rural areas. Such rules, they

said, vary slightly from one province to another and do not apply to autonomous regions for ethnic minorities. Despite the effort to curb birth rates, they added, medical help is still given those who have difficulty bearing children.

In cities, according to a recent Guangdong provincial broadcast, housing priority will be given those with no more than two children and those who marry late. Couples with one child will be given apartments for four-member families. There is to be job preference for children of families with a single son or two daughters if the children are suitable for employment. Presumably the distinction is to allow for those who try a second time to have a son, a traditional goal in this society.

The children of such small city families "need not be sent down to rural areas," according to the broadcast. If they have already been dispatched to farms they will be granted preference in returning.

Such incentives, which appear to apply nationwide, were described to American science writers visiting the July First commune near Shanghai. The visitors were told of the salary reduction for large families that will take effect next March and the allocation of additional private garden plots to those with small families.

A couple with one child is given a plot allocation for a four-member family and a small salary bonus. Those who at retirement have had no children receive additional pension benefits. This is meant to compensate for the traditional desire to produce a son who could support parents in their later years.

The chairman of the women's committee of the commune, Xie Gundi, is a forceful 34-year-old woman whose tasks include lecturing women of the commune on family planning and seeing to it that they comply.

The July First commune (named for the birthday of the Chinese Communist Party) is primarily an agricultural community of 5,016 households and 19,004 people. They are divided into 11 production brigades and 88 production teams. Most cultivate rice, wheat, cotton, vegetables, pigs, poultry (including eggs) and mushrooms.

The surplus produce is marketed in Shanghai. There are also shops assembling electric engines for irrigation pumps, furniture, boats and farm implements as well as home production of knitted and bamboo goods.

After a woman bears her first child, Mrs. Xie or one of the commune's medical workers tells her the choices of contraceptive methods, recommending an intrauterine device. The woman is told that according to the rules she may have only one more child, after four years have elapsed. Where preventive measures, fail abortions are available.

This year, as of August, the commune's 2,978 women of child-bearing age had borne 120 children and there had been 50 abortions. Of the more than 3,000 married couples, 190 had elected sterilization to prevent further births. In 95 percent of these cases it was the woman who was sterilized.

By 1970 a number of Chinese women were using monthly injections of a steroid that reportedly can cause irregular menstrual bleeding. According to Mrs. Xie, some at the commune use the injections. A small percentage of couples use a diaphragm, condoms or the rhythm method, which she termed unreliable. The paper pill, she added, is not used at all.

That pill is marketed in sheets of various colors, which indicate the contraceptive preparation with which the paper is treated. Like the wrappings of some Chinese and Japanese confections, the paper is digestible. Each perforated sheet contains 22 stamps for the 22 days of the monthly cycle on which they are to be eaten.

The vacation pill on which the most information is available contains a derivative of the steroid hormone anordrin synthesized by the Institute of Organic Chemistry in Shanghai. It is taken immediately after intercourse and again the next morning, the procedure being repeated throughout the brief vacation.

In this way the woman is not subjected during the rest of the year to the side effects of steroid hormones in the standard pills, notably occasional effects on the circulatory system, such as blood clots. Anordrin, the compound from which this pill was derived, was discovered by a French chemist 20 years ago. It was rejected in the United States as a contraceptive because it prolongs the menstrual cycle and sometimes causes vomiting, dizziness and other complaints.

The Chinese apparently consider it acceptable for limited periods. It was introduced into general use in 1974.

Clinical trials have been under way since last year on a new "morning-after" contraceptive developed at the Shanghai institute. It is a form of norgestrienone believed to prevent implantation of an embryo in the uterus. The drug is related to some under study in West Germany and elsewhere.

It can also be used to induce early abortion. For more advanced pregnancies trichosanthine is being used. This is derived from the extract of a gourd root that as early as the 15th century in China was inserted into the vagina to induce discharge of fetal membranes.

According to a professor at the Shanghai institute, two milligrams of purified trichosanthine injected as late as the fifth month of pregnancy will lead to labor and abortion within a week. It is coming into general use in areas where medical facilities are limited.

October 10, 1979

41

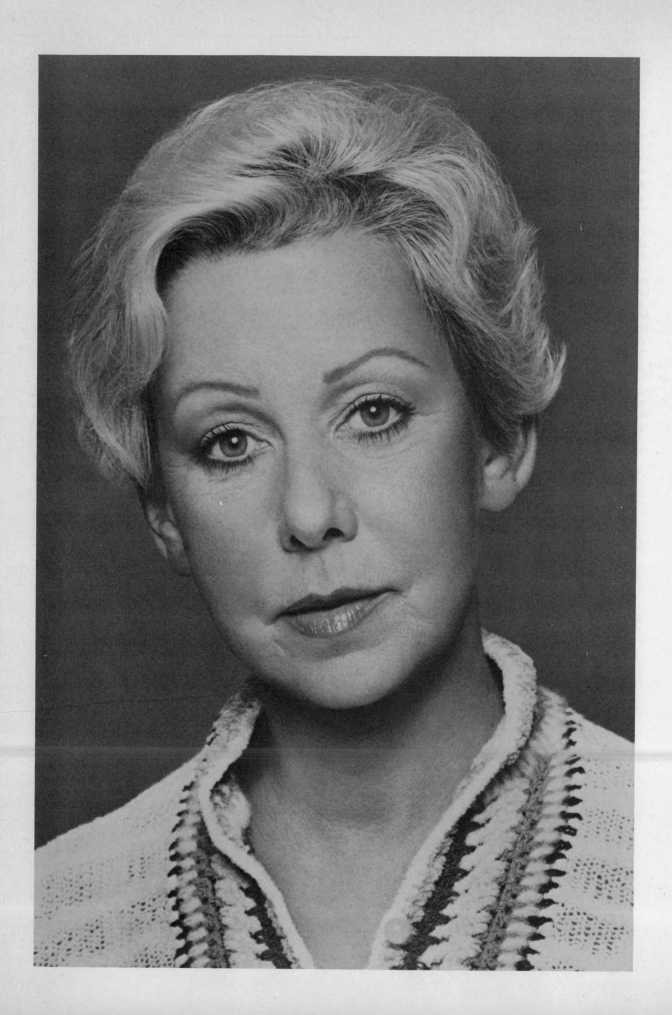

CHAPTER 4

Women: Their Changing Roles

Jane M. Byrne surprised the nation by defeating the Democratic machine in a bitter primary campaign on her way to becoming Mayor of Chicago.

One of the first women to receive her wings as an Air Force pilot, Capt. Kathy LaSauce is learning to fly jet fighters in Arizona.

THE CASE AGAINST WOMEN IN COMBAT

Should women provide manpower for the military by flying fighter planes and serving in the infantry? It sounds like 'equal rights,' but it may not work.

By George Gilder

Helmeted young women charging over beaches under fire with bayonets fixed, teen-age girls occupying oil fields in an energy crunch, Marine coeds bunking down on bivouacs in combat zones, female pilots shot down and captured behind enemy lines: Such images may seem sheer fantasy — the farthest shores of radical feminism or an arresting montage for a futuristic film. Until recently, in fact, the military services and some factions of the women's liberation movement have dismissed the idea of warrior roles for women as "unthinkable." Then, less than a year ago, the Defense Department requested that Congress end the legal ban on women in combat. Today, when the subject comes up, the word more often chosen — by high officials in the Pentagon and by outside observers of military affairs — is "inevitable."

As Congress reconvenes, current law for the Navy and Air Force still prohibits — at least formally — the use of

George Gilder is the author of "Sexual Suicide" and "Visible Man."

women in "combat roles," although it does not clearly define them. (The Army, while assuming the same prohibition, has never made it explicit in legislation.) In recent years, however, many Pentagon observers have come to believe that — because of growing personnel shortages — the United States cannot maintain an adequate combat force in any of the services without either reinstating the draft or using women more widely and flexibly than present law permits. It was primarily for this reason that, last Valentine's Day, Deputy Secretary of Defense Charles W. Duncan asked Congress to void all current restrictions.

☐

The one-time fantasy of women warriors is depicted as a quite palatable reform, made plausible by the changing nature of warfare and military technology. Women will indeed serve in "combat," it is suggested; but combat will no longer depend on bayonets and physical force — rather on lasers, microprocessors and other sophisticated devices that render obsolete the conventional images of battle. In the future, there will be no front lines.

The issue of women in battle, however, cannot be so easily rationalized or

evaded. Conventional combat and "police" activities are by no means unlikely in a world where many modern weapons are too destructive to use. All these once unthinkable fantasies, moreover, are already a daily agenda of learning in training camps across the land, where women do indeed dig foxholes, fix bayonets, fly planes, and serve on ships under "fire."

Congress has already approved the assignment of women to a wide variety of near-combat naval vessels. An initial class of 10 women flying fighter-style jets has been graduated from Air Forces courses. And thousands of female soldiers serve in Europe today, driving supply trucks and performing maintenance operations in logistical units that would bring them deep into combat areas in the event of war. Recently, these changing policies have brought forth new rules prohibiting "relationships" between persons of different ranks and different sexes; and the classic training exercise called the "slide-for-life" was restricted by the Army after a 20-year-old recruit fell to her death when her grip gave way on the high rope. The military is trying in many earnest ways to accommodate the female presence.

Pentagon officials claim that they

plan no abrupt changes regarding the use of women in combat which might endanger "the national security." Nevertheless, they do entertain the idea — as one of the Joint Chiefs of Staff said — that "women may prove to be even more effective, more aggressive than men," and they ask for more flexibility in deploying them.

In addition, the military is under constant pressure from Equal Opportunity Officers. A new alliance between Pentagon personnel administrators and policy makers and the women's liberation movement has emerged, pushing the United States military into a position where women will be so fully and flexibly involved in our armed-services organizational structure that, in a war emergency, it will be very difficult (if not impossible) to separate them out.

In a recent interview, two leading figures in the new confederation of professional feminists and military men explained this change of attitudes. The feminist was Kathleen Carpenter, a 34-year-old lawyer, formerly special counsel for employment practices at Norton Simon Inc., who now serves as Deputy Assistant Secretary of Defense for Equal Opportunity. The military man was Comdr. Richard Hunter. In recent years, he has emerged as one of the Pentagon's leading experts on personnel matters.

Both Commander Hunter and Miss Carpenter believe that the differences between the sexes are no more important, and possibly less important, than the difference between the races. "When I came into the military, blacks were treated about the way women are now," Commander Hunter said. He and Kathleen Carpenter see the matter as a simple problem of civil rights.

Although Commander Hunter acknowledged that there were some differences in physical strength, Miss Carpenter demurred. "Tests show," she said, "that while men have greater upper body strength, women have greater midsection strength." The military is changing its task structure, she maintained, to make better use of female midsection strength. This re-

Marine Robin Garrett tends a sore foot.

structuring of jobs, she said, "enriches the work experience for all."

A background study entitled "The Use of Women in the Military," ordered by Defense Secretary Harold Brown and produced last year under the leadership of Commander Hunter, offers two reasons for the growing urgency of introducing women into the battle forces: "First is the movement within the society to provide equal economic opportunity for American women. Second, and more important, use of more women can be a significant factor in making the all-volunteer force continue to work in the face of a declining youth population."

Although Commander Hunter admits that combat performance is the only fully valid standard for the quality of a recruit, he believes that possession of a high-school diploma is the best available predictor of success in combat, and that the Armed Forces Qualification Test is a valuable index of trainability. At present more than 90 percent of women recruits have high-school diplomas, compared with only 63 percent of men, and the women's test scores are approximately 20 percent higher than the men's.

However, the Army — like all the services — is restive under the new policies of the Pentagon's civilian leadership. It warns, in a rider to the Hunter report, that "we should err on the side of national security until such time as we have confidence that the basic mission of the Army can be accomplished with significantly more female content in the active force." The Hunter study responds: "Is recruiting a male high-school dropout in preference to a smaller, weaker, but higher-quality female erring on the side of the national security, in view of the kinds of jobs which must be done in today's military?"

One way or another, it is indeed questionable whether the stepped-up recruitment of women will resolve the alleged personnel shortages of the all-volunteer force (A.V.F.) or whether the removal of the combat prohibition will give women essentially equal opportunities. Although serious difficulties loom regarding reserve recruitment and overall costs, the A.V.F. is now doing quite well statistically — attracting male recruits of "higher quality" with better education and test scores for longer average enlistment periods than at any previous time in the 20th century. Moreover, both the Korean and Vietnam Wars were fought under far more stringent manpower conditions than the A.V.F. faces in the foreseeable future. In the 1950's, for example, the supply of young men was only double the military demand; through the 1990's, supply will run between four and five times the demand. Increasing rates of immigration, legal and illegal, will tend to expand further the number of eligible males.

The presence of women in the American armed services — currently nearly 7 percent and expected to rise to 11 per-

cent by 1983 — is already greater than in the military forces of any other country, including the Soviet Union or Israel. The United States Army already employs women more extensively in combat-related areas than any other national force.

The Marine Corps plans to double its number of women by 1982; the Navy, which doubled its number of females between 1973 and 1976, envisages an increase of more than 46 percent; the Air Force projects a 50 percent gain. All services are already recruiting women at a rate that is causing serious management and mobility problems, particularly among the one-third of service women who are married to service men and among the increasing numbers of military women with small children.

But even if expanded use of women were desirable to relieve recruitment and attrition costs and to fill certain technical needs, removal of the combat exclusion might nevertheless remain unnecessary and unwise. The proposed change is based primarily on a simple idea: that women could be used more efficiently if they were less restricted to traditional roles. This assumption may be false even on narrow grounds of personnel management. Although attrition rates for women are sometimes lower than for men, this advantage is totally lost when women move away from traditional kinds of work. The female attrition rate is higher than the male rate in the combat-oriented Marine Corps, and between two and five times as high wherever women work in nontraditional fields. In fact, for both men and women, attrition rates soar when conventional sex roles are ignored in the delegation of military assignments. The recent Brookings Institution study "Women and the Military" conceded in a footnote: "Of all the women who entered the electronic-equipment repair speciality in 1973 only 24 percent remained at the end of 1976, whereas the comparable figure for males was 49 percent. In maintenance, the corresponding proportions were 10 percent for women and 47 percent for men." These low rates of female retention covered all nontraditional areas and they are not improving with time.

Conversely, in the administrative, clerical, communications and medical areas, among secretaries, nurses and the like, male attrition rates were nearly double those of females. It seems clear that, at least in the military, most men and women prefer traditional work.

The military, in general, may be going too far in "objective" and rationalistic personnel policies that nevertheless defy common sense. Its tendency to view personnel "quality" almost entirely in terms of education and test scores could conceal a new form of bigotry, perhaps unintentional but certainly real. This is the fallacy of assuming that the most easily measurable of human qualities are more important than the unmeasurable ones. It may be coincidental that prominent among the victims of the new bigotry are the same dropout youths (Continued)

Above: A coed supply company of the Seventh Infantry gets a phys-ed session.

Above: Marine Kay Mesner discovers the use of camouflage in Quantico, Va.

who have always suffered. But the result is no less discriminatory. It is discrimination in favor of female credentials and test scores and implicitly against male aggressiveness and physical strength; discrimination in favor of quantifiable abilities tested in classrooms and against the intangible qualities tested in combat.

Almost every study of productivity in American industry or of success in the economy, as well as of courage in combat, has shown that the intangibles of character — enterprise, aggressiveness, drive, willingness to risk — are much more important than the measurable quotients of classroom ability.

The argument that advanced technology requires a more highly educated work force is also familiar in civilian life. Yet studies by Lester Thurow of M.I.T. reveal that the vast majority of all work skills are learned on the job, and that more advanced technology often reduces rather than enhances the need for outside education, since the necessary skills are incorporated into the machines. Many of the technical industries of Europe that compete successfully in American markets make heavy use of poorly educated aliens who do not even know the language of the country in which they work.

No major military force in the world appproaches the educational level of the American armed services. It may well be that the Pentagon should create new inducements — and suspend certain military disciplines and promotion requirements — to aid in the recruitment of highly educated specialists in certain fields. But this has little to do with the general matter of increasing the proportion of recruits with high-school diplomas, or of introducing women to combat roles.

□

The hard evidence is overwhelming that men are more aggressive, competitive, risk-taking, indeed more combative, than women. In fact, even feminist scholars Carol Jacklin and Eleanor Maccoby, chairman of Stanford University's department of psychology, acknowledge decisive biological differences between the sexes. In their scrupulously objective and voluminous text on the subject, "The Psychology of Sex Differences," they sum up their argument: "(1) Males are more aggressive than females in all human societies for which evidence is available. (2) The sex differences are found early in life, at a time when there is no evidence that differential socialization pressures have been brought to bear by adults to 'shape' aggression differently in the two sexes. (3)

Similar sex differences are found in man and subhuman primates. (4) Aggression is related to levels of sex hormones, and can be changed by experimental administrations of these hormones."

Another male trait — manifested in every human society but just as hard as aggression to measure in an exam — is aptitude for group leadership. Partly a result of greater aggressiveness and larger physical stature, partly an expression of the need to dominate (perhaps based on the neurophysiological demands of the sex act itself), males in all societies ever studied by anthropologists overwhelmingly prevail in positions of leadership and hold authority in relations with women. Steven Goldberg's rigorously argued book "The Inevitability of Patriarchy," described by Margaret Mead as "flawless in its presentation of the data," refutes every anthropological claim that there has ever existed in human affairs either a society where women rule, or a society where final authority resides with them in male-female relations.

Similarly, on the admittedly less crucial matter of physical ability, no pettifoggery about midsection strength can dispel the facts of greater male size and power. As the Hunter study concedes, "Women have about 67 percent of the endurance and 55 percent of the muscular strength of men. Even when size is held constant, women are only 80 percent as strong as men."(A recent Army study conducted at West Point showed that young women improve their strength less than half as much as men under the same regimen of physical training. Testosterone is a hormone not only of male sex drive and aggression but also of protein synthesis.)

Bureaucracies that ignore these facts of life, in favor of narrow criteria of test scores and schooling, may create in time the same kind of demoralization, restlessness and drive to early retirement that is often found in the American system of Civil Service, and indeed the same kind of disgruntlement and dissatisfaction that Kathleen Carpenter now finds in the military services and ascribes to inadequate pay and benefits. (Both modes of compensation are more generous in the military than in other branches of government.)

The increasing tendency to use academic testing methods for all personnel and promotion systems is turning untold thousands of jobs in the military and elsewhere into dead ends for all but those adept at taking tests — dead ends for all the slow-learning leaders, all the dropouts with ingenuity, inspiration and charisma, all the reasonably intelligent and exceptionally ambitious

men who have excelled in every test of combat but who know they will never rank high in a rote-administered Civil Service-type exam.

Each service plays a special part in this debate. Does the naval leadership really mean it when it claims, as did Navy Secretary W. Graham Claytor Jr. recently, that 85 percent of all shipboard roles require no particular "physical dexterity"? Perhaps the Secretary of the Navy ought at least to acknowledge that, in the heat of battle, crises often arise in which Navy personnel must switch jobs suddenly and perform different and more onerous roles, sometimes entailing acute physical stress and risk of injury.

The Marine Corps has always especially prided itself on "building men," and has traditionally attracted recruits eager to meet the challenges of stressful and blatantly virile training. But the new military, in the typical "to get along, go along" spirit of bureaucracy, has forced a steady softening of this regimen.

Even without the equal rights amendment, the Army has made women eligible for more than 95 percent of all job listings. On Sept. 8, 1977, it was announced that "unit commanders are authorized to employ women to accomplish unit missions throughout the battlefield. . . ." At the discretion of the Secretary of the Army, women continue to be prohibited from "engaging an enemy with an individual or crew-served weapon while being exposed to direct enemy fire, a high probability of direct physical contact with the enemy's personnel and a substantial risk of capture." Beyond that technical statement, anything goes.

But if the Congress feels constrained to enact a law allowing female midshipmen to undergo temporary training on ships not on combat missions, it should at the same time enact a law keeping women cadets out of foxholes and other battlefield positions, even while in training. With Equal Opportunity Officers attached to all large military units, any sensible barriers will be subject to erosion unless they are explicitly and unmistakably written into statute.

Similarly in the Air Force, 94 percent of jobs, including many noncombat pilot roles, are now open to women. If the combat restriction is removed, the Air Force will soon open all its jobs to women. But even assuming that female pilots were of equal ability in the air, many would be captured by the enemy and face the same stresses and exploitations that men know in P.O.W. camps. Is there good and sufficient reason for the United States to expose

women to such danger and abuse?

The advocates of change — from the Defense Management Journal to the Brookings report and various internal analyses for the Defense Department — regard current problems as merely transitional. As the entire society moves toward liberation from the "irrational" constraints of sex roles, so it is believed, everything will work out famously. Commander Hunter is very hopeful about the promise of radical sex-role training in the nursery schools, and Kathleen Carpenter sees inspiring progress in private industry. Jeanne Holm, a retired Air Force major general, maintains: "[The change] is inevitable . . . because that's the direction our culture is going."

All such assumptions are profoundly doubtful. Other countries have made similar efforts to force abandonment of sex roles. The result has been not an inevitable progression toward androgynous societies but — as with the Israeli kibbutzim — a return to traditional roles. The Soviet Union made great changes after the Revolution, and again after the heavy male losses of World War II. Yet today the status and earnings of women, their university attendance levels, and their work in nontraditional fields have all reverted to earlier patterns.

□

Symbols are important in society. The Pentagon might closely consider whether it wishes to become a mere bureaucracy, on the order of the Department of Health, Education and Welfare, or whether it would not do better to build on its classic role as a focus of patriotism and a strong force in the socialization and enhancement of young men. There is deep evolutionary symbolism in the warrior mystique, the chivalric tradition, the aggressive glory of the martial style. In its pursuit of the favor of a basically antimilitary counterculture that, needless to say, will protest and picket in case of war, the Pentagon may not realize how close it has come to losing its supreme and most motivating assets.

□

Like most other wars in human history, the campaign against sex roles is in reality dominated not by women but by men. It joins two fundamentally conflicting forces. On the one hand, there is the powerful and basically antiindividualistic movement toward ever more fully rationalized bureaucracy, governed by objective and quantifiable standards. (Computers prefer unisex.) On the other hand, there is the essentially individualistic movement of militant feminism, which advocates a society without sex roles in order to promote personal fulfillment. These forces actually have little in common, yet they reinforce each other. The feminists, with their rhetoric of equal rights, lend to the bureaucrats a tone of moral inspiration which their policies would otherwise lack, while the bureaucrats lend to the feminists an air of tough-minded practicality.

The consequence however, will be a rationalized, still male-dominated military bureaucracy that gives short shrift to the feminist critique of an excessively competitive, violent and hierarchical world. The women who wish to abolish exploitation and inequality may find themselves dupes of the military, acceding to men who merely wish to exploit women with ever more brutal efficiency.

The ancient tradition against the use of women in combat embodies the deepest wisdom of the human race. It expresses the most basic imperatives

of group survival: a nation or tribe that allows the loss of large numbers of its young women runs the risk of becoming permanently depopulated.

Beyond this general imperative is the related need of every society to insure that male physical strength and aggressiveness are not directed against women. All societies teach their men to avoid physically fighting with women and, most often, to avoid competing face to face with them. All civilized societies train their men to protect and defend women. When these restraints break down — as in tribes like those studied by Colin Turnbull, and the Mundugumor, described by Margaret Mead — the group tends to disintegrate completely and even to grow extinct.

The military services, however, are

unanimous in asserting that successful use of women in battlefield units depends on men overcoming their natural impulses to treat women differently and more considerately. The consequences of this latest demand for equality may be nothing more or less than a move toward barbarism.

Nevertheless, the serious subject of women in combat roles has been treated, even at the top, in a strangely light-hearted spirit. Instead of confronting the realities of battle conditions, advocates evoke cheery images of paneled offices and glistening machinery and intent young women peering through microscopes or programming computers. There are visions of strapping young corporals reacting instantly in crisis to melodiously trilled female commands; of boys and

girls who hear always — through the roar of hormones during long months at sea — the cool voices of orderly bureaucracy and cooperation. Above all, the arguments advanced for expanding the warrior role of women are as pleasantly devoid as Disneyland of any vulgar mention of sex . . . or war.

Before we move ahead on this matter, however, we should consider the idea that combat is not family entertainment — that it is a real possibility, not chiefly a job opportunity. While the Soviet Union's nearly all-male armies grow to a size three times ours and Cuba pursues its prey around the world, it would be unfortunate for American leaders to give the impression that they regard combat chiefly as an obstacle to women's rights. ◼

January 28, 1979

Chicago Is Her Kind of Town

Jane Margaret Byrne

By WILLIAM ROBBINS
Special to The New York Times

CHICAGO, Feb. 28 — Jane Byrne strolled down victory lane today, down the Magnificent Mile, the fashionable shopping stretch of Michigan Avenue. Cab drivers honked horns and shoppers shouted, "Give 'em hell, Jane!" and her bright red lipstick flashed like a Christmas wreath around her winner's smile.

Woman in the News

Other, shocked Chicagoans were still awakening to the fact that their next Mayor, all but certainly, will be a person who visits a beauty parlor. In fact, one of those was her destination today, the day after she defeated Mayor Michael A. Bilandic and the vaunted Democratic machine of the late Mayor Richard J. Daley in a bitter primary campaign.

She won the nomination, which has been tantamount to winning office, in one of the heaviest voter turnouts — 58.97 percent — in the city's history, carrying more than half of Chicago's 50 wards. [Page A17.]

It was her town today, and the 44-year-old battler — "a little spitfire," her husband fondly describes her — could afford to be magnanimous. She waved an olive branch when asked about chances for a reconciliation with the organization.

"I'm a Democrat," she said. "I believe in the Democratic Party."

Hers was a relaxed walk, unlike the recent past. She had been running for a long time, some say since she first walked into Mayor Daley's office 15 years ago to talk about a job and to become a protégé of the man she came to idolize.

Mayor Daley was her sponsor in politics and he defended her against what her daughter, Katharine, 21 years old, called "the flak from the aldermen" when she angered business constitu-

ents as his Commissioner of Sales, Weights and Measures.

Since she came out of the same Daley machine as Mr. Bilandic, she shocked the Mayor's forces when, nearly a year and a half ago, she first began to fight City Hall.

In November 1977, she disclosed to the press a memorandum, written to her earlier and notarized, that accused Mayor Bilandic of leading a conspiracy of city officials to "grease" the way for a taxi fare increase.

Mayor Bilandic seemed to have easily weathered the crisis. Lie-detector tests appeared to show that neither the accused nor his accuser was answering falsely. A grand jury failed to find grounds for an indictment.

But Mrs. Byrne, whom the Mayor dismissed, was soon to plan another attack — to take his job.

As her daughter, Katharine Byrne, recalls the episode, her mother "was disenchanted."

"She had idolized Daley," Miss Byrne, a senior at St. Mary's College of Notre Dame, said. "But the proper channels weren't working anymore."

It was "after she got fired," Miss Byrne said, and "people were coming up and slapping her on the back and saying, 'Keep fighting — we're behind you'" that Mrs. Byrne started thinking of running. Political experts gave her not the ghost of a chance.

But Mrs. Byrne outmaneuvered the machine. She wrapped about her the mantle of the late Mayor, whose memory is still revered by most Chicagoans, and her opponents, with all their cries of outrage, could not rip it away. They accused her of bad taste, but she had television clips of herself with the late Mayor and audio of his words praising her, and she continued to run them as campaign commercials.

Finally, of course, the luck of the Irish was with her, and a record snowfall buried city streets and smothered the Mayor's campaign as Chicagoans perceived a poorly administered effort

to overcome the problems.

The Bilandic forces should not have been surprised. Mrs. Byrne had been preparing herself for politics since 1960, when she joined the Presidential campaign of John F. Kennedy and attracted the attention of Mayor Daley. He was attracted by her manner — "her no-nonsense attitude and direct stare," as one writer put it — but it was not until 1964 that he offered her a job.

"I'll never forget the first time I went to his office in City Hall," she was to recall later. "It was early in 1964 and I had received a phone call at home that the Mayor wanted to see me. Well, you can imagine how thrilled I was."

Soon the Mayor became her mentor.

"He told me if I really wanted to have a career in local politics that I better get active in some grass-roots work," Mrs. Byrne recalled. "He said, 'If you don't work your way up, they'll get you in the end.'"

But Mr. Daley was helping, first with a job in a Federal antipoverty program. Later, in 1968, he named her Commissioner of Sales, Weights and Measures, and he soon discovered he had created problems for himself.

She dismissed a number of staff members for what she has described as corruption. According to her daughter, Mayor Daley called Mrs. Byrne in and asked:

"Janie, what are you doing? You've got half of this town's aldermen on my back."

Mrs. Byrne's response, her daughter said, was:

"Well, you told me to clean it up."

Soon she was enforcing compliance with meat-grading ordinances, regulations on posted octane ratings for gasoline, toy safety standards and a ban on phosphates in detergents. And Mr. Daley was getting flak from lobbyists.

Yet he found her valuable, experts here say, as a women's leader. He had her named Democratic National Committeewoman and gave her the honorary post of co-chairman of the Cook County Democratic Committee.

When the Mayor died, her Cook County Democratic post also expired.

And it is a measure of how little chance she was given in the race against Mayor Bilandic that no one has bothered to question her on even such issues as the equal rights amendment, which the Illinois Legislature has repeatedly refused to ratify. No one at Mrs. Byrne's headquarters today could say whether she was for or against it.

Mrs. Byrne, a native Chicagoan, was born May 24, 1934, and, as Jane Margaret Burke, she attended Barat College in nearby Lake Forest before marrying William Byrne, a Notre Dame graduate who was soon to become a Marine Corps pilot. He was killed in a plane crash in 1959.

Last St. Patrick's Day Mrs. Byrne was married to Jay McMullen, a political reporter for The Chicago Sun-Times, and, since then, he says, life has been more interesting.

Early this morning he, his son Stephen, Katharine, and William Burke, Mrs. Byrne's 74-year-old father, were with her as the returns came in and as it became clear, as her walk today confirmed, that Jane Byrne had carried the day.

March 1, 1979

'If anyone tries to get our rights, women will fight again,' vows a former demonstrator. A majority of Iranian women, however, couldn't care less about the struggle for equality, insisting that they are fully protected by the Koran. What the two groups share is a devotion to Islam, and the belief that 'it takes Eastern eyes to see clearly the situation for women in Iran.'

IRANIAN WOMEN: LOOKING BEYOND THE CHADOR

By Gregory Jaynes

One fine Monday morning this spring, a woman with seven children went down to a street gutter in south Teheran to do her family's wash. Balanced on her head was a plastic basket filled with dirty dishes, and flung over one shoulder was a twisted blanket filled with dirty clothes. She carried a clean, suckling infant in a sling across her bosom. A bright sun reflected up from the brown waters that sluiced through the gutter — or *jube*, as it is called — and, as the mother rinsed and wiped, sloshed and wrung, the baby tried in vain to catch his own liquid image in the curbside sink. In an hour the work was done and they returned up the hill to the tiny room the family rents. Nothing else lay ahead in the day but the soiling of more plates and apparel. The woman's name is Gholi Dahri. She is illiterate, as are about three-quarters of the 17 million or so women in Iran. She wears the chador, the crown-to-ankle veil, but not when she is washing. It would get in the way. When she is washing, she wears a scarf. She could not care less about the heavily publicized struggle for women's rights in this country, and in that regard she is in an overwhelming majority. Her highest hopes are for more food, more income and more than a single room in which to live out her life. These are the common goals of that same majority.

Gregory Jaynes, a Times reporter, has been covering the revolution and the new regime in Iran.

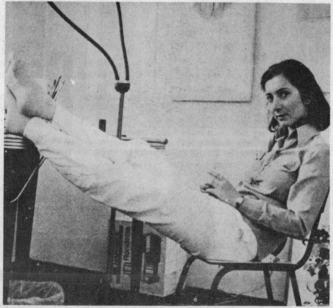

Painter Susan Kamalieh will leave Iran if Islamic decrees restrict her life style.

Light years away from Gholi Dahri, both in station and in thought, are the thousands of women who poured into the streets of Teheran one week in March to demand equality. They wore jeans that fit them snugly, or designers' dresses beneath expensive furs, and they chanted metaphorically that in the dawn of freedom, there was no freedom. They had been stirred to motion by a pronouncement by Ayatollah Ruhollah Khomeini — a call for women in Islam to wear the chador. The demonstrations begat a spate of clarifications from the Ayatollah and his government. Women should dress "decently," the clarifications said.

The demonstrations stopped. Not so much as a whisper came from the feminist camp. "And what of the women?" a friend writes me from New York. "From here, they seem to have disappeared completely."

To be sure, they have not gone away. There are today perhaps a dozen women's groups meeting in Teheran. Their leaders are thoughtful, upper-middle-class Iranians who organized and supported the chador demonstrations, then quickly pulled back when the concessions began. They are devout Moslems who say the last thing they wanted to do was divide the revolution. They feared that counterrevolutionaries would exploit the women's issue. Having made their point, they fell into quiet discussions.

There is no single person who speaks for the women of Iran, nor is there a formal group one can lean upon to check the movement's direction. "Movement" might even be stretching the happening. When the Ayatollah dropped his chador bomb, seven or eight prominent women in Teheran got on the telephone, urging friends and friends of friends to express their disenchantment. As it happened, the following day was International Women's Day, and a small observance was planned at Teheran University. From that tiny rallying point, the event mushroomed.

"If anyone tries to get our rights, women will fight again," says Mahrokh Mirmajlessi, a set designer, mother of two and wife of a doctor. Mrs. Mirmajlessi was one of the women in the thick of things last month. She feels the problem now is to make sure women's rights are not ignored in the new republic. To that end, she meets with other women, some of whom have husbands in government, all of whom have husbands sympathetic to the feminist cause. It is a subtle movement, in that men who will have some say in Iran's reconstruction begin their working day with an earful of suggestions from home and the knowledge that if forthcoming decisions rub their women the wrong way they will take to the streets again.

It is also, in the light of numbers, something of a sideshow. The Ayatollah could summon millions to a graveyard by simply saying, "Come." The women mustered 15,000 demonstrators to their most successful rally. "That is why," says Nasrim Farrokh, a musician, stage director and opera singer, "our problem is, in the main, the education of the masses. The illiterate Iranian man thinks the woman is second class. The illiterate Iranian woman accepts it.

"One of the wonderful things about this revolution, however, is that people are open to argument. I go among illiterate men, and I speak in a soft voice, and the information I give is absorbed like a drop of moisture in the desert."

There is no clear indication of what the women of Iran can expect under Islamic rule. The smart money has it that women in enlightened households will suffer no repression they are not already accustomed to, and that the strongly patriarchal society will continue among the poor. Meanwhile, as Nasrim Farrokh moves about the country on her self-appointed educational missions, women on all levels shift for themselves in ways a Westerner might find difficult to grasp.

There is a Persian word, zarangeh, that has no English equivalent. In French, it is débrouillardise. Loosely, it means everyone must try to find a better way. The Persian feels belittled if he does not hold to its standards. You see its application in the snarled traffic of Teheran.

Or on a downhill slope of the magnificent Elburz Mountains, where Susan Kamalieh, 30 years old and every bit as fetching as this fine spring morning, maneuvers out ahead of two male skiing companions and fairly flies 1,000 dropping meters, throwing up great sprays of snow. Today she has made up her mind to quit her job at a cultural center in the city and to move with her Iranian boyfriend to Paris, where they will live together. She may be gone for two months; she may be gone for good. It is her zarangeh.

Susan Kamalieh is the daughter of a middle-level bureaucrat in the Ministry of Agriculture and of a public-school teacher, a liberal couple in many ways. As a child she wore a chador only when she went to the mosque. Later she studied art and mathematics in Detroit, East Lansing and, finally, at New York University. She paints in the realist vein. She was married briefly and divorced. Her parents did not pick her husband, as is Islamic custom. For all that, she cannot live with her boyfriend in Teheran.

Nor can she live alone. Few, if any, women live alone in the capital. It simply is not done. Nor is it wise for them to venture out alone on the streets at night. Men gather around them and harass them, especially if they are not veiled. The men jeer at women out on their own, calling them cabaret dancers or whores. Single women like Susan Kamalieh live with their parents.

"The real truth is that Iranian men haven't had enough women," Susan says. "The tradition, the customs, lead to a high level of sexual frustration. If you are driving alone, they follow you. If you are walking alone, they stop you. It isn't worth the hassle doing anything independently at night in this country."

Years ago, there was a convenient Is-

lamic custom called sigha. With sigha, a man and a woman could be wed for a period of less than an hour if they wanted to. All they had to do was visit a mullah, delare their intentions to be married for a certain stretch of time, relieve their desire, and part with no stigmatic consequences from society. Widows are said to have been particularly fond of the custom. It lasted well into the 20th century before dying away.

"I would venture that every sane man in Iran is waiting to see whether the Islamic government brings sigha back," a male friend of Susan's says.

Susan has brought beer to the mountains for herself and her companions. Taking a break from skiing, they decide it would be best not to drink it in

"My children are my sole amusement," says Zahari Sajadi. A poor 24-year-old widow, she does not worry about equality, only about her children's limited future.

Imprisoned by the Shah, Ashraf Rabi (glasses) fought with the guerrillas during the revolution that overthrew his regime. She now lectures on the future of Iran.

the open. Alcoholic beverages are banned in the new republic. Back in the city, thousands of cases of liquor and wine have been smashed by the komitehs. So these skiers drink their beer in the car on the way home.

Susan paints at night, often quite late. During the revolution, she was too busy with demonstrations, but now she has returned to her art. She works in a pair of sandals, jeans and a denim shirt. Applying a stroke, she slips one bare foot out of a sandal and scratches the back of one calf with her toes.

"The revolution changed the man-woman relationship, which had been very bad," she says. "The men were horribly dominant. But there was unity

during the revolution. Now the feeling seems to have reversed itself, and I find it sad."

She drops a brush into a glass of turpentine and continues: "Everybody is afraid now that women are going to be more and more discriminated against. Maybe even pushed out of their jobs. Kept from being educated. Mainly kept from being equal to men. We need a liberal government to work for the education of the people and to bring them up to certain standards. My hope was that the revolution would bring that about.

"This is the time Khomeini has to be taken seriously. If he is given more power and more power, he can have another Savak and then we are lost again. The opposition's full blast must come now, but it isn't coming. The lower

classes, the people who fought the revolution, the people you see kissing Khomeini's picture, they are not aware of what is happening. Among other things, Islamic rules will keep women backward. I would like to live and work in Iran, but not in an Islamic republic."

□

There are, says a female professor of economics in one of Teheran's universities, no more than 200,000 women like Susan Kamalieh in the whole country. "They form a speck of the population. That's all. Just a speck," the woman says. She asks for anonymity.

"This silly issue about the chador. This silly fuss about equal rights by a

speck of the women. We must reconstruct a nation and get on with our affairs, not pay attention to fools. In Islam, the woman is equal with the man. Our task is to educate the masses. The woman is free not to wear the chador. It is her decision. But she must make that decision after she is educated about what the chador means.

"Look in the provinces. You will see women without chadors working in the fields next to the men. There is no discrimination there. It is only when they move into the slums in the cities that the woman becomes second-class. She no longer works side by side with her husband. She does nothing to win the man's respect but waits all day for him to return from work. Why bake bread when bread can be bought on the corner in the city? There is where the discrimination exists, and there is where we have to educate."

Gholi Dahri, the woman with seven children who washes in the gutter, waits all day for her husband. He is a carrier in the bazaar. When someone buys something and wants it delivered, Gholi's husband puts it on his back or in a cart, depending on the bulk, and runs it across town. He earns the equivalent of about $7 a day. Half of the five million people in Teheran earn about that much and lead similar lives.

Ashraf Rabi is not one of them. Here is a woman who actually fought — carried a rifle and fired it — in the revolution. She is a member of the Mujahadeen, the Islamic nationalist party of guerrillas. To this day, she teaches the use of weapons in the event the guerrillas will be called to arms again. The daughter of a merchant in the bazaar, she wears a chador or a scarf at all times.

Ten years ago, when she was 19 and studying physics, Ashraf Rabi spent her nights plotting the overthrow of the Government. She was arrested twice, she says, and tortured with burning cigarettes and beaten with cables. Her stays in prison were not long, though, and she returned to her underground activities. She married a fellow covert activist. Then, four years ago, a homemade bomb exploded accidentally in a house that was the headquarters of her secret group. Ashraf Rabi was the only one injured.

At the hospital, Savak agents came to ask why she would have a homemade bomb in her possession. Then, she says, they beat her until she lost the hearing in one ear. "They poured alcohol in my wounds," she says. "They put needles under my nails. They also put splinters under my nails and burned the splinters. They wanted the names of the people I had been working with, but I kept going unconscious and I did not tell them."

She was placed in Qasr Prison in Teheran and there she stayed until this year. One day, she was yanked from prison and taken to the morgue. "They rolled out a body and said, 'Who is this man?'" she recalls. "I said, 'This is my husband.' He had been shot in the street. Then they took me back to prison."

Ashraf Rabi was freed in January, when the Bakhtiar Government was still in power. Then began the fight. "It is very easy to go and die by a bullet," she continues. "My responsibility is now more than that." She lectures groups on the future of Iran. She is a stout woman with a moon face, a soft voice and a *(Continued)* sharp mind. Over tea in her father's living room, she has this to say: "The freedom we have now is so dear, so dear, that we will solve all our problems. Our largest problem is that the wrong image of Islam has been introduced to the people. The chador issue is something to sell newspapers. It is completely foolish. It is an educational thing, a thing to be studied. It is nothing that will be forced on the people. Women who study Islam will wear it proudly.

"We have more fundamental problems. We have a country, but everything is destroyed. We must reconstruct to save the revolution. We have to educate the masses and transfer our knowledge."

In Iran, sugar cubes are put into the mouth, not into the tea. Then the tea is sipped and it mixes with the sugar in the mouth. Ashraf puts a sugar cube into her mouth and speaks even more softly.

"Equality to me is a mechanical term, a procedure. Everyone should do the things she can in the best way. That is the major thing. If you don't believe this, it is as if you are expecting to hear through your eyes and see through your ears. It is impossible.

"The fundamental thing is that we want to make a society in which a human being can live in the best possible way that gives her satisfaction. In Islam the value of a woman is much, much higher than in an imperialistic country. A woman is only a product in your country. In a Communist country, a woman is only an instrument of work. In Islam, the laws of the Koran are the laws of life. The woman has always been beside the man."

I tell her that in some quarters women feel the Koran too aged a document to be taken literally.

"The Koran is obsolete?" Her voice hones up enough to cut brick. "How horrible! Saying the Koran is obsolete is like saying Newton's Law is obsolete because it was discovered 300 years ago."

She says that one day, when it is in the interest of the people, she will return to her study of physics.

□

Iranian women revolted in many ways. A flower vendor recalls giving away her flowers. "I had nothing else to give," she says. "I knew nothing else to do." A woman in a challow kabob restaurant (challow kabob, the national dish, is beef or lamb with buttered rice and a raw onion and yogurt on the side) says she stuffed sandbags with rice. "I had no sand. My rice protected our brave fighters." Mrs. Keyan Hajseyed Javadi dis-

"Everyone in my family is equal," insists this village woman.

tributed her husband's leaflets.

Mr. Javadi is an economist and a writer. He was one of the leading anti-regime writers in the Shah's time. For two years, because of his writing and her proselytizing, neither of them was allowed to work. She has now returned to teaching history and literature in a Teheran high school.

The Javadis live in a comfortable apartment in central Teheran. They have been married for 15 years and they have two daughters. "I come from a very big, traditional family," Mrs. Javadi says. "I was married at 25. We were introduced at a big family party. My husband is from a large family. In our society, if you are from a large family, you like your daughter to marry into another large family. We were married, with both families' approval, at a large family party."

Sitting in her living room, sipping tea and eating cookies she baked, Mrs. Javadi says that it takes Eastern eyes to see clearly the situation for Iranian women. She wears a green skirt of modest length, a paisley blouse, stockings and sandals. She says she derives a great deal of pleasure from caring for her husband, a pleasure that Western feminists do not seem to take to.

"I feel I have less freedom than my husband, but it is a part of Eastern society," she says. "I personally am not bothered by it. The work I do at home I love. I love to have a clean house and to serve my husband and children in the best way I can.

"Look at the Japanese housewife. You can never compare her with the American housewife. She likes to feel owned. And this silly chador thing. I personally would not

wear the chador. My husband does not care for it. It is part of the Eastern woman's nature to wear what the husband likes."

Mrs. Javadi attended schools in England, France, the Netherlands and Switzerland. She says that no Western woman possesses anything that she covets. "When," she asks, sounding exasperated, "will the Westerners understand that an Eastern woman doesn't expect so much from life. An Eastern woman pays more attention to her family. If the husband is not exploited outside the home, he will not transfer that exploitation in-

side the home. The woman is treated tenderly.

"You people seem alarmed that 95 percent of our marriages are still arranged by the families. I will say that the 5 percent who choose their own mates comprise most of our unsuccessful marriages."

Then doesn't it follow, I venture to ask, that the arranged marriages are so successful because the woman are so subservient?

"This is true," Mrs. Javadi says. "Also, the social pressure is such that people don't divorce. Also, women in Iran are more faithful. Widows tend to remarry less than widowers."

This line of conversation soon runs to the Family Protection Act, a 16-year-old law, revised in 1975, that gave a number of freedoms to women in Iran, including the right to divorce under special circumstances. The law also said that men had to have the consent of their first wife before they could take a second. When asked about the manly privilege of having as many as four wives in an Islamic society, educated women usually reply that the Koran says

'When will Westerners understand that an Eastern woman doesn't expect so much from life? If her husband is not exploited outside the home, he will not transfer that exploitation inside the home.'

that a man can have more than one wife only if he loves and cares for them equally. Since this is not possible, the educated women reply, why fret? In the villages, where polygamy is common, women look at a Western inquisitor as if he is crazy for bringing up the subject.

In any event, Ayatollah Khomeini recommended suspension of the Family Protection Act a few weeks ago. Mrs. Javadi pounces on the subject. "It was not abolished by Khomei-

Ashraf Rabi (right) teaches a woman to shoot in the event the guerrillas are called to arms again.

ni," she says. "The mullahs surrounding him were working for that. Some of the lawyers went to him and said, 'What is going on here?' I think it is being worked out. We'll see in the new Constitution what they will do.

"My husband has seen the Constitution and he believes it is a very democratic Constitution. I believe that our rights will not be ignored. I am not worried at all about my rights."

Mrs. Javadi makes reference to the fact that women in Islam can own their own property, that they do not have to share it with their husbands. I say, yes, but isn't it true that under the culture, the woman actually becomes the property of the man at the time of marriage? I quote a passage from a book by Vern and Bonnie Bullough, "The Subordinate Sex: A History of Attitudes Toward Women": "Women, like jewels, are admired and sought after, but should be protected and guarded lest they be stolen. They are property, valuable property, but really not persons, and must not take upon themselves the prerogatives of persons who are, after all, exclusively male."

Mrs. Javadi gives a large sigh, and pats her

about as confidently as if they paid property taxes. The village is by the Caspian Sea, and orange trees and lemon trees thrive in the climate.

Zahari was married to a man called Shaban, and she bore him three children: Arazoo (wish); Omid (hope); and Fereshth (angel). Every year, she and her husband earned about $700 from their lemons and oranges. The husband supplemented their income by seining for fish in the Caspian. From that, he made about $5 a day. A little over a year ago, as Shaban was carting his fruit to the city of Noshar for sale, he was run down by a car and killed.

Shaban's father, Hossein, doubled the size of his one-room house to accommodate his daughter-in-law and

and says in front of Zahari, "She is hopeless. You will not get an answer. She is too ignorant."

The translator remarks on the beauty of Zahari's children, and asks her hopes for their futures. She shakes her head. "I try not to laugh and laugh at your questions," she says. "Why do you ask me such impossible things? Yes, I want my children to grow and be educated and have a career and a good life. Wouldn't I be happy if my son could be an engineer. But this is not possible. What difference is my hope when my hope is impossible?"

A neighbor, Soghra Ali Pour, the mother of six children, puts in her own convictions: "Everyone in my family is equal to everyone else in my family. My husband of 25 years and my children and

under Islam," the Ayatollah replied. "To me, this is the kind of job that is not suitable for a woman, because a woman does not possess all the characteristics that are necessary to be a good judge. The woman has a delicate and more kind heart than a man. Therefore her judgment in important cases of the court cannot be a fair one."

Ayatollah Khomeini's wife of 50 years, Batool, who was married when she was 13 years old, says: "About internal affairs of the home, I make the decision. About outside affairs, he makes the decisions. But we always consult and discuss the problem before the final decision is taken."

Mrs. Khomeini also stresses that the Koran says that women are equal. However, the Koran at the same time says that in court cases the testimony of two women equals the testimony of one man because women are more emotional, and less capable of rational judgment. The Koran also says that if a man dies without a will, the girls in his family may split one-third of the inheritance, while the males receive two-thirds.

Just how strictly Islamic law will be applied in Iran remains unimaginably vague. Will, as many Westernized women in Teheran fear, male hairdressers be banned because a male is not allowed to touch a woman's hair? Will Pierre of the Desert be thrown on the first plane back to Paris?

And what is to be done about the gap of misunderstanding, a gap with proportions something along the lines of the Grand Canyon, between an educated Susan Kamalieh skiing on the slopes and an illiterate Gholi Dahri dragging dirty forks through the gutter? Just how far apart the two stand was dramatized during one of those demonstrations in March.

Said one veiled onlooker: "We are better than you."

"Why?" asked a smartly dressed demonstrator.

"We are pure. We helped in this revolution."

"I am pure. I also helped in this revolution. What about before and after the revolution? It is we who do the work."

"What work?"

"I am an engineer for 15 years. I have been in forests and wild places in this country where you would never go."

"So what if you're an engineer?" the veiled woman said. She fairly spat the words. "Good for you."

"Not good for me. Good for the country. I don't sit home and breed like you do." ∎

An Iranian woman's traditional head-to-toe veil contrasts sharply with her modern transporation.

breast bone as if beset by indigestion. "I doubt that you will ever understand," she says. A maid appears and collects the empty teacups.

□

Once, not all that long ago but long enough to seem as though it had never happened, Zahari Sajadi went to Teheran to be a maid. It was in 1970, when she was 15 years old. She enjoyed the city — electricity and television particularly — but then, after a year, an uncle came and said it was time to go home and be married. Dutifully, she made the trip back across the Elburz Mountains to the village of Koshesara, where belled cows amble

grandchildren. First, he lashed poles together with grass rope, the poles serving as studs, and then he fashioned mud walls around them. Now Zahari spends her days minding her children and washing their dishes and laundry in a swift, muddy creek that runs in back of the house. The father-in-law cares for everyone by picking fruit, growing rice and selling fish.

"My children are my sole amusement," Zahari says one day as a downpour plays loud on her tin roof. The children squirm about her like puppies. With the help of a translator, I dumbly ask something about equal rights for women. The village butcher, Jahangir Pasha, comes into the room

myself can afford meat only once every two months. We eat the same thing and we work the same. I ask you, where is this not balanced?"

□

What is the official position on women in Islam? Ayatollah Khomeini urges that "women be active outside the house, too, if they can. I have said many times that all our men and women should take part actively in the reconstruction of our society and our country."

Another leading religious leader in Iran, Ayatollah Shariat-Madari, was asked recently why women were not judges in Iran. "Judgment is not a simple work, especially

Prime Minister Callaghan, whom Mrs. Thatcher is challenging at the polls this Thursday, with reporters outside Parliament.

MARGARET THATCHER: A CHOICE, NOT AN ECHO

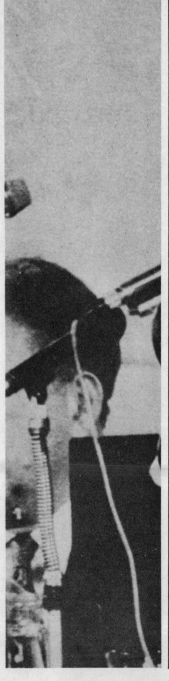

By R. W. Apple Jr.

When she opened the Conservative Party's election campaign earlier this month, Mrs. Margaret Thatcher, the party's leader, used a sentence that has become familiar to her political intimates. "We believe," she said, "that you get a responsible society when you get responsible individuals." In a number of ways, it was a comment typical of the 53-year-old woman whom most Britons expect to lead the Tories to victory in next Thursday's general election, thereby becoming Britain's first woman Prime Minister (and, in-

R. W. Apple Jr. is chief of the London bureau of The New York Times.

deed, the first woman to lead a major European nation). In that one sentence, you have the essential Thatcher — resolute, a little preachy, devoted to the essentially middle-class notion of "responsibility," and convinced in the deepest part of her being that only individuals, not governments, can achieve it.

Mrs. Thatcher — "Maggie" to her friends and to Fleet Street, "Mrs. T" to politicians outside her inner circle, and "the Blessed Margaret" to the Conservatives' resident wit, Norman St. John-Stevas — is a small, fine-boned woman, with pale blue eyes, the kind of complexion the English always liken to a rose and hair that she readily admits to dyeing blond. She has a voice that has occasionally proved a political liability. It tends at moments of stress to rise to an unpleasant stridency (Jean Rook of The Express once wrote that it sounds like "a record of a cello played at the wrong speed"). In addition, she has an

accent that is considerably more "plummy," or aristocratic, than might be expected of a person from a background as decidedly middle-class as hers. In a society that still takes accents as talismans of class, Mrs. Thatcher's prompts some people to accuse her of attempting to be something that she is not.

They are quite mistaken. Whatever gloss childhood elocution lessons and Oxford may have put on her speech, whatever lessons of dress or style she may have learned in 20 years in the House of Commons, Margaret Thatcher remains a middle-class person, knows it, talks about it and is proud of it. She stands for those values — thrift, industry, self-reliance, honesty, chastity, family solidarity — that were as entrenched in British families like hers 40 years ago as they were in families of that era in similar circumstances in Fort Wayne, Ind., or Saint Joseph, Mo. They are values that have fared poorly

Vintage Thatcher: "People want less tax taken from their pay packet . . . and they distrust and fear the power of the unions."

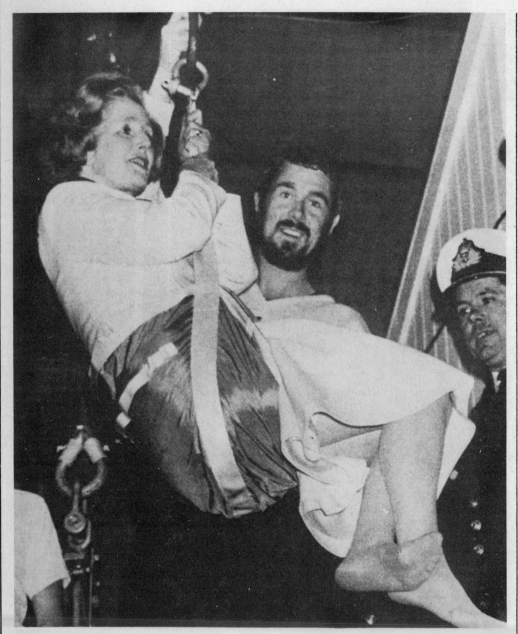

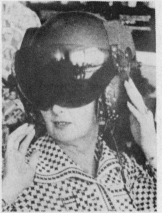

Margaret Thatcher on Britain's campaign trail (clockwise from top left): riding a jackstay at a recent boat show; trying on a new computerized helmet at an aircraft factory (data flashes to her on a small screen inside the visor); standing before a bust of Winston Churchill; visiting an old-people's home outside London.

in postwar Britain, and what Mrs. Thatcher is about, in one sense, is their restoration. One of her friends says that she is "Grantham writ large," Grantham being the market town in Lincolnshire, north of Cambridge, where she grew up; another says that "you can't understand Maggie without understanding her father." They were saying the same thing: that she was formed by a family and a community where idleness was thought evil, self-sufficiency good, and where effort brought reward.

"She is the product of what I believe you might call the heart of this country," said one of the men who would almost certainly serve in a Thatcher Cabinet, a man whose own aristocratic background could not be more different from hers. "She is the daughter of a grocer, a small, provincial shopkeeper who made his own way and looked after his family. We were a nation of shopkeepers once, but that's all gone now; we're a nation of collective enterprise, full of doubts about ourselves and our role in a world changed beyond recognition. She misses the old verities, and she thinks they are as valid as ever. Put it another way: Her heart is with the petite bourgeoisie, whose views have been out of fashion for two generations now. We have been governed either by Tory grandees like Macmillan and Heath or by the working class standing shoulder-to-shoulder in mass trade unions. These people, the middle class, have been the dispossessed, and we may be talking about half the people in the land. She is with and of them."

Mrs. Thatcher's instincts and beliefs about individuals have shaped her politics. Notice, for example, the points she made last year when she was asked what he British people wanted. They

THATCHER

wanted, she said, "less tax taken from their pay packets so that they have more of their own earnings to spend in their own way. They want to own their homes, and if they haven't had that chance themselves, they'd like their children to have it. They want to have more say in the education of their children. Our people distrust and fear the power of the trade unions, and are asking how those who wield such power can be brought to use it responsibly. Our people want governments to stick to the role of governments and not to try to do everyone else's job. Their role is, first, to keep the currency stable; second, to preserve law and order; third, to see that our country is properly defended."

No mention, you will have noticed, of all of the functions taken for granted by the British people since the political earthquake of 1945, no mention of unemployment benefits or of industrial subsidies. No mention of government as a guarantor of civil liberties. Instead, a simple, direct prescription for minimum government, straight out of Adam Smith. In a country where government is involved in everything, except perhaps the Chelsea Flower Show, it is not surprising that what Mrs. Thatcher is proposing is seen as radical, almost revolutionary.

Margaret Hilda Roberts was born on Oct. 13, 1925, above her father's grocery shop. He was what was called in those days a self-made man, who had dropped out of school at 12 to become a grocer's apprentice. In time, he came to own two shops and to serve as Mayor of Grantham. He was a teetotaler Methodist, a lay preacher, an austere man by all accounts, but fascinated by the world of public affairs and determined that his two daughters, especially Margaret, the younger and brighter, should play a larger role in it than he had. They were not poor, the Robertses, but at the beginning there was no garden, no indoor toilet and no hot water upstairs. Alfred Roberts bought his first car, a second-hand Ford, after World War II.

"It was your duty to better your lot," Mrs. Thatcher recalled several decades afterward, "and to do it through your own efforts. Everything had to be accounted for, and

my father looked with scant sympathy on people in the town who lived up to the hilt of their income, heedless of the need to save for a rainy day."

Margaret was her father's daughter. She worked hard, finishing at the top of her class at secondary school in every year but one, when she was second. Then as now, she was fastidious and unfailingly tidy, polite, well-mannered, "nice." An enterprising reporter who some years ago tracked down one of her classmates was told, "Mothers would say to their daughters, 'Why can't you be more like Margaret Roberts?'" And above all, she was self-confident almost to the point of conceit. When at 9 she won first prize in a recitation contest, the headmistress told her she had been lucky. To which she replied, "I wasn't lucky; I deserved it." Forty years later, at a press conference following her election as Tory leader, a reporter asked why she thought she had won. "Merit," she replied.

Learning that Oxford would require her to know Latin, she mastered it in a single year with the help of a private tutor. She won a bursary — a kind of grant-in-aid, something short of a scholarship — and settled into a curiously bifurcated life, dominated by chemistry and politics. She obtained a second-class degree, failing to win a first largely because she worked so hard in the Oxford University Conservative Association, of which she became only the second woman president. Afterward, she had a couple of research jobs; in one, she was set the task of finding a new adhesive that would bond a plastic to metal or wood. She turned from that to the law, eventually establishing herself as a barrister, or trial lawyer, specializing in tax matters.

But all the time, she really wanted a political career, and she began, in the time-honored Tory manner, by standing for a seat (Dartford, south of London) that was safely Labor. She campaigned with characteristic vigor, and succeeded in pushing the Conservative share of the vote from 36 percent in 1950, her first try, to 40 percent in 1951.

But the most important moment in her Dartford experience, at least in human terms, was a ride back into London with a man named Denis Thatcher, a quiet sort, 10

years older than she. He was working for the family paint firm. Margaret Roberts married him five weeks before the 1951 election and dropped out of politics for a time. In 1953, with the kind of efficiency that a career woman needs, she had twins, Carol and Mark, completing her family at a single stroke. When they were 6, it was time to look to politics again, and Margaret Thatcher, now 33, was adopted for the safe Tory seat at Finchley, a leafy suburb just north of the capital. She won handily and has won handily in every election since.

Mrs. Thatcher was an immediate success in the House of Commons, but not because she was a particularly skilled debater (she still is not in the same class as Prime Minister James Callaghan, a master of the sarcastic riposte, or Michael Foot, the Leader of the House, who scores points by the sheer power of his intellect). Her successes, like those she had made earlier in life, resulted from competence and diligence, rather than brilliance. The 1966 budget debate was an early example of her technique: She had read every major speech given in budget debates during the previous 20 years, then flabbergasted her rivals with the fluent command of relevant statistics that remains a hallmark of her public performances. Both civil servants and party researchers came to expect demands from her for far more thorough briefings than they were accustomed to.

In 1970, the Tories returned to power, and she entered the Cabinet — only the second woman to hold such a position in a Conservative Government. For the next four years, she presided over the Ministry of Education, a crucial testing ground between tradition and progressive thought. The biggest public row during her tenure followed her withdrawal of free-milk allowances, which led hecklers to taunt her wherever she went ("Thatcher, Thatcher, milk snatcher!"). A public-opinion poll in January 1972 showed her to be the least popular member of the Government. She was reproached by newspapers and by opposition politicians for being out of touch and out of reach.

Curiously, that episode has faded from public memory. What is remembered more than anything else is her battle against comprehensive secondary schools on egalitarian grounds. Labor wanted to do away with the grammar schools, which are paid for

with public money, but chose their students on the basis of intellectual potential, and replace them with comprehensives, which all students would attend. It was the same sort of argument that has revolved around such American schools as the Bronx High School of Science. Mrs. Thatcher defended the grammar schools, which had opened the high road to success for many boys and girls of modest means, including Margaret Thatcher. She delayed the implementation of the comprehensive plan in many areas, arguing that all parents should have the breadth of two choices open to those who could afford private education. She also cut the budgets of universities so that she could put more money into primary education, and she refused, despite heavy pressure, to raise the age at which children were legally permitted to leave school.

During her 20 years in Parliament, the subjects that have engaged her interest have been predominantly domestic and most often linked to financial questions. Her voting record over two decades shows not the slightest deviation from Tory orthodoxy, although she has been tougher than many Conservatives in the area of social reform. She voted to restore the death penalty, for example, in 1964, 1966 and 1969, and in a more limited form in 1974. She failed to vote on homosexual-law reform and she opposed the most important reform of abortion laws.

As the abortion vote may suggest, she is no feminist, despite her pioneering role as a woman in the upper reaches of British politics. Although 60 years have passed since Nancy Astor became the first woman to sit in the House of Commons, women have remained a tiny minority in what still operates much like a St. James's men's club. Of the 635 members of the last Parliament, only 27, or slightly more than 4 percent, were women. There is only one woman Minister in the outgoing Cabinet, Mrs. Shirley Williams, and there would probably be only two in a Thatcher Cabinet — Mrs. Thatcher herself and Sally Oppenheim.

But there is no recorded instance of Mrs. Thatcher's berating her party for not selecting more woman candidates. In this area as in others, she seems to believe that if she can succeed without special consideration so can everyone else, though Mrs. Thatcher's path to the top has undoubtedly been hindered by the resentment of male colleagues and, more important, by the fear of male colleagues that voters will oppose her because she is a woman. After one of her television appearances, one still hears almost as much commentary about her hair style, her clothes and her accent as about what she actually said.

She has drawn the ire of many in the women's movement. The journalist Katharine Whitehorn, for example, once wrote that Margaret Thatcher acts more like the wife of a politician than a politician, with "the glassy smile" and the "conventionally boring clothes." Even to "the less radical women caught up in the groundswell of the feminist tide who would in theory most welcome a woman leader," Miss Whitehorn added, "she is exactly the sort of woman they loathe: the archetypical Tory lady."

To all of this, Mrs. Thatcher has reacted with equanimity. She concedes, without apparent bitterness, that if she loses this time she will not get another chance to lead the Conservatives to victory, partly because she is a woman, and she concedes that she might lose some votes for the same reason (even though one of the most thorough recent polls suggests that she will not). Her way of dealing with the problem is not to try to submerge her womanliness: Indeed, she has emphasized, in newspaper and magazine interviews, that she is soft and feminine despite the toughness and resilience that she demonstrates in her public life. She told Women's World magazine: "There are times when I get home at night and everything has got on top of me, when I shed a few tears silently alone. I am a very emotional person. I'm a romantic, you know, at heart, and believe wholeheartedly in love. Anyone who doesn't must be terribly unhappy." She told Kenneth Harris of The Observer, when he asked about her husband: "Denis is absolutely marvelous. He behaves as a sort of shock absorber. . . . I certainly do talk to Denis — about everything. But it's not only the talk. It's that he protects me, he really does." And she talks constantly about her own knowledge of the problems of household budgets and about what she discovers in the stores when she goes shopping.

In fact, however, she is not the archetypical Tory lady by any means. She has far more ambition, far more steel, far less inclination to defer to men. In politics, she has said, "If you want anything said, ask a man; if you want anything done, ask a woman." The phrase applies with some precision to Mrs. Thatcher's greatest triumph to date, her election as Tory leader in February 1975. Having lost two elections in a row, having won the ideological enmity of some of his colleagues by his flip-flop on economic policy and the personal enmity of others by his icy manner, Edward Heath was ripe for defeat. Everyone was talking about it, but the likely candidates to challenge him chose for one reason or another not to do so.

Margaret Thatcher had no qualms. Beginning with the support of a handful of M.P.'s who carried little weight but knew a good deal about modern media techniques, Mrs. Thatcher quickly built up support across a broad spectrum within the party. Then she did something unusual in British politics: She marched into Ted Heath's office and told him, firmly but politely, that she would oppose him. "These things are usually done by stealth," says David Howell, one of the ablest of the young Tory M. P.'s. "It was a very, very courageous thing for her to do, and typically straight." On the first ballot, she knocked Mr. Heath out of the contest, to the astonishment of most of those who voted for her, and, on the second, she won a decisive overall majority.

She has shown the same boldness in staking out controversial positions for her party when the opportunity pre-

sented itself. Last year, for instance, she took a strong stand on immigration, asserting that whites in certain cities and towns in Britain felt "swamped" by Indians and West Indians, and promising to do something about it at a time when party policy on the subject had not been decided. It caused her some difficulty within the Shadow Cabinet, and the Conservative election manifesto is less committed than she was. She was denounced in some quarters as a racist. But she succeeded, in a single statement, in identifying her party with the millions of Britons who do feel threatened by immigration and who long for the racial homogeneity of the good old days.

But Mrs. Thatcher can also be cautious. One of her fellow members of the Shadow Cabinet commented that "her basic instinct is caution, because she is surer of where she wants to go than how to get there." She demonstrated this quality in the episode that gave her what she had waited for so long and so impatiently, the opportunity to lead the Tories into a general election. She had submitted a motion of no confidence once before, last year, and suffered an embarrassing defeat when Prime Minister Callaghan patched together a Liberal-Labor alliance at the last minute. This time, she waited until the key minor parties had committed themselves — until the Scottish Nationalists had put forth a no-confidence motion of their own, until her whips reported to her that the Northern Ireland members were likely to go along with her, until David Steel of the Liberals had issued a statement of support. The newspapers accused her of timidness, but her caution proved justified when she won by only a single vote and only because a Labor M.P. was so ill (he died a few days later) that he could not leave his bed to cast the vote that would have saved Mr. Callaghan once again. Similarly, she agreed to a late-starting campaign, with fewer than 14 days outside her own constituency, when it became clear that she was beginning with a substantial lead. And even though she had said last September that "I think I would gain from a TV debate with Mr. Callaghan," she refused to engage in one when he agreed.

Perhaps the most famous parliamentary wisecrack ever made about Mrs. Thatcher came from Labor's bully boy, Denis Healey, the Chancellor of the Exchequer. He called her "Ted Heath in drag." Like Mr. Heath, Mrs. Thatcher does have trouble projecting warmth to groups of any size: She tends to come across as distant and scolding. But unlike her predecessor, she relates well to individuals, especially those whom she has

At funeral of her close adviser Airey Neave, killed by terrorists.

known for some time. She is an excellent street campaigner. Her staff dotes on her, and obscure backbenchers speak gratefully of the time she spends listening to their views and inquiring about their well-being. "She remembers who has a bad back, whose wife has been sick, who is having business problems," one of them said, "and she always asks. The result is that she commands the loyalty of the back benches through affection, where Ted got it only by cracking the whip."

She is so single-minded — "It is as if she had taken holy orders," said a supporter in her constituency — that some of her friends wonder whether she might not crack under the strain of running the Government and commit serious errors. "Prime Ministers need things to take their minds off the job," said an old Tory who has known several, "and she doesn't have them. She doesn't drink with the boys, or the girls, either, for that matter. She doesn't tell jokes. She may relax, but I've never seen it in the 15 years that I've known her." Mr. Heath had his music; Mr. Wilson his coterie of chums outside politics; Mr. Callaghan his farm. Mrs. Thatcher likes music and played the piano and sang in choirs as a young woman, but she is seldom seen at Covent Garden or Glyndebourne these days. And when she reads, according to one of her closest friends in the House, she al-

most always chooses books that relate to her work, or something by the conservative philosopher Friedrich von Hayek, or the "Gulag Archipelago."

If Mrs. Thatcher wins the election, her first test will be the formation of a Cabinet. Her selections will be important, because a British Prime Minister, unlike an American President, shares power with Cabinet colleagues. Mrs. Thatcher has had a mixed record in shaping and leading the Shadow Cabinet. "She isn't jealous of other people's gifts and talents," one colleague said, "but she talks too much. She comes into the conversation too early, gives her view — bang — and that pre-empts free discussion. As you know, there's no subject on which she has no view. She will have to get over that if we're in Government."

It is not that Mrs. Thatcher is intellectually arrogant, but that she sometimes speaks in the first instance out of a rather rigid instinctive sense of what is right — her Margaret-the-moralist view, one young Scottish M.P. calls it. She offended some senior American officials when she visited Washington in 1977 by precisely that sort of performance, telling President Carter abruptly, "Foreign policy is simply a matter of national self-interest." But if she can be persuaded to stop, to listen to contrary arguments and to analyze a situation — and, by all accounts, she often can — she is ready to change her view. A case in point came in the heated discussions over the Conservatives' position on Rhodesian sanctions. Her heart lay with the white Rhodesians, and she followed it for a time. But Lord Carrington and other Tory foreign-policy experts persuaded her that the abandonment of sanctions would have dire effects. She not only changed her mind but brought most of the front-bench party leaders with her; one exception was young Winston Churchill, the great man's grandson, of all people, and she sacked him.

John Biffen, a Tory thinker who has worked closely with both Mrs. Thatcher and the rightist Enoch Powell, compared the two in this way: "Both are intellectually potent — God, yes — but with Powell you were always sure of the context in which he was thinking. In other words, he was always logical. She isn't; her ideas often flow from her empathies, and that introduces some inconsistencies." Would she make a good Prime Minister? "No one, absolutely no one, can answer that question, not even Mrs. Thatcher herself."

One of the inconsistencies that most perplexes those who watch Mrs. Thatcher is this: She presents herself as an ideologue, with a set of ideas she wants desperately to put into effect, yet she has behaved with evident pragmatism on many occasions. She often says that she hates the word "pragmatic," because "it implies that there are no values worth fighting for," and she told an interviewer not long ago that she would not try to assemble a Cabinet of "peo-

ple who represent all the different viewpoints in the party." "We've got to go in a clear direction," she said to Kenneth Harris of The Observer. "As Prime Minister, I couldn't waste time having any internal arguments." Yet those who served under her in the Education Ministry remember that she decided between preserving grammer schools or closing them on a case-by-case basis rather than attempting to enforce a sweeping policy line. One of the members of her present personal staff commented: "Water will always flow downhill, but it goes around obstacles and impediments. Mrs. T is like that. Now you tell me, is she a pragmatist or not?"

If she really means to exclude from the Cabinet those whose views are not completely in concert with her own, she may be in trouble within her party from the start, since she's been known to have differences of one degree or another with most of the party's leading Cabinet candidates. Then too, if she really means to push ahead with strictly her own vision of the British future totally without regard to shifts in the political weather, she would almost certainly find herself in confrontation with the trade unions and perhaps with other important segments of British society. That is what brought the Heath Government down only five years ago.

The three most important things that Mrs. Thatcher will attempt to do as Prime Minister, if she reaches her goal, are to reduce income taxes, restrain the labor unions and normalize relations with Rhodesia. She may, in the end, be dissuaded from the last by the

argument, which some of her advisers are prepared to make, that she would be risking the break-up of the Commonwealth.

On income taxes, she has said, "You've got to change the balance between the distribution of wealth and the creation of wealth. That means you've got to have as your first priority the reduction of personal tax." The underlying assumption is that high tax rates make it pointless for either workers or managers to do their best and are therefore the root cause of "the British disease" — low productivity, low rates of investment, economic stagnation. Grine Worsthorne, the iconoclastic Tory columnist, suggested recently, it is quite possible that "instead of reaching to [tax cuts] as an opportunity for greater activity, people would seize upon them as an excuse to do even less." There may be far less latent initiative in this country than most people suppose, Mr. Worsthorne argued. "What if the Socialist lid is removed and very little capitalist steam bursts forth?" In that event, Mrs. Thatcher would have achieved little.

Even if Mr. Worsthorne's dire view is incorrect, Mrs. Thatcher's plans to cut income taxes would surely run into resistance if, as seems inevitable, they were accompanied both by an increase in sales taxes and a sharp cutback in some social-welfare services. The likely result would be a rise in trade-union militancy at the very time when she was attempting to dampen it through a ban on secondary boycotts, restrictions on closed shops and reductions in strike benefits.

Emboldened by the poll data showing that fully two-thirds of all Britons believe that the unions have too much power, Mrs. Thatcher is determined to crack down on them. She thinks that "some of the people in positions of influence in the trade unions are out to destroy a free society, and the weapon they are using is intimidation. We shall have to stand up against those elements." Strong words. But stand up to them with what? A prolonged industrial crisis, no matter how noble the motive that brought it on, would bring down upon Mrs. Thatcher the wrath not only of the public but also of industrial leaders, bankers and investors. Prime Ministers are held responsible for strikes and stoppages (viz., Mr. Callaghan this past winter), which is precisely the reason they try not to antagonize the unions.

In a country whose economy has failed to respond to the nostrums of a succession of postwar Prime Ministers of varying ideological hues, a country that has been in slow but steady economic decline since the shattering experience of World War I, it is difficult to believe that Mrs. Thatcher's simple formulas will suffice. It is hard not to conclude that she has overgeneralized from her own personal experience, and to wonder, along with one of her colleagues in the Shadow Cabinet, "whether this woman of sharp focus and narrow vision, this not-so-eloquent true believer in a world of eloquent skeptics, can save us from going down the drain." That, no less, is the task she has set herself. ∎

April 29, 1979

Right of Women to Self-Defense Gaining in 'Battered Wife' Cases

By WAYNE KING

"It was the most brutal act in recent history," said David Berg, the lawyer who defended Diana Cervantes Barson in her trial for the murder of her common-law husband in Houston. "And it took the jury only one hour and 40 minutes to acquit her."

Miss Barson shot the man, cut his body into five pieces, put it in garbage bags, loaded it into the trunk of her Cadillac and drove to California to ask her relatives for help in disposing of the bags. The police found her in an orange grove in San Bernardino, her wrists cut in a suicide attempt and some of the bags nearby.

The Houston jury's acquittal of Miss Barson, Mr. Berg said, was "a clear-cut

case of affirmation of a woman's right to physical self-defense."

Miss Barson was acquitted largely through her emotional recounting on the stand of two years of beatings by her husband, ending in three days of threats and abuse with loaded revolvers and an ice-pick.

Her trial is a dramatic example of the growing and effective use around the nation of the "battered woman" defense in cases of homicide.

In the most recent such case, a jury of four men and eight women in Charleston, S.C., found Cynthia Hutton, 22 years old, not guilty of murder last month. She shot her 24-year-old husband in the back with a 12-gauge shotgun, but testimony corroborated her account of repeated beat-

ings over a five-and-a-half-year marriage, and the jury acquitted her.

Some such cases never reach trial. In New Orleans, 24-year-old Viola Williams was charged with first-degree murder after she shot a male friend 13 times with a .38-caliber revolver, reloading twice in the process and shooting him 11 times in the back.

Charges were dropped after the authorities learned that she had suffered a decade of abuse, including slashing with a knife, beating with a baseball bat and having her face pushed into a hill of red ants.

Of the 13 shots, Assistant District Attorney Sheila Myers said that "the first two were in self-defense and the rest for what he'd done to her for years."

Whether wife-beating constitutes 'a killing excuse" or affords the victim "the right to kill," as some suggest, is a matter of dispute.

Prosecutors tend to question such defenses, while feminists and others contend that acquittals in such cases constitute a simple recognition by jurors that the use of force, even deadly force, is justified to ward off a beating when a pattern of repeated abuse is established.

Last year, the Center for Constitutional Rights in New York published a guide that the organization said was "intended to aid attorneys representing women who commit homicides after they have been been assaulted or after their children have been molested or abused."

25 Cases Are Described

The guide noted: "Women charged with homicide in response to abuse formerly pleaded guilty or pleaded insanity and were routinely convicted. They are now speaking out about their circumstances, describing the reasons for their actions, and asserting an equal right with men to defend themselves."

The article named 25 abused women from all over the country who killed their assailants, usually the husband. Nine were acquitted on the ground of self-defense or won reversals of their convictions, one was not prosecuted, one was found not guilty by reason of insanity, two were given probation, nine were convicted (some have appealed) and three were awaiting trial when the guide was published.

"I'm a born skeptic," said Joseph Murray, chief of the homicide unit in the Philadelphia District Attorney's office. "A lot of these things aren't the way the wife says; then again, a lot of them are. How many of these women charged with homicide are actually bruised? Maybe five to 10 a year."

Mr. Murray said he had seen no increase in the number of women who struck back against their husbands or male friends but had noticed "a more liberalized treatment toward the woman who does strike back."

In Berkeley, Calif., a 46-year-old nurse killed her husband in a scuffle over a pistol he had threatened her with. She was convicted of involuntary manslaughter and was sentenced to "house arrest" — she cannot leave the county and must report daily to a probation officer. It was believed to be the first such sentence in the nation's history.

'Permission to Fight Back'

Beth Doolittle, a 26-year-old "street fighting" instructor at the Marin County Rape Crisis Center, said that women today have "more societal permission to fight back."

"The image of Fay Wray cringing under the grip of King Kong is being replaced by images of women fighters like the Bionic Woman," said Miss Doolittle, adding that she taught women "to fight dirty — kick below the belt." Legal attitudes also appear to be changing.

In a Detroit case, for example, Judy Hartwell was charged with murder after stabbing her husband five times and leaving him to die on the kitchen floor. Judge Victor Baum, in his instructions to the jury, said, "A wife has the right to use whatever force necessary to resist her husband's forcible sexual attack."

The Hartwell case, which resulted in acquittal, was hailed by women's rights groups as something of a landmark in determining a woman's right to defend herself against her husband.

Less clear-cut was a case in Ingham County, Mich., involving Francine Hughes, who, as a result of years of "physical and mental torture," waited until her husband fell asleep and then set the house afire. She was found not guilty by reason of insanity.

Marjory Cohen, a Detroit lawyer with extensive experience in such cases, said that, paradoxically, the notoriety of these two cases resulted in harsher attitudes toward women accused of killing men who beat them.

"There's a real backlash, at least on the part of the public," she said. "People think any woman can kill her husband and get away with it."

Because of the absence of consistent long-term statistics, there is no reliable way of determining whether there has been an increase in the tendency of women to fight back. However, a study of woman inmates in the Cook County Women's Correctional Institution in Chicago in 1975 and 1976 found that the major reason given by 132 women awaiting trial for manslaughter or murder was retaliation for, or defense against, physical abuse.

Of the 132, 53 had killed their husbands or male friends. Twenty-seven used knives and 26 used guns; in some cases, the gun had earlier been used to beat the woman.

Every one of the women said they had called the police on numerous occasions after being beaten. Every woman said she loved the man she had killed, and in almost all cases, alcohol was said to be the cause or a factor in the abuse.

Twenty-five of the 53 women are free on bond, four have been found not guilty, four are on work release and three were remanded to the Department of Mental Health. Seventeen were found guilty and served or are serving average sentences of two to five years; usually, they are eligible for parole after one-fourth of the sentence has been served.

May 7, 1979

HIGH COURT UPHOLDS A CIVIL SERVICE EDGE FOR WARS' VETERANS

WOMEN'S MOVEMENT SET BACK

7-to-2 Ruling Finds Job Preference for Service in Military Is Not Sexually Discriminatory

By LINDA GREENHOUSE
Special to The New York Times

WASHINGTON, June 5 — The Supreme Court ruled today that a state can hire veterans in preference to nonveterans without unconstitutionally discriminating against women, even though the policy may "overwhelmingly" benefit men and have a "devastating impact" on women's Civil Service job opportunities.

By a vote of 7 to 2, the Justices reinstated a Massachusetts law that gives any veteran who passes the Civil Service test a right to be hired before any nonveteran, regardless of test scores. The law had been struck down twice on sex-discrimination grounds by a Federal District Court.

The Federal Government, 45 states besides Massachusetts and many local governments offer veterans some form of preference in employment, typically a five- or 10-point bonus on the Civil Service test. Today's decision, upholding one of the most sweeping preference laws, would appear to shield the milder laws from constitutional challenge as well.

War Experience Required

Most of the laws require the veteran to have some wartime experience, although that is generally defined rather loosely. New York, New Jersey and Connecticut are among the states with some form of veterans' preference.

As expected, the decision was praised by veterans' organizations and denounced by women's organizations as "devastating" to efforts to end sex discrimination. [Page A17.]

Another ruling of the High Court affected women; the Justices held that the Constitution gives an employee the right to sue a Congressman for damages on grounds of sex or race discrimination.

In a third case, the Court reaffirmed its support of the exclusionary rule, which bars the use of illegally obtained evidence at a trial. [Page A17.]

The Massachusetts law was challenged by Helen B. Feeney, a 54-year-old former Civil Service worker who was denied several chances for promotion when veterans with lower test scores were given the jobs she sought.

The decision, Personnel Administrator v. Feeney (No. 78-233), is a setback for the women's movement, which has made the elimination of veterans' preference laws a top priority. The National Organization for Women, the League of Women Voters and a number of other groups filed briefs in behalf of Mrs. Feeney, who was represented by the Massachusetts chapter of the American Civil Liberties Union.

While the Justices today did not specifically address the question of whether veterans' preference laws discriminate against blacks — 92 percent of the nation's 30 million veterans are white — the Court's analysis of the kind of evidence needed to prove unconstitutional discrimination makes a successful race-discrimination challenge to the laws extremely unlikely.

The equal-protection clause of the 14th Amendment, Associate Justice Potter Stewart wrote for the majority, "guarantees equal laws, not equal results."

In other words, he said, it was not enough for Mrs. Feeney to show that the Massachusetts law has the effect of keeping out of the best Civil Service jobs the 98.8 percent of Massachusetts women who are not veterans.

Rather, she had to show that the law — the first enacted in the 19th century to help Civil War veterans — had a "discriminatory purpose" — that the state intended to keep women out of the Civil Service.

It is the majority's definition of "intent" that is likely to give today's opinion a continuing impact beyond the veterans' preference issue.

The majority opinion was joined by Chief Justice Warren E. Burger and Associate Justices Byron R. White, Lewis F. Powell Jr., Harry A. Blackmun, William H. Rehnquist and John Paul Stevens.

Associate Justices Thurgood Marshall

and William J. Brennan Jr. filed a dissenting opinion, written by Justice Marshall. They said the fact that Massachusetts had enacted an absolute veterans' preference, when it could have chosen a less sweeping approach to helping veterans, "evinces purposeful gender-based discrimination" sufficient to meet the Court's test of intent.

"Where the foreseeable impact of a facially neutral policy is so disproportionate," the two Justices wrote, "the burden should rest on the state to establish that sex-based considerations played no part in the choice of the particular legislative scheme." Massachusetts could not meet such a burden, they concluded.

The majority's rejection of the idea that the "intent" standard can be met by showing that a challenged policy had a discriminatory result that could have been foreseen is likely to prove important in other types of civil rights suits.

Plaintiffs in school integration cases, for example, often try to meet the intent test by showing that a school board should have foreseen that certain policies would probably enhance segregation. Some Federal courts have accepted this argument; the Supreme Court has never explicitly ruled on it.

The Federal Government itself has adopted this "foreseeable result" theory in a number of school cases. However, the Justice Department's friend-of-the-court brief in today's case argued against applying that theory and urged the Justices to uphold the Massachusetts law.

"When the district court found that the purpose of the veterans' preference statute was to aid veterans and not to injure women, that should have been the end of the matter," the Justice Department brief said.

June 6, 1979

THE CASE FOR MOTHER'S MILK

Despite mounting evidence that nursing is beneficial to both mother and infant, a woman's choice is complicated by social and economic factors.

By Robin Marantz Henig

The Williston Park Community Pool in Mineola, N.Y., had no doubt been the scene of more daring displays of flesh than Barbara Damon's. Mrs. Damon, wearing a one-piece bathing suit and draped in a towel, was breast-feeding her 2-month-old son, Michael, when the lifeguard confronted her, saying she was offending the other sunbathers. "He told me that I could nurse my baby in the restroom," Mrs. Damon said. "I said that the bathroom was not an appropriate place to nurse the baby."

Because Mrs. Damon continued to nurse Michael in public, the pool manager canceled her family's membership. The incident, brought to national attention by a suit the Damons filed in 1978, highlighted the dilemma facing many new mothers today, whether the benefits of breast-feeding warrant the extra effort it takes to nurse a child in contemporary America.

The number of women choosing to breast-feed has grown dramatically in recent years. According to a study made by a formula manufacturer, in 1971 the proportion of new mothers who left the hospital breast-feeding their babies was 23 percent; by 1977, the proportion had grown to 43

Robin Marantz Henig, an editor at BioScience magazine, is writing a book on aging and the brain.

Iowa City firefighter Linda Eaton was suspended for breast-feeding while on duty.

percent, and it's still climbing. The message of academic pediatricians and nutritionists

has finally sunk in: Mother's milk is usually far superior to infant formula for a developing child.

However, even in the face of mounting evidence showing the benefits of breast-feeding both to mother and child, the decision to nurse may be a difficult one for the modern woman to make. Nursing is inconvenient, especially for the woman who works. The recent case of Linda Eaton, an Iowa City, Iowa, firefighter, exemplifies the problem. Mrs. Eaton was suspended from her job because she breast-fed her son while on 24-hour shifts. She nursed him during her "per-

sonal time," time male firefighters use to shower, eat, sleep or play cards. According to the city manager, she was suspended for refusal to follow orders. During a nationally televised news show before her suspension, Mrs. Eaton was shown nursing when the fire alarm sounded. She handed her baby over to his baby sitter, smiled at the cameras, and was the first person on the fire truck.

Other considerations that may complicate a woman's decision to nurse are hospital obstetrical procedures — which often interfere with successful nursing by sedating the mother during delivery and by separating her from her baby during its first days of life — and environmental pollution, which can make nursing dangerous, since some toxins may accumulate in breast milk. The woman's decision may also be influenced by her social status, educational level and economic standing.

☐

Breast-feeding does not necessarily come naturally — even in the world of nature. Zoologists have found that chimpanzees need the instruction of other female chimps before they can successfully nurse their young. Interference in social groupings among animals — such as separation of a mother goat from her kid and the other goats during the first hour after birth — can hamper an animal's ability to nurse.

In humans, production of milk begins as a signal of the infant's need. As the baby sucks at the breast, the nerves of the mother's nipple send a message to the hypothalamus, an endocrine gland at the back of the brain. The pituitary gland, in turn, releases two hormones — oxytocin and prolactin. Oxytocin, the same hormone that doctors give to induce labor, stimulates contraction of the small saclike cells behind the areola, the dark area encircling the nipple. This forces the milk from the glandular tissue of the breast through the nipple, and initiates the flow of milk. At the same time, prolactin, the other hormone released by the pituitary, stimulates the production of milk through its action on the milk glands in the breast.

Almost anything can interrupt this delicate process. One study, conducted in early 1948 by Drs. Niles and Michael Newton in Philadelphia, showed that a woman who had nursed successfully for seven months could be so distracted by a tickle, a splash of ice water or a tricky mathematical problem that her milk flow virtually stopped. Dr. Mavis Gunther, a British pediatrician, found that even ideas could interfere with the flow of milk. Watching women using a breast pump to empty their breasts because their infants were too sick to nurse, Dr. Gunther said she could see the ebb and flow of the milk supply change with the topic of conversation.

There are also social and psychological obstacles. If a woman believes that for some reason she cannot breast-feed, she may think herself into a physiological state in which she cannot. But the fact remains that almost all women who want to breast-feed can. Small-breasted women can nurse as well as large-breasted women, since it's the amount of glandular tissue, not fat tissue, in the breast that determines the ability to produce milk. Large-breasted women won't increase their breast size by nursing; pregnancy, not nursing, is what accounts for the change in shape and texture of the breasts of women. Certain breast conditions associated with problems during nursing — such as mastitis, blocked milk ducts, breast engorgement or sore and cracked nipples — can usually be overcome with some simple techniques. And once an adequate milk supply has been established — usually after one month or so of frequent feedings at the infant's demand — most doctors agree that a daily bottle can be given in the mother's absence without interfering with milk flow, milk production, or the unique immunological protection afforded by human milk.

"Human milk is for human infants; cow's milk is for calves," said the late Dr. Paul Gyorgy, who spent more than 50 years studying the difference. As scientists discover more and more components of mother's milk, evidence mounts that the liquid contains just about everything an infant needs, not only for growth and nutrition but for

coping with the harsh realities of life outside the womb.

Studies comparing the health of breast-fed and bottle-fed babies show that infants who nurse are less likely to have respiratory infections, allergies, diarrhea, gastrointestinal infections, eczema, ear infections and iron-deficiency anemia.

This is not to say that bottle-fed babies are unhealthy. As researchers learn more about what's in mother's milk, industrial scientists are able more closely to mimic it in new, "humanized" infant formula preparations. But even formula manufacturers agree that infant formula will always be second best. Among the compounds found uniquely in human milk are lipase, an enzyme that metabolizes fat; taurine, an amino acid involved in bile acid metabolism that may also help transmit nerve impulses in the brain; lysosome, an enzyme that fights bacteria; and lactoferrin, a protein that deprives certain bacteria, most often staphylococcus and E. coli, of the iron they need to survive. Breast milk also contains white cells that are better able to absorb bacteria than white cells found in blood. These cells, or macrophages, are transferred directly to the infant, and they engulf and kill any bacteria with which they come in contact. And the particular qualities of human milk — its low phosphate, low protein and high lactose content — create an acidic environment in the infant's alimentary canal that stimulates the growth of the so-called friendly bacteria and inhibits the growth of bacteria that could be harmful.

Some scientists think the immune protections provided through breast milk may last far beyond the weaning stage. At the Harvard School of Public Health, Drs. Isabelle Valadian and Robert B. Reed are following up a study initiated nearly 50 years ago on child health, growth and maturation, in Boston. A preliminary report indicates that adults who had participated in the study are today less likely to develop allergic reactions, such as eczema, rhinitis and asthma, if, as babies, they were breast-fed for at least six months.

Other components of breast milk also help to make for a healthier baby — and, perhaps, a healthier adult. Valadian and Reed of Harvard found that 30-year-olds who were breast-fed up to the age of 6 months have significantly lower serum cholesterol levels than adults who were breast-fed for two months or less. The difference may be due to the higher concentration of cholesterol in human milk than in cow's milk, a concentration some scientists believe may induce cholesterol-metaboliz-

ing enzymes to develop and lead to more efficient use of cholesterol in later life.

Mother's milk is, of course, not perfect. Some environmental pollutants have been shown to collect in breast milk, particularly such fat-soluble toxins as the agricultural chemicals DDT and PCB DDT has been banned in this country, but harmful chemicals still persist in the waterways and food sources of some regions. In a recent policy statement, the American Academy of Pediatrics, which favors breast-feeding, cautioned women who are in high-risk occupations or live in areas where the chance of exposure is great, to have their breast milk analyzed for toxins before they begin nursing.

"Many compounds and drugs are also found in human milk," notes Dr. William Weil, chairman of the department of human development at Michigan State University, who testified before a Senate committee in 1977 about the hazards of contaminated mother's milk. "Many medications — anticonvulsants, antibiotics, sedatives — appear in breast milk, but by and large they are found in low quantities that are not considered disadvantageous to the child. Still, their presence could set up allergies and sensitivities to the drugs in later life."

The American Academy of Pediatrics has also recommended that nursing women refrain from taking specific drugs that might harm their infants: antithyroid compounds, antimetabolites, anticoagulants and cathartics. Some pediatricians tell nursing women to stay off the Pill, which interferes with the process of lactation. And they encourage moderation — or, if possible, abstinence — in smoking cigarettes and drinking coffee and alcohol.

☐

If mother's milk is so good for her child — providing immunological protection, good nutrition, and possibly prevention of future problems like obesity, allergies, cavities and maybe even coronary artery disease — why isn't every young mother breast-feeding?

One reason is that many women simply do not know about the advantages. Doctors are reluctant to suggest breast-feeding, for they fear stirring up feelings of guilt and failure if the patient chooses the bottle. From a doctor's perspective, bottle feeding is easier. "The obstetrician's definition of a 'successful outcome' for his patient — the woman — is an uncomplicated delivery and an easy postpartum recovery," says Dr. Myron Winick, director of the Institute of Human Nutrition

at Columbia University. "And the postpartum course is greatly simplified when the woman bottle feeds." The pediatrician, too, is apt to prefer treating a bottle-fed baby, Dr. Winick says, in that he knows exactly how many ounces per day of exactly what formula the infant is getting.

By not advocating a specific feeding method, a doctor may be helping the woman to chose the one with which she is most comfortable, and a happy, relaxed mother feeding her child infant formula is better than a nervous mother trying to feed him breast milk. On the other hand, women may be making decisions about infant feeding without adequate information.

The women who are leading the movement back to nursing are those who have sources of information other than their obstetricians: better-educated, higher-income women who read books and magazines, attend natural childbirth classes, and tend to have the personal and social resources needed to make their own decisions. Among studies of young women in the university communities of Boston and East Lansing, Mich., where the women or their husbands were graduate students or faculty members, the proportion of breast-feeders was found to be about 75 percent. And in a poll of alumnae of Swarthmore College, nine out of 10 respondents had breast-fed.

☐

Breast-feeding also has a connection to income. In Alabama, the proportion of middle-class mothers who breast-feed is twice that of lower-class mothers, and these figures are confirmed in other scattered surveys. Poorer women, who tend to have fewer sources of information and more problems to cope with, rely on the bottle because it is practical. With a bottle, anyone — grandmother, baby sitter, older child, father — can feed the baby.

Obstetric procedures in most modern hospitals, especially the larger, understaffed public hospitals serving many of the poor, tend to work against the ability to breast-feed. The first days after birth are crucial to building up both an adequate milk supply and a comfortable relationship between mother and child. But most hospitals separate the mother from her infant for all but feeding time, and feedings are scheduled at four-hour intervals — even though a breast-fed baby needs to nurse every two or two-and-a-half hours in the first few days.

When a woman is heavily sedated during childbirth, it may be days before her infant is alert enough to suck vigorously at the breast. In 1961, pediatrician Dr. T. Berry

Brazelton at Children's Hospital Medical Center in Boston found that heavy medication with barbiturates during delivery impaired the success of breast-feeding for as long as five or six days. On the first day after birth, 65 percent of lightly medicated mothers had "effective" feedings, compared to just 30 percent of heavily medicated mothers.

Once a woman goes home with her baby, breast-feeding becomes even more traumatic. The new mother is frequently unable to tell if her child is getting enough milk. Added to this natural uncertainty are the problems a nursing woman encounters when dealing with the outside world — especially if she is a working woman.

"We haven't adjusted our society to the needs of the breast-feeding woman who works," says Dr. Winick of Columbia. "How do you breast-feed if you're working in an office or a factory? We're just not geared for that, and we haven't yet said as a

society that the advantages are so great that we have to make special provisions for it."

With some ingenuity, breast-feeding and employment *can* mix. Working women can be taught to pump their milk by hand during the time they're away from home, to keep milk production stimulated. The milk may either be kept for the next day's feeding or the baby may be fed formula while the mother is away from home.

The frenetic pace of life today seems out of step with the measured, tranquil image of breast-feeding. What woman has time to nurse her baby five or six times a day, 20 or 25 minutes a feeding? Yet women who have nursed successfully, even the busiest among them, say they wouldn't miss the experience for the world. Not only is breast-feeding surprisingly convenient, they say — no bottles to clean, no formula to warm up, no equipment to tote — but it is fun: a sensual, satisfying experience that women look for-

ward to as a haven in a hectic day.

The mother also benefits physically from the process. Hormones involved in milk production and lactation help shrink the uterus, utilize the fat deposits of pregnancy, prolong the postpartum period of infertility, and may even have a tranquilizing effect. Sex researchers William Masters and Virginia Johnson found that women who breast-feed resume an active sex life more quickly after delivery than do women who bottle-feed, and may in fact reach levels of excitement that exceed their prepregnancy states. Several investigators have also noted a decreased incidence of breast cancer among women who breast-feed, but confounding factors, such as the woman's age at childbirth and the number of children she has, make such findings difficult to interpret.

To allow women to savor this unique aspect of mothering, though, will take a reexamination of our society's values and priorities. Work schedules for

new mothers will have to become more flexible to allow for longer lunch hours and more frequent breaks; feelings about the sight of bare breasts and nursing infants will have to be placed in perspective; information about the best way to breast-feed, and signs to assure a mother that her infant is being well fed, will have to be transmitted more effectively through obstetricians, pediatricians, nurses and even informal gatherings of nursing women. Only then will large numbers of women be able to experience that special kind of maternal pride. "The greatest part of breast-feeding," says Dr. Marvin Cornblath, a pediatrician at the National Institute of Child Health and Human Development in Bethesda, Md., "is when a mother comes in with her healthy, thriving baby and says, 'Look at that kid! Look how much he's grown since he was born! And all that came from me.' " ■

July 8, 1979

OH, THE CAPTAIN, SHE'S A LADY

By Grace Lichtenstein

No question about it; the khaki-colored ship was headed straight for us, portside, its two sets of .50-caliber mounted machine guns pointing skyward out of the fore and aft turrets.

"All deck force to the fantail," our skipper shouted over the intercom, the command carrying over the noise of diesel engines grinding to slow. Within seconds, seamen and petty officers were scurrying around the stern.

What was happening that hot July morning 15 miles north of Havana was an international incident at sea — the kind that doesn't necessarily make waves but could easily make headlines if mishandled. The Cape Current, a 95-foot United States Coast Guard cutter on patrol, was about to turn over a disabled small craft, its four Cuban occupants and their two dogs to Cuban gunboat No. 138, after rescuing them a day earlier off remote Cay Sal island in the Bahamas. The rendezvous, which had mysterious overtones, was among the very, very few carried through between this country's oldest naval service and a military vessel of its nearest Communist neighbor.

Almost as dramatic was the fact that it was occurring during only the second regular patrol of the Cape Current under its rookie captain, Lieut. (j.g.) Susan Ingalls Moritz, one of but two pio-

Grace Lichtenstein is currently working on a new book about women adventurers.

neering women to have been given command of a United States patrol ship.

The gunboat and the Cape Current performed a wary pas de deux in the slightly swelling waters. They settled in formation side by side, stern to bow, a few hundred yards apart. "I wish I had my camera," the captain said, her hands almost absent-mindedly maneuvering the throttles. Her manner was businesslike, nearly brusque. In her working uniform of dark blue pants and light blue shirt, an impressive set of keys dangling from her belt, her hair softly framing a girlish face free of makeup, she seemed at the same time younger and older than her 25 years. At this rather tense moment, nothing in Lieutenant Moritz's demeanor on the bridge betrayed the apprehensive "Uh-oh!" that flickered through her mind when she realized that the "fishing boat" she had been told to expect at the rendezvous was actually a ship built for combat.

The Cubans sent out a speedy runabout to aid in the transfer. While the coastguardsmen loosened the lines that hugged the pleasure craft to the Cape Current's side, Petty Officer Jim Sherburne helped Faustino Revilla and the three women in his party climb up out of the cramped crew quarters below deck, where they had been resting. Mr. Revilla, sporting a pair of American jeans (a gift from one of the crew) excitedly hailed the gunboat in Spanish. Then he shook hands with Chief Petty Officer Bob Walden, called a heartfelt *"Muchas gracias"* up to Lieutenant Moritz and climbed down a rope ladder into his boat. *"De nada,"* came the reply from the bridge. As soon as the Cubans grabbed the pleasure boat's line, the American skipper thrust the

throttles forward.

The whine from the Cape Current's engines broke the tension. "Go, Capt'n!" bellowed Chief Walden. "Smoke 'em down!" chimed in the others. "Show off! . . . Whoo-e-e-e!" To the cheers of her all-male crew, Susan Moritz rocked the Cuban Navy with the wake from her spunky ship, churning north toward home at a full 20 knots. Before the gunboat was too far away, the seamen and the captain got out their cameras and snapped some parting shots. The rendezvous was the highlight of an otherwise ordinary patrol for the cutter.

☐

In fact, though, nothing had been quite ordinary aboard the Cape Current since last June 8. That was the day Susan Moritz assumed her duties as commanding officer at the cutter's home port, Fort Lauderdale, Fla.

A veteran of four years in the Coast Guard, she had attended Officer Candidate School and had served in major posts on the Gallatin, a much larger vessel based in New York, prior to landing the Cape Current assignment. About the same time, another woman, Lieut. (j.g.) Beverly Kelley, was put in charge of a cutter in Hawaii, but her ship was in dry dock for repair for weeks — so, for the record, Lieutenant Moritz was the first woman to captain a regular patrol. The Coast Guard, the most peaceable of the military services, had thus taken the lead in giving women front-line positions of authority, rather than desk jobs or support commands. Moreover, Lieutenant Moritz's assignment is especially high-profile: the Straits of Florida are a cauldron of licit and illicit activity requiring Coast

Guard attention. How she performs during her two-year tour of duty is bound to have some impact on how interested parties, from the Pentagon on down, judge the efficacy of women as top brass.

The mission of the seven 95-foot cutters that work out of the Seventh Coast Guard District in Florida is, on paper, the same as that of any in the Atlantic or Pacific fleet. These intermediate-sized ships, usually carrying a crew of 13, handle search and rescue (SAR) cases — helping pleasure or commercial craft that have either run out of fuel, are leaking or are otherwise disabled. They also help evacuate injured persons at sea and occasionally clean up oil spills or other marine environmental pollution.

In recent years, however, the most publicized mission of the Florida-based Coast Guard ships has been "law enforcement" — code words for drug-smuggling or Cuban-refugee escort cases. Since Miami has become the Number 1 port of entry for South American drugs, the Coast Guard has been increasingly called upon by Federal agencies to seize drug-laden boats. The crew of cutters like the Cape Current board and search suspicious boats, acting either on their own strong hunches (like literally smelling a pleasure boat that reeks of marijuana a half mile downwind) or upon orders from Miami headquarters. The service has the legal authority to arrest alleged smugglers at sea. Normal procedure upon finding suspected dope on a boat is to place the occupants under arrest, put some petty officers on board to guard them, and then escort the captured vessel to the nearest port. Any foreigners involved are turned over to the Immigration Service; American smugglers, to the Drug Enforcement Agency; and the boats and contraband, to United States Customs.

Coast Guard officials are the first to play down their anti-drug-smuggling capability. "For every boatload of pot we catch, there are hundreds of tons more that get through. It's a big ocean," a Miami spokesman said. On the other hand, they are the first to publicize any good-sized seizure. The message is not lost on Coast Guard crews. They know drug busts are regarded as the glamorous side of their work. Not long ago, one Florida-based cutter was fired upon by the irate crew of a suspected drug boat, although such shooting incidents are rare. Like other ships that have made pot seizures, the Cape Current has decaled on her single smokestack a green marijuana leaf, right next to the Coast Guard emblem. With each new bust, cutters add a black chevron.

Someone stenciled the first chevron beneath the Cape Current's leaf during our patrol. The week before, on Lieutenant Moritz's first patrol as captain, the ship had tagged a cabin cruiser 75 miles southeast of Miami that was packed with 3,000 pounds of marijuana. It was hardly one of the bigger hauls, but when the Cape Current crew brought the cruiser into Miami, they were congratulated by members of the Coast Guard Congressional Oversight Committee, who happened to be touring in the area.

Lieutenant Moritz tends to enjoy law-enforcement cases more than SAR missions because they are more unpredictable. "You never know what's going to happen." Her crew members are more cynical, even ambivalent. "Hell, I don't think marijuana should be illegal," one of them said. Then, too, many of the tips prove false, and a few men wonder privately if any boats in trouble have been neglected while they chased suspected smugglers.

"The ones we catch are the dumb ones anyway," claims Chief Walden. Some boats are so overloaded with marijuana that they start sinking and are forced to call the Coast Guard for help. Meanwhile, big dealers, reported to be using destroyer-sized ships equipped with radar, are said to ply the Florida coast undetected.

Drug busts or no, it was evident from the first day of this second Moritz patrol that we were in for an eventful voyage. At 7 A.M. the cutter's skipper had trotted up the gangplank at the Fort Lauderdale dock. "Good morning, Capt'n," Chief Walden greeted her. "You need a hand, or can you make it yourself?"

She made it herself, all the way up to the bridge for a last-minute check of the Cape Current's cache of weapons — several .50-caliber machine guns, M-16 automatics and .45-caliber handguns. Then she gathered her crew on deck for the customary briefing. "We're going out to Great Isaac and down to Orange Cay," she announced, referring to two tiny Bahamian islands, east and south of Fort Lauderdale. "The seas are two to four feet, winds 10 to 20 knots, so make sure everything is stowed for sea."

Climbing back to the bridge, Lieutenant Moritz took the wheel, cautiously easing the ship out of the busy harbor. A message over one of the half-dozen two-way radios within her reach re-

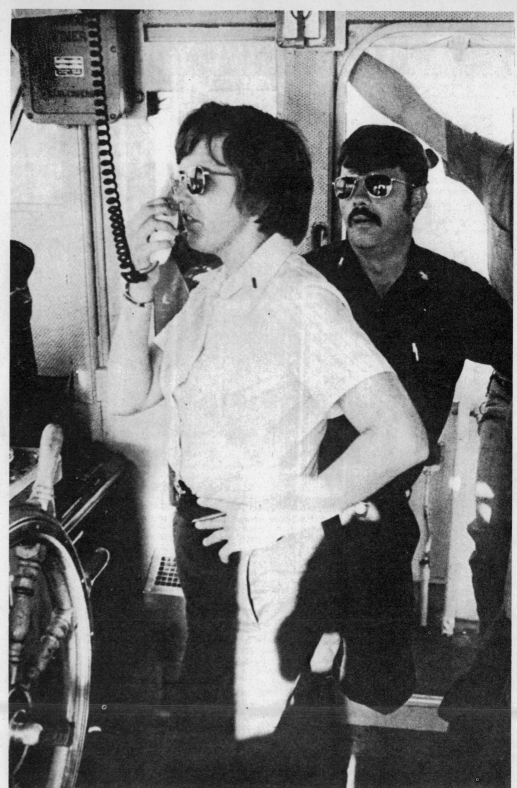

Ship-to-shore conference: Lieut. Susan Moritz consults with headquarters on how to deal with four Cubans found adrift off a

called her to shore almost immediately. Headquarters had some fresh "law-enforcement" advisories. It was simpler to go back to port to discuss them over a regular telephone than to have them transmitted laboriously in code over the open ship-to-shore radio wavelengths. After its false start, the Cape Current sailed out again, saluting Lauderdale's posh beachside high-rises

with a mighty foghorn blast. Soon, the high-rises were minuscule boxes on the horizon behind us and the sea was a deep 300 fathoms blue.

At midmorning, a motorboat sputtered toward us, almost out of gas. On orders from Lieutenant Moritz, crew members winched the rubber pontoon Avon dinghy, which sits at the ready on deck, over the side. The petty officers

in the boarding party strapped on side arms for the trip, a routine precaution, however unsuspicious the encounter. Then the Avon party sped off to the motorboat, whose gasping engine they refreshed with three gallons of gas.

Lieutenant Moritz gave a running commentary on the routine SAR case, which took 20 minutes in all, over the radio to Miami headquarters. I asked

remote island in the Bahamas. In the picture above, she assists one of them aboard her cutter. Later the group was turned over to a Cuban gunboat outside Havana harbor.

about the gas. Policemen, after all, do not give out free government gas to stalled motorists on freeways. She shrugged noncommittally. "The alternative is towing them in, and that would take us an hour and a half the wrong way."

Over lunch in the air-conditioned galley below deck, one coastguardsman speculated about the boat he had just

fueled. "I think they were after dope. They said they were going fishing, but there wasn't any tackle on board."

The rest of the afternoon was hot and noisy. Ever-present coffee cups chattered on the galley tables from engine vibrations. Logy from the sun and the motion of the boat, now matter how much coffee they drank, the crew napped between watches. A 95-footer

has no room to spare for amenities, so the crew's quarters — racks of two- and three-deck mattresses crammed between up-   *(Continued)* right foot lockers — were definitely steerage class, although air-conditioned. When not dozing, the men preferred to sprawl in the bucket seats of the Avon, or under the shade of

the bridge or stand in the bow and catch the spray. The captain, the chief and a senior petty officer rotated four-hour watches at the helm.

During her breaks, Lieutenant Moritz wrote in the ship's log, helped make regular radar or loran (long-range radio aid to navigation) checks on the ship's course, and read "A Man Called Intrepid" in her plain cabin. A platform bed dominates the room, which also contains a file cabinet, some book shelves, a desk with a small safe on top, a chair and a chart of the Straits of Florida tacked on the wall. The room has no porthole, but it does have about three square feet of open floor space, two and a half more than the only other single cabin, which belongs to the chief petty oficer.

□

Late that afternoon, the captain and I wandered toward a quiet part of the bow to talk about why she had chosen such a distinctive career.

When she was growing up in Lynn, Mass., Lieutenant Moritz's initial ambition had been to be a veterinarian. "Then I wanted to be an astronaut," she said, "but my brother told me they didn't take girls. My senior year in college [heading toward a degree in marine fisheries] I went to Woods Hole, Mass., and had the opportunity to go on an oceanographic cruise. I discovered I really enjoyed sailing."

She paced the deck, hands in pockets. "I did consider briefly, and I do mean briefly, the Navy. But I learned what my job would be and I wanted something more operational, smaller, more personal." She settled on the Coast Guard right after graduation from the University of Massachusetts, and spent the required 17 weeks at Officer Candidate School in Yorktown, Va. At the time, the Coast Guard still had a policy of not allowing women at sea, but enough rumors about imminent changes made her confident of a job on a ship eventually.

She got her wish aboard the Gallatin, following a stint in Long Beach, Calif., as security manager and assistant branch chief of intelligence for that district. On the Gallatin she served as a deck-watch and weapons officer — she is an expert shot — getting her first chance to command men at sea. During that time, a higher-ranked officer at the same base caught her attention. He was Comdr. George Moritz, about whom she will say little except that he is now being transferred to duty in Miami. A small photo of her bearded husband, taken on their scuba-diving honeymoon in Bermuda last spring, sits on a shelf above her bed on the cutter. Her priority is to stay based close to her husband, but beyond that she's not sure what her ultimate goal is. "After two years at this," she says, "I'll be ready for a desk job."

Had she experienced or did she expect any problems as a female captain? She'd had that question before. Her answer was a firm negative. "I've found a few people who are totally opposed to women at sea, but even they're willing to work it out." Her all-male crew seem receptive to her, although they weren't wild about

the needling they got from other guardsmen when her assignment was announced. Most of her men are in her own age range and accustomed to women holding executive positions. If Lieutenant Moritz had had any run-ins with her crew, she wasn't forthcoming. "They know who I am; I'm just learning who they are," she said crisply.

Just then, Chief Walden, an easy-going older salt of 34, who set the crew's tone in dealing with the skipper — forthrightly, but with respect for her rank — paged her from the bridge. A coded message had come from Miami. There was animation in her voice when she returned: "You'll be happy to know we should have some action for you by dawn. A law-enforcement case. It's one of your 'Do not pass go, do not collect $200' deals.'" With that, the Cape Current surged full speed south-southeast past Bimini toward Cay Sal island.

□

The Miami report: four Cubans fleeing their own country to the United States in a small boat had had engine trouble and were stranded off Cay Sal, a patch of Bahamian real estate a long way from nowhere. The Bahamians, never friendly with the Cubans, had asked the U.S. Coast Guard to collect their unwelcome visitors.

When we arrived off Cay Sal at breakfast time the next morning, sure enough, an open pleasure boat was floating in the emerald shallows. Lieutenant Moritz sent the Avon to check out the situation. Chief Walden returned with the only man in the Cuban party. Climbing up to the bridge, his hair bleached and lips puffed from too much exposure to the sun, Faustino Revilla wore a long-sleeved shirt, swimming trunks and a hangdog expression. He uttered just one word in English: "Dramamine."

Nobody on board the Cape Current, including the skipper, spoke more than a few words of Spanish. As she handed Mr. Revilla a bottle of antiseasickness pills, Lieutenant Moritz requested the Key West Coast Guard station to put a Spanish translator on the radio, pronto. There followed a brief exchange between a land-based translator and Mr. Revilla. Then the Key West operator gave the captain the news.

It was all a mistake, caused by the language barrier. The Bahamians, who did not speak Spanish either, had misunderstood. The Cubans were lost all right, but they had no intention of fleeing to the United States. They just wanted to go home to Cuba. Would the Coast Guard please take them?

Susan Moritz caught her breath in surprise. "Request immediate phone patch with District Op Center," she snapped over the intercom. Once in touch, she reported tersely that the Cape Current had obviously been sent 160 miles out of its way through a misunderstanding. What was she to do with the Cubans now?

The Miami center, incredulous, told her to await new orders. "We might drop anchor," she said, a touch of resignation creeping into her voice. "This could take hours." It did.

The disabled boat, with Mr. Revilla's mother and two young women still on board, was brought alongside the Cape Current. Bedraggled from four days of drifting, the women climbed on board, clutching small religious statues. But the circumstances of the Cubans' voyage were somewhat mysterious. Why, for instance, was there a five-day-old issue of The Miami Herald on their boat? Why did the boat have a current Florida registry sticker?

The Miami Operations Center, in classic bureaucratic style, made each Cuban individually pronounce over the intercom his or her desire to go to Cuba, not the United States. Then headquarters issued contradictory instructions, first telling Mr. Revilla his only two choices were to go to Miami or to be left stranded at Cay Sal. Seconds later, the Op Center changed its mind and told everyone to stand by again. This "law-enforcement" adventure was becoming a boring bilingual farce. Susan Moritz vowed to sign up for a course in Spanish as soon as she returned to Fort Lauderdale.

Without Miami's approval, Lieutenant Moritz was powerless to move. Her crew was exasperated. The Revillas were still seasick. Taking a longing glance overboard, the young skipper sighed, "Boy, would I love to jump into that water for a swim!" After lunch, that's exactly what she and the crew did, leaping off the rail of the bridge with delighted yodels. (The skipper, donning snorkeling gear and flippers, was ostensibly investigating a suspected crack in the hull.) The swim restored spirits considerably.

At 3:30 P.M., Miami headquarters finally relayed a decision. The Cape Current was to tow the Revilla boat to a rendezvous 15 miles outside Havana harbor — 110 miles and 10 hours sailing time away. There, we would deliver the Revillas and their boat to their Cuban countrymen. No reason was given for the decision.

(A week later, in a brief statement that still left many questions unanswered, the Coast Guard said that a check of the Florida registry number had turned up no evidence that the boat was stolen. However, the boat had been in Key West in partially disabled condition some days before winding up off Cay Sal island, which was probably how The Miami Herald found its way aboard. After receiving assistance in Key West, the boat had departed for Cuba and run into trouble again when we picked it up. The Coast Guard, in consultation with Cuban authorities finally decided to have the Cape Current tow the Revillas

to Cuba, rather than leave them near Cay Sal to be collected, because such a move would "favor relations" between the two countries. Without knowing so at the time, the cutter had become part of an American good-will mission.)

The voyage to the rendezvous took almost twice as long as it should have because the line towing the pleasure craft kept coming loose. Not until nearly lunchtime the following morning, day three of the patrol, did the Cape Current deposit the Revillas near the gunboat and veer north. The next stop — Key West, for refuelling and a brief five-hour shore leave.

"Tell me," the skipper asked Chief Walden, "was it always so busy on board this boat?" "Not until you came," the chief answered, laughing. Besides hauling in the marijuana boat the previous week, the Cape Current had, on the same patrol, helped a Russian ship evacuate seriously burned people from a third ship that had caught fire in the middle of the ocean. Throughout these happenings, crew members said, Susan Moritz, though green, had kept her cool and had earned their respect, despite the occasional X-rated sexist remarks they made behind her back.

It was dinnertime as we skirted the Key West beaches. Lieutenant Moritz called a meeting on deck. "I thought I'd leave it up to you how long we spend here," the skipper said. She carefully pointed out that if the crew, wanted to

Over the rail for a swim in the Atlantic

Entering day's doings in the ship's log.

frolic ashore for more than five hours, the Cape Current, the only available cutter in the vicinity, would become "a prime target for a SAR diversion," which would delay our scheduled 8 A.M. docking back at Fort Lauderdale. The guardsmen got the picture.

"You guys did a good job these past few days and I want you to know it," she added. "Seeing a Cuban gunboat that close is a rather rare occasion. Incidentally, based on our last drug case, Admiral Stabile [commander of the Seventh District] sent his regards for a job well done."

Thus commended, the crew roared off in a van borrowed at the Key West Coast Guard station to celebrate. Several beers later, a number of the men plus the skipper regrouped at Sloppy Joe's, a huge bar that sells T-shirts with Ernest Hemingway's face on them. Still more beers later, shortly after Lieutenant Moritz had mentioned her hobby of making New England tombstone rubbings, the party came to an abrupt halt. A seaman arrived to inform the captain that they had just gotten a SAR call after all. Back it was to the cutter. We splashed through the night to the other side of the Cay Sal Banks to rescue a foundering yacht.

As we steamed toward the mainland, yacht in tow, Lieutenant Moritz sloughed off some of the inner tension remaining from the Cuban operation by talking about the question of women commanding men. Gender, she felt, has little to do with it. "Some people are natural-born leaders. Others have to work at it. I have to work at it," she said straightforwardly. "It's getting people to do things and making it as pleasant as you can. For example, when we get back in, we'll be on Bravo Two status [ready to take off on patrol again on short notice]. But at the same time, we've been out for days. The guys are tired. So I'll give them a day off, but they'll know they'll have to be ready to respond quickly if necessary."

She did not have any profound insights to offer on the subject of female versus male leadership. But not many military officers are given to deep meditation. Besides, she pointed out, "I haven't been in charge very long." She agreed that the precise lines of command, based on rank, have made it simpler for her to handle men who are older and more experienced than she. But how had she felt the first time a crewman addressed her as "Captain"? She grinned. "Strange!"

For this grand title and the long hours it requires, Lieutenant Moritz gets paid a paltry $12,000 a year, plus a food allowance. Only because her husband's salary as a commander is much higher could they afford to put a down payment on a house in Hollywood, Fla. Some of the seamen, who make far less, live on board ship even in port.

Home port, nevertheless, was on everyone's mind as we dropped off the yacht near Miami and surged up the coast. "Channel fever!" hissed Tim Jinkins, one of the seamen, impatiently. With the Saturday sun disappearing into a cloud bank over Miami Beach, we pushed hard until the Fort Lauderdale highrises once again came into view. The Cape Current had covered 735 nautical miles. Susan Moritz had logged four and a half more days' experience on the not-so-high seas in the service of her country.

"All hands, boarding stations!" she cried, nosing the ship toward the dock. The hands at the bow applauded. Sunday was going to be a banner day off for everyone. The skipper dropped her reserve long enough to confide this would be a special day off for her, too. Her husband was also returning from patrol that weekend. Soon, they would be reunited — for the first time since their honeymoon 10 weeks ago. ■

August 26, 1979

Conference Examines Pornography as a Feminist Issue

By LESLIE BENNETTS

One by one the images flashed onto the screen, each accompanied by gasps and exclamations from the audience: a naked woman gagged and bound, hanging from the ceiling by her wrists, her back flayed with whip lashes; a woman's body being fed into a meat-grinder; the dismembered body of a woman as a chicken on a plate of cacciatore, covered with tomato sauce — or blood.

Culled from magazines, record album covers, billboards and other sources, the pictures made up a slide show that began a two-day conference on women and pornography held this weekend at Martin Luther King Jr. High School in Manhattan. Sponsored by an organization called Women Against Pornography, the symposium was billed as the "East Coast Feminist Conference," although the audience of some 700 women came from as far away as Ireland and England, in addition to cities around the United States.

A Feminist Issue

They had gathered, according to Dolores Alexander, one of the founders of Women Against Pornography, "to claim pornography as a feminist issue of national proportions."

"Pornography is antiwoman propaganda," she said.

Many among the audience who arrived unconvinced of such assertions had changed their minds even before the opening slide show had ended; the presentation left most of the women profoundly shocked. Nevertheless, during the lectures, workshops, panels, films and discussions that followed, a good deal of time was spent on defining such basic questions as what is pornography. Most viewed it not as an isolated phenomenon, but as one of many cultural manifestations of male hatred of women that interact and reinforce each other, all having far more to do with violence than with sex.

"The word pornography in its very origins means 'writing about women captives or slaves,'" explained Gloria Steinem, the feminist and editor of Ms. magazine. "Erotica is something quite different, portraying love as something chosen. Pornography is not sex, and sex need not be violent or aggressive at all. It is violence and domination that are pornographic."

As for why the women's movement was tackling the pornography issue at this time, she added, "It's organic, in a way: having untangled sex and violence on the issue of rape, which came first because it's direct physical violence — as opposed to pornography which is indirect — then it's a rational progression toward untangling sex and violence in pornography. Pornography is the instruction; rape is the practice,

battered women are the practice, battered children are the practice."

The lives of many of the women at the conference had, as they saw it, been scarred by the influence of pornography. During a "speakout" session, many told harrowing stories of being molested as children by their fathers or brothers, who had often prepared for such events by reading pornography. "I see pornography and incest as closely related," said Katherine Brady, a participant. Although her family was "middle-class, moral religious regulars in the Midwest," she said, her father had molested her from the time she was 8 years old until she was 18. "As a child, the only time I ever saw pornography was with my father; the pictures he showed me told me it was okay for men to do whatever they wanted to women."

The fact that some women are stimulated by sadomasochistic images of sex was discussed in several workshops as a product of prevailing social values. "We've eroticized being submissive," said Robin Morgan, a poet and feminist writer. "As for the part of one which may respond to pornography, I acknowledge it the way I acknowledge scar tissue — not to affirm it, but to say it exists."

The speakers at the conference — who included such writers as Lois Gould, Shere Hite, Barbara Seaman,

Alix Kates Shulman and Barbara Grizzuti Harrison — suggested a variety of different reasons for what was generally viewed as "the explosion of pornography in the last few years," as Susan Brownmiller, a founder of Women Against Pornography, put it.

Climate of Acceptability

Sexual liberation has created a climate of acceptability for this explosion, some said. A backlash against the women's movement was seen as another factor in what appears to be the increasing violence portrayed by much current pornography. The increase in child abuse was suggested as related: "Part of the backlash is men who are unable to relate to women as equal human beings turning to little girls," said Dorchen Leidholt, a founder of Women Against Pornography. "Kiddie porn is an important genre, and 25 percent of all rape victims are under the age of 12; half are under 18."

First Amendment Implications

One factor inhibiting activism against pornography has been concern over the First Amendment implications of such a crusade, an issue which is hotly debated among civil libertarians concerned about threats to freedom of speech, whatever form such freedom may take. Most feminists at the conference saw that as a "a false issue," as Miss Steinem put it.

"My position on pornography is not contrary to my lifetime position as a civil libertarian," asserted Bella Abzug. "I do not believe it is necessary for us to interfere with anyone's constitutional right to produce pornography. But that doesn't require us to encourage and assist in the proliferation of pornographic materials on the streets and in the stores."

Among the tactics suggested for combating pornography were boycotts of supermarkets and stores selling

Susan Brownmiller, left, and Bella Abzug

such magazines as Hustler, which had published as its cover the picture of the woman being fed into a meatgrinder; civil disobedience used to interfere with patronage of pornography shops or the showing of pornographic movies; enforcing and extending public nuisance statutes to help curtail the display of pornographic materials; and generally raising public awareness so that consumption of pornography is socially ostracized. A march on Times Square is also planned for Oct. 20.

Some among the audience had their doubts about the feasability of such goals, given the fact that pornography is a $4 billion a year industry, according to conservative estimates. Avalon Krukin, a guidance counselor who works with children, said, "It's a

worthwhile goal, but I don't know whether it can be accomplished."

At the least, most participants felt the consciousness-raising goal of the conference had been achieved. Jeri Levitt, a recent college graduate from New Jersey, concluded, "This was an educational experience for me. I didn't know whether, as a liberal, I was going to come out of this conference feeling that taking a position on pornography was justified or unjustified. But seeing the extent to which violence is being used against women in this way, I now feel very strongly about it. We have to educate each other about this kind of violence. It's not just happening on 42nd street, it's affecting all of us in the bedroom."

September 17, 1979

Pontiff Reaffirms Opposition to Women as Priests

SAYS CLERGY CANNOT RESIGN

By FRANCIS X. CLINES
Special to The New York Times

CHICAGO, Oct. 4 — Pope John Paul II came to the heart of the country today, emphatically reaffirming in his crosscountry tour that the Roman Catholic priesthood required a lifelong commitment from men and was not a proper calling for women.

In a frenetic day of prayer, teaching and touring, the Pope raced in 12 hours from Philadelphia in the urban East to a solemn mass in a vast Iowa farm field, then here to the commercial capital of mid-America.

Moving by plane, by helicopter, and by motorcades that often passed crowds in a blur of schedule haste, John Paul maintained the patient, polite, smiling demeanor that has been dominating national news all week.

Stresses Themes of Rural Life

In his frenetically paced day, the Pope stressed various themes — the church loyalty of Ukranian Catholics, the virtuous perseverance of inner-city Puerto Ricans, "the more human dimension" of America's rural life, which he extolled at a tiny country church and for a throng in the Iowa field.

But his main address of the day was his firm, patriarchal emphasis on the re-

strictions of the priesthood — a major issue in the church in America in recent years, in which thousands of priests have returned to secular life.

Obviously speaking to the pressure for reforms, including the ordination of women, the Pope drew standing applause early in the day from thousands of priests and nuns at Philadelphia's Civic Center when he reaffirmed the priesthood as a celibate male preserve.

"The church's traditional decision to call men to the priesthood, and not to call women, is not a statement about human rights, nor an exclusion of women from holiness and mission in the church," he declared.

In mid-sentence, at the words "and not to call women," the Pope was interrupted by applause from his highly attuned audience of professional religious. The gathering of religious, about 10,000 priests seated on the main floor of the center and about 2,000 nuns seated in the balcony, followed his every word.

The faces of many in the crowd indicated that nuns as well as priests seemed heartened by the Pope's words.

Rejects Female Priest Issue

In rejecting the criticism that the church's ban against female priests creates a human rights issue, the Pope declared: "Rather, this decision expresses the conviction of the church about this particular dimension of the gift of priesthood by which God has chosen to shepherd his flock."

The Pope did not elaborate on his comments about the role of "holiness and mission in the church" for women. Rather, he focused on shoring up the resolve of the church's priests, many of whom have seen colleague after colleague quit holy orders to marry and to build secular a career.

In his year in office, the Pope has not approved one request to leave the priesthood; Pope Paul VI approved an average of 2,000 a year.

"Priesthood is forever," Pope John Paul declared, repeating this in priestly Latin for emphasis — "Tu es sacerdos in aeternum."

Instantly, there was applause rolling back from the elaborately vested front rows of men, more than 100 ranking prelates seated to the side of a red-carpeted altar where the Pope celebrated mass.

"We do not return the gift once given," the Pope said of the calling to priestly life. "It cannot be that God who gave the impulse to say 'yes' now wishes to hear 'no.' "

He then restressed the need for priests to remain celibate in order to "express the totality of the 'yes' they have spoken to the Lord."

The Pope and his entourage of three planes of church aides and news reporters left Philadelphia at midday. He flew to Des Moines, Iowa, where he found the farmland mellowing in the autumn harvest. He also found the usual massing of church and government officials, police and school children welcoming him. This scene of cheering mobs was repeated here this evening as the Pope arrived for another full round that will feature a special session of the United States Bishops's Conference.

In his four-hour trip to the farm fields, Pope John Paul visited an old rural Irish parish, St. Patrick's, and celebrated mass in a 600-acre bowl-like clearing of farmland packed with scores of thousands of Middle Westerners in waving, wheat-like abundance. In a field nearby, colts frisked and rolled in the grass.

On the rolling, sun-warmed land, the Pope gleamed in his vestments on a regal-looking, raised altar.

John Paul spoke in his homily at St. Patrick's as "one who has always been close to the earth." He told the farmers, who are well into the age of agribusiness and mammoth machinery, that their land was a special gift from God.

Attending the mass was Joe Hayes, the parishoner who had invited the Pope to St. Patrick's. This farmer now has been invited to the White House this weekend, where Pope John Paul is to meet President Carter.

Here tonight, after another fast-paced motorcade, Pope John Paul again addressed groups of religious, but he did not emphasize the strictures of the priesthood. Instead, he celebrated their love of Jesus. "Brothers," he said to one group, "Christ is the purpose and the measure of our lives."

John Paul finally rested here after dinner with John Cardinal Cody, ending a long day's journey across America.

At the Philadelphia gathering of clergy, where the Pope stressed holy orders, one young student for the priesthood, Edward Ulrich of Devon, Pa., applauded and cheered. "It was great," he said of the address. "The Pope spoke directly to each one of us as individuals and as a group."

Across the way, Sister Mary Jude, of the Sisters of Jesus Crucified, beamed and applauded. Asked about the notion of women priests, Sister Mary Jude said, "That's being overdone by a minority. A majority of us do not want to be priests."

Visit to Ukranian Cathedral

Earlier, the Pope had visited the Ukrainian Cathedral of the Immaculate Conception in Philadelphia, part of the church's Byzantine tradition. This branch of the faith does permit married men to become priests in Europe, but not in the United States.

Here, the Pope did not discuss the priesthood issue. Rather, he praised the Ukrainian church's cultural distinctiveness and said this was beneficial, and "far from being a sign of deviation, infidelity or disunity."

In contrast to all the police escorts, smoothly-run entourages, and mass events directed minute-by-minute for his benefit, the Pope seemed relieved earlier today when the microphone failed during his address to the clergy in Philadelphia.

A covey of robed aides fumbled for a replacement and, once restored to full amplification, the Pope said: "I see sometimes in United States something may be not O.K."

October 5, 1979

FEMINISM TAKES A NEW TURN

By Betty Friedan

In California last month, I went into the office of a television producer who prides himself on being an "equal opportunity employer." His new "executive assistant" was waiting for me. She wanted to talk to me alone before her boss came in. Lovely, in her late 20's or maybe 30-ish, "dressed for success" like a model in the latest Vogue advertisement, she was not just a glorified secretary with a fancy title in a dead-end job: The woman she replaced had just been promoted to the position of "creative vice president."

"I know I'm lucky to have this job," she said. "But you people who fought for these things had your families. You already had your men and children. What are we supposed to do?"

She complained that the older woman vice president, an early radical feminist who had vowed never to marry or have children, didn't understand her quandary. "All she wants," the executive assistant said, "is more power in the company. . . ."

A young woman in her third year of Harvard Medical School told me, "I'm going to be a surgeon. I'll never be a trapped housewife like my mother. I would like to get married and have children, I think. They say we can have it all. But how? I work 36 hours in the hospital, 12 off. How am I going to have a relationship, much less kids, with hours like that? I'm not sure I can be a superwoman."

In New York, a woman in her 30's who has just been promoted says, "I'm up against the clock, you might say. If I don't have a child now, it will be too

Betty Friedan, a founder of the National Organization for Women, is currently working on a book about changing sex roles and the aging process at Columbia University's Center for Social Sciences.

late. But it's an agonizing choice. I've been supporting my husband while he gets his Ph.D. We don't know what kind of job he'll be able to get. There's no pay when you take off to have a baby in my company. They don't guarantee you'll get your job back. If I don't have a baby, will I miss out on life somehow? Will I really be fulfilled as a woman?''

An older woman in Ohio reflects, "I was the first woman manager here. I gave everything to the job. It was exciting at first, breaking in where women never were before. Now, it's just a job. But it's the devastating loneliness that's the worst. I can't stand coming back to this apartment alone every night. I'd like a house, maybe a garden. Maybe I should have a kid, even without a father. At least then I'd have a family. There has to be some better way to live. A woman alone. . . .''

□

With the same mix of shock and relief with which the women's movement began in the 1960's, feminists at the end of the 1970's are moving to a new frontier: the family. Another corner in history is turned, I believe, with the convening of a National Assembly on the Future of the Family by the N.O.W. Legal Defense and Education Fund at the New York Hilton tomorrow.

It's hardly new for women to be concerned with the family. But aren't feminists supposed to be liberating themselves *from* the family? Isn't the women's movement supposed to be trying to destroy the family with the Equal Rights Amendment? And isn't the family, after all, the last bastion of conservatism? Is the women's movement surrendering, then, to the forces of reaction by retreating to the family? Or is feminism truly entering a new stage?

I think, in fact, that the women's movement has come just about as far as we can in terms of women alone. When I think back to the explosion of the women's movement at the opening of the decade — thousands of women marching down Fifth Avenue on Aug. 26, 1970 on that first nationwide women's strike for equality, carrying banners for "Equal Rights to Jobs and Education," "The Right to Abortion," "24 Hour Child Care" and "Political Power to the Women" — our agenda then seems so simple and straightforward.

Ten years later, though the women's movement has changed all our lives and our daughters take their own personhood and equality for granted, they — and we — are finding that it's not so easy to live with — or without — men and children solely on the basis of

that first feminist agenda. The great challenge we face in the 1980's is to frame a new agenda that makes it possible for women to be able to work and love in equality with men — and to choose, if they so desire, to have children. For the choices we have sought in the 70's are not as simple as they once seemed. Indeed, some of the choices women are supposed to have won by now are not real choices at all. And even the measure of equality we have already achieved is not secure until we face these unanticipated conflicts between the demands of the workplace and professional success on the one hand, and the demands of the family on the other. These conflicts seem insoluble because of the way the family and workplace have been structured in America.

The second feminist agenda, the agenda for the 80's, must call for the restructuring of the institutions of home and work. But to confront the American family as it actually is today — instead of hysterically defending or attacking the family that is no more, "the classical family of Western nostalgia," as Stanford sociologist William J. Goode calls it — means shattering an image that is still sacred to both church and state, to politicians on the right and left. And dispelling the mystique of the family may be even more threatening to some than unmasking the feminine mystique was a decade ago.

For instance, a White House Conference on Families was supposed to have been held this year, but it was suddenly canceled as too "controversial" when the experts assembled to plan it began facing the facts about American families today. The flak started when participating Catholic priests discovered that the eminent black woman coordinating the conference was herself divorced and raising her family as a single parent. And when it was revealed that fewer than 7 percent of Americans are now living in the kind of family arrangement to which politicians are always paying lip service — daddy-the-breadwinner, mother - the - housewife, two children plus a dog

and a cat — the White House decided to call the whole thing off.

According to Government statistics, only 17 percent of American households include a father who is the sole wage earner, a mother who is a full-time homemaker, and one or more children. (And one study found that one-third of all such full-time housewives planned to look for jobs.)

28 percent of American households consist of both a father and a mother who are wage earners, with one or more children living at home.

32.4 percent of American households consist of married couples with no children, or none living at home.

6.6 percent of American households are headed by women who are single parents with one or more children at home.

0.7 percent are headed by men who are single parents with one or more children at home.

3.1 percent consist of unrelated persons living together.

5.3 percent are headed by a single parent and include relatives other than spouses and children.

22 percent of American households consist of one person living alone (a third of these are women over 65).

As Muriel Fox, president of N.O.W.'s Legal Defense and Education Fund, put it in her charge to the Family Assembly discussion leaders: "Our assembly will accept — rather than deny — the fact that 93 percent of American families today fit patterns other than the traditional one of a breadwinning father, a homemaking mother and two or more dependent children. We will accept the inevitability of continuing future change in the relationships and roles of men, women and children within families. And we will seek new responses to the conditions that are cause and effect of such change.

"We do not share the frequently voiced opinion that American families are in a state of hopeless collapse. People are living together in new combinations for the intimacy and support that constitute a family — unmarried adults with or without children, single-parent families, multigenerational communes, various new groupings of the elderly.

"The future of the family is an overriding feminist issue."

☐

When feminists proclaimed "the right to choose" a decade ago, we meant the right for a woman to decide whether or when to have a child and to control her own reproductive process — a more basic right, as the Supreme Court ruled, than many of those spelled out in the Bill of Rights by and for men. Lately, feminists upholding "the right to choose" have been fighting a desperate holding battle against the "right to life" or "pro-family" forces whom politicians from President Carter on down have tried to appease by barring Medicaid funds for poor women needing abortions.

But what is beginning to concern me even more today are the conflicts women now suffer as they reach 30 or 35 and cannot choose to have a child. I don't envy young women who are facing or denying that agonizing choice we won for them. Because it isn't really a free choice when their paycheck is needed to cover the family bills each month, when women must look to their jobs and professions for the security and status their mothers once sought in marriage alone, and when these professions are not structured for people who give birth to children and take responsibility for their upbringing.

My own feminism began in outrage at the either/or choice that the feminine mystique imposed on my generation. I was fired from my job as a reporter when I became pregnant. Most of us let ourselves be seduced into giving up our careers in order to embrace motherhood, and it wasn't easy to resume them. We told our daughters that they could — and should — have it all. Why not? After all, men do. But the "superwomen" who are trying to "have it all," combining full-time careers and "stretch-time" motherhood, are enduring

such relentless pressure that their younger sisters may not even dare to think about having children.

The many women struggling with the conflict between careers and children cannot be dismissed as victims of their mother's expectations, of the feminine mystique. Motherhood, the profound human impulse to have children, is more than a mystique. At the same time, more women than ever before hold jobs not just because they want to "find themselves" and assert their independence, but because they *must*. They are single and responsible for their own support, divorced and often responsible for their children's support as well, or married and still partly responsible for their families' support because one paycheck is not enough in this era of inflation. In all, 43 percent of American wives with children under the age of 6 are working today, and by 1990, it is estimated that 64 percent will have jobs — only one out of three mothers, approximately, will be a housewife at home.

Yet the United States is one of the few advanced nations with no national policy of leaves for maternity, paternity or parenting, no national policy encouraging flexible working arrangements and part-time and shared employment, and no national policy to provide child care for those who need it.

It seems almost obscene, as the United Nations' "Year of the Child" draws to a close, that the United States, the richest of all nations, should be spending less on child-care programs now than it did 10 years ago. But, all in the name of "preserving the family," Presidents, both Democratic and Republican, have ignored the need for such services, available to parents at all income levels, as they are in countries much poorer than ours, with a sliding scale of fees according to a family's ability to pay.

Dr. Sheila B. Kamerman, professor of social policy and social planning at the Columbia University School of Social Work, investigated the child-

care arrangements of 200 white and black mothers, half of them professional or executive women, the other half low-paid, unskilled workers, all of them working full time, with at least one child under the age of 5 to care for. According to Dr. Kamerman, their children "may experience three, four or even more kinds of care in an average week, as they spend a part of the day in nursery school, another part with a family day-care mother (or two different such women) and are brought to and from these services by a parent, a neighbor, or some other person.

"Aside from the difficulty of obtaining good child care, the greatest difficulty for these women is the rigidity and unresponsiveness of the workplace. Beginning hours at a job often conflict with conventional school opening times; the lack of any benefits for maternity leaves, or for sick leaves to care for an ill child, creates financial and emotional stress for families dependent on the income of two wage earners."

Confronting the crisis in child care, the Family Assembly is dispensing with rhetoric and theoretical arguments in favor of concrete proposals: new options, child-care solutions that have worked in other countries, child-care services that could be provided by a company or union in the actual workplace, home-based child care, commercial child-care services. The Amalgamated Clothing and Textile Workers Union, for instance, has been writing into union contracts provisions for a well-run child-care nursery on factory premises as a basic employee benefit. The Stride Rite Corporation has been running such a child-care center at its Boston plant for eight years now, and parents who live nearby but do not work at the plant can use the facility, too. In Alabama, a community center for senior citizens doubles as a child-care center. Another new concept is a "neighborhood network," a volunteer referral service whereby families pass on to new parents child-care arrangements that have

worked for them.

Dispensing with the obsolete question of whether or not women should work, the Family Assembly is also trying to "balance the demands of workplace and family." Officials of A.T.&T., American Can Company and industry's Committee for Economic Development have joined Stanford sociologist William J. Goode and Caroline Bird, author of "The Two-Paycheck Marriage," in proposing "work-pattern innovations" to preserve "the quality of family life in an industrial and postindustrial society [in which] home and job can no longer be regarded as separate worlds." More and more companies are finding that "flextime" actually increases productivity and profits while reducing absenteeism. "Flextime" is the new system whereby everyone works at the office or factory during a midday core of hours, but arranges starting, leaving and lunch times according to individual needs. Some mothers and fathers start work at 10 A.M., after dropping their kids off at school; others skip lunch hour and leave at 3 to be home after school; others do a week's work in four days in order to have a long weekend for themselves. American Can, for instance, even provides "flexible benefits" — health-insurance, vacation and sick-leave packages tailored to individual family needs.

☐

It seems clear to me that we will never bring about these changes in the workplace, so necessary for the welfare of children and the family, if their only supporters and beneficiaries are women. The need for such innovations becomes urgent as more and more mothers enter the workplace, but they will come about only because more and more fathers demand them, too.

And men are beginning to demand them already. After all, they've long been subject to the same pressures that increasing numbers of women are experiencing.

A Vietnam veteran, laid off unexpectedly during the energy crunch by the airline with which he thought he had a lifetime career, tells me, "There's no security in a job. The dollar's not worth enough anymore to live your life for. I'll work three days a week at the boiler plant and my wife will go back to nursing nights, and between us we'll take care of the kids. We're moving to where we can raise our own food and have some control over our lives. But what if she gets laid off?"

"I thought seriously of killing myself," says a St. Louis man who was forced to resign at the age of 59 when the company he headed was taken over by a large conglomerate. "I saved up the arsenic pills I take for my heart condition. How could I live without that company to run, my office, my staff, 600 employees, the wheeling and dealing? But then I realized how much of it I'd really hated: the constant worry, getting in at 6 A.M. to read the reports of six vice presidents, fighting the union to keep wages down and being hated, and knuckling under to people I despised to get accounts. The only good thing was knowing I'd made it to president of a company, when my father never got past stock clerk. Now I want to work for myself, to live, to enjoy the sunsets and raise begonias. But my kids are gone now, and my wife started her career late, and all she wants is to get ahead in the agency. It's unthinkable somehow, what's ahead in life...."

A younger man in Rhode Island quit his job at a bank to take care of the house and kids and to paint at home, while his wife, sick of being a housewife, was happy to get a lesser job at the same bank. "I think you're going to see a great wave of men dropping out," he told me. "All we've been hearing for years now is, 'What does it mean being a woman? How can she fulfill herself?' But what does it mean being a man? What do we have but our jobs? Let her support the family for a while, and let me find myself...."

A young man refuses an extra assignment, which would mean working nights and traveling weekends, on top of his regular job. It doesn't matter that it might lead to a big promotion. "We're having another child," he tells his boss, "and I'm committed to sharing the responsibilities at home because my wife's going to law school at night. It hasn't and won't interfere with my job — you were more than satisfied with my last report. But I'm not taking on anything extra. My family is more important to me."

"That man isn't going to get far," his puzzled boss tells a

colleague. "Too bad: He was the pick of the litter."

The colleague asks, "What if they all start acting like that? Where are we going to find the men to run the economy, for God's sake, if they all start putting their families first?"

As we move into the 1980's, it becomes clearer that the women's movement has been merely the beginning of something much more basic than a few women getting good (men's) jobs. Paradoxically, as more women enter the workplace and share the breadwinning, their family bonds and values — human values as opposed to material ones — seem to strengthen. The harassed working mothers and their husbands in Dr. Kamerman's study of families place more importance and reliance on these bonds, not only with each other and their own children but with their own parents and other relatives, than do comparable families conforming to the traditional housewife-breadwinner image. The increasing tendency for young men to refuse corporate transfers and to put more emphasis on their self-fulfillment and "family time" as opposed to "getting ahead" has been reported by Yankelovich, Gail Sheehy in Esquire and even Playboy magazine.

When his wife is also earning money and his identity and standard of living do not depend entirely on his paycheck, a husband is something more than just a company man. Such a man has more freedom and opportunity to develop human values and to share the reality and responsibility of parenting. The Wall Street Journal reports that, of 300 recently relocated executives surveyed by a Chicago management-consulting firm, the great majority said that their most important consideration was not the job itself but "winning family consent."

"Tomorrow Begins Today," a study issued by the Security Pacific National Bank of Los Angeles, states, "Currently, 80 percent of all families earning $20,000 or more are two-wage-earner families.... Two-wage earners place less emphasis on careers and increased value on leisure activities, child care and household services. Career roles are less instrumental than the search for self-identity and good health."

Ah, yes — good health. One of the most remarkable results of the women's movement has been the unprecedented new vitality and growth experienced by millions of women who have defied the deterioration, depression and despair that used to be considered "normal" symptoms of aging in women.

Twenty years ago, the mental health and general well-being of women were much

worse than those of men in every age group after 20. Today, however, a repeat of the famous Mid-Manhattan Study shows that women no longer deteriorate in middle age. Indeed, women aged 40 and 50 are just as healthy as they were in their 20's and 30's, and much healthier than middle-aged women were 20 years ago. But such improvement has not been noted among men. The scientists who conducted these studies suspect that a crucial factor was the women's movement. It seems to have been a fountain of youth for women as we have moved to our new self-respect and put new energies and new talents to use, directing our aggression outward, even when the obstacles have been tough, instead of inward, against our own bodies, as despairing women used to do after their early peak of marriage and childbearing.

The widening discrepancy between men's life expectancy and women's — now roughly 77 years for women, 69 for men — makes it urgent that men break through their conventional sex roles as so many women have. Some men are already doing so, and the increasing awareness of men's midlife crises may be a harbinger of more changes to come. Furthermore, the discrepancy should alert women to the dangers of adopting too closely the obsessive careerism that has made so many men die prematurely of stress-induced heart attacks and strokes.

Unfortunately, according to Dr. Alexander Leaf, Chief of Medical Services at Massachusetts General Hospital, more young women than ever before are being admitted there with heart attacks, though hard data are not yet available nationally. And cancer experts have been noticing lately that as more young men stop smoking, more young women are starting. ("You've Come a Long Way, Baby!") A repeat of a massive mental-health survey originally conducted 15 years ago by the National Center for Health Statistics shows that today more women in their 20's and 30's are suffering from stress. Women's equality will have been for nothing if its beneficiaries, by trying to beat men at their own old power games and aping their strenuous climb onto and up the corporate ladder, fall into the traps men are beginning to escape.

The young men of the "counterculture" of the 1960's and 70's were rebelling against the great pressures to devote their

whole lives to money-making careers — just as women in the feminist movement rebelled against the great pressures to devote their whole lives to husbands and children and to forego personal ad-

vancement. But to substitute one half of a loaf for the other is not an improvement. Why should women simply replace the glorification of domesticity with the glorification of work as their life and identity? Simply to reverse the roles of breadwinner and homemaker is no progress at all, not for women and not for men. The challenge of the 80's will be to transcend these polarities by creating new family patterns based on equality and full human identity for both sexes.

☐

The willingness, finally, of the modern women's movement to come to grips with the practical problems of the family has inestimable historical significance. Reviewing the history of the original feminist movement and why it failed to alter the lives of most American women, historian William O'Neill in his book, "Everyone Was Brave," concluded that the trouble was rooted in the movement's unwillingness to tackle the problems of the family. Most of the early American feminists were either young single women opposed to marriage and the family or else married professional women who didn't have children or preferred to concentrate on loftier issues, such as the vote. They assumed that winning suffrage would automatically usher in equality and purify society. Yet, wrote O'Neill, since the masses of American women were married or wanted to be, the only way that equality between the sexes could have been achieved was through a "revolution in domestic life" reconciling the demands of family and career — a revolution that the first feminists never attempted — which is why they fizzled out after the vote was won in 1920.

A few years ago, it looked as if the same thing might happen again. The popular (and unpopular) image of the modern feminist was that of a career "superwoman" hellbent on beating men at their own game, or of young "Ms. Libber," agitating against marriage, motherhood, sexual intimacy with men, and any and all of the traits with which women in the past pleased or attracted men.

But the nine founding mothers of N.O.W. who are participating in the Assembly on the Future of the Family preached a feminism that was always rooted in personal truth. For us to have ignored or attacked men and families would have been a lie. We averaged more than two children apiece. Men were members of N.O.W. from the very beginning. Muriel Fox, a high-powered public-relations executive, brought her two kids along to N.O.W.'s founding conference in 1966, and her surgeon husband put them to bed in the hotel while she stayed up all night getting out our first press release. The feminists who have followed us in the movement's mainstream speak today of their own evolving families. Eleanor Smeal, the brilliant housewife president of N.O.W., found that commuting between Washington and her family in Pittsburgh didn't work, so her metallurgist husband and two children will join her. He will change jobs.

It seems to me you can trust feminists — or any other "-ists," for that matter — only when they speak from personal truth in all its complexity. Such truth is never black or white. The image of "women's lib" as being opposed to the family was encouraged by women locked in violent reaction against their own families and identities. Their anger was real enough, but their rhetoric denied other elements of their personal truth.

When extremists — both feminists and antifeminists — perpetrate the myth that equality means death to the family, other women have a hard time figuring out what their real options are — and their own real feelings. So Phyllis Schlafly and Marabel Morgan make a lot of money pursuing their own careers, going around the country lecturing women that they don't need equal rights, just husbands to support them — which they'll allegedly stop doing when the Equal Rights Amendment is passed. But the underlying reality is no different for the bitterest feminists and most stridently fearful defenders of the family. None of us can depend throughout our new, long lives on that "family of Western nostalgia" to meet our needs for nurture and support, but all of us still have those needs. The answer is not to deny them, but to recognize and strengthen new family forms that can sustain us now.

In the years since my divorce, I seem to have created a new extended family that consists not only of my own children but also of friends on whom I can always depend for a bed and a loving ear. Although I live alone in my little house in Sag Harbor, my

family stretches from California to Cambridge. And there is an authentic note of family reunion as we who founded N.O.W. a generation ago reconvene for the Assembly on the Future of the Family, our sons and daughters at our sides, our hands extended to men. As our children grow up and leave home, as more of us struggle with the problems of husbands' strokes and heart attacks, and our own illnesses and loneliness, we admit our needs and vulnerabilities to each other. We get love and support from this family that was knit together in our battle for equality. The movement itself has become "family," and now we call each other longdistance just to say "hello" as families do.

My personal concern for the 1980's is the possibility of and need for new patterns of intimacy and growth, love and work in the third of life that most women — but not yet most men — can now hope to enjoy after 50. What social policies and institutional innovations will enable us more easily to enjoy vital, integrative lives after childbearing, making our own professional and personal choices, and receiving as well as giving family support to each other and our children? As our population becomes proportionately older, the dominant social revolution of the 1980's will almost certainly be made by men and women in this new third of life. For they will be us, demanding our own voice and participation in society instead of the diminished opportunities, low status, dehumanizing discrimination and passive roles as patients that our society now imposes on the aging.

For instance, the assembly asked architects to come up with new kinds of communal housing for people who live alone and for older couples rattling around in drafty houses too big to keep up with the children gone — housing which provides everyone with her or his own private living space but also offers new kinds of shared space for eating, socializing and doing chores. There are also proposals for the same kind of mortgage financing now available only to

young families, and for re-education loans, like the G.I. Bill, for the new third of life after 50.

Robert N. Butler, director of the National Institute on Aging, has been complaining for some time that most of our policies on aging are based on research conducted on men — and men in pathological decline, outside the family setting — whereas most of the people actually living in old age are women. There is some speculation that women survive better and longer than men because they forge more human and familylike connections even after childrearing. As more men share that early nurturing, will they also develop new survival skills?

□

To some it may sound strange for a feminist like myself to be arguing so passionately for the importance of families. Such arguments have even been dismissed by some radical feminists as "reactionary family chauvinism." But it may very well be that the family, which has always been considered the bastion of conservatism, is already somehow being transformed by women's equality into a progressive political force. For when men start assigning a higher priority to their families and self-fulfillment, and women a higher priority to independence and active participation in "man's world," what happens to the supremacy of the corporate, bureaucratic system?

Some recent management studies, for instance, indicate that the corporate policy of frequently transferring executives and demanding that they work nights and weekends is not really necessary for the work of the corporation, but that, by estranging them from their communities and families, it serves to make executives corporate creatures, "company men." Will women renounce their bonds and their power within the family in order to become "company women"? Some already have, but in most instances, women's equality, in the home

and in the workplace, strengthens the family and enables it better to resist dehumanization. Families are easier to control when women are passive and dependent. One of the first acts of the Nazis when they took over Germany was to take away some of the rights of women.

Can American capitalism accommodate a strengthened, evolving American family? Why not? Despite the rhetoric, the family has never ranked high on the American political and economic agenda, except as a unit to which to sell things. The business of America, as everyone knows, is business, and until recently it's been man's business. Now that women are beginning to have an active voice in the economy and politics, the nation's agenda may begin truly to include the family. Not just because women insist — they don't have that much power yet — but because men have a new stake in the family. The new sharing of parenting and the envy many men are beginning to express now of women's liberation suggest that the family, instead of being enemy territory to feminists, is really the underground through which secretly they reach into every man's life. The new urge of both women and men for meaning in their work and life, for love, roots and family — even though it may not resemble the ideal family that maybe never was — is a powerful force for change.

I'm not even altogether sure that the women's movement as such will be the main agent of this next stage of human liberation. But if we don't want to retreat — with women and men withdrawing into tired, lonely disillusionment and backlash — we must somehow turn this new corner toward the family of the future. Women must now confront anew their own needs for love and comfort and caring support, as well as the needs of children and men, for whom, I believe, we cannot escape bedrock human responsibility. ■

November 18, 1979

CHAPTER 5

Black Africa

Bishop Abel Muzorewa, head of the biracial Government of Zimbabwe Rhodesia, reached agreement with rival guerrilla factions and British mediators on a transitional formula for his country's new constitution.

United Press International

Rhodesian Whites Vote to Accept Limited Rule by Black Majority

By JOHN F. BURNS
Special to The New York Times

SALISBURY, Rhodesia. Jan. 30 — White Rhodesians voted overwhelmingly today to accept limited black rule of this southern African nation, which has been dominated by whites since the first colonial settlements in the 1890's.

Under the relentless pressure of a guerrilla war, the 95,000 eligible voters approved a constitutional plan that provides for a black-led government with extensive safeguards for whites, including nearly a third of the Cabinet posts for the first five years. The new government is scheduled to be chosen April 20 when the 2.5 million adult blacks can join whites in a one-man, one-vote election.

With 37 of the 54 districts reporting, nearly 85 percent of those voting supported the plan. With the turnout running at close to 70 percent, the result was hailed as a victory by Prime Minister Ian D. Smith, who gave up his 14-year-old struggle to maintain white rule and negotiated the constitutional plan with three moderate black leaders.

Visiting a polling station, Mr. Smith said that a strong "yes" vote in the referendum would strengthen his hand in appeals to the United States and Britain for support against the guerrillas, who have vowed to fight on until any black government that emerges from the April election is overthrown. The Communist-backed guerrillas, calling the Smith accord a "sellout," are intensifying a conflict that already is taking more than 500 lives a month.

British-U.S. Plan Is Rejected

A plan offered by Britain and the United States that would require an internationally supervised one-man, one vote election and the replacement of the existing army with a force based on the guerrillas has been rejected by Mr. Smith. After a British negotiator reported this month that prospects for agreement between the warring parties were practically nil, Prime Minister James Callaghan, with the Carter Administration's support, shelved plans for a conference of all the parties to the dispute.

Blacks, who outnumber whites by more than 27 to 1, have been given no opportunity to vote on the agreement between Mr. Smith and his black partners, Bishop Abel Muzorewa, the Rev. Ndabaningi Sithole and Senator Jeremiah Chirau. But unofficial soundings among urban and tribal dwellers have suggested that many foresee a civil war between the black factions allied to Mr. Smith and the guerrillas of the Patriotic Front.

The guerrillas, loyal to Joshua Nkomo and Robert Mugabe, the co-leaders of the Patriotic Front, appeared to have made no effort to disrupt the voting today. But the Government took no chances, posting

Associated Press

BASIC TRAINING: A white Rhodesian soldier being upbraided by a black noncommissioned officer recently at the Llewellin Barracks in Bulawayo, Rhodesia. The soldier was one of 600 conscripts — about half of them black — that arrived at the camp for basic training. The Rhodesian Army has become a very integrated force as a result of a drive last year to conscript blacks in an effort to expand the army for the war against nationalist guerrillas.

troops and police reservists at the polling stations and moving pupils to classrooms away from voting booths in schools to protect them against bombs.

Farmers in the eastern districts, where the guerrillas have been active, drove to the voting places in armored cars, rifles and machine guns at the ready against ambush. Elsewhere, voters living along rural roads formed convoys. Most of those voting appeared somber, reflecting the foreboding with which most whites regard black rule.

'We May as Well Go Along'

"The way I look at it, the situation is hopeless if we reject the agreement, and only slightly less hopeless if we accept it," said Robert Cross, an automobile mechanic who braved a driving rainstorm to

vote in Salisbury. "I figure we may as well go along with Smithy, and pray that he can somehow pull it off."

The view was widely shared. In a three-week campaign across the country before the vote, Mr. Smith, who built his political career on policies that discriminated against blacks, made no secret of his chagrin at having to cede leadership of the Government. "This is the most difficult exercise I've ever undertaken," he told a crowd of hecklers at one rally in the capital. "I'm trying to sell you something I've tried to avoid."

Two right-wing groups, the Rhodesian Action Party and Save Our Nation, mounted a drive for rejection of the new constitution, arguing that it would plunge the country into chaos and result in a an eventual takeover by the guerrillas. "If you vote 'yes' Mugabe and Nkomo will dance in Cecil Square," the Action Party said in full-page advertisement yesterday in The Herald, the country's leading newspaper. "Better they dance in Moscow."

The National Unifying Force, a white liberal group that has opposed the plan as too restrictive for the black majority, formally adopted a neutral stance in the referendum campaign. But its leaders argued that the only way to avoid a civil war that would pit the Patriotic Front against "private armies" loyal to Bishop Muzorewa and Mr. Sithole was for Mr. Smith to hand over control of the country to Britain, renouncing the rebellious independence he declared in 1965.

Mr. Smith, subjected to frequent vituperation by right-wingers who once were his staunchest supporters, answered bluntly. He said the country lacked the manpower to turn back the guerrillas, 12,000 of whom now roam the country, and declared that the only hope for whites in the long term — he put it at "50-50" — was that the United States and Britain would lift economic sanctions and recognize the new black-led government.

At a rally in the northwestern town of Karoi, he acknowledged for the first time that the Government had made contingency plans with the South African Government in the event that the quarter of a million whites remaining here have to flee.

The 59-year-old white leader also

argued that the constitutional plan was "the best deal available," leaving what he described as considerable influence to whites. Besides guaranteeing whites 28 of the 100 seats in the Assembly, and a proportionate number of seats in the Cabinet, the plan effectively bars any major changes in the white-led civil service, armed forces, prison service, police and judiciary for the first decade after a black prime minister takes power. Nationaliization of the 50 percent of the farming land that is white-owned is also effectively barred.

Mr. Smith ridiculed a proposal for a return to legality under Britain, calling the British-American plan for a settlement with the guerrillas a formula for chaos.

January 31, 1979

Zaire's Rich Mineral Area Is Racked With Poverty

Sygma

Young Zairians marching in support of President Mobutu Sese Seko in a recent demonstration in town of Lubumbashi

By JOHN DARNTON
Special to The New York Times

LUBUMBASHI, Zaire — Giant billboards extolling President Mobutu Sese Seko, the incarnation of the central Government 1,500 miles away, loom at every major traffic circle. He is "our only hope," the "single man we all follow for unity, tranquillity, security," they say.

In the early morning, under skies already darkening for the daily downpour, several dozen workers from the Société Générale d'Alimentation, the supermarket chain owned by the President's uncle, gather in a public square under a peeling flagpole. With a listless air they perform 15 minutes of exercises and slogan-singing called "salongo," a compulsory exhibition of public spirit.

When General Mobutu toured this dilapidated provincial capital recently, he was greeted by crowds of respectable size. They applauded under the importuning of the youth branch of the single party, the People's Movement for the Revolution.

These superficial signs of homage are

75

misleading. Shaba Province, formerly Katanga, the mineral heartland of sprawling Zaire, the seat of the secessionist dream that tore apart the former Belgian Congo in the 1960's and the scene of two rebel invasions from neighboring Angola over the last two years, still seethes with discontent.

Industry Is Barely Creeping

For lack of spare parts and basic materials, its factories that still function do so at under 50 percent capacity. There is a severe shortage of gasoline; Zaire can afford only 60 percent of its required fuel, and 90 percent of that is reserved for the capital, Kinshasa. In Lubumbashi, the three working traffic lights blink down at streets largely devoid of traffic.

Unemployment can only be guessed at among the province's three million or so people, but it is estimated at 80 percent. Over the last year the price of food has doubled. The essential commodities that can be found, from explosives to mine the copper at nearby Kolwezi to the fruit and vegetables at the three high-priced restaurants, come directly from black Africa's sworn enemies, Rhodesia and South Africa.

Missionaries and others who travel the all-but-impassable back roads report an alarming increase in malnutrition. Children in particular are showing the telltale swollen bellies and red hair that bespeak the protein-deficiency disease called kwashiorkor. "Frankly," said a Belgian financier who has lived here 15 years, "the population is on a starvation budget. We are sitting on top of a volcano."

The streets of Lubumbashi are patrolled by Moroccan soldiers in khaki battle fatigues and Senegalese paratroopers with speckled camouflage uniforms. The 1,500 Moroccans and 600 Senegalese are the main components of the Inter-African Force, which also includes a token 140 soldiers from Togo and Gabon and 80 medical corpsmen from the Ivory Coast. The force, drawn from French-oriented African countries, replaced the Belgian and French paratroopers who repulsed rebel forces attacking southern Shaba from bases in Angola last May. In that invasion, 94 whites were massacred in

The New York Times / Feb. 4, 1979

Kolwezi and over 1,000 blacks were killed.

The Moroccans patrol frequently, scouring the countryside in helicopters for signs of rebel activity. But this appears to be largely for morale and discipline since there have been no sightings since July, when a small band of rebels turned cattle rustlers were found south of Kolwezi.

In theory the foreign troops are here to fend off another invasion and are to remain only until Belgian and French advisers finish training Zairian troops. Now that Angola, as part of a normalization of relations with Zaire, has removed the rebels it once harbored in the border area, the immediate threat of a sudden full-scale attack has diminished. Instead, the foreign forces appear to have taken over the basic functions of law and order.

"They're a thousand times better than the Zairian Army," said a Greek mechanic, waiting at an impromptu roadblock where drivers' licenses were checked by three Moroccan soldiers. "They're disciplined and they don't steal. They're the only thing protecting us from the Zairian troops."

Whites in Lubumbashi and elsewhere, traumatized by the events in Kolwezi and worried by the rising crime rate, frequently voice the pledge that they will leave as soon as the Inter-African Force departs, which could be within six months. There are perhaps 3,000 to 4,000 whites in Shaba, compared with 13,000 when the country became independent in 1960, and their number is steadily dwindling.

The blacks have a somewhat more ambiguous attitude. Some appear to regard the Moroccans as an occupying force, but an equal number hate and fear the Zairian soldiers and regard them as occupiers too. The Zairian Army is paid irregularly and scantily, so its undisciplined troops customarily extract bribes and even rob.

Disadvantages — and Advantages

"We are bothered that a country like ours, 17 years after independence, needs to rely on a foreign army," a well-respected teacher commented. "But from the point of view of security, when you see a Moroccan, you're not afraid he is going to stop you and shake you down just for the fun of it."

Shaba, whose rich deposits of copper, cobalt and other minerals provide about 60 percent of Zaire's foreign exchange, has received little in return. Over the last decade the central Government has not built a single school, hospital or road.

The ill will toward the central Government is held in check by strict security measures. In August, under the guise of a program to clear the streets of beggars and the unemployed, the Government conducted a campaign in which over 2,000 people were rounded up and shipped in American-supplied C-130 transport planes to what was termed an agricultural training center in Kivu. Actually, the place was more like a detention center. There have been reports that many of those sent to the center, among them lawyers, other professionals and suspected dissidents, died for lack of food and medical care. In September those remaining were released.

February 4, 1979

Zaire Could Be Very Rich But Now It Faces Ruin

By JOHN DARNTON

KINSHASA, Zaire — Several years ago, during his "authenticity" campaign to wipe out the colonial past and help Zaire rediscover its African heritage, President Mobutu Sese Seko outlawed Christmas. Nationalistic symbolism was meant to disarm leftist intellectuals at home and make African nations forget that General Mobutu came to power with the aid of white mercenaries, the United States Government and the Central Intelligence Agency. But, like much

else in Zaire, "authenticity" failed. So, last December, President Mobutu reinstated Christmas. Zaire's 26 million people had little to celebrate.

The country is on the brink of ruin. Shaba Province, invaded in 1977 and again in May 1978 by rebel exiles from Angola, is still patrolled by 2,500 troops from other African nations. The threat of invasion has receded, but local insurrection cannot be ruled out. Last week the Belgian Government, at Zaire's request, sent in 250 paratroopers ostensibly for training. Belgium was reportedly concerned about the safety of thousands of Europeans in Zaire, including

Abbas/Contact; Alon Reininger/Contact
Pro-Government rally in Kinshasa and President Mobutu Sese Seke.

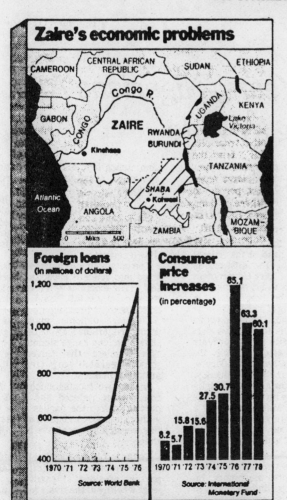

Zaire's economic problems

Foreign loans
(in millions of dollars)

Source: World Bank

Consumer price increases
(in percentage)

85.1

63.3
60.1

30.7
27.5

15.8 15.6

8.2 5.7

1970 '71 '72 '73 '74 '75 '76 '77 '78

Source: International Monetary Fund

many Belgians, but observers noted that European countries concerned for their nationals customarily urge them to leave, rather than sending in troops to protect them before trouble has started.

There is real cause for discontent throughout Zaire. The living standard has dropped to an all-time low. Expatriate businessmen who left after their farms and businesses were nationalized in 1973 and 1974 are being called back to repossess them. Officials of the International Monetary Fund and Western creditor nations are assuming control of banking, finance, customs and transportation. Zaire owes $3.5 billion it cannot repay on schedule. With no foreign exchange to buy spare parts or basic materials, small industries are closing and big ones are operating at 50 percent capacity. The infrastructure is collapsing: roads between provincial capitals are no longer drivable, half the locomotives are incapacitated and Air Zaire cancels more flights than it runs.

Zaire's 60,000-man army and paramilitary gendarmery are a sham. Underfed and underpaid, they are driven to extortion and outright theft. One diplomat estimates that when Shaba was last invaded, 85 percent of the looting in Kolwezi was done not by the rebels but by Zairian soldiers who occupied the town after it was retaken by French and Belgians. Observers agree that corruption is endemic, starting at the top, where the President, according to diplomats and businessmen, rakes off 10 percent on most international business deals, all the way to the bottom, where primary school students pay $5 for a passing grade. The former Minister of Culture sold treasures from the National Museum. The Postmaster General plunders money orders from the mail. The generals sell gasoline intended for army trucks. The doctors sell hospital medicines on the black market.

While the politically connected elite think nothing of spending $200 on dinner, the masses are close to starvation. A confidential report from an international relief group calls it "the worst rural malnutrition in decades." Belgium's Foreign Minister was quoted last week as saying that the paratroopers were sent out of fear that 3 million hungary Zairians in Kinshasa might turn on 30,000 well-fed Europeans. Inflation has zoomed since 1972, sometimes running above 80 percent. Since November, when the Government began an unannounced devaluation that has reached 50 percent, prices have doubled. The cost of a 50-kilo sack of rice or manioc, the staple food, now exceeds the average monthly salary. Farmers have abandoned cash crops for subsistance agriculture. In Kinshasa, the jails are so crowded with thieves that they sleep on the floor in shifts. Diplomats are shocked to see people fainting in the streets.

Zaire's slide into bankruptcy and economic chaos is not merely a product of one-man misrule; the events that turned what should have been Africa's richest nonoil-producing country into one of Africa's poorest are not that simple. Zaire borrowed heavily between 1973 and 1975, when copper, its main export, was selling at record high prices. But only 2 percent of the loans were used to develop agriculture. Fifty percent went for grandiose transportation and power projects, ill-conceived and soon mismanaged, but tempting to an emerging nation that was striving to establish itself in world councils. Many of these projects were conceived on French and American drawing boards. A French firm constructed the monumental $60 million Voix de Zaire, a futuristic communications complex — and turned off the elevators when payment was delayed. An American company is constructing a $600 million, 1,500-mile power transmission line from a new dam on the Congo River to copper-laden Shaba. The electricity could have been purchased from nearby Zambia at a fraction of the cost, but the Government was attracted to

a scheme that would give it control over its richest and most rebellious province.

As early as 1974, Zaire began falling into arrears on debt service payments, but new loans were still forthcoming. When they came due, the United States and France were at the front of the repayment line, getting more than half the $820 million Zaire repaid between 1970 and 1977. In the early 70's, the average terms extended to Zaire were much harsher than those granted to 30 other countries with a per capita gross national product of less than $250. The banks that were scrambling over one another to lend money then ($561 million worth), are now balking at lending an additional $220 million. Into the breach have come the World Bank and the International Monetary Fund. Next month the fund is expected to order a stabilization plan to grant stand-by credit and unlock the money markets but, like the currency devaluation, the plan's immediate effect will be to further lower the standard of living.

The rescue of the economy could of course be the rescue of President Mobutu. He is an isolated figure, unpopular at home, ignored by much of Africa, disparaged by the East and embarrassing to the West. But he has a genius for cutting down challengers. The one sign of potential opposition was an incident which reportedly occurred in Kinshasa several weeks ago. (Western diplomats are still not convinced it did happen, although they say contacts in the secret police have verified it.) Two men were reportedly found crucified in an empty lot, with labels identifying them as "thieves of the people" and a list of political accusations against the President.

February 11, 1979

After a Wait of 13 Years, Nigerians Are Preparing to Vote

By CAREY WINFREY
Special to The New York Times

LAGOS, Nigeria, March 5 — Between now and Oct. 1, Nigeria's nearly 50 million voters are expected to elect the country's first civilian government in 13 years.

If history is any guide, the campaigns will range from raucous to riotous as the nation gives vent to political enthusiasms all the more tumultuous for having been suspended so long by military decree. But now it is quiet.

The reason is that Lieut. Gen. Olusegun Obasanjo's military Government has yet to set the dates for the elections or, for that matter, to announce the number of times voters will go to the polls to elect the nearly 2,000 civilians who will govern them from statehouse to presidential mansion.

Nor has the Government announced its position on the latest political controversy, over the form the expected presidential runoff will take. Once the field of five candidates has been reduced to two through a popular election, will the runoff take the form of another popular election or will it take place in the House of Representatives? Because the electoral decree is ambiguous, the debate has raged.

With so many important questions still

to be answered, the presidential campaigns have yet to shift into high gear. One candidate, Chief Obafemi Awolowo, has returned to his village and vows not to campaign until the dates have been announced.

"Everybody asks the same thing: When is the date for the election?" said Alade Odunewu, the head of the federal electoral commission. The answer, he added, is when the military Government tells him to set it, an event now not expected until after the new budget is signed at the beginning of April.

Until February, most knowledgeable Nigerians expected the first of five elections to be in April with voting for state legislatures. Other elections, for the House of Representatives, the Senate, state governorships and the presidency, were expected to follow at three week intervals. Now virtually no one expects the first election before May, with June a more likely date and July or even August possibilities. "Everybody's guessing, but nobody really knows," said a professor at Lagos University.

The Government is waiting for assurances that the electoral machinery, including some 125,000 polling booths, is in place, no small task in a country the size of Texas and California combined with

communications and tranportation that can most charitably be described as inadequate.

"With all the technical difficulties this country has in every sphere," said Stanley Macebuh, chairman of the editorial board of The Daily Times of Lagos, "the federal election commission needs all the time it can get."

The potponements have yet to be translated into concern about the election itself. "The elections will come off unless the sun stands still," said a newspaper editor here, expressing a common view. A Western diplomat made the same point. "Nigerians may not be the greatest shakes on long-range planning," he said. "Nevertheless the Government has shown a real ability to put things together at the last moment."

In the two months since the election commission reduced the field of 52 parties to five the parties have put together slates of candidates and in the last few weeks platforms have begun to take shape.

To the left is the Socialist-oriented People Redemption Party. Aminu Kano, the party's presidential candidate, is the least wealthy of the rivals for his nation's leadership and has a long record of dedication to the poor. He has his greatest support in the northern city of Kano, Nigeria's second largest city. However, this Moslem constituency, divided as it is by two other northern candidates, is not considered likely to carry him into the presidency.

Obafemi Awolowo, the candidate of the Unity Party of Nigeria, is a strong contender. As a member of the Yoruba tribe, the nation's largest, he begins with a household name and a broad ethnic base of support. His weakness in the north is offset by the speed with which he associated himself with popular issues: universal free education and medical treatment. His party espouses a progressive program of development of roads, schools and government services.

Though Awo, as he is called, turns 70 years of age this week, his age represents little handicap in a nation that generally reveres its elderly.

Even more advanced in years is the 74-year-old first President of Nigeria, Dr. Nnamdi Azikiwe, who came out of retirement to challegnge his longtime rival, Chief Awolowo. As the oldest practicing politician in the country, Zik, as he is called, cannot be taken lightly. A coalition with Aminu Kano is a possibility and it could propel him to the nation's leadership despite frail health and his pronounced identification with the past. His Nigerian People's Party is basically centrist and advocates a continuation of the progress made by the military.

Like Aminu Kano, Waziri Ibrahim, a wealthy Kanuri businessman, has his base in the Moslem north. He made his fortune supplying arms to the federal Government in its war against secessionist Biafra and is identified with Arab business interests. His candidacy must be counted a long shot but he has made dramatic gains in the last month.

As the candidate of the National Party of Nigeria, the nation's most broadly based party, 58-year-old Shehu Shagari would have to be considered the frontrunner if the election were held tomorrow, though he himself is far from a charismatic campaigner.

Supported by Business Community

He has support and financing from the business community and his moderately conservative campaign emphasises stability and the status quo.

What the five candidates share, beyond ambition, is age and experience. A local magazine characterized them as old wine in new bottles and lamented the fact that no younger men qualified as candidates.

A university professor said that the candidates' age, their rigidity and their identification with the past virtually guaranteed a return of military rule within two years. But others are not so sure. "It would have been illogical to expect younger chaps to come into the mainstream at this time," said Stanley Macebuh. "They'll get their chance four years from now."

March 6, 1979

CAPITAL OF UGANDA TAKEN BY INVADERS; NEW REGIME FORMED

RESIDENTS WELCOME SOLDIERS

Broadcast by Amin From a Secret Site Denies Statement That Government Has Fallen

---

By JOHN DARNTON
Special to The New York Times

NAIROBI, Kenya, April 11 — An invading force of Tanzanian soldiers and Ugandan exiles captured Kampala, the capital of Uganda, early today as thousands of residents poured out of their homes shouting and weeping for joy.

The attack, which met only scattered resistance and came after five and a half months of sometimes desultory war, effectively toppled President Idi Amin and ended his eight years of murderous rule over Uganda's 12.5 million people.

The Ugandan exiles announced tonight the formation of a provisional Government, headed by Yusufu K. Lule, a 67-year-old academician who was chosen only last month as chairman of the new Uganda National Liberation Front. [Page A16.]

Amin Broadcast Denies Defeat

In a dramatic and bizarre twist late tonight, President Amin raised a defiant voice in a broadcast thought to have originated somewhere in Uganda. He said: "I, Idi Amin Dada, would like to denounce the announcement . . . that my Government has been overthrown. This is not true."

The President, whose voice was readily identified by correspondents here, accused the invaders of murdering civilians and destroying private property. He insisted that his forces controlled 90 percent of the country and added:

"The Ugandan Armed Forces must not surrender their arms to any rebellion. We are in full control all over Uganda. I am speaking as President of the Republic and Commander in Chief of the Armed Forces."

Despite his assertions, Tanzanian and Ugandan exile sources and military analysts did not believe that he had enough troops to hold out for long.

There was no definite word on where the Ugandan dictator was. But several reports from Kampala, including one from a Cabinet minister who denounced the regime in its final moments, said he had been seen fleeing yesterday evening toward Jinja, an industrial city about 50 miles east of Kampala where the remnants of his loyalist troops have gathered.

Frightened residents said the troops were desperate, exhausted and hungry and had begun looting. They were thought to have little fight left in them.

Many reports of looting as well as of the jubilation in Kampala, the capital, were provided over the telephone by residents and diplomats, including one ambassador who sat down to tea with a Tanzanian commander, and by a Western correspondent who went with the front-line Tanzanian troops.

During the attack on the city, whose population of 350,000 has been greatly reduced by several weeks of shelling, mobs of civilians turned their vengeance on isolated pro-Amin soldiers, beating at least 10 of them to death. There were also many reports of widespread looting by civilians, who have been low on food and other essential goods for weeks.

The New York Times / April 12, 1979

Kampala fell to invaders, and Idi Amin reportedly fled toward Jinja.

Downtown shops were rapidly emptied, and in some cases even the store shelves were taken. Tanzanian soldiers, outnumbered by the frenzied populace and reluctant to fire on them, were unable to stem the plunder.

Hatred for Nubians Is Shown

Ugandan exiles in Nairobi noted that after President Amin expelled 40,000 Asians in 1972, many of Kampala's stores were handed over to his favorite group in his tribally based Government, Nubians from southern Sudan, Zaire and elsewhere. The Nubians, who were also a hard core in the 20,000-man army and the security organizations, are hated as foreign mercenaries by many Ugandans.

"There's a belief that all this wealth belonged to the Nubians," said one Ugandan. "So now the people are paying them back."

The attacking force, which has had Kampala under siege for a week, practically walked in. Approaching in darkness from the south and west, they ran into pockets of Government soldiers and sporadic sniper fire. With tanks and AK-47 rockets, they easily overpowered the Ugandan positions.

The populace greeted the invaders as liberators. The troops were mobbed as the people cheered and chanted, "Nyer-ere up, Amin down," and "Thank you, Nyerere," referring to President Julius K. Nyerere of Tanzania.

Tanks Are Decked with Flowers

An American free-lance journalist, Tony Avirgan, who accompanied the Tanzanian 12th Division, the first to enter the downtown area, said the celebrators hugged the troops, lifted them in the air and decked their tanks with flowers. So wild was the reception, he reported, that the Tanzanian commanders were worried that the army could not move around to wipe out the remaining pockets of resistance.

It was announced that Tito Okello, the commander of the invading forces called the Uganda National Liberation Army, had arrived in the capital. Commander Okello served under President Milton Obote, who has found refuge in the Tanzanian capital since his overthrow. He is close to President Nyerere, and many people had assumed that President Nyerere would push to reinstate him.

As the invasion gathered force, however, the Tanzanian Government prodded 18 often-feuding Ugandan exile groups to form a common political and military front. On March 23 they met at Moshi in northern Tanzania and elected a provisional 11-man committee to oversee a period of interim rule leading to elections. Conspicuously, the committee, broad based both tribally and ideologically, did not contain a single overt Obote supporter.

Mr. Obote, whose own brand of authoritarianism is not fondly remembered by many Ugandans, is believed to be eager to pick up the reins of power. A substantial number of the 3,000 or so exiles who joined the Tanzanian force of 4,000 to 6,000 soldiers were initially Mr. Obote's supporters, and this could create tension in the fledgling Government.

New Leader Widely Respected

Since President Nyerere will now have an extraordinary say in Ugandan affairs, Mr. Obote's chances of returning to office will depend to a great extent on whether the Tanzanian President still backs him. Most observers believe that Mr. Nyerere, who has already run the risk of being condemned by other African nations for interfering in the affairs of a neighboring state, will not want the visibility that would come with hand-picking the country's next President.

Mr. Lule, the country's new leader, is widely respected as a figure of unity and comes from Uganda's largest tribe, the Paganda, which has not held power since independence in 1962. But he does not have the forceful personality that may be required to steer Uganda through an un-certain period of tribal tension.

Economically, the country that Winston Churchill once called "the pearl of Africa," is in ruins. It has no foreign exchange, its basic installations and facilities are crumbling and its currency has dropped to one-tenth its face value. Only high world market prices for its main export, coffee, kept the country afloat last year, and now that too is falling. Western powers that broke with the Amin regime are standing by to rush in aid once a new Government is in place to administer it.

Embassy in U.S. Occupied

Special to The New York Times

WASHINGTON, April 11 — In a festive mood, two dozen anti-Amin Ugandans occupied their country's embassy today.

Mahmud Musa, chargé d'affaires of the old regime, said he had agreed to add an extra lock to the front door so that both parties would have to be present to open the door. Mr. Musa said that when accredited representatives of the new regime arrived, he would turn the building over to them. ——

U.S. Welcomes New Regime

WASHINGTON, April 11 (AP) — United States officials welcomed the ouster of President Idi Amin today and said the United States planned to establish a normal relationship with the new Government quickly.

April 12, 1979

Ugandan Exiles Establish Provisional Government

United Press International

Photograph made available by The Tanzanian Daily News is said to show a Ugandan colonel, loyal to President Idi Amin, after his capture in Kampala. The other prisoners are described as being Libyan soldiers.

By CAREY WINFREY
Special to The New York Times

DAR ES SALAAM, Tanzania, April 11 — Yusufu K. Lule, a 67-year-old academic who describes himself as apolitical, announced here tonight that he would assume the presidency of a provisional Ugandan Government.

Mr. Lule, the chairman of the Executive Council of the fledgling Ugandan National Liberation Front, said in a statement that he would also be Minister of Defense and Commander in Chief of the armed forces. The statement named a 14-member Cabinet that included most of the members of the front's Executive Commmittee. The Government plans to fly to Kampala tomorrow from Dar es Salaam..

In a broadcast Mr. Lule pledged to restore democracy and promised elections in Uganda as soon as conditions permit. Ugandans have not voted since 1962.

Mr. Lule's calm, scholarly voice contrasted sharply with the kind of speeches that Ugandan listeners had come to expect from President Amin.

In the broadcast, Mr. Lule also announced that all foreign exchange transactions would be temporarily frozen, and he called on civil servants to go to work tomorrow.

Mr. Lule has previously said that the front, which was formed three weeks ago at a meeting of 18 Ugandan exile groups, planned to establish an interim Government for up to two years.

At that meeting in Moshi, Tanzania, 300 miles north of here, plans were laid for a Parliament of 70 to 90 members, a Cabinet of not more than 20 ministers and an Executive Council. The meeting was orchestrated by Julius K. Nyerere, President of Tanzania, who encouraged Ugandan exiles of all political persuasions to attend.

Conspicuously absent was Milton Obote, the Ugandan President overthrown by Idi Amin eight years ago. Mr. Obote, who lives in exile here, commands an estimated 2,000 rebels inside Uganda and, until the creation of the front, was widely assumed to be preparing to return to power. Although Dr. Obote and Mr. Nyerere are close friends of long standing and share a socialist ideology, the Tanzanian President has insisted throughout war that his troops were not fighting to restore Dr. Obote to the presidency.

A spokesman for the front said five administrators were already at work in areas of Uganda captured earlier from President Amin's forces.

The spokesman said the administrators had been telling Ugandan citizens they had nothing to fear despite having worked for President Amin. He quoted one administrator as having said: "Nobody is going to be victimized. Only criminals have anything to fear and they will go through the law courts. The structure of the legislature and the judiciary are still there. It is just a question of rebuilding them."

April 12, 1979

RHODESIANS STRIKE AT ZAMBIAN CAPITAL IN BID TO GET NKOMO

REBELS IN BOTSWANA ALSO HIT

10 Reported Killed, 12 Injured in Raid in Lusaka — Guerrilla Leader Escapes Injury

By JOHN F. BURNS
Special to The New York Times

JOHANNESBURG, April 13 — Rhodesia sent raiders into neighboring Zambia and Botswana today to attack command centers of guerrillas who have vowed to disrupt elections in Rhodesia next week. At least 10 people were reported killed.

In a 3 A.M. strike into Lusaka, the Zambian capital, Rhodesian commandos disguised as Zambian soldiers wrecked several buildings in an attack apparently aimed at capturing or killing Joshua Nkomo, one of the two principal Rhodesian guerrilla leaders.

A bungalow where Mr. Nkomo lives was destroyed in the raid, but the guerrilla leader escaped injury. At first, Zambian officials and guerrilla aides said that Mr. Nkomo had not been at home during the attack, but Mr. Nkomo insisted later that he had escaped during a gun battle.

The report that at least 10 people were killed and 12 wounded in the Lusaka raid was made by Zambian officials, but they did not say whether any senior aides of Mr. Nkomo were among the casualties. There was no official word on Rhodesian losses, but some sources in Salisbury, Rhodesia's capital, said the raiders returned without casualties. They were ferried across the Zambezi River in Land-Rovers on pontoons.

Other Rhodesian troops, dressed in the uniforms of the Botswana Defense Force, drove into Francistown on the border at 1 A.M. and captured 14 of Mr. Nkomo's aides. The officials were in charge of coordinating the passage of guerrilla recruits and refugees from Rhodesia through Botswana into Zambia.

At about the same time, Rhodesian attackers destroyed a ferry that crosses the Zambezi between Botswana and Zambia at Kasungula, the point where Rhodesia, Botswana and Zambia meet. A Rhodesian military communiqué issued in Salisbury said the ferry had been carrying war materials for the Zimbabwe People's Revolutionary Army, the military wing of Mr. Nkomo's Zimbabwe African People's Union.

No Comment on Objective

A Rhodesian military spokesman refused to say what the principal objective of the Lusaka attack was. But it appeared to be similar to that of raids into Mozambique over the last three years, when military units attempted to seize or kill commanders of the other main Rhodesian guerrilla force, which is led by Robert Mugabe. On each occasion the raiders struck at camps where they believed guerrilla commanders were present, only to find that their intelligence information had been faulty.

There was speculation in Rhodesia today that the attacks into Zambia and Botswana were part of a military offensive aimed at blunting guerrilla plans to disrupt elections for a black-led Government scheduled to begin on Tuesday. The elections, involving 2.8 million black voters and more than 100,000 whites, are to continue for five days under one of the most elaborate security operations ever mounted in Rhodesia.

The latest drive on the guerrillas began last month with a series of raids against camps and other targets in Zambia and Mozambique. However, military spokesmen have said that three bombing raids this week against targets in Zambia, which left 350 people killed or wounded, had come too late to have any effect on rebels already inside Rhodesia preparing to disrupt the vote.

A possible motive for today's attacks could have been to demoralize guerrillas led by Mr. Nkomo and encourage them to defect, as several hundred insurgents have already done under a Government amnesty program. Another motive could be to avenge the deaths of 107 Rhodesian civilians in two Air Rhodesia airliners shot down by Nkomo forces in the past six months.

The attack on Lusaka lifted sagging morale among the 250,000 Rhodesian whites, some of whom compared it to the Israeli raid on Entebbe Airport three years ago. The military command gave few details, but reports from Lusaka indicated that the raiders were probably members of the Selous Scouts, a unit that operates independently of the army's main force, specializing in assaults across the border.

Witnesses in Lusaka said that some of the raiders entered the Zambian capital in Land-Rovers identical to those used by the Zambian Army, and bearing Zambian markings. Others were reported to have parachuted onto a golf course close to the Nkomo residence.

The attackers were described as blacks. Most members of the Selous Scouts are blacks, but whites blacken their faces for raids across the border. Witnesses said the raiders were dressed in Zambian uniforms.

Later, a Zambian official surveying the remains of Mr. Nkomo's bungalow pointed out that the raiders must have assume that their barrage of rockets, grenades and small-arms fire would kill the nationalist leader.

In recent months, Mr. Nkomo has been alternating his overnight stays between the bungalow and other houses in the Zambian capital. But he said after today's attack that he had been in the bungalow at the time, escaping during a 45-minute gun battle.

After wrecking the Nkomo home, the attackers destroyed several residences of other officials close to the guerrilla leader. They also destroyed Liberation Center, an office complex on Lusaka's southern outskirts that houses the Nkomo wing of the Patriotic Front, whose other wing is headed by Mr. Mugabe.

April 14, 1979

Muzorewa Party Wins in Rhodesia With Bare Majority in Parliament

Black Premier For Rhodesia

Abel Tendekayi Muzorewa

Special to The New York Times

SALISBURY, Rhodesia, April 24 — Almost 15 years after Prime Minister Ian D. Smith took over the leadership of their nation and embarked on a collision course with the world, Rhodesians today learned the identity of the black leader who will soon be charged with resolving the seemingly intractable guerrilla war that the white leader will leave behind.

Man in the News

In character and political style, Bishop Abel Muzorewa is a complete contrast to Mr. Smith. In place of the stubborn adherence to unfashionable convictions that has marked Mr. Smith's tenure, Bishop Muzorewa is viewed among many of his countrymen as a hesitant, sometimes vacillating figure whose capacity for toughness in difficult situations is open to doubt.

A Popular Following

The Bishop may have compounded those doubts tonight when he gave an interview to a local newspaper in which he admitted to having "mixed feelings" about the election victory that will install him as Prime Minister some time late next month or early in June. The Herald, the country's principal newspaper, quoted the Bishop as having said that he was "frightened at the immense responsibility" that the task would place on his shoulders.

Among whites, concerns about the 53-year-old prelate have been allayed by Mr. Smith's announcement that he intends to remain in the Government, in one of the Cabinet seats reserved for whites, until Britain and the United States recognize the black-led administration Bishop Muzorewa will head. Among blacks, Mr. Smith's plan has raised fresh doubts about the authenticity of the "black majority rule" Bishop Muzorewa has promised them.

Largely because of his image as a peacemaker, fostered by his position at the head of one of the country's smaller Methodist denominations, the Prime Minister-designate has attracted a popular following matched by few if

The New York Times/Maggie Steber
Bishop Abel Muzorewa in Salisbury after his victory.

any of his rivals in nationalist politics. But to the guerrilla leaders who will now be challenging him, the dapper cleric with the awkward public manner may well appear an easier opponent than Mr. Smith. "The little man should go back to his pulpit," Joshua Nkomo, one of the guerrilla leaders, said contemptuously last month.

From a childhood spent barefoot in the mountains along the Mozambique border, he seemed destined to spend his days as evangelist among the poor and illiterate people who constitute the bulk of tribal dwellers. From their thatched hut, his peasant parents sent him to the local mission school. From then on, for nearly 40 years, he was never far from religious studies and missionary work.

Nowadays, he traces the first stirrings of his political consciousness to an incident in 1957 when he heard Winston Field, later Prime Minister, make "a remark about Africans not deserving to go to heaven" in a speech to an audience that included black farm workers. But it was not until 1971 that he emerged as a political figure, after detained leaders like Mr. Nkomo saw his position as Bishop of the 55,000-member United Methodist Church as an ideal platform from which to lead opposition to unpopular constitutional proposals agreed upon by Britain and Prime Minister Smith.

When protests organized by the Bishop caused Britain to abandon the plan, he was launched on the path that will make him Mr. Smith's successor in the Prime Minister's whitewashed mansion here. Benefiting from his reputation as a peacemaker between warring factions of the nationalist movement and showing a personal ambition that surprised rivals like Mr. Nkomo, he built the organization developed during the 1971 campaign into the political party, the United African National Party, that swept last week's vote.

The biggest crowds in Rhodesian history — up to 250,000 — have flocked to his rallies in Salisbury, caring less about his unimpressive speechmaking than their sense that he could somehow end a war that has taken the lives of more than 14,000 people, mostly blacks. Although he has confessed to disliking clerical collars, he invariably wears one on public platforms, a powerful symbol in a nation that is one of the most heavily Christian in Africa.

On the most controversial issue in nationalist politics, the future role of whites, he has been more forthright than any other moderate. At rallies, he invariably regales blacks with his experiences in the black-ruled nations farther north, telling them how the departure of white communities has left the people without enough food and the hotels without soap and light bulbs.

Among the nationalists, he is the most flamboyant in his personal style. His wardrobe is filled with London-tailored suits, often offset with two-tone shoes. He likes to carry a fly-whisk, the symbol made famous by Jomo Kenyatta of Kenya, and lately he has added to his sartorial reputation by appearing in an array of vividly colored dashiki robes acquired elsewhere in Africa or specially tailored for him here.

In an interview that attracted wide attention here recently, he said that he meditates every morning at 6 A.M., and acknowledged an unusual caution about food, including a refusal to eat fish unless it is served with the head on.

"Otherwise," he said, "I just don't know whether it could be a snake, which tastes the same." Friends say he also worries a good deal about assassination, a constant threat in Rhodesian nationalist politics.

Abel Tendekayi Muzorewa was born on April 14, 1925, near the Old Umtali Mission, an American Methodist settlement.

From 1958 to 1963, he studied on a scholarship at Methodist colleges in Fayette, Mo., and Nashville, Tenn. He was consecrated Bishop of his church's Rhodesian wing in 1968. He and his wife have five children, including two sons who work as political aides and bodyguards. For recreation, the Bishop likes to relax on his farm at Dowa, near his birthplace, tending chickens and pigs.

April 25, 1979

Idi Amin: A Savior Who Became The Creator of 8 Years of Horror

By JOHN DARNTON
Special to The New York Times

KAMPALA, Uganda — When Tanzanian soldiers and Ugandan exiles marched into Kampala on April 11, they found the carnage left by the institutionalized brutality of a state gone insane.

At Makindye military-police barracks, the Tanzanians found concrete prison walls splattered with blood. Stashed on one side were crude instruments of death, including car axles, sledgehammers and machetes.

At the pink-stucco headquarters of the State Research Bureau, the dreaded secret police, they found underground cells packed with corpses of prisoners, slain in a final bloodletting by the fleeing captors.

Miraculously, a handful of prisoners survived; their jailers, who had long ago stopped feeding them, had forgotten they existed. The survivors had listened for days to the screams of fellow prisoners being strangled and hacked to death and had managed to live by drinking their own urine and eating dead bodies.

Absolute Hold Over His People

The horrors of Idi Amin's Uganda were the product of a megalomaniacal ruler who was hailed as a savior when he came to power in a coup in 1971 and who gained an absolute hold over his people.

From interviews with Ugandans who experienced President Amin's rule and from documents retrieved from the ransacked offices of his Government, significant features emerge about life under his eight-year regime.

Three Security Organizations

No one knows, and no one ever will know, the number of people who died under President Amin. Estimates from defecting ministers and from human-rights organizations like the International Commission of Jurists and Amnesty International range from 100,000 to 300,000. But they are no more than guesses.

Thousands perished in prison cells run by three separate security organizations that sprang up in 1971. But thousands more died in massacres by assassination squads and army units in remote villages.

Because Uganda, like much of Africa, has a traditional system of extended families, in which cousins twice removed are regarded as brothers, it is likely that by the end of the Amin regime almost every person in the country had felt the loss of a close relative.

President Amin tried to disassociate himself from the butchery. He was cautious in giving execution orders, speaking in northern dialects understood only by his henchmen, or in euphemisms, such as, "Give him the V.I.P. treatment."

But there is no doubt that the trail of blood leads directly to him. Even as the Tanzanian invaders were closing in on Kampala, a State Research Bureau squad was cruising northern towns in a yellow-and-green Mercedez minibus with an assassination list from the President's office.

But some Ugandans seemed to absolve the dictator and held officials around him responsible for the murders and economic chaos. This accounts for petitions from within the army and for numerous anonymous letters to the "Life President," pleading with him to step in and correct the situation.

Periodic Purges of Officials

It was an impression that President Amin fostered to remain in power, with periodic purges of top officials, promises of reform, "love and reconciliation," and displays of largesse.

"Up until the end," a Ugandan businessman said, "some people believed that he had the power to save them, if only they could get to him and explain what people were doing in his name."

When Mr. Amin came to power in January 1971, reportedly with the help of the British and the Israelis, he was a hero. The regime of Milton Obote that he overthrew had alienated much of the country with its increasingly oppressive policies, a proclaimed "move to the left" and the dismantling of four traditional kingdoms. These included the revered Kabakaship of the Baganda, Uganda's largest tribe.

"There was dancing in the streets," recalled Jack Kironde, a building contractor. "Everyone ran outside and rejoiced."

President Amin promised that his was a "caretaker government" that would soon return to the barracks. But within five months he had suspended political activities, abolished all rights of assembly and free expression and reinstated detention without trial.

Mass Killings of Officers

At the same time there were purges of army officers from the Acholi and Langi tribes, which President Amin regarded

Ugandan family at a refugee camp in Kenya where they are waiting for permission to return to their country.

as supporters of Mr. Obote, with mass killings in barracks at Makindye, Lira, Mbarara, Masindi and elsewhere. By the end of the year, the killings had spread into the lower ranks of the army and the civilian sector, with wholesale massacres in Acholi and Langi villages in the north. The rest of the country, unaffected, appeared unconcerned.

Rapidly, however, President Amin was transforming the army. Nubian recruits from the southern Sudan were trucked across the border with the lure of good wages and easy plunder. Top positions were also given to members of President Amin's own Kakwa tribe. The changes in the complexion of the army meant that all of Uganda's 28 tribes that did not come from President Amin's native West Nile Province came under the gun.

Three security organizations came into being. One was the military police. When it was taken over in late 1971 by Maj. Hussein Marella, it turned into a terror squad and its barracks at Makindye was nicknamed "The House of Death."

Most Feared Unit in Early Years

Another was the public safety unit, so called because it was set up to deal with armed robbers and empowered to shoot them on sight. Under Ali Towelli, like Major Marella a Nubian, it became the most feared unit in the early years. At its headquarters at Naguru, just outside Kampala, prisoners were bludgeoned to death in a courtyard in full view of a neighboring high-rise building.

The third, and most powerful of all, was the State Research Bureau. It was housed in Kampala's residential Nakasero Hill on a bougainvillea-draped street next to the President's lodge and connected to it by an underground tunnel. Under Farouk Minawa, it specialized in intelligence, infiltration, assassination and the extraction of information through torture. The French Ambassador, whose residence was next door, said that he could judge the political climate of the country from the screams he heard during the night.

It has been estimated that the three security organizations contained about 3,000 officers and agents at any one time. There was a rapid turnover, however, as graduates went on to run operations in the army and elsewhere, so that the total of the state's enforcers was many times that number.

In August 1972, President Amin appeared at a barracks at Tororo and announced to the troops that God had appeared to him in a dream the night before and ordered him to expel the 50,000 Asians in the country. The Asian exodus, which applied even to those holding Ugandan citizenship, ushered in the country's economic ruin. The re-allocation of property was chaotic, with shops being given over to illiterate soldiers.

Essential goods, such as salt, sugar, soap, cooking oil and gasoline, disappeared. Inflation galloped. More destructive still was the growth of "magendo," an all-inclusive term for corruption that includes kickbacks, profiteering, black marketing, hoarding and outright theft.

The example was set at the top: President Amin proudly drove a Cadillac and Mercedez-Benz that he seized from a multimillionaire Asian family. In the lower ranks, soldiers stopped Asians on the road and stripped them of their possessions. By mid-1973 the army turned the same tactics against other citizens. Houses of prominent officials who disap-

Camera Press

Former President Idi Amin

peared were taken by officers. Buses were stopped and their occupants beaten and robbed. To refuse could mean arrest or sometimes instant death.

As the spiral of lawlessness and violence grew, civilians began engaging in killings by proxy. They found that an anonymous letter or tip to a spy could eliminate a competitor.

"There were a lot of what might be called personal killings," said a doctor at Vulago Hospital. "If I was in the same trade as you, I would have you killed. Or if I wanted your car, or we shared a girlfriend, or we were up for the same Government post, someone makes an allegation against you of some sort and you just disappear."

Military Control Was Effective

President Amin's power base was the army, and in a country where citizens do not own guns, control through the military was effective. It came to an end only when the alliance of the military and Moslem northwestern tribes came apart in early 1978. That led to internal mutinies, which eventually prompted President Amin to search for an external enemy and to invade Tanzania last October.

The horrors of the regime were well publicized abroad, and they were known to most of its people. A blackout on news from foreign sources was clumsily enforced; banned periodicals, including Time and Newsweek, were obtainable under the counter. Thousands tuned in surreptitiously to nightly programs on the Voice of America, Radio Deutschwelle and, above all, the British Broadcasting Corporation.

The people learned to talk in code. "We developed a kind of language," said Father Joseph Bragotti, a missionary priest. "If someone said, 'We had a thunderstorm last night,' you knew there had been a lot of shooting."

"We knew much about what was going on," said Sister Mary Nives Kizito, a soft-spoken 34-year-old nun whose mission house is located down the street from the State Research Bureau.

Tales of Torture Documented

A few months ago, on a trip to Nairobi, Sister Nives picked up a copy of Drum magazine and read an account by four es-

caped soldiers of the tortures inflicted at the secret police center. But this did not prepare her for what she found shortly after Kampala was taken by the Tanzanians.

At 7:30 in the evening, she heard sobs and screams from the road outside. There, she found a 14-year-old boy, talking incoherently. Crawling behind him, naked and "reduced to a skeleton," was his father, who later died. They had emerged from two weeks' imprisonment in the center.

The next morning Sister Nevis and two other nuns went into the building. Stepping over corpses that blocked the passageways, they discovered the underground cells and the suriving prisoners. With the help of Tanzanian soldiers, nine prisoners were carried upstairs; three of them died on the way.

It was not necessary to live on the same street as the State Research Bureau to be well informed. Kampala, though a capital city, has only 350,000 residents. It is essentially a small town where news travels quickly along the sidewalks.

When a prominent official was abducted, frequently the license number of the car that carried him off, usually beginning with the "UVS" registration of the State Research Bureau, would be supplied by sympathetic witnesses. Telephone calls to well-connected officials would give a hint about his fate. If they went unanswered, it was a sign he would not return.

Most of Kampala's residents knew about the Namanve Forest, a Government reserve eight miles out of town that was used as a dumping ground for bodies. In Jinja, Uganda's second largest city, hundreds of workers who commuted daily over the Owen Falls Dam saw the remains of floating corpses that had been thrown to crocodiles.

The Government made desultory efforts to cover up reasons for the disappearances. President Amin, in long, rambling speeches, variously blamed guerrillas fighting for Mr. Obote, bandits or unnamed "imperialists."

The Government flaunted its arbitrary powers in things large and small. Mail censorship and phone tapping were carried out openly, with envelopes ripped apart and resealed and phone lines going dead as soon as talk became controversial.

The radio warned that people hiding guerrillas "will lose children and never see them again." The security squads performed abductions in daylight and in public.

Uganda's Chief Justice, who had released a British defendant for lack of evidence, was dragged off the bench of the High Court while horrified judges looked on. An eminent doctor, in whose house an anti-Amin soldier had taken refuge, was pulled out of an operating theater as he was performing surgery. A prominent politician was seized at a cocktail reception in the International Hotel and trundled down the stairs, screaming,

"They're going to kill me."

The vice chancellor of Makerere University disappeared, but only after it had been announced on the radio that he was missing. A former Minister of Public Works, whose arrest touched off a small riot when friends tried to prevent it, was reported by the radio as having fled to Tanzania.

Not surprisingly, the fact that such murders and kidnappings were so blatant increased their power to cow the people.

"It was difficult for anyone to put up any resistance because these people were ruthless," said Odongo Okino, an accountant in the Ministry of Agriculture. "Any time anyone tried, it led to more killings. I've seen the way these people kill. They enjoy it. They open you up like a fish."

Mr. Okino spent 115 days in Makindye prison in mid-1971. On his first night, he said, he peered through a peephole and saw 35 soldiers in an adjacent cell killed. "They herded them into one corner and shot them," he said. "Then they finished them off with pangas," African knives.

Mr. Okino, who was eventually released when a commanding officer agreed to check his record and found no offense written there, was asked how he managed to survive. He laughed and said: "Instead of talking, I developed a big beard." After his release, he said, "I kept myself quiet and out of sight."

Apolitical and Passive

The reaction was typical. Ugandans remained apolitical and passive. They sought spiritual refuge in the church; they viewed the Nubians as "foreign mercenaries" and acted like a people under occupation. "You had to conduct yourself in such a way as to survive," explained W. Senteza Kajubi, the vice chancellor of Makerere University.

There was no anti-Amin underground, no organized opposition, no anonymous pamphlets denouncing the regime. There was one anti-Government demonstration in eight years. It was launched by students from Makerere, after a student was shot and killed by a policeman who was trying to rape his friend. It was quickly broken up.

"We were like chickens of the Government," said one Ugandan. "If they fed us, we grow fat. If they want to kill us, they slit our throats."

Outsiders sometimes wonder why the people did not rise up against President Amin, or why no one managed to step out of a crowd and put a bullet through his head. The answer is complex.

Some Ugandans point to a cultural tradition of obedience and fealty, deeply ingrained from subservience to the ancient kings, some of whom celebrated their ascension to the throne by slaughtering hundreds of slaves.

Others cite the quirks of President Amin's own personality. Cunning and changeable, talented in the art of survival, he appeared invulnerable and almost omnipotent, especially to the thousands of Ugandans who believe in witchcraft and magic. He also had a genius for political showmanship.

"I remember a rally in Gulu," said Father Bragotti. "It was at the peak of the massacres. There must have been 10,000 people in this stadium who hated his guts. He drove in with only four soldiers and the place fell totally silent.

"They put on a war dance called the Otole. They came charging up to him with spears brandishing. I was sure that one of them was going to thrust it in his chest. Then he asked for a spear and a shield. He went into the dance with them, and the whole place just melted."

A third explanation lies in the capricious nature of the violence. When a man left home in the morning, there was no certainty that he would return in the evening. Law-abiding or not, rich or poor, all were vulnerable. Remaining "quiet" was no guarantee of safety. Cells were filled with prisoners who simply happened to be in the wrong place at the wrong time.

Prisons Reflected Outside Society

The prisons themselves were like miniature versions of the society outside, a universe of incomprehensible unpredictability. Some prisoners were marched outside to be shot, others to be released. While most died, some lived. They escaped when guards were drunk or they bought their way out. Sometimes, when a prisoner's family paid "ransom" for his release, he would be shot instead. Ugandans were living in a netherworld of rumors and horror stories.

"Over and over," said a local priest, "you would hear 'So and so is gone; so and so is dead; so and so has disappeared.' But you had no facts. Only rumors."

A Kampala businessman was asked what his lasting impression of the years under President Amin would be. "I was afraid all of the time," he replied immediately. He paused, then added, "There were no rules."

April 30, 1979

A Key to the Future for Ugandans: Restoring the Nation's Moral Fiber

By CAREY WINFREY
Special to The New York Times

KAMPALA, Uganda — Rich in people, minerals, climate and the fertility of the soil, Uganda stands a far better chance than most developing countries of rebounding from the eight years of terror, mismanagement, corruption and tribal favoritism under Idi Amin.

Whether it does will depend on the still undiagnosed dimensions of the wounds to the moral fiber of the nation of 12.5 million.

"Unless we succeed in moral reconstruction," said Andrew Adimola, Minister of Reconstruction and Rehabilitation in the new Government, "the physical reconstruction will be useless."

"The Ugandan is a certain type of African," said J. S. Magoba, who heads the United Nations Educational, Scientific and Cultural Organization's office here. "You slap one cheek, he will put up the other. Even in the worst days they walked the streets, saying hello, good day, how are you? We have lived under conditions worse than any other nation on earth. And we managed."

Virtually no one starves in Uganda. "You throw a seed and a few days later it's a plant," said Richard N. Posnett, a former British colonial administrator who is Acting British High Commissioner. Ninety percent of the people live in rural areas, eating what they raise.

"If this had not been the case," said Sam K. Sabagereka, the new Minister of Finance, "we would have had a major catastrophe on our hands."

Foreign Reserves Exhausted

Cotton and coffee used to provide substantial foreign exchange, but production of all cash crops — coffee, tea, sugar, cocoa and particularly cotton and tobacco — fell when President Amin stopped paying for them and farmers turned increasingly to growing food. Mr. Sabagereka is confident, however, that "the resources are still there and we could be exporting again within six months."

The sooner the better. Foreign reserves are nil. Because the Amin Government, which was ousted early this month, invested most of its reserves in a top-heavy and demonstrably ineffectual military establishment, the economy is bankrupt. Although vegetables have begun to appear in markets, there are almost no essential commodities.

Copper mining, which in 1965 accounted for $25 million in foreign exchange, plummeted when the dictator removed several managers. In addition, Uganda boasts largely undeveloped deposits of beryl, columbite, limestone, rock phosphate, tin and tungsten.

Because millions of dollars in currency were printed unsupported by reserves, inflation is estimated at 700 percent over the last year. A gallon of gasoline on the black market — the only place it can be bought — costs $120. Mr. Adimola, the Minister of Reconstruction, said the new Government would soon introduce a new currency.

"All Amin did in eight years of power," said the Rev. Joseph Bragotti, a Roman Catholic priest, "is bleed the country dry."

A civil servant commented: "Social services have gone down the drain. Industrial machinery has not been replaced or repaired and many factories are on the verge of collapse from sheer neglect." At Kampala's Mulago Hospital, equipment has not been replaced in the 16 years of its operation.

Estimates of how long it will take to get the economy moving again vary from six months to three years. Mr. Adimola predicted that it would take "one year to stand up and a second year before we start taking a few steps forward."

Mr. Sabagereka has announced that he will seek $2 billion in foreign aid. A member of the American delegation that visited Kampala said the requests he saw totaled $900 million, which seems within reach.

The United States, Britain, France, Canada and West Germany and a host of private relief agencies have expressed willingness to help. Private oil companies, which cut off supplies when Field Marshal Amin ran up a bill of $14 million, say they are willing to sell oil immediately and wait for back payment. The United States Congress is expected to repeal a trade embargo.

Economic recovery, necessary as it is, will count for little if spiritual recovery does not accompany it. "Magendo" — rampant corruption — and ethnic animosities remain.

"There will have to be a period of psychological rehabilitation," said Minister of Justice Daniel W. Nabudere. "The people will have to be re-educated. Democratic principles have to be redeveloped. Without democracy you can't talk of law, and without national unity there will be chaos."

From his bare office, where even the chair cushions were pilfered, Mr. Sabagereka, the Finance Minister, gestured at piles of rubbish on the streets, the residue of a wave of looting by Kampala residents after the capital fell to an invading force of Tanzanians and Ugandans opposed to the Amin dictatorship. "What you see on the surface is better than what you don't see," he said glumly.

Damage to the National Outlook

The head of the Ugandan Red Cross, Silvano Katuma, agreed. "Much more damage has been done to the outlook of the people themselves," he said. "Our scholars are no longer proud of their alma maters. Our tradesmen are no longer trustworthy. No one is proud of what he has been doing."

Tribal and religious animosities pose special problems for the future. Many go deep into the past. Uganda can be seen as two nations, north and south, divided by the Nile. To the north are the warrior animist tribes dominated by Acholis and Langis; to the south are the predominantly Christian Bantu people, dominated by 2.5 million Bagandas. North and south, Moslems constitute less than 10 percent of the population.

The Bagandas, who trace their roots to a highly developed kingdom in the 16th century, have long dominated intellectual and professional life, while the Acholis and Langis, with a warrior tradition, ran the army. Milton S. Obote, the first Prime Minister after independence from Britain in 1962, who was forced out of the presidency by Field Marshal Amin in 1971, is a Langi.

A Moslem from the small northern Kakwa tribe, Field Marshal Amin brought Kakwas and Nubians, many from the Sudan, with him into power. Because he hated and feared Acholis, Langis and Bagandas and distrusted Christians, he intensified the traditional hostilities by persecuting and killing them.

The number of recent revenge killings may never be known, but there is all-too-abundant evidence that retreating troops loyal to the dictator have massacred Bagandas, Acholis and Langis. As the anti-Amin forces advanced there were widespread reports of reprisal killings of Moslems, Nubians and Kakwas.

Yusufu K. Lule, the new President, in his inaugural address, appealed to Ugandans "not to indulge in evil acts of revenge against the regime we have just removed" and called for a return to sanity.

With calm repeated calls for restraint, national unity and the rule of law, the new Government has set a tone that most Ugandans have echoed with enthusiasm and relief. But the Government will ultimately be judged by its success in solving problems.

The potential for divisions within the leadership is worrisome. Formed at Moshi, Tanzania, only five weeks ago from a group of exiles embracing philosophies that range from Marxism to monarchism, the Government lacks a cohesive ideology.

Pretenders to power exist outside as well as inside the Government, not the least of whom is Dr. Obote, who controls some 2,000 Ugandan soldiers in the country from his exile in Tanzania.

While the Government has consistently played down the former President's importance — Mr. Lule characterizes him as one of 12 million Ugandans — few political analysts expect him to remain on the sidelines for long. "Obote will come back," a member of the Cabinet said. "He has supporters in the new Government." The timing of his arrival may determine the role he will play.

Mr. Lule, who is 67 years of age, has pledged elections within two years and, failing disaster, can probably rely on the support of most Ugandans for that time.

While the threat of a coup d'état exists, the general feeling is that it recedes with each passing day. While the new Minister of Commerce, Robert Serumaga, is optimistic, saying "Our best protection against dictatorship is that we have seen it with our own eyes," Father Bragotti, who remembers the rejoicing in the streets when Field Marshal Amin took over, tempers his hope with historical perspective. "If this Government doesn't work," he said, "a few years from now some general will come up and say, 'I'm going to save you,' and we'll be back where we started."

May 1, 1979

Hard Times Follow Ghana Coup

By CAREY WINFREY
Special to The New York Times

ACCRA, Ghana, July 25 — This has been a difficult summer for Ghana. The coup d'état on June 4 that overthrew Gen. Fred W. Akuffo has been followed by the arrest of 150 people on charges of corruption, the deportation of more than a dozen foreigners, and the execution of eight high-ranking military officers, including General Akuffo and two other former heads of state.

These events and a rash of little mutinies in some army units and police and fire departments have taken Ghana to the threshold of anarchy and jeopardized its long-awaited return to civilian rule. In an interview here, Flight Lieut. Jerry J. Rawlings, the half-Scottish leader of the coup who has become chairman of the ruling Armed Forces Revolutionary Council, insisted he would hand over the Government to elected civilians by Oct. 1. But Western diplomats estimate the chances of that happening at no better than 50-50.

Ghana's economy, stagnant before the coup, has further deteriorated. A revolutionary council decree that goods be sold at reduced, "controlled" prices has drastically limited the amount of food and other commodities in the marketplace. Undisciplined soldiers reinforce the edict by beating merchants they find selling goods at the old prices.

Lines stretch for blocks at stores selling bread and milk. Vehicles line up for as much as two miles to buy gasoline at the few stations still dispensing it.

And because of excoriations by Flight Lieutenant Rawlings, about half of Accra's 3,000 Lebanese, who virtually controlled the city's economy, have fled the country. A much-needed $22 million installment of an International Monetary Fund loan is now in question.

Most Are Unperturbed

But despite the turbulence, uncertainty and deprivation, most of the citizens of this drab city of overgrown lawns, peeling paint and potholed streets are going about their business seemingly unperturbed, even with renewed vitality and patriotism.

For in spite of a witch-hunt atmosphere in which businessmen, bankers, civil servants and the owners of expensive automobiles are all suspect, and many fear for their lives, Flight Lieutenant Rawlings's campaign to eradicate corruption through a "moral revolution" appears to have won wide favor among an overwhelming majority of the people. At the time of the executions, crowds gathered in the streets to chant, "Kill, kill, kill." And when the 32-year-old pilot, who is blessed with swashbuckling good looks, appears in public, he is mobbed and cheered.

"All the popular support you currently enjoy stems not just from the fact you have succeeded," read an open letter to the flight lieutenant in an Accra newspaper the other day, "but from the fact that the entire country is agreed that the thieves should not be allowed to get away with the booty."

Western diplomats here say Flight Lieutenant Rawlings is idealistic and well intentioned, if politically unsophisticated. They also consider him the most moderate of the 15 members of the revolutionary council, which includes seven noncommissioned officers and one private. Flight Lieutenant Rawlings acknowledged in the interview that he approved the executions of his former superiors only after his own life was threatened by other members of the council.

Judgment Is Questioned

While few doubt his sincerity, many question his judgment. One diplomat called his program "naïve populism" and described the flight lieutenant as "a real romantic idealist."

"He's a naïve, honest guy who says poor people have been victimized and it's time to spread the bounty. How to do it? He hasn't a clue."

And a friend of Flight Lieutenant Rawlings said: "I know Jerry is honest. I just wish he knew that being honest is not enough."

Both diplomats and Ghanaian intellectuals say the situation remains so volatile that anything can happen, including another coup d'état.

There is concern over how much control Flight Lieutenant Rawlings exerts within the revolutionary council. There are rumors of council dissatisfaction with him, growing factionalism and increasing worry among the members about what will happen to them once they no longer control the levers of power.

Motives of Others Questioned

"If it were just Jerry, I wouldn't worry," his friend said. "But there are others seeing themselves on the front pages, their names mentioned. They're the ones I'm worried about. They're beginning to believe the people who tell them they're the best thing that ever happened to Ghana."

It was in the still hour before dawn on Monday eight weeks ago that a handful of enlisted men overpowered guards at a military jail where Flight Lieutenant Rawlings was being held on charges of leading an abortive coup three weeks earlier.

A little after 6 A.M., astonished radio listeners heard the breathless pilot proclaim he had taken over the government to cleanse it, but that civilian elections scheduled for June 18 would go ahead as planned. Before the day was out, at least 13 people, including the Army Chief of Staff, would die in skirmishes.

A few days later, the revolutionary council announced that the return to civilian rule, scheduled for the first of July, would be postponed for three months to allow for a "housecleaning." In a runoff election this month, Hilla Limann, a former Foreign Service officer, defeated Victor Owusu for the presidency. In a

Associated Press

Flight Lieut. Jerry J. Rawlings

separate interview, Mr. Limann said that he did not doubt that he would be inaugurated Oct. 1 and he vowed "to continue the cleaning-up process."

Trials Are Condemned

The hasty, secret trials and executions that followed the takeover brought condemnation and appeals for mercy from Secretary General Kurt Waldheim of the United Nations and from many countries.

Nigeria, whose military leaders also plan to hand over their government to civilians on Oct. 1, registered its displeasure by halting oil deliveries to Ghana. As a result, more than half of Accra's motor vehicles are now out of circulation.

"The few positive economic steps taken under Akuffo — the tightening of the budget, the I.M.F. loan, — have been reversed," said an American economist here. On the other hand, he pointed out, "prices have come down, people are queuing up to pay their taxes, merchants are voluntarily reducing their prices."

In an hour-long interview conducted in his office at Burma Camp, an army base a few miles outside Accra, Flight Lieutenant Rawlings insisted repeatedly that the June 4 takeover averted an uprising that would have cost many more lives. "What happened in Iran could have happened here," he said. "A potentially explosive situation existed."

He said that even if the revolutionary council could not accomplish all it had set out to do by Oct. 1, it would still hand the

government over to Mr. Limann and other elected civilians. "We can only hope," he added, "that we would have laid the foundation that would make it possible for the incoming government to continue, to see the necessity of having to continue the cleanup."

Flight Lieutenant Rawlings said there would be no more executions and hinted that some of the life sentences might be reduced once the public hunger for vengeance abated.

He denied that he had singled out the Lebanese community for special persecution. "We've got good Lebanese, hardworking, honest Lebanese citizens in this country, and they can stay and continue with their honest work," he said. "But there are corrupt ones who are going to be dealt with. There are corrupt Ghanaian citizens who are going to be dealt with so it is not just a question of Ghanaians and Lebanese. We are trying to restore something very healthy in this country which will make it impossible for this kind of blatant exploitation to continue."

Asked if the price of that restoration might be more violence and instability, Flight Lieutenant Rawlings replied that it would be "a mistake to think we don't question the morality of having to run a country using the barrel of the gun." "None of us like this responsibility."

"We are not crazy," he added, his deep voice rising. "We are not mad. We are very sane people and know what is right and what is wrong."

The flight lieutenant, who wore an open-necked blue air force shirt over tailored blue trousers, toyed with silveredged aviator glasses and spoke with passion. He said his reference to Ghana going "the Ethiopian way" in his June 4 broadcast did not mean he admired either Marxism or Communism. "I wouldn't want to use those words because I don't know much about them."

"All I know," he said, "is that all of you are on the moon. You are using the capitalist system, they're using the Communist system, but you're both on the moon. So it means that down the line there must be something in common and I think it is some degree of efficiency, selflessness, dedication toward a cause."

Then leaning forward and gesturing with his hand for emphasis, he added: "This is what's been taken away from us. We've got to get it back. And it hinges on morality and values, the restoration of the values that we've known."

July 30, 1979

Visiting Zulu Leader Has Evoked Black Hostility

By LINDA CHARLTON

Gatsha Buthelezi, a black South African political leader, is not popular with the ruling white regime, but he is the chief minister of a Zulu "tribal homeland" that the Government is trying to persuade him to accept as an independent state.

"They're caught," said a laughing Chief Buthelezi, explaining that if he was "head of a state which is going to be independent," they must let him have a passport.

Chief Buthelezi, who is on his eighth visit to the United States, has consistently refused to accept "homeland independ-

ence" for the fragmented Zulu territories, holding out for majority rule in the country as a whole. While this has not endeared him to the Government, his preference for working for change from within has earned him the hostility of other black Africans.

He said in a speech to the Foreign

Policy Association in New York City yesterday that his espousal of foreign investment in South Africa has also brought him criticism from some church groups and liberals in the United States. His reply, he said, is that his people must survive.

'Live to Struggle for Justice'

In a 1977 speech on investment, given at the opening of a chemical plant, he said: "We accept investment because we need jobs so that we and our children may live to struggle for justice."

Chief Buthelezi criticized Western "meddling" in southern African affairs, later defining the word as "anything the West tries to impose on us." In his speech yesterday, he said: "Blacks in South Africa are not the pawns of games played by the Western Five. South Africa isn't a major piece on a board with which to maneuver Communism into a stalemate."

He said it was the ordinary black people of South Africa, not "international strategists," who would determine that nation's future. "Because an apartheid regime has rendered us voiceless," he said, "the world behaves as though we are powerless and inconsequential."

The 50-year-old chief is the president of Inkatha, a "liberation movement" that he says has nearly 300,000 members, and his speech stressed the "dominant role" of the group in giving expression to "the aspirations of ordinary people."

'Shattered' by Young Affair

In an interview after his speech, he was asked if he envisioned himself eventually as the leader of a multiracial South Africa under majority rule. He said "not necessarily," but he appears to have some political ambitions.

Chief Buthelezi said he was "really shattered" by the recent resignation of Andrew Young, the United States representative at the United Nations. He said Mr. Young "had given credibility to your foreign policy, had won the hearts of blacks in Africa for America" and "spoke in such forthright terms." But he said it did not matter whether Mr. Young's successor was black or white, "if he builds on his framework."

Chief Buthelezi calls himself a pragmatist and a realist. Speaking of the rights of white South Africans, he said, "They have just as much right as Americans have to be here. It's a fair truth." And he believes in working with whites who oppose the Government's apartheid policy. "One can't spurn their hands," he said.

"Foundations built now across the color lines," he said, "will help prevent ultimate race war, or at least limit the violence." He said the the questions he is asked most often concern the inevitability of a bloodbath and that he has no absolute answers, except that a state of war already exists. "A process of escalation is all that we are talking about now," he said, later adding that this escalation will mean open warfare but that "I don't have to be paralyzed until that day comes."

He said he wa committed to nonviolence, until it comes to accepting the fragmented Zulu territories as an independent homeland. "If they force at the point of a gun," he sid, "then nonviolence, which I espouse, will cease to be a noble cause."

August 23, 1979

Zimbabwe to Be Nation's Name

SALISBURY, Zimbabwe Rhodesia, Aug. 26 (Reuters) — Prime Minister Abel T. Muzorewa has announced that Rhodesia will be dropped from his country's name before the constitutional talks begin in London on Sept. 10.

The country will then become known as Zimbabwe, Bishop Muzorewa said yesterday at a rally in Fort Victoria.

The announcement means that for the first time since 1890, all reference to Cecil John Rhodes, the British explorer who organized the first white settlements, will be eradicated from the region's name.

Zimbabwe has long been the name used by African nationalists. The name is taken from the ancient Zimbabwe ruins, regarded as a symbol of the country's African past before the arrival of white settlers.

Mr. Smith and Bishop Muzorewa and other black leaders agreed on the name Zimbabwe Rhodesia as part of a compromise on the new constitution that brought black majority rule to the country this spring.

August 27, 1979

Oil Price Rises Put Kenya's Economy in Jeopardy

By CAREY WINFREY
Special to The New York Times

NAIROBI, Kenya, Aug. 19 — "The Arabs tried to make slaves out of us 500 years ago," a Kenyan journalist said here the other day, "and now they're trying to do it again."

His hyperbole was aimed at oil price increases and their likely effect on Kenya's economy. In eight months, two increases by the oil producing countries have nearly doubled the price of a barrel of oil delivered from the Persian Gulf to Kenya's port city of Mombasa. The average price last December: $12.70. The average price today: $23.50.

The new costs mean that a third of Kenya's foreign exchange earnings must now be spent on oil, as opposed to a quarter a year ago, thus deferring badly needed rural development projects designed to coax more food out of marginal soil, build new roads and schools and bring electricity to vast regions of the country.

A Government economist pointed to new statistics that predict a $2.5 billion budget deficit over the next five years and a decline in per capita income. "We are more than worried," he said. "Say what you want, it's bad." Referring to a red volume fresh from the bookbinder that bore the title "Development Plan 1979-1983," he added, "This is already out of date."

The plan, the fourth since independence in 1963, focuses on resettling African farmers on unused or mismanaged land and on major development programs in water and agriculture. Although Kenya remains an agricultural country, only about 10 percent of its total land area has high potential, with another 5 percent suitable only for raising animals.

The economist estimated that about 30 percent of the programs in the new four-year plan would have to be scrapped, though he and others say it is too early to know which ones will be sacrificed.

Because of a quadrupling of prices by the Organization of Petroleum Exporting Countries in 1973, the third four-year development plan, published that year, was also out of date before the ink was dry. Only about 40 percent of the projects outlined in the third plan, most of which dealt with agricultural development, were ever realized.

Like most of Africa south of the Sahara, Nigeria being the outstanding exception, Kenya imports all of its oil, the source of 80 percent of its energy. This year Kenya will use 12 million barrels, up 6 percent over last year.

More Expenditure Than Income

Also like most African countries, Kenya spends more than it earns. Because of unseasonal rains and a drop in world prices, a sharp decline in income from coffee and tea cut foreign exchange earnings by 40 percent. While income went down, spending on defense and energy went up. In one year, Kenya's balance of payments plummeted from a $286 million surplus to a $210 million deficit.

"The balance of payments gap is going to get a lot worse," says an American economist here. "Money they would have been putting into development they're now going to use to pay the Arabs." Or, as Mwai Kibaki, Kenya's Vice President and Finance Minister, said in a speech here recently, the more that is spent on oil "means less for everything else."

"For every dollar increase in the price of a barrel of oil," says Tom S. Tuschak, a Canadian on loan to the Government as an adviser on energy, "Kenya would have to increase exports by one and a half percent just to stay even."

For coffee and tea, which bring in $60 of every $100 earned abroad, Mr. Tuschak's formula would mean production increases of about 30 percent. But coffee and tea producers say that because of the excessive rains earlier this year they will be lucky to match last year's low levels, much less increase production.

The affect of oil price increases on tourism, Kenya's other major foreign exchange earner — it brought in $135 million last year — is no more encouraging. Higher oil prices mean fewer travelers, higher air fares and increased costs for going on safari.

Lacking coal or oil deposits, Kenya must import some 90 percent of its energy needs, including hydroelectric power from Uganda. Only 20 percent of the oil Kenya imports is used in private automobiles. The rest is absorbed by essential services, transportation of goods and by industry and the armed forces. The logical cutback, in oil used by the armed forces, loses support the higher one goes in the Government.

Kenya has taken some measures to conserve foreign exchange and to cut fuel consumption. Foreigners are forbidden to purchase plane tickets with local currency; all importers must deposit a fee of from 40 percent to 150 percent of the value of the goods to be brought in, and vehicle registration fees have been raised proportionately.

But Western economists here say Kenya could be doing more, particularly by passing more of the energy costs along to the consumer. Although the price of crude oil has gone up almost 90 percent since the beginning of the year, the price of a gallon of gasoline in Kenya has gone up only 20 percent, to about $1.80 a gallon.

Tanzania Introduces Rationing

In neighboring Tanzania last week, President Julius K. Nyerere introduced gasoline rationing, allocating private vehicles between 15 and 18 gallons a week based on the number of passengers.

"Pricing in Kenya," says an economist here, "has not been totally cognizant of these problems."

Another economist points to a pervasive faith shared by many Government officials that as long as the country remains politically stable and run along capitalist lines, Western nations will bail it out financially.

Kenya receives more than $200 million in aid each year from the United States, Britain, West Germany and the Scandinavian countries, and loans from the International Bank for Reconstruction and Development, the International Monetary Fund and many nations. But in April Kenya was forced for the first time since independence to go to the European commercial market to close the gap on a more than $2 billion annual budget. An agreement to borrow $200 million from a consortium of European and Japanese banks at an interest rate of 14 percent was signed last month.

Next month, President Daniel arap Moi will take his quest for aid to Saudi Arabia. Although the purpose of the trip is to secure a low-interest loan, it is being billed as a mission of "good will," a commodity in short supply here at the moment.

As Joe Kadhi, writing in The Sunday Nation of Nairobi, said the other day: "Among the members of the neck-strangling club called OPEC, are our 'brothers' from Arab countries and other African nations such as Nigeria. Whenever we meet their brothers in other forums, such as the Organization of African Unity, they do everything possible to make us support their cause of backing the Palestinians. But apparently when they meet in their other club — OPEC — they tend to forget our brotherhood."

August 27, 1979

South African Might Review Ban On Interracial Sex and Marriage

CAPE TOWN, Sept. 26 (Reuters) — Prime Minister P. W. Botha has offered to consider changing the South African laws barring sex or marriage across the color line.

His statement startled politicians and churchmen today. The Immorality Act and the Mixed Marriages Act are widely regarded as pillars of the apartheid policy of separation of the races, and Mr. Botha is bound to face stiff opposition to changes in them from a powerful right-wing faction in the Government that has vowed to block any measures encouraging racial mixing.

But at a congress of his ruling National Party last night, Mr. Botha said the Government was prepared to consider suggestions for changing the acts. "No act can be regarded as a sacred cow," he said. "I will not tolerate any laws on the statute book that insult people."

'Amenable to the Suggestion'

Although there were biblical examples of mixed marriages, the Prime Minister said, from a practical point of view they remained undesirable. "But I concede there is a problem where people really love each other and want to get married," he added.

Turning to the Immorality Act, which outlaws sexual intercourse between people of different races, Mr. Botha said:

"Immorality of any sort must be fought in a Christian country.

"If we can improve the Immorality Act," he said, "not only in connection with people of different races but in all areas, I will be amenable to the suggestion."

More than 10,000 people are estimated to have been convicted under the Im-

morality Act since 1950. Policemen acting on tips from neighbors stage regular raids on mixed couples, forcing their way into bedrooms to obtain evidence. Dozens of couples are convicted annually, and some are sentenced to jail terms.

Leaders of the Dutch Reformed Church, the Methodist Church and the Council of Churches welcomed Mr. Botha's remarks, which follow a series of moves to improve race relations in South Africa while maintaining the basic principles of separate development. Yesterday the Government granted union rights to all black workers in a gesture to the country's voteless majority.

Mr. Botha is the first Prime Minister to question the laws that make sexual relations between the races a crime. Two years ago John Vorster, Mr. Botha's predecessor as Prime Minister, vowed that the statutes would remain after a Government panel warned that their abolition would lead to what it called "an orgy of sexual excess."

September 27, 1979

89

London Negotiators See New Hope for a Rhodesia Pact

By WILLIAM BORDERS
Special to The New York Times

LONDON, Sept. 26 — As British negotiators met again today with representatives of the warring factions in Zimbabwe Rhodesia to hammer out complex constitutional changes, a new mood of optimism seemed to have overtaken the two-week-old peace conference here.

Although all sides continued to stress the significant difficulties that lay ahead, leaders of the three delegations, the Zimbabwe Rhodesians and the Popular Front guerrilla alliance as well as the British, almost simultaneously made more hopeful statements than before during the conference, the ninth major diplomatic attempt to solve the Rhodesian problem since the colony declared independence from Britain in 1965.

Bishop Abel T. Muzorewa, the black who was recently elected Prime Minister of the war-torn country, said that the conference had already been "a triumph for democracy," and Health Minister David Zamchiya, a member of his delegation, hailed the "substantial progress" at the meetings today, saying: "If the present pace is continued, we should be able to conclude this part of our exercise by the end of the week."

Mugabe: Somewhat Optimistic

At the United Nations in New York, Lord Carrington, the British Foreign Secretary, said, "I dare to hope that the moment may not be too far away when the British Government will be able to grant legal independence to Rhodesia," and predicted that before too long it would be joining the world organization.

Even Robert Mugabe, co-leader of the Patriotic Front, said, "I am optimistic, to some extent."

When the conference opened Sept. 10 there were widespread predictions that the front might pull out of the talks at an early stage on the ground that they were irrelevant. But in a radio interview here today Mr. Mugabe echoed the sentiment of the other particpants with this assessment: "If we continue as we have started, the possibility of reaching a settlement is there."

Referring to the complex diplomatic matters yet unresolved, Mr. Mugabe added that he would be more optimistic after the start of discussions of an interim arrangement on the transition from the present Government to whatever form of government emerges from the conference.

The guerrilla front, which has engaged the Salisbury Government in a long and costly war, has consistently taken the position that the real work of the conference will be in setting the rules for the transition, not in revising the Constitution. In particular, sharp differences are expected on the question of control of the armed forces during the transition, a matter especially important to the white minority, whose members are already emigrating at the rate of a thousand a month and who might leave in a flood if they felt less safe than they do now.

The Patriotic Front dismissed out of hand a cease-fire appeal yesterday by Bishop Muzorewa's Government, made on the ground that the two sides had reached broad agreement on the general principles of a new constitution, After Foreign Minister David Mukome said that "the people of Zimbabwe, who are dying daily on account of the front's atrocities, would be happy to see the front stop this senseless procrastination," Mr. Mugabe replied: "We didn't come here for verbal exchanges. This isn't the time for scoring points."

In Mr. Mugabe's radio interview, with the British Broadcasting Corporation, he acknowledged his potential rivalry with Joshua Nkomo, co-leader of the Front. The two men are in London as equals, sharing the chairmanship of the front's delegation, but in the event of new elections Rhodesia, one of them would presumably have to be selected as leader. Asked about this, Mr. Mugabe replied: "I am prepared to be what the people want me to be. I am prepared to play whatever role the people assign to me."

September 27, 1979

Parley Worries White Rhodesian Farmers

By CAREY WINFREY
Special to The New York Times

BEATRICE, Zimbabwe Rhodesia, Sept. 26 — Collin J. Cook came home here yesterday after three weeks he spent in Britain for a medical checkup and a close look at the talks that may determine his future.

Today he inspected the wheat on his 3,400-acre farm, checked on the progress of construction of five handsome new brick pigstys and worried aloud about developments at the constitutional conference being held in London.

"I'm very pessimistic about the outcome," he said, expressing a view widely shared by the 160 other white farming families surrounding this pastoral community about 35 miles south of Salisbury, the capital. "I just can't see Carrington achieving a compromise between people with avowed Communistic feelings and those who believe in democratic goverment." Lord Carrington, the British Foreign Secretary, is presiding at the conference.

Like most whites in Zimbabwe Rhodesia, the 54-year-old farmer is disappointed in Prime Minister Abel T. Muzorewa, not so much for his acceding to British proposals to reduce the power of the white minority as for seeming to do so without exacting concessions in return. He calls Bishop Muzorewa's actions "appeasement."

'A Very Great Mistake'

"I think he's made a very great mistake," Mr. Cook said as he surveyed a seamless green expanse of wheat. "He put all his cards on the table. Whatever game you play, poker, gin, you don't do that. He's made concessions he can't now retract."

He added that he thought Bishop Muzorewa should have agreed to the reduction of the number of seats reserved for whites in Parliament from 28 to 20 only if the British would agree to lift economic sanctions.

"We've reached a state where the economy is really suffering," said the British-born architect-turned-farmer, who came here in 1953. He explained that sanctions had reduced his income by one-third and had driven the cost of imported implements up to five times what they were a few years ago. The farm operates on a profit margin of about 8 percent. "Unless something is done quickly," he asserted, "there will be a great outflow of Europeans from the country."

In the last two years, eight white families have left Beatrice following the ambush killings of one couple by members of the Patriotic Front guerrilla group. In addition, the guerrillas burned down the houses of two other families and made rocket attacks on three others.

Angry Reaction Recalled

"After one of the attacks," Mr. Cook said with a quiet smile, "they left a note that said they were going to kill us all and rename the street 'blood river road.' That made most of us just want to get out and eliminate them."

When he was asked about the Smith & Wesson pistol he wears at his waist, his eyes twinkled. "I hope I kill a 'terr' or two before I'm finished," he said. "Terr" is an abbreviated form of "terrorist."

While Mr. Cook and others say they are willing to make concessions to achieve economic relief through the lifting of sanctions, he draws the line at what he considers questions of survival. "If it means giving up the white control of the armed forces and the police, I say 'no,'" he said. "You can't compromise your police security with the P.F."

He said he was convinced that the neighboring "front-line states" — Tanzania, Botswana, Angola, Zambia and Mozambique — were trying to force a Communist government on Zimbabwe Rhodesia and then use this country as a "jumping-off board for the same thing to happen in South Africa."

He said that if the Patriotic Front was given any kind of role in the Government, he and his wife and their three children,

two of them grown, would have no choice but to leave the country, even though that would mean forfeiting a quarter of a century of work and a farm worth hundreds of thousands of dollars. "They have said their avowed intention is that whites will have no further say in the government, in the armed forces, and that they're going to take over the land," he explained. "So how can we stay?"

He and his fellow farmers are also dubious about the prospects for any new

election, which under the British proposals would follow a cease-fire in the war. He insists that it was the white farmers

who made possible the election that put Prime Minister Muzorewa into power this year by organizing farm workers and local people in the countryside, assuring them there would be no repercussions if they voted. They will not do so again, Mr. Cook declared.

September 27, 1979

A Quiet Leader For Nigerians

Shehu Shagari

By PRANAY B. GUPTE
Special to The New York Times

LAGOS, Nigeria, Oct. 3 — Nigeria's new civilian President, the first after more than 13 years of military rule, prides himself on being a "contact person," someone who meets scores of ordinary people every day and inquires about their troubles.

Man in the News That, associates say, is among the problems that 54-year-old Shehu Usman Aliyu Shagari will now have to cope with. Noting that he spent more than 14 hours on Sunday, the day before his inauguration, receiving visitors, many of them peasants from distant parts of the country, they say that as President Mr. Shagari will have to learn to budget his time better.

But since the ceremonies on Monday that formally made him President, there has been no letup in the stream of visitors, although for the moment at least they have been of a different sort — politicians from his National Party of Nigeria and applicants for Government positions. Yesterday, his first working day as President, Mr. Shagari, who described himself as the "reluctant candidate" during the campaign, dealt with visiting politicians for more than 12 hours.

The President, a slight man who wears eyeglasses and tends to gesticulate as he speaks, is described by associates as displaying calm and confidence in both private and public regardless of the problems he faces. And the tasks before him, they say, are enormous.

Party Far Short of Majority

At the head of the list is how to get a legislative program through a new American-style Senate, where his party holds 36 of the 95 seats, and through the 449-member House of Rep-

Associated Press

Has always considered himself a conciliator and a healer.

resentatives, where only 168 members support him.

Mr. Shagari — pronounced sha-GA-ri — has received informal pledges of support from others, but the opposition to him is expected to be led by the Unity Party of Chief Obafemi Awolowo, who had sought through court action to block him from assuming office, charging that the legal conditions for his election had not been met.

Expressing hope that he and the legislators can work together, Mr. Shagari said at a news conference, "The President has the whole nation as his constituency, and he should be above any petty differences between the various sections of the community."

His problems include how to meet the rising expectations of the 80 million Nigerian people, most of whom are poor even though the nation is rich in oil.

Mr. Shagari has let it be known that he is keenly aware that he is taking office at a time when some people abroad, especially in the West, are skeptical that a civilian Government

can succeed in running a country where regional and ethnic differences have sometimes threatened to smash national unity, as in the period from 1967 to 1970 when a Biafran secessionist movement was put down.

"I think it is high time Nigerians sat down together to build a stable nation," he said recently.

Mr. Shagari says that he was "pressured" into running for President by associates who cited his experience in previous Governments as Minister of Public Works, Finance, Economic Development and Internal Affairs.

For the last three years he served as chairman of Peugeot Automobile Nigeria Ltd., a post to which he was appointed by the military administration of Lieut. Gen. Olusegun Obasanjo.

Reared by His Older Brother

Shehu Shagari was born on April 25, 1925, in the northern state of Sokoto. His father, Magaji Aliyu, died when Shehu Shagari was 5 years old, and the future President was reared by his older brother, Muhammadu Bello. Mr. Shagari was educated at Kaduna College, where he received a bachelor's degree in science, and he subsequently went to Mecca, the Islamic holy city in Saudi Arabia. A Sunni Moslem, he has made the trip several times since.

After graduation from college, he taught science for 13 years in various schools and became increasingly involved in local politics. In 1954, he won a seat in the federal House of Representatives and in 1959, a year before Nigeria's independence from Britain, he was named Parliamentary Secretary to the Nigerian Prime Minister.

Besides his service in government, Mr. Shagari also was a member of the assembly that drafted the American-style Constitution that went into effect on Monday.

The new President is known as an intensely private family man. His Government biography notes only that he was married in 1946 and that he has five children.

He has been acclaimed by literary critics in Nigeria and elsewhere in West Africa for his poetry, but this is available only in the Hausa language.

October 4, 1979

There's Still a Uganda

By Jonathan Power

LONDON — In the 1880's, the British fought tooth and nail to get hold of Uganda, the point of entry to the head waters of the Nile. Now nobody wants it. There are no Cuban soldiers there, no Russian technicians. If ever there was the archetypal "not-wanted," "no-strategic-value" nation, Uganda is it. As far as 99 percent of the world is concerned, Uganda can rot in its Tanzanian juice. The great hurrah that went up when Idi Amin was finally toppled by invading Tanzanian troops in April has subsided into an embarrassed murmur.

The United Nations High Commission for Refugees appealed for $13 million to help resettle the quarter of a million displaced persons and refugees. So far, only three donations have been made, by Denmark, Switzerland and the Netherlands, totaling $750,000. Yet the Commission's appeal last year for aid for the more strategically important Ogaden, after the Somali-Ethiopian war, and its appeal this year for aid to Indochina have been generously met.

Britain has offered an emergency grant of $4 million, a paltry sum that compares unfavorably with what Britain has saved since it cut off its aid to Amin. The West Germans, French, Dutch, European Economic Community and Organization of Petroleum Exporting Countries have all announced modest aid programs. Little if any of this has been actually spent.

The United States until last month was constrained by legislation outlawing aid to Uganda. The law has now been repealed and an aid appropriation of $6 million has been announced.

The Soviet Union has offered a couple of hundred scholarships but no cash.

The American and Soviet cases aside, the reason given for the go-slow is the anarchy inside Uganda. Aid cannot be used in a situation where looting is still rife, where the police force cannot stop random murders, where Government ministers don't know what their colleagues or subordinates are up to, where even the President, Godfrey L. Binaisa, has only a rough idea of what is going on.

The caution thus makes a certain amount of sense. Yet the logic of it leads one into a classic Catch-22 situation. For without aid, there can be no effective, well-trained police force, without emergency relief there can be no coming to grips with the destabilizing refugee crisis, and without a sense that outside countries are prepared to support the post-Amin regime, it is all too easy for internal forces that do not like the Binaisa Government to undermine it.

What Uganda needs now more than anything else is an injection of outside confidence. How could this be done? First and foremost is the need to establish law and order. In the corridors of the Lusaka, Zambia, conference of the Commonwealth heads of government in August, the Tanzanians and Kenyans were touting the idea of a Commonwealth defense force to replace the Tanzanian conquering army. The Tanzanians frankly acknowledge that their troops are not behaving particularly well in their role of occupiers. The Kenyans, Uganda's other principal neighbors, are convinced that the longer the Tanzanians stay, the surer becomes the return to power of Milton Obote, whom Idi Amin overthrew. Mr.

Obote, they have convinced themselves, would join with Tanzania in organizing an anti-Kenya socialist East African front.

Despite the odd confluence of opinion, none of the other Commonwealth countries were interested. The Nigerians, the most likely African candidates, who only recently had their fingers burned when they sent peacekeeping troops into Chad, said no. The British, who in 1964 sent in troops to help Mr. Obote put down an army mutiny, also refused.

It was an opportunity missed. Here was the chance to show the world the value of benign military intervention: going in when invited to avoid political disintegration and perhaps the likelihood of a local military buildup at a later date.

Law and order aside, Uganda needs to come to grips with its omnipresent black market. While basic commodities are scarce the ripoff merchants work their will. A recent study made by the Commonwealth Secretariat estimated that to buy the essential goods to rebuild the economy and undermine the black market, $900 million was needed over a six-month period. With less than $150 million promised and little of that delivered, the gap between need and practice is about as wide as the Nile is long.

Left to its own devices the only course for Uganda is downhill. No outside experts, no foreign businessmen, no one but the brave and balmy will be prepared to go to Uganda and help. The West, its governments and its news media spent eight years denouncing Idi Amin from the safety of the sidelines. Now that he's gone, they should make sure the chaos he created goes too.

Jonathan Power is a columnist for The International Herald Tribune.

October 5, 1979

Ghana Looking To New Leader For Better Life

By PRANAY B. GUPTE
Special to The New York Times

ACCRA, Ghana, Oct. 7 — Jamestown is a neighborhood like dozens of others in this West African city. Small but tidy shacks are squeezed behind narrow, rutted roads, and in scores of tiny stalls mounds of smoked fish and kenkey, a delicious snack made of ground corn, are displayed for sale.

Such neighborhoods far outnumber the enclaves of affluence in Accra. It is in places like Jamestown that most is being expected of the country's first civilian Government after seven years of military rule, and it could well be that in neighborhoods like these that the life span of President Hilla Limann's administration will be determined.

"We all have very high hopes for Limann," said Victoria Yanke, a young mother of two, as she kneaded dough for bread that she would sell to customers who form long lines outside her home on Zongo Street. "The question we have is whether he will get more flour for us, and more food. Will we get electricity in our homes?"

These are not questions that President Limann and his People's National Party, which won 71 of the 140 seats in Parliament, can answer easily. The inaugura-

tion of the Government on Sept. 24 has opened the floodgates of popular expectations.

Income Declines Sharply

Dr. Limann, a former diplomat, is taking over a country of 10 million people whose average income is declining alarmingly. The World Bank recently revised Ghana's per-capita income down to $380, from about $450. There was virtually no growth in the gross national product last year. Cocoa production, the mainstay of the economy, is expected to fall to a record low of 265,000 tons this year. As recently as a decade ago, Ghana produced 500,000 tons annually.

"The unfair thing about all this is that everybody, rather unrealistically, expects Limann to get things moving fast," a British diplomat said.

An American diplomat echoed his sentiments. "The next six months are going to be extremely critical for the Limann Government," he said. "There may very well be a false sense of security now. It is very difficult to say if Limann will be in office six months from now."

Some Ghanaians, like Leslie Nyarko, a young medical technician, feel that in neighborhoods such as Jamestown, frustration may explode into rioting if the Limann Government does not insure that enough food and essential consumer items like soap are supplied soon.

"We waited for things to get better under Acheampong, we waited under Akuffo and we waited under Jerry Rawlings," Mr. Nyarko said, puffing an American cigarette that he had bought on the black market. "Tell me, how long can we wait?"

Leaders Are Executed

Gen. Ignatius K. Acheampong seized power in January 1972 from the civilian Government of Prime Minister Kofi A. Busia. Excessive spending on public projects and corruption sent Ghana into a severe economic decline from which it has yet to recover. Lieut. Gen. Frederick W. K. Akuffo, who overthrew General Acheampong in 1978, took some measures to improve matters, such as reducing Government expenditures, devaluating

the currency and balancing the budget.

But General Akuffo was overthrown in June by Flight Lieutenant Rawlings and his youthful Armed Forces Revolutionary Council, and analysts here said that the Akuffo program of stabilizing the economy quickly came to a halt. Eight persons, among them Generals Acheampong and Akuffo, were executed.

The Rawlings regime was in office barely four months before it handed over power to the elected administration of Dr. Limann. About the only thing the military group did in the economic sphere was to impose price controls. Market women were flogged in Accra's squares for alleged price gouging, and many foreign traders, mainly Lebanese and Indians, were frightened into shutting their shops.

"Thus while Rawlings was gathering popular support through his punishment of alleged offenders, he was also in effect creating a situation where already scarce goods were being driven away from the market," said Elizabeth Oheme, editor of the Government-owned Daily Graphic.

Priorities Are Listed

The 52-year-old Dr. Limann has been in office for two weeks now and he has offered few specifics about his economic programs. There is general agreement

among politicians and diplomats about what his priorities should be.

Among them is the revitalization of the cocoa industry, with increased incentives for farmers whose production slackened in recent years because they were not paid enough, the easing of price controls for essential goods so that more appear in the market, and insuring that the supply of new money is kept down and that the budget is balanced.

Still another priority, politicians and diplomats say, should be foreign aid. Little is coming now, largely because of uncertainty among Western donors about the political situation here.

In conversations with Ghanaians, considerable wariness was evident about what the military might do in case the new civilian regime falters. Although Flight Lieutenant Rawlings and his group are reportedly being given diplomatic or military positions abroad, no one in this capital is ruling out a return to power by other elements of the armed forces, particularly if continuing shortages of food and other products result in rioting in Accra and elsewhere.

"Come back in six months time," said Leslie Nyarko, the medical technician. "Then we'll know how we're doing."

October 10, 1979

GUERRILLAS ACCEPT RHODESIAN CHARTER PROPOSED BY BRITISH

THEY MAY REJOIN TALKS TODAY

By YOUSSEF M. IBRAHIM
Special to The New York Times

LONDON, Oct. 18 — The Patriotic Front accepted Britain's proposals for a new Zimbabwe Rhodesia constitution today, and Britain responded by indicating that the guerrilla alliance could resume its role in the three-sided negotiations tomorrow.

Two days ago the British began talks with the delegation from Salisbury headed by Prime Minister Abel T. Muzorewa after the guerrilla group balked at constitutional proposals pledging future Zimbabwe Rhodesia governments to compensate whites for seizure of their lands.

The return of the guerrilla leaders to the conference table appears to end, for the time being at least, the crisis over the compensation issue.

The guerrillas had insisted that Western governments, principally Britain and the United States, assume the financial

burden of easing the whites out of their thriving farmlands and turning the land over to black Zimbabwe Rhodesians.

The deadlock was broken by a general agreement between Britain and the guerrillas that any new nationalist government, in which the guerrillas expect to play a major part, will not be responsible for making these payments, estimated to exceed $2 billion.

However, Britain, which is the chairman of the conference, still refused a firm commitment to any specifics, stating only that it would help in putting together a multinational fund with the participation of the World Bank, the European Economic Community, other international institutions and the United States.

Conference sources here suggested that while the talks will move on tomorrow to the so-called second phase of transitional arrangements, the constitutional issue is being "papered over" and might still present the conferees with a problem in the end. "It is too early to call this a breakthrough," one source here said, suggesting that more time will be needed before any optimistic pronouncements on the talks can be made. The divergence of views was made clear this afternoon when Lord Carrington, the British Foreign Secretary, issued a statement announcing that the Patriotic Front, following a meeting with him this afternoon, had dropped its objections on the knotty issue of land compensation and had agreed to return to the talks. He neglected, however, to report that the guerrillas had strongly restated their views on

land settlement as an integral part of their constitutional compromise.

The British omission led to an angry response from Dr. Eddison Zvobgo, the spokesman for the Patriotic Front, who released the full text of the message delivered to Lord Carrington by the guerrilla co-leaders, Joshua Nkomo and Robert Mugabe, in their 15-minute meeting this afternoon. "Since when did the British start taking responsibility for distributing Patriotic Front statements?" Dr. Zvobgo said to reporters in the press center here. "This was taken totally out of context."

The text of the Patriotic Front statement tied any final settlement to resolution of the land issue, and asserted that the front had been promised Western participation in land programs. "We have now obtained assurances that depending on the successful outcome of the conference, Britain, the United States of America and other countries will participate in a multinational financial donor effort to assist in land, agriculture and economic development programs," the guerrilla group stated.

Analysts here suggested that the British Government hoped to use the issue of compensation as a lever in the second phase of talks on transition of power to moderate the Patriotic Front's demands for a larger share in any new coalition government.

The analysts also said that neither the British nor the American Government had discussed in any detail the size or shape of the land fund, which has become a central issue in these talks.

October 19, 1979

93

Uganda, 'Pearl of Africa,' Battling Legacy of Amin

By GREGORY JAYNES
Special to The New York Times

KAMPALA, Uganda, Oct. 18 — Uganda, said Winston Churchill, is the pearl of Africa.

"You dig a hole here and something grows," said David Macadam, an International Harvester farm-equipment salesman and a farmer himself. "Pitiful though it is now, it remains the pearl of Africa."

The country's fertile soil and friendly citizens leap first to the tongues of foreigners who know it well. The third subject, inevitably, is whether Uganda, now six months and one week out from under Idi Amin, can overcome problems that appear insurmountable.

"Nothing quite works right here anymore," said a top-ranking civil servant. "Do you see any refugees returning? Of course not. I suspect it is the other way around; people are getting out if they can."

There are no hard statistics, but world relief organizations anticipated the return of about 80,000 refugees and so far this has not happened. Doctors have been particularly shy about returning; there were 1,650 in Uganda in 1970 and last month only 510.

A 'Sickening' Reaction

"There are exiles," said Dr. John Davies, the administrator of Mengo Hospital in Kampala, "doctors who have come back and taken one look and decided they needed 'further study' elsewhere. I find it sickening."

Uganda has 76 hospitals with 15,567 beds to care for the country's 13 million people, but medical manpower needs are "staggering," according to a recent study. At Kampala's Mulago Hospital, with 1,547 beds the largest in the country,

there are 191 key posts unfilled. "What's the use?" a doctor at Mulago asked. "We have no medicine to administer anyway."

Dr. Davies, whose own little missionary-run hospital has been without running water for months, said the "vaccine situation is terrible, terrible."

"We've got nothing," he added. "Antitetanus serum is what we need. We're having people here die of tetanus. It's running about one death a week."

Tanzanian Soldiers at Airport

Other problems are apparent to visitors from the start, at Entebbe Airport, situated on a spit of land jutting into Lake Victoria. There, rusted antiaircraft guns are aimed at the sky and Tanzanian soldiers doze at noonday in the shade under the wings of a Hercules C-130. Nearby is a Boeing 707 with three flat tires and faulty engines.

"The Government brought in a friend of mine to repair that 707," said Hermann Marcksheffel, a building contractor. "He was inspecting it and taking pictures of it and the soldiers arrested him. He spent two days standing up in a cell because there were so many cockroaches and spiders on the floor. He will never be back."

Few airlines fly into Entebbe, ostensibly because of a lack of firefighting equipment. Ugandan Airlines' own ragtag fleet of planes takes off with irritating infrequency. On many days ticketed passengers wait up to eight hours only to be told the flight has been canceled and they must make the 25-mile pothole-ridden return trip to Kampala on a bus with hard wooden seats — a brain-jarring journey.

In the terminal a dozen clocks perpetually read a few minutes past 12. A Swiss expert was called in a month ago and re-

paired and set them. He was unable to collect his pay, so he stopped the clocks one by one, locked their faces down tight and flew away.

Thief's Body Left in Street

In the capital there are always fresh stories of crime. The body of a suspected car thief is left in the streets three days as a warning. A man suspected of murder is stoned to death in the central market, his body covered with dried banana leaves and set afire.

Now and again there are stories of good deeds from the new Government. One minister took a bus ride that was supposed to cost 75 shillings — about $7 — and was charged 300. "He went to the manager," an aide reported, "and he told him, 'Just once more and you won't know what hit you.'"

From the corners of the country come tales of pillaging by remnants of Mr. Amin's army. From the seat of government, the building where a new 127-member Parliament recently convened, comes talk of impending trials of 3,000 "sycophants of the former regime."

"You've got practically every kind of Ugandan you can think of in the new Government," a Western diplomat observed. "I expect there will be some pretty bitter political rivalries before the elections in 1981."

"It may take 100 years to get it right politically," said Keith Bennett, a British electrician. "However, the wife won't hear of living elsewhere. We put in five years in Tanzania, two years in Kenya. I ask her if she wants to go with me on business trips to Nairobi and she says thanks all the same, she'll stay put. It's the pearl of Africa."

October 25, 1979

Rhodesia Parley Reaches Accord Over Transition

Only Terms of Cease-Fire Remain for Agreement

By R. W. APPLE Jr.
Special to The New York Times

LONDON, Nov. 15 — Britain and the rival factions in Zimbabwe Rhodesia reached agreement today on arrangements for a transition from guerrilla war to all-party elections, increasing the possibility of an early end to the 14-year-old crisis.

In 10 weeks of talks at Lancaster House here, the British Foreign Secretary, Lord Carrington, has persuaded the Patriotic Front guerrilla alliance and the biracial Government of Bishop Abel T. Muzorewa to accept both an independence constitution and a transition formula. All that remains is bargaining over the terms for bringing the armed combat to a halt.

Although the cease-fire talks, which will begin tomorrow, could prove difficult, Lord Carrington said in an interview tonight that he hoped to complete them "quickly — in a matter of days."

Historic Success for Tories

By itself, today's agreement constituted an historic success for the Foreign Secretary and for the Conservative Government of Prime Minister Margaret Thatcher. The talks have succeeded in part because of Lord Carrington's firm management and in part because of the changes in the status of the participants.

"Even if we can't get an agreement on the cease-fire," said a ranking British official involved in the talks, "we have done far more than we dreamed possible. The front is committed to so much now — the constitution, the transition — that they would be hard put to sell an election boycott to the rest of the world. If we carry out what we have agreed to in good faith, even if the front walks out at the last minute, we will have a fair part of Africa and much of the world with us." Under the current settlement plan, blacks would hold 80 of the 100 seats in Parliament and would have the power to amend the constitution.

To gain the assent of the guerrilla alliance for the transition proposals, Britain had to agree to add only one sentence to the plan it put forward on Nov. 2. The sentence read: "The Patriotic Front forces will also be required to comply with the directives of the governor." The front took the sentence as evidence that their forces would have equal status with those of the Salisbury regime during the electoral campaign; the British took it as a reinforcement of the authority of the interim governor, a Briton who will be named shortly.

Equal status had been the main sticking point for Joshua Nkomo and Robert Mugabe, the co-leaders of the front. They felt that they must have some acknowledgment that their armed struggle had been a factor in bringing the parties to the conference table, and sought some protection against repression by the Muzorewa military forces.

Front's Spokesman Jubilant

"After the governor arrives in Salisbury there will be no terrorists," a jubilant Willie Musarurwa, the front's spokesman, told reporters. "We will become a legal force in the country. This is to us simple recognition of the contribution which young men and women in Zimbabwe have made to bring about peace, progress and democracy in that country."

As things turned out, the formula evolved by President Kenneth D. Kaunda of Zambia last week after three days of talks with Mrs. Thatcher, and put forward by the front yesterday as a compromise gesture, was a smokescreen behind which the front could retreat. Mr. Kaunda suggested a complicated series of supervisory bodies, among other things, but none of the provisions of his plan were included in the agreement reached this morning.

The breakthrough at the conference came in the following exchange between Mr. Mugabe and Lord Carrington:

MR. MUGABE: In the light of the discussions we have had as a result of President Kaunda's proposals to the Prime Minister, if you are prepared to include the Patriotic Front forces in paragraph 13 of the British paper, we are able to agree to the interim proposals, conditional on a successful outcome of the negotiations on the cease-fire.

LORD CARRINGTON: I can confirm that your forces, and the Rhodesian se-curity forces, will be under the authority of the governor. The Patriotic Front forces will be required to comply with the directions of the governor.

'A Total Capitulation'

A spokesman for Prime Minister Muzorewa, who accepted the British proposals last week, described today's agreement as "a total capitulation by the Patriotic Front to the original British position." Privately, British officials saw things the same way.

But the official British spokesman, Nicholas Fenn, issued a notably even-handed statement. "Lord Carrington is delighted with this development," he said. "He welcomes Mr. Mugabe's statement and pays tribute to the wisdom and statesmanship of both delegations. Both have made difficult decisions which have enabled the conference to move forward to negotiations on a cease-fire."

One important point dealing with the transition was left unresolved. Britain has suggested a transition period of two months; the front wants at least six months. According to conference sources, however, it was agreed secretly some days ago that the transition period would last two months after the beginning of the cease-fire. The election would therefore take place two months after the fighting stops.

The effect of that decision was to shift the dispute to the question of how much time would be needed to bring about a cease-fire. Mr. Mugabe and Mr. Nkomo have been arguing for two to three months, Lord Carrington for no more than 10 days or two weeks.

Continuing disagreement over the length of time that must elapse before the elections reflects differing imperatives. Britain's main concern is that the agreements so painstakingly reached here might come unraveled if the voting is delayed for too long; the front wants time to get all of its widely dispersed supporters back into Zimbabwe Rhodesia.

Britain Will Send Governor

If an agreement is reached, a British governor will be sent to Salisbury under a law that was rushed through Parliament on Monday and Tuesday. The leading candidate to be governor is Lord Soames, the Tory leader in the House of Lords and the 59-year-old son-in-law of Sir Winston Churchill. He would be accompanied by civil servants, police officials and a limited number of soldiers, with Sir James Haughton, a retired senior British police official, responsible for overseeing the Rhodesian police.

Participants in the conference said three elements appeared to have brought about the breakthrough after a prolonged deadlock.

The first was intense pressure exerted on the front by Mr. Kaunda, who desperately wants a settlement in Zimababwe Rhodesia so that he can rebuild Zambia's ravaged economy. His country has suffered not only from raids by the Salisbury Government's forces, but from disruption of transportation networks by the guerrilla struggle.

The second element was Lord Carrington's decision to push enabling legislation through Parliament before he had a transition agreement. This served as a warning to the Patriotic Front that he would conclude a separate deal with Bishop Muzorewa if the front did not come to terms.

Finally, the British told the guerrilla leaders that one of their main demands, a monitoring force made up of observers from Commonwealth nations, would be met by Lord Carrington in the cease-fire negotiations, if that stage was reached. In the words of the same official, "We let them have a glimpse of the goodies awaiting them."

November 16, 1979

President Honecker Was in Ethiopia Last Week

East German Afrika Korps: Force to Be Reckoned With

By JOHN F. BURNS

SALISBURY, Zimbabwe Rhodesia — Joshua Nkomo calls them "the best friends" his Rhodesian nationalist army ever had. The approbation is shared in many African countries with Marxist or left-leaning governments and by Mr. Nkomo's counterparts in other insurgent movements. In South Africa, Prime Minister P.W. Botha and his generals regard them as the most dangerous harbingers of a "Communist onslaught" they perceive as organized by the Soviet Union and its allies.

The men who inspire such contrary feelings belong to a new "Afrika Korps" that may have a more lasting impact than Field Marshal Erwin Rommel's World War II forces. Some 5,000 to 10,000 East Germans have spread their influence from Ethiopia to Guinea to to the "front-line" states facing the white minorities in Southern Africa. Although small compared to the Cuban legions that fought in Angola and Ethiopia and shored up other shaky African governments, East Berlin offers skills and an estimated $200 million a year, much of it military aid. Last week, the East German President, Erich Honecker, flew to Africa for the second time this year, to put a formal seal on military and other aid extended to Ethiopian President Mengistu Haile Mariam.

To those who feel threatened, the East Germans are viewed much as the Cubans — as proxies for the Soviet Union, advancing Moscow's ambitions without risking a direct clash East-West clash. Much of the stimulus and some of the money has come from Moscow, but the effort also serves East German interests. The German Democratic Republic has found in Africa a means of winning acceptance and prestige in a world that was largely hostile for two decades after its founding in 1949.

When the path to international favor was blocked by the superior economic strength of West Germany, East Berlin portrayed itself as the natural ally of ex-colonial peoples and insisted that Bonn had inherited the 19th century imperialist tradition of German Chancellor Otto von Bismarck. Beginning in 1953, with a trade agreement with Egypt and another with Guinea in 1958, there were 10 years of signing commercial pacts, starting cultural centers and "friend-ship organizations." The German Communists also offered help to black liberation movements, including those that later gained power in Angola, Guinea-Bissau and Mozambique, and others still fighting in Rhodesia, South-West Africa and South Africa.

In 1969, the Sudan was the first in Africa to extend diplomatic recognition. As relations between the two Germanys improved, so did the East's diplomatic prospects. Other Africans established formal ties and a pattern of aid was set in agriculture, education, medicine and technology, matched everywhere with military and security aid. In 1973, East German military advisers were spotted in Brazzaville, the Peoples Republic of the Congo capital, a showcase for Soviet bloc aid. Also headquartered in Brazzaville, the Popular Movement for the Liberation of Angola, Agostinho Neto's Marxist insurgents, received Cuban and East German military aid, which brought Mr. Neto to power in 1976. As Angola's first President, he installed East Germans in supervisory positions in agriculture, education and medicine, and in organizing and training his army and police. Mr. Neto's death in Moscow in September opened up new opportunities for East German influence among contending political forces. In Mozambique, where Frelimo guerrillas had received similar help, the East Germans were the largest Soviet bloc contingent to move in after independence. For President Samora Machel, like his counterparts elsewhere, the East Germans trained security men and the secret police. They operate from one of the largest embassies in Maputo, matched in Africa only by the East German mission in Ethiopia.

There, the Marxist Government that took power after the overthrow of Emperor Haile Selassie has relied heavily on East Germany for weapons, tractors, textiles and medical supplies, as well as training for its militia, journalists and party officials. The importance of the relationship was suggested when President Mariam interrupted a speech in Addis Ababa to announce the death in East Berlin of Werner Lamberz, a Politburo member who had quarterbacked the East German support for Cuban and Ethiopian troops who repulsed an invading Somali army in the Ogaden war.

East German support for guerrilla movements seeking power in the white strongholds of the south is growing monthly, reflected in annual visits by the Defense Minister, Gen. Heinz Hoffmann. Last February, President Honecker also visited Mozambique and Angola. Their beneficiaries include the African National Congress of South Africa, which maintains one of its most important offices in East Berlin, the Angola-based guerrillas of the South-West Africa People's Organization, and the Zambia-based army of Mr. Nkomo, the Rhodesian nationalist leader. East German experts assist these groups in organization, training and in planning guerrilla operations. A large contingent of the 2,500 "advisers" stationed in Angola, as well as some of the 1,500 in Mozambique, supervises training. They have assisted Mr. Nkomo's Zimbabwe People's Revolutionary Army with training in flying MIG interceptors, and operating tanks, heavy artillery and ground-to-air missiles such as those which shot down two of Air Rhodesia's Viscount airliners.

Since Mr. Honecker's visit, East German military advisers in Angola and Zambia, reportedly have been spotted in forward camps advising on guerrilla deployment into Zimbabwe Rhodesia and South-West Africa. Last year, they played an even more adventurous role — planning and supplying logistics for the (unsuccessful) invasion of Zaire's copper-rich Shaba Province, launched from Angola.

The new assertiveness worries military planners in Pretoria and Salisbury. Contempt for their insurgent enemies is tempered by respect for what these might become, with a few hundred East Germans providing more effective leadership. But to venture further, Mr. Honecker would have to involve his men in fighting, an excessive risk in East Germany's careful calculus. More likely, the "Afrika Korps" will remain just off stage, involved too obliquely to provoke a superpower clash but close enough to make a crucial difference to the insurgent cause.

November 18, 1979

CHAPTER **6**
Values
Americans
Live By

The town of Crested Butte, nestled below the mountain of the same name, is a last frontier now threatened by massive industrialization.

Susan Cottingham

A TINY TOWN BATTLES A MINING GIANT

Villagers in Crested Butte, Colo., are out to keep AMAX from developing a molybdenum mine that they say would destroy the wildernesss and their way of life.

By Roger Neville Williams

Crested Butte sweeps its rocky ridge into the Colorado sky, high above the tiny hamlet which has taken its name. More than a butte, it is a 12,000-foot peak with a saw-toothed crest, a mini-Matterhorn challenged only by the supine Mount Emmons across the valley. The two mountains stand like bookends above the plain.

In between lies "The Butte," an ex-coal-mining town of 1,000 people, many from New York, California and Atlanta, who have trimmed out the old village in Disneyesque fuchsias and lavenders while adding stately Victorian buildings of their own. The entire town is a national historic site, almost entirely surrounded by National Forest land.

Three miles to the north is ex-Secretary of the Army Howard (Bo) Callaway's ski area, on the rim of the vast cirque edging the valley, where Bo and his brother built their condominium village. It is a tasteful, modern ski complex. Crested Butte and the ski mountain lie some 25 miles south of Aspen as the bullet flies, or 225 miles from Aspen by paved road and mountain pass.

The townsite is in a remote, enchanting setting at the end of the road in one of Colorado's last unspoiled valleys. Crested Butte lies at the end of the road of some people's lives as well, a last frontier now threatened by massive industrialism unlike anything Western Colorado has ever seen.

AMAX Inc., formerly American Metal Climax Inc., a giant mining corporation with more than $3 billion in assets, wants to construct a billion-dollar molybdenum mine two miles from town, and the urban refugees who have settled permanently in Crested Butte, and who see themselves as preservationists of the surrounding wilderness and of the historic landmark to which they have brought a renaissance, are not too happy about it.

When I drove up the 28 miles from the sleepy town of Gunnison, past the abandoned, lichen-covered homesteads,

Roger Neville Williams is a freelance writer living in Telluride, Colo.

past the ranchers' proud turn-of-the-century homes, I was met at the plain of the North Valley by hundreds of poky Herefords coming down the road, prodded along by mounted cowboys in snow-covered chaps — a real cattle drive. This is the Old West, still mythic in its appeal, a tranquil region of range and mountains, relatively unscarred by its one compact Victorian mining town and scarcely marked by the recreation boom of the last 15 years.

In 1977, two miles from Main Street and 11,000 feet up on Mount Emmons (known to locals as "Red Lady Mountain"), prospectors for AMAX sank their core drills into what is thought to be the world's richest known deposit of molybdenum ore. Colorado has two other larger deposits, at Henderson near Berthoud Pass, and at Leadville, both being mined by AMAX's Climax Molybdenum Division, but neither ore body compares to that of Mount Emmons, which has been estimated to have a market value of $7 billion. Worldwide demand for the "gray gold" is rising at 7 percent a year, and Climax Molybdenum — which provides nearly half of the free world's supply — wants the Red Lady's moly.

In addition to the mine on the mountain, AMAX has proposed a mill and tailings pond (slime dump) of staggering proportions, all of it to be set down on surrounding National Forest and ranchland that has never known modern industrialization. The mine and mill would employ 1,300 people, cost as much as $1 billion, and require more than 2,000 construction workers during the building phase.

Crested Butte's Mayor Mitchell (in wheelcair) leads a march in Aspen against AMAX's plan to mine "The Lady," Mt. Emmons.

John Meislahn

About to fly over Mount Emmons to survey the mine site (below), David Leinsdorf plays David to AMAX's Goliath.

Of the 165 million tons of ore estimated, conservatively, to be inside Mount Emmons, 164 million tons of it would end up *outside* the mountain, on the surrounding land. There will be a tailings pond, a useless, caustic, chemical-laden sludge that's left after the moly concentrate has been extracted, covering 3,000 acres, 30 times the size of the town of Crested Butte. Gunnison County's total population, currently 11,000, would more than double over the next six years. Bumper stickers seen around town say: "DON'T CLIMAX IN THE BUTTE." All anyone can talk about in town is "the mine." Crested Butte is a town fighting to save itself.

So what is molybdenum, aside from being nearly unpronounceable? (It's been called "Molly Be Damned.") It is a silver-gray metallic element with a melting point of 4,730 degrees Fahrenheit that is used to alloy most steel products, to provide durability as well as resistance to corrosion and severe temperatures. It lightens steel, while hardening it, and it is found in everything from tools and engine blocks to steam turbines, jet engines, pipelines and nuclear reactors.

Molybdenum has been used in armor plate and cannon since 1894. In 1905, a German company discovered the world's largest deposit at Climax Station, near the town of Leadville, Colo. The company, Metallgesellschaft, changed its name to American Metal Co. when World War I came along, but that didn't prevent the War Minerals Board from seizing its stock and placing it under the trusteeship of Henry

Morgenthau and Andrew Mellon. A rival company, Climax Molybdenum, was then formed to provide moly for Allied war instruments. By World War II, molybdenum was the country's No. 1 strategic metal, and troops were garrisoned near Climax to guard the mine and mill from possible Axis sabotage.

When AMAX officials came to Crested Butte a year ago to announce their find, the last thing they expected to encounter in a tiny, rural community was a phalanx of resident New York lawyers, who, along with just about everyone else in town, want to keep the Butte and its surroundings as they found it.

There is New York University Law School graduate Myles Rademan, the town planner who came to the Butte six years ago and says: "You just can't imagine the size of this project, its magnitude. When you have 200 or 300 *million* tons of tailings, you're talking about a retaining wall the size of the Aswan Dam or bigger. Areas under consideration for tailings dumps are among the most beloved and beautiful parcels of land in the world. What we're dealing with here is the disemboweling of the West."

There is Gil Hersch, a former New Yorker who now owns and publishes the weekly Crested Butte Chronicle. He told me: "I'm opposed to the mine. We're killing our planet. I'm interested in the major paradigms in the world right now, between the traditional capi-

talistic outlook of more-makes-better, as opposed to the New Age culture which says we've got to live in harmony with nature. AMAX came in here with the statement that they were going to make this the first in a new generation of mines. We asked, 'What is a new generation of mines?' Is it that it's just 15 years newer? Or is it the difference between the Industrial Revolution and the New Age Revolution?"

And there is County Commissioner David Leinsdorf, son of conductor Eric Leinsdorf. He was recently re-elected by playing David to the AMAX Goliath. He advocates a much smaller mining operation than AMAX envisions. A Columbia Law School graduate, ex-Nader Raider and former Justice Department attorney, he moved to the Butte in 1971 and became Colorado's youngest county commissioner in 1974, acceding to a post normally reserved for conservative, native-born ranchers.

The stakes are high: A $7 billion mineral deposit versus the priceless value of wilderness; technological progress and modern industrialization versus a ranching and resort economy; transient workers and "boom town" greed versus a tiny, stable community practicing, for the most part, a kind of voluntary simplicity.

Recreation provides Gunnison County's chief income, as a million and a half campers, hikers, fishermen, backpackers, hunters, jeepers and skiers visit Gunnison National Forest each year, and one might expect the resort industry to lead the fight to preserve the integrity of this region of unparalleled natural beauty. But that's not the case.

"I wish they'd found their moly somewhere else, but I'm not going to lie down in front of any trucks to stop AMAX. It's not my style," said Bo Callaway in his office overlooking his ski lifts. "I support the free-enterprise system, and I'm not comfortable telling a mining company when and where they can mine. If AMAX goes the extra mile and operates openly, I'll continue to work with them."

As President Ford's campaign manager, before he resigned over the "Callaway Affair" and moved permanently to Crested Butte, Bo was the enemy of the area's environmentalists. He proposed a bigger ski area, on another mountain, and was accused by Myles Arber, the former editor of The Chronicle, NBC News and Senator Floyd Haskell, the Colorado Democrat, of using his Cabinet post in Washington to lever the required special-use permit from the Forest Service. He handed me a copy of Harper's with the July 1977 cover story absolving him of any wrongdoing, his sweet vindication.

Money is money, and $7 billion is sitting inside Mount Emmons. Already it has had its effect — just lying there: $75,000 condominiums now sell for $200,000 and Crested Butte lots, which brought $10,000 last year, are going for $25,000 today. Terry Hamlin, a former head of the ski patrol, ski-area marketing manager, real-estate agent and a Colorado native, is a case in point. The young, person- *(Continued)*

CRESTED BUTTE

Continued

able ex-ski bum is now manager of local affairs for AMAX. "A man on the way up," taunts Gil Hersch in The Chronicle. A shotgun blast shattered the windows of Mr. Hamlin's AMAX office last fall, but he was away at the time.

Hamlin drove me up to the mine site on Mount Emmons, two miles above town on the Kebler Pass Road. Already, 135 men and women are employed here in "exploration," bearded young geologists examining core samples, miners cleaning out and shoring up the drifts (tunnels) of the abandoned Keystone Mine. The tunnels of the defunct lead-zinc mine happen to lie directly beneath the huge molybdenum ore body and are being used for exploratory core drilling.

In 1977, AMAX took the mine over from a Wyoming firm, fully aware that the mine's 100-year-old tailings dump had collapsed, polluting Coal Creek, Crested Butte's primary water supply. Under a Colorado Department of Health cease-and-desist order, AMAX had to spend $800,000 to clean up the previous mine owner's mess and stabilize the old tailings pond. Terry Hamlin handed me a colorful, 18-page AMAX brochure that promised to bring brown trout back to Coal Creek, concluding, "AMAX is committed to the people of Colorado to help maintain the air, water and natural environment that make the good life here what it is."

"Next summer we're building a $2 million water-treatment plant up here to completely end the Keystone drainage contamination of Coal Creek," Hamlin explained. "In the past, at the historic stage of the development of the West, mining companies were not concerned with the environment. But AMAX is. We have to be." It was a theme I was to hear again and again from AMAX officials.

But no one on the Town Council has bought AMAX's expensive public-relations arguments. "It seems highly unlikely that the mine and the town can be compatible, or that recreation and mining can be compatible," says William Mitchell, 35, Crested Butte's Mayor and chairman of the Town Council. "I'm very anxious about it. This is a national playground and it belongs to everyone. These are public lands. Those of us who've consciously given up conveniences and have adopted a harder way of life as a trade-off for the natural wonders surrounding us have become trustees of a magnificent wilderness."

Mayor Mitchell — he goes by his last name alone — has been using his "situation," as he calls it, to attract attention to Crested Butte's plight. After a 1971 motorcycle accident badly burned his body, hands and face (now restored by plastic surgery) and a subsequent airplane crash fractured his spine, Mayor Mitchell is now a paraplegic — in what he calls "the least wheel-chair accessible town in America," Mayor Mitchell says, "I'm way too selfish for self-pity. If I hadn't got burned up or been in a crash, I would not have

enough money to be Mayor of Crested Butte, since the job pays only $25 a month and I'm at it full time." Mayor Mitchell's dynamism, charisma and tough leadership qualities, combined with his ex-radio announcer's voice, keep one from noticing his "situation."

"He's a one-man media event," says Myles Rademan, noting the national publicity afforded Crested Butte under such headlines as "Paralyzed Mayor Battles Mine Owners." When President Carter invited Mayor Mitchell to Washington to advise the President's Committee on Employment of the Handicapped, the Mayor dropped by the Interior Department to see Secretary Cecil Andrus, a man who, he notes, launched his career stopping a molybdenum mine at White Cloud, Idaho.

Mayor Mitchell confronts and confounds AMAX's media blitz (full-page ads in the local papers, a monthly sheet called The Moly News featuring columns by company executives, and a comic strip called "Miss Moly") with a tireless campaign of his own. In recent months, he has testified before the House Subcommittee of Mines and Mining, and has met with Paul Ehrlich, John Denver, Robert Redford, Jack Anderson and David Brower, president of Friends of the Earth. He once threatened to address an AMAX stockholders meeting, until AMAX chairman Pierre Gousseland and the then-president of the Climax Division, Jack Goth, invited him, along with Messrs. Leinsdorf, Callaway, Rademan and the Mayor of Gunnison, to visit the home office in Greenwich, Conn., last year.

"We can't give in to the divine right of ownership by mining companies," says Mayor Mitchell. "I'm fighting for every protection I can get for this valley. We can understand the value of Yellowstone and Yosemite — there are probably tons of mineral deposits in Yosemite and a decision was made in 1890 to preserve it — but what is the value of this valley? There have been more than 1,500 mining claims filed in this county this year. What they want to do is make Gunnison County one large open pit mine."

Last November, Mitchell was unhappy to see Colorado's environmentalist Senator, Floyd Haskell, a Democrat, defeated by William Armstrong, a Republican Congressman. Armstrong had been listed by Environmental Action as one of Congress's "Dirty Dozen" for his sorry environmental voting record while a Congressman in an adjoining district. But Mayor Mitchell is heartened by the position of Democratic Governor Richard Lamm, who recently counseled him: "[AMAX's] admission price to this community is going to be very high."

The result of Mayor Mitchell's activism has been an enormous amount of unprecedented cooperation between a giant corporation and a small community. Despite a unanimous vote by the Town Council on Feb. 20 to oppose the mine, the Mayor knows that it is hard to stop a billion-dollar-a year cor-

poration from doing what it wants. Messrs. Leinsdorf, Rademan and Hersch, lawyers all, also know that the Mining Law of 1872 guarantees mining companies access to all mineral deposits under public lands. And so they are fighting what the Mayor calls a "gentlemen's war."

☐

I was unprepared for all this cooperation. I had half expected to find a local Monkey Wrench Gang loose in the woods, pouring Karo syrup into bulldozers, dynamiting construction sites, rolling boulders down on workers' pickups. In fact, there has been some of that: a D-7 Caterpillar was riddled with rifle bullets, gas tanks have been sugared, and there was the shotgun blast through Terry Hamlin's window. A resident of Ohio Creek, an area threatened by the proposed tailings pond, said at a recent public hearing that if AMAX came to his valley, he would oppose the company by force, if necessary. AMAX, meanwhile, proudly points out in The Moly News that its security guards remain unarmed. So far.

reviewing proposed ski-area development, but it has never been applied to mining. The process addresses such problems as housing, transportation, schools, sewage, air pollution and environmental impact. The C.R.P. is the brainchild of Mr. Leinsdorf's old Columbia Law School classmate Harris Sherman, director of the State's Department of Natural Resources and a Lamm appointee. AMAX has signed an agreement with the Forest Service, Gunnison County and Mr. Sherman's agency to participate in the C.R.P. It is an important breakthrough in a state that has probably been the least progressive in the West on questions of land use planning and mining controls.

Mr. Leinsdorf led the fight two years ago to have the Board of Commissioners adopt the county's Land Use Resolution, one of the most innovative documents of its kind in the country. He also insisted that AMAX and the Forest Service participate in the C.R.P. Mr. Leinsdorf has taken on corporate giants before: He was a co-author of the book "Citibank" for Ralph

---

'I think we're lucky to be in a place where the community is very sophisticated. I'd rather have a lot of sophisticated enemies to work with than a bunch of dumb friends,' says Stanley Dempsey, AMAX's vice president for external affairs.

In an attempt to anticipate problems before they happen, Messrs. Leinsdorf, Rademan, Hersch, Callaway and Mitchell and a talented young planner for Gunnison County, Jim Kuziak, are attempting an untried and radical method of dealing with the massive mining operation. As Mr. Leinsdorf explained it: "I'm not as concerned with the environmental consequences as I am by the boom-town syndrome, the spin-off that comes from rapid industrialization and a high rate of growth: social and economic problems, unemployment, crime, drug abuse, alcoholism, child abuse, transient workers. So we're trying to see to it that the development that comes does not overwhelm us. It's called the Colorado Review Process, involving AMAX, the Forest Service, the Colorado Department of Natural Resources and the County."

The Colorado Review Process (C.R.P.) was devised for

Nader, a critical look at the First National City Corporation. When the Crested Butte delegation went to Greenwich to see AMAX's top executive officers, Mr. Leinsdorf opened the meeting by telling a story about his first job, in the Antitrust Division of the Justice Department.

"I was given a filing cabinet full of files on a possible restraint-of-trade and monopolistic-practices case." he began. "I was told to examine all the evidence and either bring charges against the company or close the case. That company was American Metal Climax. I was to see if a company which produced 70 percent of the molybdenum in the United States was engaging in unfair practices." As the executives tensed up, Mr. Leinsdorf laughed and added, "I found none and closed the case. But now, 10 years later, 2,000 miles west of New York City, I'm involved with AMAX once again. Which just goes to

At Climax, Colo., another big AMAX moly-ore mine is causing an entire mountain to collapse.

'To hell with the mine,' says town councilman Kirk Jones, a hard-drinking construction worker. 'Let the moly stay there. Maybe in 30 years they'll have the technology to take it out without making a mess.'

show, there really is no running away."

Mr. Leinsdorf believes that it is possible to make the Mount Emmons mine a model project. He sees the joint public review as an opportunity to develop a new process that could apply to the whole spectrum of natural-resource development throughout the West: uranium, copper, shale oil, gas and coal. In that spirit, AMAX has contributed $100,000 to the Gunnison County Planning Department, toward a computerized planning system. But Mr. Leinsdorf also told me, "If AMAX is not willing to build the mine on a scale and a schedule that maximizes the benefits to the county, I'm prepared to fight them and vote no on their zoning-change permit. The only way to keep the tourist and ranching economy from being overwhelmed is to do it on a smaller scale."

At the first C.R.P. hearing at Gunnison High School last September, every opinion on the mine was heard in the long, emotionally intense and crowded meeting. AMAX officials listened and answered questions very carefully.

AMAX emphasized employment and taxes, while the local governments pointed to environmental degradation, social upheaval and irreparable harm to the tourist industry.

Although public sentiment at that particular meeting was predominantly antimine, a few retired Crested Butte miners spoke in favor of it, remembering the prosperous coal-mining days of the 30's and 40's. One Gunnison businessman said Crested Butte had an "obligation to humanity to supply molybdenum to the world." A native Crested Butte resident stood up and shouted, "I can't eat scenery!"

At AMAX's Climax Molybdenum Division offices in Denver, I spoke with Stanley Dempsey, AMAX's vice president for external affairs. A jovial, thoughtful man of 39, Mr. Dempsey has taken on the burden of getting the mine built. In the past year, he has spent a great deal of time in the Butte, where he is respected for his candor and sincerity.

"I think we're lucky to come into a place like Gunnison County, where both the tradi-

tional community and the new-life-style community are very sophisticated. Despite a strong heritage against a lot of government control, the county has had the political will to pass a strong Land Use Resolution. Few places have that on the book yet. I'd rather have a lot of sophisticated enemies to work with than a bunch of dumb friends."

Stan Dempsey was in charge of environmental controls during the construction of the gigantic Henderson molybdenum mine, a project now hailed in the mining world, and by conservationists as well, as the most environmentally sound mine and milling operation in the world. "I don't worry about the physical impact, I think we can handle that," he says. "The growth-management issue is the challenge. That's why we really believe in open planning. We realized that the public was not going to put up with the kind of mining that's been done in the past and we know that we have to solve the housing and transportation problems which are the prinicpal constraints on mining."

AMAX, at Mr. Dempsey's

suggestion, recently organized an educational "boom-town tour" for 36 town and county officials to see what was going on in other mining towns. AMAX picked up the entire tab — charter planes, hotels and meals — for the tour of six Colorado and Wyoming boom towns recently "impacted" by energy and mineral development, including the infamous Gillette, often described as an "aluminum ghetto." Myles Rademan called me when he got back; the trip had backfired somewhat. "What we saw out there was a real mind blower. It was frightening. We will have to take a quantum leap to avoid becoming like them."

Stan Dempsey does not try to defend AMAX's environmental record at its Climax operation near Leadville, America's second-largest hard-rock mine. Nor does AMAX's public-relations director, Terry Fitzsimmons, who represents the third-generation of a Leadville mining family — his grandfather worked for Horace Tabor (of "The Ballad of Baby Doe" fame). Mr. Fitzsimmons said, "At Mount Emmons we plan to build a mine in keeping with the values of today. When Climax was built, the last thing miners or anyone else cared about was the environment."

In the Butte, it's the *first* thing anyone thinks about, and the residents continue to worry about their mountain's collapsing. "Subsidence," mining men call it, and people in the Butte don't want Mount Emmons to "subside," to crack and slip and cave in, as Bartlett Mountain did at AMAX's moly-ore mine at Climax.

One particularly concerned Butte citizen is Susan Cottingham, 30, the chairman of the legal committee of the High Country Citizens Alliance. This 140-person group devoted to preserving the quality of life in the mountains is challenging the idea that mining is the highest and best use of the land.

The mine is not unstoppable, she told me. "AMAX has been

stopped from doing projects in Wyoming and Tennessee. If a lot of aggressive and very intelligent people can't stop it here, the West might as well roll over and die. The technological dream of a 'model mine' belies human nature. I like a lot of the AMAX front men, but I don't believe their promises. I guess I'm an environmental paranoid. But then just look at their record."

Gil Hersch, down at his cramped Chronicle office, did exactly that last September. He reprinted a devastating report by some Tennesseans that carefully documented AMAX's coal, zinc, lead and copper operations in six states. "[AMAX has] a record of noncompliance with the law and of disregard for local citizens," the report stated. "Moreover, AMAX has shown a pattern of aloof resistance toward public officials, employees or ordinary citizens who have called upon the company to match its image with its performance."

In Colorado, AMAX has fought all "severance tax" proposals (taxing minerals as they leave the ground) — even though the money would be allocated to "impact mitigation," something to which AMAX says it is dedicated.

Chuck Malick, 28, is president of the High Country Citizens Alliance. The leather-worker and shopkeeper came with his wife to Colorado seven years ago as part of "the hippie dream." Standing by a crackling Ashley stove in his barn-wood paneled living room, I asked him: "This was a mining town. What's wrong with its becoming a mining town again?

"We're not against mining," he explained. "The Keystone Mine operated here until 1975 and employed about 50 people. I'd welcome several small mines. Our concern is that the public should have something to say about the management of resources on public lands, especially in a fragile area like this. What the nation is being asked to do in the West, by mining and energy corporations like AMAX, is to sacrifice our national playgrounds. The old adage that there's always someplace else to go is just not true anymore.''

Then Mr. Malick added, ''All this environmental stuff is not just flowers and trees. You kill the recreational trade and I'm out of a job. And right now there is no unemployment here, but in mining boom towns the unemployment figures are around 11 or 12 percent. We don't need that.''

The High Country Citizens Alliance has been successful in persuading the Forest Service, which has traditionally been overly cooperative with mining and timber interests, to consider a ''no-action alternative'' in its environmental-impact statement — that is, no mine. The Forest Service also listened to the Alliance's and to County Planner Kuziak's objection to AMAX's plan to build two 16-by-16-foot drifts (tunnels) to the ore body next summer; it persuaded AMAX to put the tunnels ''on hold'' for now. It is the company's first setback.

Newly appointed Crested Butte Councilman Kirk Jones, a hard-drinking construction worker, claims that he represents 85 percent of the town's residents when he says, ''I'm a consequence of the 60's and I'm tired of hearing about Leinsdorf's 'model mine.' I say to hell with the mine. Let the moly stay there; it could be a national reserve. Maybe in 30 years they'll have the technology to take it out without making a mess.''

Myles Rademan says, ''We're talking about our home here. I see this as a community struggle, not as an environmental fight.'' Which is why he led a trip to see Orville Schell in Bolinas, Calif., last summer with Messrs. Mitchell, Kuziak, Leinsdorf and Terry Hamlin, the AMAX public-relations man. Mr. Schell's book ''The Town That Fought to Save Itself'' chronicles a small community's battles with developers for control of its charming Pacific Coast town; and it's a kind of bible in Crested Butte. Myles Rademan wanted the group to meet the author, who commiserated with them. ''Here was another community refusing to accept every indignity thrust upon it', says Mr. Rademan.

However, as a town planner, Mr. Rademan believes cooperation with AMAX is the town's best defense. ''We're in a position of having to optimize something we don't really want,'' he explains. ''From a professional standpoint, it's exciting. This project could be a model for how you industrialize sensitive areas, as energy and mining companies seek to take over the West. We can set a new level of corporate responsibility.''

Corporate responsibility notwithstanding, Stan Dempsey wants to build his mine. But he's not sure it's economically feasible for AMAX to do the project on a scale any smaller than originally proposed. After two years of studies and talk, the big company is beginning to dig in its heels as it attempts to meet an already delayed timetable for construction (work is scheduled to begin in 1981).

And the Crested Butte Town Council won't listen any longer to AMAX's plans for a mine and mill the size the company wants. They are standing fast, as they made clear two weeks ago in their resolution calling for ''a stop to the proposed activity on the Mount Emmons project . . . until a real benefit to Gunnison County and the nation can be shown.'' When Mayor Mitchell gaveled the unanimous vote, the first time town officials went on record against the mine, 250 people attending the meeting whistled and cheered, as television cameras recorded the event.

The gentlemen's war is heating up. On Washington's Birthday, 30 members of the High Country Citizens Alliance ski-toured over the range to Aspen and held a street protest against AMAX. Last weekend, two commissioners from adjoining Pitkin County led a delegation from Aspen and cross-country skied back over the high-mountain route to Crested Butte, where they held a rally in opposition to the mine. It's not clear now what is going to happen to Gunnison County and the town of Crested Butte, but the people on the front lines in this fight to determine the future of some of our most beautiful Western lands know what they don't want. The lines are drawn in a battle that will be repeated in dozens of Crested Buttes throughout the country in the 1980's. ■

March 4, 1979

OPTING FOR SUICIDE

Once faced with cancer, the author planned suicide. He explains why, and argues for the right of choice.

By Edward M. Brecher

No doubt, many people who face a lingering death — from cancer, say, or some other disease — contemplate suicide. But by the time they are ready to take action, it is often already too late. Closely monitored by family and caretakers, they lack access to any effective means of suicide. They may

Edward M. Brecher is a long-time writer on science and medicine.

also be physically or mentally incapacitated. Their last-minute, improvised suicide attempts are quite likely to end in failure, humiliation and, perhaps, additional pain. This is the case history of a patient who firmly believes in advance planning. I am that patient.

I received a diagnosis of cancer of the colon, and a recommendation for prompt surgery, on June 17, 1978, shortly before my 67th birthday. As every prudent person should, I had given some thought during the preceding years to the advisability of suicide in such a contingency, and I thought fur-

The author's three sons supported his suicide planning. He is seen here with the eldest, Earl.

ther about it following my cancer diagnosis. As matters turned out, suicide proved unnecessary in my case, but I regard my suicide planning as a durable asset, a resource upon which I can draw should the occasion arise in the future.

I first began to think about the problem in October 1961. My wife and I and our 15-year-old son were hiking through New England's glorious autumnal foliage near our home when our aged German shepherd, Rufty, suffered a heart attack and lay writhing in the leaves, clearly in great pain. I hurried home and phoned a veterinarian, who arrived quite promptly. We found Rufty cradled in my son's arms.

"As soon as this convulsion is over," the veterinarian said, "I'll put Rufty out of her pain."

"No, don't wait," my son replied. "Do it right now."

The veterinarian extracted a syringe from his black bag, inserted it in a vein, and pushed the plunger. Rufty relaxed, and a moment later it was over. That moment was fraught with emotion. I looked straight at my son and blurted out what was in my heart:

"I hope someone will be as kind to me when my turn comes."

A few years later, the father of a friend of mine received the same kindness. The old man lay terminally ill on his sickbed late one night, surrounded by half a dozen members of his family. The 75-year-old physician who had attended them all, and had delivered several of them, was sent for. He arrived, examined the patient briefly, then filled his syringe and gave an injection.

"Is that to ease the pain?" someone asked.

"No, that was to ease the passing," the physician said. My friend likes to tell the story as an example of how fortunate his New England village was to have a horse-and-buggy doctor as recently as the 1960's.

My late wife, Ruth Brecher, who was also my collaborator as a writer, developed inoperable cancer in 1965 and died after a 19-month illness. Her care was managed at home, in accordance with her personal wishes, including adequate doses of morphine and sedatives as often as she requested them. I was deeply grateful to the physician who abided by her desires — but I resolved to follow a different course if I myself were ever in a similar situation.

I was vividly reminded of these earlier matters in 1975, when the newspapers carried accounts of the deaths of Dr. and Mrs. Henry P. Van Dusen. Dr. Van Dusen, an eminent theologian, was for many years president of the Union Theological Seminary in New York. Years before their

death, he and his wife had entered into a suicide pact, resolving to depart from life — voluntarily and together — at an appropriate time. When Dr. Van Dusen was 77 and his wife 80, they agreed that the time had come. Together they swallowed what they assumed were adequate doses of sleeping medication. Mrs. Van Dusen died the next day; but Dr. Van Dusen was taken to a nursing home, where he lived on for 15 days. He had failed in what was probably the most important action of his entire life.

Even so, the Van Dusens were fortunate. Many of those who attempt suicide are discovered, rushed by ambulance to the nearest hospital, and subjected to stomach pumping or other demeaning and painful emergency procedures. These are often, alas, successful.

Among those moved by the Van Dusen episode, and by even more harrowing accounts of botched suicides and successful resuscitations, was a California psychiatrist. He wrote a letter to the editor of Psychiatric News proposing a "Hemlock Society" that would "dedicate itself to providing information and personal counseling to those giving serious consideration to suicide." I promptly wrote him requesting membership in the society, but it was never formed.

"I regret to say that my own interest in this project waned," the psychiatrist subsequently wrote me. "It immediately became apparent to me that, if I pursued the matter in this community, I would be identified as 'the prosuicide psychiatrist,' or something equally controversial. . . . Also, the local district attorney's office . . . advised me that I and all persons in a 'Hemlock Society' would be subject to prosecution if we were seen in any way as 'aiding and abetting a suicide.' Consequently, I had to move other projects higher on the priority system.

"I continue to believe that there are a large number of thoughtful and concerned people who would support such a movement if certain others provided the leadership."

The basic principle of such an organization, it seems to me, can be simply and clearly stated: Nobody who doesn't want to die of cancer, or of any other disease, should have to die of that disease. Alternative modes of death should be available.

I had more important business than suicide planning, however, during the period immediately following my own cancer diagnosis. My first need, as I saw it, was to determine whether I could continue to enjoy life — or whether a cancer diagnosis in itself was enough to sour the weeks and months ahead. Accordingly,

WHEN NOT TO SAVE A LIFE

The October 1978 issue of Hospital Physician, a medical journal, presented the case of a suicide attempt by an 80-year-old woman who had for two years been living in a nursing home. She suffered from glaucoma, which had almost completely blinded her, and from cancer of the colon, for which she was receiving chemotherapy. Her husband was recently dead. Under these circumstances, I regard her suicide decision as wholly rational.

Moreover, her decision was not a whim of the moment. To relieve her chronic pain, and perhaps to mitigate the side effects of chemotherapy, she was being given hydromorphone, a morphine-like drug. In order to save up a week's supply of hydromorphone tablets she suffered through 168 hours of uninterrupted pain. Then she swallowed her hoard and went into coma.

Surely she had earned her right to die. But instead of being allowed to do so in peace, this aged patient was rushed to a hospital emergency room and subjected to a variety of procedures to save her life, including the intravenous injection of naloxone, a powerful morphine antagonist. The naloxone worked and she was returned to the nursing home — still suffering from glaucoma and from cancer.

Those who rushed this unconscious patient to the hospital were, no doubt, primarily responsible for the tragedy. But should the staff of the hospital emergency room share their guilt?

Emergency-room physicians explain that their role is to save lives. In a situation where minutes may mean the difference between death and life, they cannot pretend to be gods, deciding on the spur of the moment whom to save, whom not to save.

I dissent. Emergency-room physicians must make equally difficult death-or-life decisions of many other kinds. In fact, the staffs of a number of hospitals are currently amending their procedures to provide that life will not be artificially maintained in cases where the outlook is hopeless and neither patient nor family seeks prolongation. But this newer philosophy has not yet been applied to hospital emergency rooms or to suicide attempts of the terminally ill — the patients for whom the new policy is most clearly warranted.

I am confident that sound procedures can be developed to prevent many resuscitation disasters. I believe that tampering with life and death — whether it takes the form of killing a patient who wants to live or resuscitating a patient who yearns to die — is an intolerable affront to our humanity. Even if revised emergency-room procedures save only a handful of patients like this one from the horror of resuscitation, the effort will have been worthwhile. — E.M.B.

two days after the diagnosis I took off (as previously scheduled) for a human-sexuality workshop on Cape Breton Island in Nova Scotia.

My two weeks there were among the richest and most rewarding of my life, in a dramatic setting, surrounded by exciting people, dealing with fascinating subject matter, and even including what I have always considered the ultimate of life's enjoyments — a full-fledged falling-in-love experience. Far from souring life, the cancer in my colon added a unique zest to those two weeks.

On my return home, I consulted with my three grown sons, all of whom were nearby. They, like me, vividly remembered their mother's last 19 months.

"Whatever happens," I assured each of them, "I am not going to die of cancer." All three fully understood my meaning, and were unreservedly supportive.

The question of whether to

accept or refuse cancer surgery came next in my thinking. This is perhaps the most important decision cancer patients must make; yet few of them give it even a moment's consideration. They simply do what they are told to do. I think the medical profession is at fault in this respect. The surgical option is too often presented to the patient as if it were not an option at all but a foregone conclusion. Many patients, indeed, are "given the rush act" so that they won't have time to think things through. Thus, at the very beginning of their careers as patients, cancer victims are deprived of an invaluable privilege of the human condition: the opportunity to plot their own course. Once they let others make that crucial decision for them, they are already reduced to a dependent status.

When a cancer patient does weigh the options, however, it immediately becomes apparent that refusing surgery is utterly irrational if the only alternative is to let nature take its course and to die miserably after a protracted period of suffering. A feasible alternative to surgery must be available if the choice is to have any meaning. That feasible alternative, I believe, is to enjoy life to the full as long as possible and then to terminate it in your own way at a moment of your own choosing. Accordingly, before I could decide whether or not to accept surgery, I had to make sure that suicide was an available option. I devoted my next two weeks to "suicide shopping."

Since guns, knives, dangling from ropes and jumping from high places are not my style, I soon narrowed my alternatives to three: an overdose of sleeping medication, intravenous injection of a rapid poison such as cyanide or nicotine, and carbon monoxide (automobile exhaust).

I was doubtful about carbon monoxide because of the Federal air-pollution regulations designed to curb automobile exhaust gases; but an engineer friend of mine assured me that there is still plenty of carbon monoxide in the exhaust from even a 1978-model car. It just takes a bit longer to be effective.

A physician friend of mine was also helpful. He offered to write a prescription for a month's sleeping medication, cautioning me that it should be taken under circumstances which would prevent my being discovered while still alive and having my stomach pumped.

The offer of sleeping medication is a traditional ploy used by some physicians to retain control of suicide-prone patients. The offer makes it almost certain that the patient will return for the prescription before taking action — and the physician, instead of supplying the medication, can then institute appropriate antisuicide measures. I was confident, however, that in my case the offer was made in good faith.

□

It was at about this point that I made a most welcome discovery about attitudes toward suicide. In all, I discussed my plans very frankly with nine women and men. Eight of them, including my three sons, were fully supportive. The ninth, while not in agreement, made no effort to dissuade me. Attitudes toward suicide among civilized people, I concluded, are rapidly changing, although the change has not yet been publicly noted.

Another welcome discovery followed. I had always thought of suicide as an inevitably lonely experience — best performed, perhaps, deep in the woods near my home or in some distant motel room. I was therefore deeply moved when a young registered nurse who was very close to me said: "Let me be with you and give you the injection. I love you and I'd consider it a privilege."

A woman my own age made a similar suggestion — that I come to her island in the Caribbean, where I would be surrounded by affection.

I was even more moved when one of my sons approached me with another suggestion. He reminded me that many years earlier, when his mother and I used to read out loud to the children each night after dinner, we had read them Plato's account of the death of Socrates, describing Socrates surrounded by his intimate friends and enjoying their company to the full until the moment came for him to drink the hemlock.

"I have two requests," my son continued.

"First, I don't want you just to disappear. I want to be able to say goodbye. Don't go off to a meeting somewhere and not come back.

"Second, when you are considering alternatives, I wish you'd consider the Socratic alternative — inviting in the friends you really want to be with for a last evening or even a weekend."

My first reaction was one of shocked amusement. How could I possibly decide whom to invite to such an occasion? And think of the distress of those who were not invited! Surely that would be the ultimate snub. I reached no decision then on whether or not to follow the Socratic precedent, and I still have not, but I have decided I will not be alone.

By the end of my two-week suicide shopping period, I felt confident that I could rely on any of the three courses mentioned above, and that I could count on the support of those near and dear to me. I was ready for the decision about surgery. I chose to accept it.

Actually, I chose a modified form of conservative surgery and found a surgeon who agreed. The hospital consent form was altered to eliminate the patient's traditional consent to "any other procedures which may prove necessary." But that is another story. Fortunately, the surgery was suc-

cessful, and I now expect to die of a heart attack like any other self-respecting, cigarette-smoking, beef-eating, 67-year-old American male.

□

I have given some thought also to the ethical problem of accepting the help of others or of involving others in my planning — since aiding and abetting a suicide might result in their being prosecuted for murder.

Quite simply, the risk of being charged with murder would certainly not deter me from aiding someone I love if I deemed it appropriate. I would want to be sure, of course, that this was a considered decision rather than a momentary whim on the part of the person seeking my help. And the circumstances would have to be such that I understood and approved the decision. My decision to help would be easier to reach if I felt in addition that I would behave similarly in like circumstances. Since I am prepared to help others close to me despite the risk to me, I see no ethical objection to my accepting the help of others, despite the risk to them.

As a practical matter, the risk of prosecution is small if plans are prudently laid and skillfully executed. This is particularly true in terminal cases; for even if by mischance the facts become known, only an exceptionally strict prosecutor would willingly prosecute for murder someone whose only offense was helping a dying friend in dire need. Such prosecutors, I trust, will sooner or later — let us hope sooner — vanish from our society. Until then, prudence and caution should of course be exercised, both by those planning suicide and by those assisting them.

Even when prosecutors do intrude, juries have a way of demonstrating their sympathy and human concern. Among many cases, I need only cite that of George Zygmaniak of Monmouth County, N. J., who broke his neck in a motorcycle accident in 1973. Paralyzed from the neck down, he begged his brother to help him die. Four days after the accident, George's brother killed him with a shotgun. Following a widely publicized trial for murder, the brother was acquitted by a jury. Acquittal in a murder trial, it should be remembered, can be handed down only by a unanimous jury of 12 individuals.

My own plans, incidentally, took account of the possibility that I might at some point — following a stroke, for example — be unable to give consent

or take action myself. Those close to me agreed that, in that event, my present request for help would be taken as a continuing request.

It is unfortunate that each patient must do his own suicide shopping without even the most rudimentary guidelines from the medical profession. Most practicing physicians are themselves poorly versed in the advantages and disadvantages of various forms of voluntary life-termination available to laymen. Thus, all of us, quite unnecessarily, run the risk of suffering the fate of Dr. Van Dusen, or even worse.

As a modest first step, I suggest the preparation and widespread distribution of a pamphlet for laymen entitled "How Not to Commit Suicide," designed to warn against inadequate and inappropriate measures. If readers learn incidentally about some of the preferable ways, so much the better. I would like to write such a pamphlet, but I have not found a physician knowledgeable in such matters who is prepared to collaborate. A recent book — "Common-Sense Suicide" by Doris Portwood, published by Dodd, Mead — is clear-eyed and honest, but is not a "how-to" guide.

I think it is high time that those who think as I do about suicide "come out of the closet." I am confident that even a modest band of women and men prepared to speak out frankly on the suicide issue can have a healthy impact on suicide law and on public attitudes toward suicide. My own experience indicates that coming out of the suicide closet can be a rewarding rather than a harrowing venture.

The suicide routes available and acceptable to me depended, obviously, on my personal tastes and preferences and on the particular people who are close to me. Others must lay their plans in the light of their own circumstances and resources. Many people, moreover, have deep religious, ethical or emotional objections to suicide by any means at all. But for a large and, I believe, growing portion of society who deem suicide vastly preferable to prolonged and unnecessary suffering, I strongly recommend advance planning. And the planning should be done years in advance of the occasion, so that the available alternatives can be unhurriedly explored and prudently weighed. ■

March 18, 1979

NOW, VIETNAM VETS DEMAND THEIR RIGHTS

David Christian, one of the most decorated soldiers of the war, at home with his mementos; and (at left) with bride, Peggy, in 1969.

We're a very patriotic group. We still believe in symbols, in the flag, in America. I'm against anyone who tries to taint us for that. Most families in this county contributed two or three guys to the war. Most of these guys feel they participated in something wrong. They feel scolded. Their peers moved ahead of them. They came back and people said: Block out Vietnam — what you did was worthless. It twisted a lot of guys.'

By Bernard Weinraub

Going to war is a landmark experience in the life of an individual, an episode of tremendous importance, but in the case of Vietnam you learned very quickly to repress it, keep it secret, shut up about it, because people either considered you a sucker or some kind of psychopath who killed women and children.
— ROBERT MULLER, a former Marine lieutenant, who led an ill-fated assault on a hill near Conthien in 1968.

In the first formal job interview of my life, with a large Washington law firm, I was asked whether I committed any war crimes in Vietnam. It blew my mind.
— JOSEPH C. ZENGERLE, a West Pointer who is a former special assistant to Gen. William C. Westmoreland.

All my friends — we were kids together, we grew up together, we went to Vietnam together. They don't know what's happened. They can't figure it out. These are confused and broken guys. . . . It breaks my heart.
— DAVID A. CHRISTIAN, a Vietnam veteran who was one of the most decorated soldiers of the war.

They are in their 30's now, many of them married, the fathers of young children. They have been home for 10 or 11 years, having slipped back quietly into society without victory parades to welcome them, without brass bands or cheering crowds or even recognition. They are, in many ways, a perplexed and splintered generation, a group of men, mostly blue-collar, who have, until recently, buried their experiences of the Vietnam War, refusing to run for office or lobby through veterans' organizations — like veterans of past wars — or make political demands as a group to make amends for the education and unemployment and even psychological difficulties that have crippled them. What has dominated the mood of Vietnam veterans has been a passion for isolation and anonymity.

After nearly a decade, the Vietnam experience and its aftermath have unleashed a wave of award-winning films, such as "Coming Home" and "The Deer Hunter," and some first-rate novels, including "Fields of Fire," by James Webb, Tim O'Brien's "Going After Cacciato," Larry Heinemann's "Close Quarters," and Frederick Downs's "The Killing Zone," as well as Michael Herr's "Dispatches." But beyond this, the veterans themselves — following the nation's upheaval and trauma — are now struggling to organize, to assert themselves for the first time, to come to terms with an experience that many of them had sought to erase.

The Carter Administration, mindful of the potential political clout of Vietnam veterans, has proclaimed May 28

Bernard Weinraub is a reporter in the Washington bureau of The Times.

Robert Muller, who was shot leading a Marine assault and paralyzed from the chest down, at home with his wife, Virginia; inset, on leave on Waikiki Beach in 1969.

You know, I remember joining the Marines and standing in my dress whites and hearing "The Star-Spangled Banner" and crying like a baby. I cried out of pride. We were so idealistic.... My dream was a fraud. For a lot of guys, the foundation in our life was shaken. We were told we were fools.'

to June 3 as Vietnam Veterans Week. The gesture, while welcomed by Vietnam veterans, is viewed as more symbolic than substantive, and they point out that the Administration has so far failed in any real way to relate to their needs. On Capitol Hill, a group of 20 Democrats and Republicans has organized Vietnam Era Veterans in Congress.

More significantly, Bobby Muller and others have formed Vietnam Veterans of America, to press for jobs, education, health care and counseling. They believe they must fill a void left by the Establishment veterans groups that have for the most part, consciously and unconsciously, scorned the younger men who returned home battered and alienated from a losing war.

Perhaps now the passage of time has begun to sweep away the humiliation of the nation's longest and most unpopular war. Perhaps, as several Vietnam veterans claim, the nation is finally set to confront the issue of Vietnam, address the meaning of the war, explore its implications, face what it did to a young generation. The United States sent nearly 2.8 million soldiers to Vietnam. As Lawrence M. Baskir and William A. Strauss pointed out in their provocative book "Chance and Circumstance," "Fifty-one thousand died. . . . Another 270,000 were wounded, 21,000 of whom were disabled. Roughly 5,000 lost one or more limbs; a half million were branded as criminals, and millions had their futures shaped by the threat of going to war."

A truly detailed assessment of Vietnam veterans is impossible, but it is known that their rate of divorce, suicide and mental breakdown is unusually high compared with the rest of the population. Underemployment is a severe problem — 26 percent of Vietnam veterans are earning less than $7,000 a year. The number of veterans who have earned college degrees has proved disappointing, largely because, until 1974, there were limited education benefits available to them.

Out of the plethora of statistics about the needs of Vietnam veterans, out of the stereotypes that have emerged — the smoldering, dangerous Vietnam vet on the edge of going beserk — many, perhaps most of them, have readjusted, but with some difficulty. Taking several steps away from the statistics and the stereotypes, however, looking closely at three disparate veterans, one finds that the emotional depth of the Vietnam experience and its aftermath remain remarkably powerful.

Bobby Muller

He still remembers the choking dust, the heat, that engulfed his unit that April afternoon in 1969 as the Marine colonel shouted over the radio: "Take the hill! *Take the hill!*" It was outside the refugee village of Cam Lo. Leading a South Vietnamese unit up the hill, Lieutenant Muller, backed by 10

tanks, called in artillery and then jet fighters that pounded the hilltop. But the North Vietnamese squad, dug in, sprayed machine-gun fire across the hill, thwarting any advance by the Americans and the South Vietnamese.

It was twilight, and the colonel shrieked one more time, "Take the hill!" and Muller and his bedraggled, terrified South Vietnamese troops began walking up slowly, and then. . . .

"I caught a bullet through the chest, through both lungs, severed the spinal cord, and right out my back. I saw, suddenly, a kaleidoscope. Fragmented colors. Somebody was hitting me with a sledgehammer — that's what it felt like. Totally numbing. Then I'm looking up at the sky and it's getting dark and I say, 'I'm dying.' I didn't believe it. 'I'm gonna die on this ground.' And I felt myself recede from consciousness, and that was it.

"I should have died on the hill. The kind of wound. Where I got it. I was incredibly *lucky.* And when I woke up on the hospital ship, it was unbelievable. When they came around and told me I was going to be a paraplegic, my response, really, was, 'So what?' Man, was I lucky."

It is 10 years later now and Bobby Muller, who is paralyzed from the chest down, speaks obsessively about the unhealed wounds of Vietnam, about the silence and guilt — until lately — that afflicts the neglected veterans of an unpopular war. "Only now, after years of feeling betrayed and humiliated, after years of trying to disavow our experience, are veterans trying to understand what happened to them, trying to do something about their problems," says Muller.

"We still haven't had an accounting," he says. "Why did it happen? People always say it was a waste, but, when you ask why it was a waste, you get 50 different stories. Let's talk about it. You cannot take 55,000 lives away — and that amount of money spent — and never give an accounting for everything that happened. *Why did Vietnam happen?*"

He is sitting now in his new office on Eighth Street in the northeast section of Washington, a quiet, predominantly black street where Muller is struggling to form the Vietnam Veterans of America — the first organized group to press the case, solely, of Vietnam veterans. It is a task that was made difficult by what has been the lack of political clout of Vietnam veterans, by the preoccupation of members of the older groups, such as the Veterans of Foreign Wars or the American Legion, seeking higher retirement benefits or higher pensions from Congress. Such a preoccupation is a source of deep resentment on the part of the younger men, who are seeking psychological and health care, or an education or jobs.

Beyond this, Muller and other veterans contend that the White House, although giving "lip service" to the vets, has brushed them aside politically and symbolically. Fewer than half a dozen of 700 key political appointments by the Carter Administration have been given to Vietnam veterans, and Muller points out that for nearly two years the White

House domestic-policy aide whose responsibilities include handling the problems of Vietnam veterans has been a 30-year-old former teacher, Ellen Goldstein, who had Democratic Party ties and no direct experience with Vietnam.

It is widely known that last autumn Stuart E. Eizenstat, the President's adviser for domestic policy, strongly recommended that Mr. Carter personally announce before a group of veterans his proposal for $250 million in benefits for those who served in Vietnam. But Gerald M. Rafshoon, Mr. Carter's public-relations aide, argued that the President should avoid meeting the group because it would probably attack the program as inadequate. Vice President Mondale made the announcement, and Muller and other Vietnam veterans were insulted.

"At the White House there's not a Vietnam-era veteran dealing with Vietnam problems," he says. "It's unbelievable. How can you be formulating a policy for nine million guys and not have these guys represented?"

Although Bobby Muller was a strong voice in Vietnam Veterans Against the War, he has a mild contempt for some radicals and antiwar types who, for example, deplore "The Deer Hunter" because it depicts the Vietcong and North Vietnamese in a cruel light. "You get out of the movie what you bring to it," he says. "It's not a great movie; it's a good movie. It's not a literal interpretation of the war but — the Russian-roulette scene, that's a great metaphor. War is a numbers game. You're plucked out. It doesn't matter how bright or brave you are. If they reach out and grab you, that's it."

Muller grew up in Whitestone, Queens, went to Great Neck South High School and the State University at Cortland, where his dominant interests were wrestling, cross-country and track. "Going to Vietnam was the expected thing to do, at least in my circles," he says. "I joined the Marines principally because I was going to class one day and there was this guy, this recruiter standing in dress blues, sharp as a tack. I went up to him and I said, 'Hey, talk to me.' And this guy, he was beautiful, he said, 'If you want to join the service and you want to do a job and you want to bust heads, then join the Marines. It's a unit of pride and distinction.' "

As with many G.I.'s, doubts about the war were brushed aside by Lieut. Bobby Muller of the Second Battalion, Third Regiment, Third Marine Division in northern I Corps, near Khesanh and the Ashau Valley. Survival was the key.

After he was wounded, he spent a week on the hospital ship, the U.S.S. Repose, and then six weeks at the St. Alban's Naval Hospital in Queens. It was only after being transferred to the Kingsbridge V.A. Hospital in the Bronx that Muller faced the nightmare that still terrifies him.

"From the time I got shot I never cried," he says. "The hospital ship, the naval hospital — the doctors and nurses and attendants cared. When I got to Kingsbridge Hospital and saw that this was going to be the place where I was

going to be staying, it so overwhelmed me that I broke down and cried. My mother broke down and cried. It was overcrowded. It was smelly. It was filthy. It was just disgusting."

He grips his wheelchair and speaks in a choked voice. Drainage bags attached to the sides of men's beds often went unemptied and overflowed onto the floor. Rats and mice climbed on the beds at night. The patients were Vietnam wounded as well as old-age veterans who stared vacantly at the ceiling all day.

"My best friend killed himself in that ward," he says. "You want to know how a quadriplegic kills himself? He has his brother help him. Right off the top of my head, five guys I know killed themselves there.

"On my first day in the hospital, I asked for a wheelchair, and they said, 'We can't get you one,' and I went crazy. 'What do you mean you can't get me a wheelchair?' I was a quiet guy, always, but when they put me in a corner like that I began to scream."

An evening attendant told Muller he was too busy to help the veteran get from his bed to his wheelchair, and that he had to stay in bed after 4 P.M. each day. A psychologist told Muller that he should slip off into a corner and cry for himself.

"The doctor came by the second day with a little card and said matter-of-factly, 'You've got a total severance of your spinal cord, and you'll never walk again.' Period," Muller recalls. "I was aghast. I had spoken to neurosurgeons at St. Albans. They said you couldn't say anything with certainty. Then here comes this jerk . . . ."

In 1970, Life magazine did a cover story about Muller's ward — an article that resulted, according to Muller, in a new paint job, some alterations, but no real change for several years. Kingsbridge was scheduled to be demolished and replaced by 1978 but the new building may still not be ready before the end of this summer.

Following his one year in the hospital — as he realized that "Vietnam veterans were getting a raw deal" — Muller rented an apartment in Great Neck, L.I., joined Vietnam Veterans Against the War and attended Hofstra Law School. His initial obsession, he says, was the inadequacy of the V.A. system, whose $20 billion-a-year medical, education, housing and welfare programs are heavily weighted in favor of the established veterans lobby, supported by veterans-affairs committees in Congress.

Initially Muller, who lives on Long Island with his wife, Virginia, was legislative director of the Eastern Paralyzed Veterans Association, but, discerning quickly that Vietnam veterans had unique needs, he has struggled to form the new Vietnam Veterans of America. "You know, I remember joining the Marines and standing in my dress whites and hearing 'The Star-Spangled Banner' and crying like a baby," he says. "It was 1967. I cried out of pride. We were all so idealistic."

"My dream was a fraud," he says quietly. "For a lot of guys, for me, the basic foundation in

VETERANS

Joe Zengerle at home with his son, and (left, center) in Vietnam a decade ago with General Westmoreland.

our life was shaken. We came home and were told we were either fools or killers. Guys came home after this significant life experience and they've never had the opportunity to talk about it. It's torn people apart. It's unnatural. . . ."

Joe Zengerle

Ten days after returning home from Vietnam, Joe Zengerle slipped out of his apartment on Connecticut Avenue while his wife, Lynda, slept and drove his car to southeast Washington, a tough, high-crime area. It was after 2 A.M.: "I mean, I wanted somebody to mug me, to try to hurt me, and I wanted to kick the hell of them," he says quietly now. "I wanted to release something. I wanted to release certain emotions. I walked the street for hours," he says, smiling. "Nobody attacked me." *(Continued)*

"I came home and sat on the edge of the bed with my wife and I began talking and telling her that there were times that I was very scared. I heard a helicopter go by and I began telling her that I was petrified of helicopters. I never admitted that before to anyone, myself included, and suddenly my hands began to shake and I began crying uncontrollably for an hour. It was my way of getting back. The next day, I was sitting on a footstool in the living room, listening to Simon and Garfunkel, and Lynda asked me what I was going to do with my life and I said I didn't know. She said she wanted to go to law school and I said, 'I guess I'll go to law school with you, too.' "

On the face of it, Joe Zengerle's wartime experience — and his postwar career — are rare among Vietnam veterans. A West Pointer, class of 1964, an outstanding, gung-ho graduate of airborne school, Zengerle was assigned for six months to serve as a special assistant to General Westmoreland and his deputy, Gen. Creighton Abrams. As an intelligence aide, he briefed both men on incoming communications from Washington, went on private missions for the two generals, traveled with them and served, in effect, as an executive assistant. On the second portion of his one-year tour in Vietnam, Zengerle was assigned to an intelligence unit of the Americal Division in northern I Corps. He submitted his resignation from the Army in May 1968.

Zengerle rapidly immersed himself in a career that seems golden. He graduated from the University of Michigan Law School in 1971, worked for the prestigious Washington law firm of Arnold & Porter, and served as a clerk to Judge Carl McGowan on the Court of Appeals and then to Chief Justice Warren Burger on the Supreme Court. He worked for Justice Burger in 1973-74, the year of what he calls "the Nixon case."

Although he then had his pick of Washington law firms, Zengerle selected a medium-sized one, Shea & Gardner, partly to liberate himself from the crushing hours of high-pressure firms that provide Washington lawyers with rich fees and impoverished family lives. His wife and two small children, he says, are the most important elements in his life.

Slowly, in ways that still confuse him, Zengerle has become deeply immersed in the Vietnam-veterans issue. His neighbors, his law colleagues, many of his friends never served in Vietnam. "I had virtually no contact with people from Vietnam for several years. I buried Vietnam in my mind."

At the 10th reunion of his West Point class in 1974, Zengerle began talking to his classmates, reminiscing about the men who fell in Vietnam, recalling the problems of coming home and seeing friends who had fled the war.

"I was wondering about a coherent principle to tie this reunion together and, I guess, I had a sense that it fractured a generation into at least two camps and that the fracture line certainly has temporary, perhaps permanent, implications," he says. "Camps were created that had — and still have — a disproportionate effect on our generation.

"On the one hand, you have the people in Vietnam. Some of them were rabid militarists who thought the greatest thing in the world was to get into a gunship and hover over a village and pull the trigger. Some of them were the most rabid saboteurs of the military you could find — they would frag officers, subvert the war effort, do anything to undermine the military. There were guys who loved the war and guys who hated the war and a hell of a lot of guys who just wanted to live through it and come home. But they were part of one camp."

The other camp, he says, consists of the loosely allied political and ideological movements of the 1960's — the consumer, environmental, civil-rights, women and antiwar movements. "I think those people — many of them running the levers of government now — made certain moral judgments about those serving in Vietnam, created a certain climate of hostility, created a sense that people who went were either evil or stupid," he says. "And that was such a pervasively held view that it didn't have to be expressed anywhere, on the record, in writing. It was just felt by a lot of people."

Zengerle is sitting in a restaurant, gripping a linen napkin in a tight fist. "Look, I lost three of the closest friends I've ever had in my life in Vietnam," he says. "War — any war — tends to create an environment in which the things you've experienced are indelibly etched in your recollections — the friendships you've had, the affairs you've had, the situations you've gotten into, whatever dangers you've faced.

"Well, we, as a group, were denied the opportunity to share that experience, to talk about it, review it, have some degree of self-acceptance," he says. "There's this inability to gain value from this experience through open discussion. O.K., it may have been a bad experience. But we were entitled to a climate that allowed us to identify and secure to us the value of a bad experience, whatever value there may have been. And I think they took it away from us by their moralistic judgments."

Joe Zengerle is caught between two worlds, and he knows it. He lives in Washington, D.C., and the luncheons and cocktail parties that he attends are cluttered with colleagues and neighbors whose decision to avoid the war grates on him increasingly. He compares the Vietnam veterans with Jews — he converted to Judaism after marrying his wife. "Like Jews, Vietnam veterans accepted the perception handed down to them that they were inferior or not worthy or they could be trod upon," he says. "Then came the holocaust and the birth of Israel."

He is tentatively planning to take a year or two and work full-time for Vietnam veterans. Although he is a Democrat who campaigned actively for Jimmy Carter, Zengerle expresses anger at the "natural discrimination against Vietnam veterans" within the Administration, the "paltry" programs for veterans, the failure to appoint any more than a half-dozen veterans to Administration jobs out of 700 high-level posts.

After Jimmy Carter was nominated, Zengerle got an inkling of the future Administration's point of view toward Vietnam veterans. He approached Dr. Peter Bourne and asked him about working on "substantive issues" during the campaign. Bourne, Zengerle says, was "extraordinarily condescending" and said with obvious lack of conviction that he would try to come up with something. Zengerle, at that point, walked out angrily. He feels now that Bourne, who was an antiwar activist, resented him because he had worked for Westmoreland.

"Last summer, almost 10 years after I first saw Vietnam, I had lunch with a young woman who had barely been in high school in 1967," he says. "She asked me if it was true that soldiers my age had borne the brunt of the war. I told her that three of my closest friends, including my plebe roommate, had been killed in action. Another person with us, a man closer to my own age, remarked: 'Yeah. And a lot of North Vietnamese died, too.'

"In no way did my answer to the woman's question deserve the implication in that remark that I or my dead friends were responsible for the war," he says. "Their sacrifice does not even establish that they supported the war policy. Some things should be put behind us; others should never be put before us in the first place."

Dave Christian

It is shortly after 7 P.M. at Dave Christian's home in Washington Crossing, Pa., just outside Levittown. A half-dozen Vietnam veterans stand and salute the flag as the meeting of the David Christian post of AMVETS begins.

The post president, Joseph Kennedy, a burly, gentle man who was seriously wounded in World War II and suffers headaches and tremors and collects disability payments now, asks for a minute of silence in honor of the war dead. Kennedy's son, Joey, is a paraplegic, crippled in Vietnam.

Dave Christian — perhaps the most decorated soldier of the Vietnam War — sits on the sofa in the living room of the big split-level wood and brick house. The litany of the men facing him is poignant, bewil-

Christian's Purple Heart and Distinguished Service Cross.

dering: "Dave, Dave, I keep running into closed doors. . . . We're now on food stamps. . . . I'm too old now, I'm 32, I don't like to train with 20-year-olds. . . . Sure there are CETA jobs around, but, man, I don't want to be a dishwasher or cook. . . . I am just disillusioned. . . . How can I support three kids on $3 an hour. . . . It sickens me, the guys who went to Canada and jail, those guys are the heroes. We're nothing. . . . Dave, I got a job literally picking up dead cats and dogs from the highway. Couldn't take it after a while but, Jesus, that does terrific things to a

man's ego, you know? . . . Dave, I just can't plug into the system. . . ."

Dave Christian listens impassively. He is a handsome, boyish figure, blond and blue-eyed, speaking softly and dramatically. "Look, you're veterans, be proud of it, use it, don't be ashamed," he says. "You're somebody, somebody to be respected. There is leverage. If employers give you a runaround, get back to me. You're under affirmative action programs. Don't get beaten down, man. . . ."

He grew up on welfare in Levittown, Pa., his father an alcoholic who walked out on Dave's mother, leaving her to raise three small boys. "It was just so painful," Dave Christian says quietly now, sitting in his home after the meeting. "Kids made fun of us — the color shoes we wore. The welfare. My mother bought clothes for us at the Salvation Army, hand-me-downs."

It was — and is — a working-class community whose men worked in the steel mill, a patriotic blue-collar community whose sons went to Vietnam because their fathers went through World War II. "We're a very patriotic group. We still believe in symbols, in the flag, in America," he says. "I'm against anyone who tries to taint that for us. Most families in this county contributed two or three sons to the war. Many of them lost their sons, or had sons badly wounded.

"Most of these guys still can't figure it out. 'What happened?' These guys have been on a 10-year guilt trip. They feel they participated in something wrong. They feel scolded. Their peers moved ahead of them. They came back and high-school kids jeered at them. Some of these guys had the lives of men in their hands, had awesome responsibilities, dealt with millions of dollars in machinery. They wanted to keep that momentum going. They came back and people said, 'Forget it. Block out Vietnam. What you did was worthless.' It twisted a lot of guys."

Christian, whose two brothers were wounded in the Army, enlisted in the military at the age of 17. "I joined the Army only for upward mobility, not for war or that stuff," he says. "I wanted to get the G.I. Bill."

He sought out perhaps the most dangerous assignments of the conflict, long-range reconnaissance patrols into Laos and Cambodia and North Vietnam to uncover North Vietnamese base camps. He was wounded on at least eight different occasions — "Actually those were suicide missions," he says now. In October 1968, his patrol was trapped in a North Vietnamese base camp in Cambodia and, by the time an American unit broke through to save the patrol, Christian was nearly unconscious. He had been stabbed, shrapnel wounds covered his neck, his thighs and crotch were bloodied by the explosion of an antitank weapon. As he was evacuated to Japan, his wife, Peggy, received a telegram from the Defense Department saying that her husband's death was imminent.

After recovering in an Army hospital in Japan, Christian returned to Vietnam and served in the Central Highlands. During a North Vietnamese attack, he was accidentally struck by na-

palm — friendly fire — and suffered second- and third-degree burns over his entire body. Christian was eventually evacuated and was hospitalized for nearly two years at Valley Forge Army Hospital outside Philadelphia.

His weight dropped from 170 to 90 pounds. He was given the last rites of the Catholic Church twice. He caught a respiratory disease, melioidois, that ate away portions of his lungs. "It was horrible," he says. "The shots were painful. The burns were painful. You could stick your fingers through my tongue, it had so many holes in it."

Although scars from the napalm attack cover most of his body — and he is usually in some discomfort and pain — Christian's endurance and recovery were based on a powerful psychological and physical need to survive. "I justified the war in my mind in order to survive," he says. "If I agonized about the war, if I felt what I did was wrong, it would have twisted me. No guilt. I justified everything I did. I know I killed people who had mothers and wives and children like me, and *I did not feel guilty about it* because if I didn't kill them, they sure as hell wanted to kill me. There is a self-destruct mechanism about many G.I.'s; you got so much abuse and guilt that you had to go into a decline," he says. "I just lived day by day, not thinking about the war, just surviving; gaining a pound was a big deal. . . ."

Christian was twice recommended for the Medal of Honor. Among the combat decorations he earned were the Distinguished Service Cross, two Silver Stars, two Bronze Stars, seven Purple Hearts, the Air Medal for 25 combat assaults from a helicopter and two Vietnam crosses for gallantry.

He completed his studies at Villanova University in 19 months, and then worked for a direct-mail advertising company in Pennsylvania. In 1978, he took a job with the Department of Labor in Washington, working on veterans affairs, and last January — amid some controversy — he was removed from the consultancy post because, he said, he had been "too visible" and "too frank" in the job and had "made people in the front office uneasy" by his speeches to Vietnam veterans in which he urged them to find jobs, to organize politically, to take advantage of Government training programs and to take pride in having served in Vietnam. In the ensuing publicity, the embarrassed Labor Department said that his one-year consultancy had merely expired and that it was making efforts to find him another job. He is now working on veterans' issues at the Labor Department's Philadelphia office, a job he prefers because of its proximity to his home, his wife and three children.

"I really don't know how I survived," said Christian, who, at 20, was the youngest captain in the war. "It was a class war — there weren't any rich or college-educated boys fighting with me — but I didn't come back bitter. I figured life was just getting knocked down and getting back up again. I didn't shift the blame on anyone else. I just took it upon myself to deal with life." ■

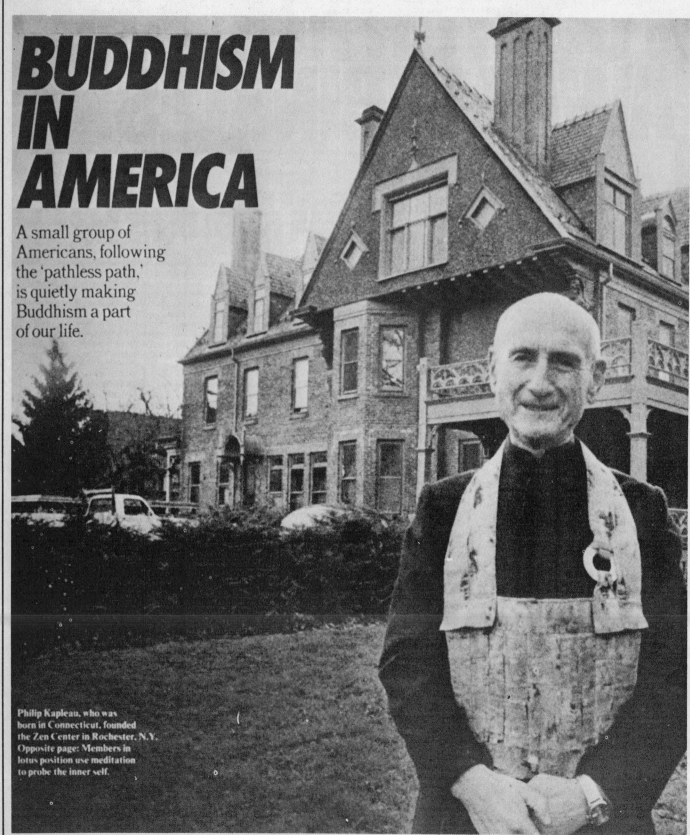

BUDDHISM IN AMERICA

A small group of
Americans, following
the 'pathless path,'
is quietly making
Buddhism a part
of our life.

Philip Kapleau, who was
born in Connecticut, founded
the Zen Center in Rochester, N.Y.
Opposite page: Members in
lotus position use meditation
to probe the inner self.

Bob Adelman

By George Vecsey

The father's words reflected the pain in his face, as he talked about his teenage son.

"At first he said he was an atheist and didn't want any part of religion," the father said. "But lately he has been saying he's an agnostic, so maybe that's promising. Sure, we would have liked him to follow our example."

These words of disappointment could

George Vecsey covers religion for The New York Times.

have been spoken by millions of Christian or Jewish parents whose children showed no interest in the family faith. In this case, however, the father was an American-born converted Buddhist.

Yet, as disillusioned as it may sound, the father's lament may be an indication that Eastern religion has reached a new stage in America, a stage in which many "founding" fathers and mothers are secure enough in their religious beliefs to want to see them adopted by their children.

Eastern religion has enjoyed several periods of growth among non-Oriental Americans since the end of World War

II — the first with the Beat generation in the late 40's, then a larger wave with the peace-and-love generation of the mid-60's, which sought self-awareness as one antidote to the Vietnam War and general disillusionment with American institutions. Most of the wanderers of those generations have moved on to the next stimulus by now, their interest in meditation and harmony either diverted or disrupted.

But as the 1970's make their last inscrutable stand, a smaller core of converts to Buddhism is building distinctly American communities combining the teachings of Zen with the demands of contemporary American life. In clusters of varying size across the country, the largest of which are centered in Rochester, New York City, Los Angeles and San Francisco, an estimated 500,-000 Americans are quietly practicing some form of Buddhism.

□

Perhaps the best example of community-building among contemporary Buddhists is the Zen Center on the comfortable east end of Rochester, where about 300 members, who live in houses and apartments throughout the city, are finding a Buddhist focus for their lives.

The Zen Center consists of two large old houses, now joined by an archway, with a peaceful garden and meeting hall out back. Just inside the front door, a wall of photographs resembles a suburban community-house bulletin board. There are snapshots of a Buddhist Thanksgiving (since many Buddhists are vegetarian, the turkey is released in a game preserve rather than eaten); snapshots of children wearing goblin costumes on Halloween; snapshots of the center's softball team.

Every Sunday, 150 to 200 people walk or bicycle to the center (they are discouraged from driving cars for environmental and good-neighbor reasons) for the 8 A.M. meditation. They are almost all Caucasians, ranging in age from 18 to well over 65; most are trim and healthy looking. As they arrive at the front door, they are already withdrawing into the silence of the next few hours, taking off their shoes, putting on brown robes, finding a spot in one of the austere but airy rooms where they can assume a full lotus position. With eyes half-closed, they will empty their minds of distractions.

The meditation is broken into three half-hour segments, punctuated by walking meditation exercises through the house. After the meditation, the group gathers to chant — in English — and listen to the words of the center's *roshi*, or "venerable teacher," Philip Kapleau.

The natural feeling of the rooms in these houses is enhanced by the woodwork, which has been stripped of paint. Carpets and simple furniture complement the plants arranged in front of every window. The food, shared by the staff following the meditation session, is magnificent — steaming dishes of grain, fresh vegetables from a farmers' market, homemade applesauce and breads, cheeses, hearty soups. ("We don't let new people work in the kitchen," says one of the center's monks. "We believe you need experienced, calm people there. Good karma creates good food.")

In the center, there are statues of Buddha in prominent places, of course, but little else speaks of Buddhism's Eastern origins or the 13 years Philip Kapleau spent in Japan. It is his objective to help people make Zen an American way of life rather than a foreign religion, to "keep the spirit of Zen but give it an American dress. If Zen Buddhism is going to grow in this country," says Roshi Kapleau, "we must get rid of foreign cultural encrustations."

□

Philip Kapleau is one of 10 or 12 bona fide roshis teaching in America today. (Since there is no central Buddhist organization, roshis are known mainly by reputation.) He was born in 1912 in Connecticut and worked as a court reporter for the United States at the Nuremberg trials and later at war trials in Japan. He says he was impressed with the way the Japanese dealt with their role in the atrocities of the war compared with the way Christian Germany dealt with its past. He was introduced to the Buddhist notion of "karmic retribution" — for inflicting pain, a harvest of pain will be reaped — and determined to explore Buddhism. In 1952, with no knowledge of the Japanese language, he gave up his job in America and sought a Japanese master, eventually studying under three, one of whom was Dr. D.T. Suzuki. In 1965 he decided he had "grown stale" and came home to America — a 53-year-old man who shaved his head and considered himself ready to share Zen Buddhist teachings with others.

He established his first center in Rochester in 1966 with a nucleus of 20 people. Today the Zen Center has American affiliates in Denver, Chicago and Boston. There are four centers in Canada, one in Mexico City, one in Costa Rica and one in Poland; the Zen Center's estimated membership is 750 worldwide.

The center supports itself with membership dues ranging from $20 to $50 a quarter; fees for training programs and seminar workshops; voluntary contributions, and royalties from Roshi Kapleau's three books. The Rochester center has a staff of 28, none of whom, including Roshi Kapleau, earns a salary.

Today, Roshi Kapleau sees more older people moving toward Buddhism. In the 60's it was mainly the young. "They were disillusioned and in despair, many of them on the verge of suicide," he says. "They felt the Establishment was insincere, not meeting genuine needs. A lot of young people were in drug trouble, looking for something. Many of them had no skills and could be called irresponsible.

"Young people wanted something simple, something worthwhile. I had to show them they could change their karma [a Buddhist and Hindu term meaning a self-caused inevitability]. To do so meant to take responsibility for their own lives. I had to show them enlightenment came from within."

Buddhism does not seek converts; proselytizing is foreign to it. "In Zen, the idea is almost to push people away," says Roshi Kapleau. "A person needs real ardor and perseverance to attain a higher order of enlightenment."

Roshi Kapleau tells this story: "A young man once came to a venerable master and asked, 'How long will it take to reach enlightenment?' The master said, 'Ten years.' The young man blurted: 'That long?' The master said, 'No, I was mistaken. It will take you 20 years.' The young man asked, 'Twenty years?' and the master said, 'Come to think of it, maybe 30.' "

□

Buddhism eludes easy definition. A group of psychiatrists once engaged Roshi Kapleau to explain Zen to them. He obliged by silently munching a ripe banana. When a student asked what it all meant, the roshi rubbed the banana peel in the student's face and said, "You have just witnessed a first-rate example of Zen. Are there any questions?" Most of the psychiatrists said they still didn't get it and pressed their speaker for a definition. Roshi Kapleau thought awhile and said, "All right. Zen is a flea copulating with an elephant."

Knowing that the roshi feels about definitions, I nevertheless persisted in asking him for one.

"A pathless path that has to be walked, not talked about," he said. "A path between two extremes, self-indulgence and self-mortification. The realization that we exist on two levels — relative to time and space, and transient, beyond time and space. Does that do? Leave it blank or answer it yourself."

According to a recent survey, there may now be two million adults in this country involved in some form of Eastern religion. In a Gallup survey of "experiential" movements — groups stressing personal spiritual discovery — this total fell behind other movements such as transcendental meditation (six million), yoga (five million), charismatic renewal (three million) and mysticism (three million).

In her book, "Buddhism in America," Emma McCloy Layman estimates that 500,000 native-born Americans have made some form of personal commitment to Buddhism, but, she adds, "The influence of Buddhism in America has been and will continue to be greater than membership figures would lead us to expect, and it is here to stay." Dr. Layman notes that meditation practices have already been integrated into some psychiatric methods and some Christian worships, and she says Buddhist concepts of "suffering, impermanence and nonself have already been widely accepted." She predicts many people will be influenced by Buddhism to seek a "simple life style."

"If you are in touch with yourself through meditation, you can influence other people," Roshi Kapleau says. "If you are in touch with deeper forces, you can propel these forces. You can be a powerful transmitter."

□

Ruth Sandberg's face is serene; her body is as lithe as a teen-ager's, but she is a grandmother who spends six months a year in Florida and six months teaching hatha yoga in Rochester.

"When my husband died, I was still searching," she says. "It was my good karma that a friend gave me a clipping about the roshi's first book. I read it, and as soon as I could, I hopped on a plane to Rochester. The minute I walked into this center, I knew I was home. Sitting on that cushion, meditating, I lost all my hang-ups."

During a Saturday workshop, Mrs. Sandberg demonstrates exercises to assist in meditation. To a novice whose muscular legs are resisting the lotus position, she points out sweetly: "An inflexible body is the sign of an inflexible mind."

□

"Pain" is a word frequently used by Buddhists around the country. At the Zen Center of Los Angeles, a tranquil priest named John Daishin Buksbazen told me how the pain of the politics and the assassinations in 1968 sent him to meditation. At the Tibetan Buddhist Nyingma Center in Berkeley, a gracious student, Kimberley Bacon, told of the emptiness that led her around the country, searching for something. And many can tell stories like Zenson's, a 30-year-old priest in the Zen Center.

Zenson grew up in a small town outside Rochester. He hung posters of Mickey Mantle on his wall, played sports and experimented with drugs in the 60's. He went to Denmark for a while, afraid of being drafted. He had abandoned his parents' Christian beliefs early on.

One day a  (Continued)

"We believe we need calm people in the kitchen. Good karma creates good food."

BUDDHISM

Continued

friend told him about the Zen Center. His first response to meditation was the pain of a posture that was impossible for his leg muscles, thickened from years of playing sports. It took him six months to stretch his ligaments so he could sit in the full lotus position, legs crossed, each foot resting on the opposite thigh. Sitting longer and longer, hands in his lap, eyes downward, counting his breaths, he learned not to think about sports or sex or the traffic outside the building or dope or any of his troubles. He learned simple woodworking and clerical tasks, and he learned to take his meals in silence, as is the custom.

"I can only say what it has done for me," Zenson says. "Since I came here, I have had a sense of oneness with the world. It's ongoing, ever-deepening."

Two years ago he was ordained as a priest, shaved his head and adopted the name of Zenson — son of Zen. (Roshi Kapleau gives names to his priests and monks that approximate Western sounds.)

"My parents came to my ordination," Zenson says. "I once heard one of my relatives say, 'He's such a nice boy. How did he get mixed up in that Buddhist thing?' They didn't understand. Buddhism was the cause of my being a nice boy. I think they were surprised when they came here. They expected me to be into some dark, weird place. Look around you. This is a nice place, isn't it?"

Meditation is only one part of the path to enlightenment. Another important part is the dialogue with the roshi, what Roshi Kapleau calls a "knee-to-knee, eyeball-to-eyeball confrontation," neither intellectual nor spiritual, but rather like a joust using water pistols and beanbags as well as lances and spiked clubs.

In his recently published book, "Zen: Dawn in the West," Roshi Kapleau writes that it is his job to "take away much that is foreign to your true nature: the sticky beliefs, chesty opinions, petty rationalizations, illusory ideals and deluded thoughts, all of which imprison you as in a cocoon."

Descriptions of these meetings with the master, called *dokusan,* remind me of the old joke about the seeker of truth who struggles for 20 years to find the ultimate guru on top of a mountain, only to be given

the advice that "life is a fountain." When the visitor protests the shallowness of this advice, the guru asks, "You mean it's not?" After spending two days around the unpretentious Roshi Kapleau, I began to see a level of truth in the joke.

One time a student asked him, "What is Buddha?" The roshi replied, "Who are you?" The roshi agrees this might sound like pretty weak stuff, but he adds the student "came to awakening as a result of this exchange."

Although a roshi is allowed to do some outrageous things (tease, taunt, whack with a ritual stick or smack the student on the side of the head), the roshi is not above criticism. (Roshi Kapleau's books include anecdotes about pupils who whacked their masters back.) Roshis have been known to be wrong. "The role of the teacher is to preserve the student from the teacher's influence," says Roshi Kapleau. "You cannot lean on the teacher. He will push you away." Each student is also given a riddle, called a *koan,* to ponder for five or 10 years. When he autographed a book for me, the roshi inscribed: "To George: In the seas birds fly, in the air fish swim."

Deeper Zen meditation comes in four- and seven-day periods of seclusion called *sesshin,* in which fasting and silence produce alternate moments of despair and joy. With no deity to cling to, seven days of *sesshin* are a plunge into the inner self.

At a recent *sesshin,* the members were resting in the garden when they heard the unmistakable sounds of a prowler being chased by police in the next yard. None of the meditators stirred for fear of losing the intensity needed for a seven-day seclusion.

Some Westerners question whether Buddhism is a religion at all. As it moved from India to China to Japan and then to the West, it has constantly acquired divergent, even opposite, ways. There are celibate priests, married priests, divorced priests. (Roshi Kapleau is married. His wife lives with their teenage daughter in Toronto and leads a spiritual group under an Indian teacher.) There are hermit sects and worldly sects. There is no god or dogma. But since it has its priesthoods and its rituals, since it deals with the meaning

of life and death, since it fills the same deep needs that Judaism and Christianity fill for others, since it proposes a life of spirituality, it fits all other criteria of a religion.

All the forms of Buddhism stem from one man, an Indian named Siddartha Gautama, who enjoyed a 45-year ministry six centuries before Christ. His thousands of disciples called him the Buddha, a Sanskrit word for "enlightened one."

Different Oriental cultures, different teachers, have created vastly different forms of Buddhism, many of them practiced in the United States by people of Chinese, Japanese, Korean and Tibetan ancestry and culture. For Americans of European background, the interest has most often been in Zen Buddhism, an adaptation of part of a Chinese word for meditation, *ch'an.* Zen is not a worship of a god, and it is not an intellectual philosophy but rather a series of disciplines that includes sitting in a meditative position for long periods of time. Very long periods of time. For a deeper understanding, a sense of humor is also useful.

Some Buddhists feel Christians and Jews can comfortably adapt most Buddhist teachings. C. T. Shen, a Chinese-born industrialist who sponsors many Buddhist activities in the New York area, says Christ and Buddha had "similar visions" but that Buddha had more time to complete his vision because he lived 80 years and Christ lived only around 33.

□

The Death of Master Hofuku (Pao-fu)

The master called his monks together and said, "During the last week my energy has been draining — certainly no cause for worry. It's just that death is near."

A monk asked, "You are about to die. What meaning does it have? We will continue living. And what meaning does that have?"

"They are both the Way," the master replied.

"But how can I reconcile the two?" asked the monk.

Hofuku answered, "When it rains it pours," and wrapping his legs in the full lotus, calmly died.

*—"The Wheel of Death," by Philip Kapleau**

One of the central beliefs of Buddhism is that life endures over a longer cycle than mere

* Reprinted by permission of Harper & Row, © 1971 by Philip Kapleau.

Chores done in the neighborhood enhance a sense of humility

birth and death. As Roshi Kapleau wrote in "The Wheel of Death," human life has "an unseen subterranean existence and appears at other places in other times and other shapes. The lesson from the deaths of some Zen masters is that dying is just another step, a time of ceremony and insight."

One of the most persistent criticisms of Eastern religions is that they lack the activism of Western groups — the philanthropic, educational and community-spirited zeal of the major Western religions.

In his recent book, "Turning East," Dr. Harvey Cox, the Harvard theologian, praised many of the techniques and results of Eastern religion, but said he was suspicious of "inward" movements that were uninvolved with other humans who had not yet reached, or sought, their brand of enlightenment.

Nevertheless, he described Oriental meditation as "a step toward escaping illusion and ego, and toward seeing the world of impermanence and suffering for what it is." He called Christianity "one pole of the dialectic of action and

repose, being and doing," and he predicted that Christianity would act as a "prism" to "transform" Buddhism.

Most Buddhists have heard Dr. Cox's criticism of inward movements and have an answer. "We are a lot like the Quakers," says Dr. Michael Taizen Soule, a biologist and full-time official at the Los Angeles Zen Center. "Quakers do more good than anyone, but they do a lot of work on themselves first. If you have a beneficial impact on yourself, after a while you can't help but act properly."

"When one purifies himself," says Zenson, the priest at the Rochester center, "one purifies society. You start where you are. You beautify your building. We won an award for the best restoration work in our part of town, and we didn't even ask to enter the competition. We help clean up the city. We live in harmony with our neighborhood."

What has Buddhism accomplished in America? Some members are sure the Vietnam War ended because of the moral pressure vibrated outward by meditation sessions. Some Buddhists feel Ameri-

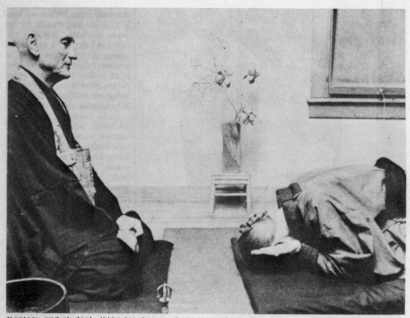

Kapleau and student: "The teacher must preserve the student from the teacher's influence."

cans have toned down their lives, cleaned up their towns, evaluated what services and efforts were necessary, because of Buddhist thinking.

The writing they cite most frequently is the book by E. F. Schumacher, "Small Is Beautiful," and the chapter "Buddhist Economics," in which the author writes, "Consumption is merely a means to human well-being; the aim should be to obtain the maximum of well-being with the minimum of consumption."

"You can inflame people's minds subconsciously," Roshi Kapleau says. "When we hold a *sesshin*, our friends as far away as Poland say they can feel the vibrations across the ocean. That is the basis of prayer. People who are developed in Zen can cast positive forces into the world. They can influence millions of people. It is happening already."

Roshi Kapleau cites an old Buddhist adage: "A small master hides himself in the mountains; a great master hides himself in the marketplace." The era of seclusion is over for many Buddhist groups, which are finding their mission right out in the bustle and, some would say, hustle of the marketplace.

The face of American Buddhism includes self-improvement courses, publishing ventures and Tibetan self-healing techniques offered by the Nyingma Center in Berkeley; photography and cultural studies and advice on how to care for the dying, all offered by the Los Angeles Zen Center; college degrees offered by the Naropa Institute, an unac-

credited college in Boulder, Colo., which is as close in spirit to Woodstock as it is to the Orient.

The Tibetan teacher Tarthang Tulku Rinpoche urges his disciples at the Nyingma Center to make their wisdom available to the world. Thus, a handsome 24-page brochure

□

boasts "There's something for everyone" at the institute. He insists these courses are offered not to evangelize but as a way of sharing wisdom for its own sake.

Native-born blacks and whites are also joining more traditional Oriental groups. Paul Kennedy, a 23-year-old Irish-American Rutgers graduate, recently became the first non-Oriental monk ever ordained by a Chinese Buddhist group, when he joined the Temple of Enlightenment in the Bronx.

Most Oriental Buddhist groups have cultural programs for children to keep their Japanese or Chinese or Korean traditions, yet their worship services closely resemble Protestant church services, with Western-style

'People who are developed in Zen can cast positive forces into the world. It is happening already.'

hymns and altars. These Oriental-American groups have almost nothing to do with groups like the Zen Center of Rochester, which are trying to shed all "foreign cultural encrustations" on the path to enlightenment.

□

Many of the frustrated young people who discovered Buddhism in the 60's are now sober not-so-young couples of the late 70's. Many are raising modest families of a child or two. Their worldly skills are producing decent incomes, and the American-born Buddhists are trying to build communities.

The neighbors on Arnold Park in Rochester seem to have accepted the quiet people who come to meditate, but some citizens in New York City, Iowa and Minnesota have, within recent months, stalled land sales to Buddhists. Roshi Kapleau has taken great effort to avoid the child-nabbing charges occasionally leveled against some cults, and he recently received approval from a Rochester minister before the minister's 17-year-old daughter was allowed to visit the center. The Japanese-born roshi in Los Angeles, Taizan Maezumi, tries to visit families of his members when he travels across the country.

The Buddhists seem to know they are not for every American, and indeed they have no room for every American. Meditating requires intense motivation. The ones who

come have some real desire for spiritual, moral, philosophical or intellectual uplifting.

Dedicated Buddhists feel their way of life offers something important to Americans who can understand it. Joseph Lippman, a former Congressional aide who came to Buddhism through his two sons and is now the dean of the Nyingma Institute in Berkeley, says he believes life is going to get more painful in America — both spiritually and physically. He says Buddhism can teach people to live with less, to be content with themselves, to cope with the diminishing pleasure and greater pain that will accompany the last quarter of this century.

Can these perceptions be transmitted from one generation to another? Unlike Catholicism, with its first communion at the age of 7, or Judaism with its bar and bat mitzvah at the age of 13, Buddhism has no rites of passage for children.

Richard and Vicky Wehrman moved from Missouri to be close to Roshi Kapleau, to raise their children as Buddhists, if possible, in a polyglot society.

"My daughter is going to Catholic school this year because it was the best alternative," says Mrs. Wehrman. "She takes religion 45 minutes a day, but they're not trying to convert her. And as Buddhists, we are wary of setting up programs for our children, but lately my daughter has come to the center for some of the celebrations." Buddhists tend to want their children to be exposed, not indoctrinated.

"When we first moved into our neighborhood, we used to sit on the floor to eat, but the children felt strange about bringing their friends home, so we got chairs. We try to have a middle-class style for them, but we do have altars around the house. I don't think anybody says anything to the children about being different."

Actually, some Buddhist parents say the most noticeable aspect of their Buddhism is their vegetarianism. But even that is becoming commonplace in schools and offices these days. Though not all Buddhists are vegetarian, those who are like to take some credit for the rising consumption of fruits and vegetables.

The Los Angeles Zen Center has a day-care center for children, and allows the children to improvise their own Buddhist rituals; other groups are leery of early exposure be-

cause of memories of their own Jewish or Christian childhoods. The Rochester group has been concentrating on American-style holidays (New Year's Eve parties, with a traditional Buddhist driving-out-of-ghosts; Halloween; Thanksgiving).

Some parents do not transmit their Buddhist beliefs to their children. Alan Temple, a leader of the Rochester center who lives and works in a suburban neighborhood, says he is "hurt" that his 22-year-old son has no interest in Buddhism.

Other families have more success. The Wehrmans try not to appear "unusual" to their two teen-age children, but they do conduct meditations in their home, and they see signs of passing the practice to the next wave.

"My son resisted for a long time, but lately he has been coming around," Mrs. Wehrman says. "The other day we were all cross with each other for some reason. A lot of bad vibrations in the house. My son said, 'Mom, we'd better have a sitting.' So he put on a robe and got a cushion and he began to sit. He only lasted five minutes, and we all lasted longer than he did, but we got good karma back into our house."

Sylvan Busch, a commercial artist who is president of the Zen Studies Society in New York City, says his 27-year-old son has just started to attend the center's meetings. Mr. Busch, who has been practicing Zen Buddhism since 1958, says he never tried to influence his son's choice. "I knew if he was curious he would look in."

Of the future of Zen Buddhism in America, Mr. Busch says: "We have made a good start. But with pockets of 200 people here and 200 people there, we can't really be considered established. Bringing a new religion to a country as big as the United States is not easy. But we are doing a lot of good; we are putting down roots."

Roshi Kapleau has been thinking about putting down new roots as well. He has recently decided to build a new headquarters on 200 acres in Denver. A female disciple, Toni Packer, will run the center in Rochester as an affiliate. "Most Zen teachers move around," he says of his decision. "There is no concept of an irreplaceable guru. Part of Zen training is not to let people get dependent on you. They must stand on their own feet." ■

June 4, 1979

A NEW LOOK AT LIFE WITH FATHER

Researchers have lately probed the father-infant relationship and found few significant differences in the way children relate to fathers and to mothers. Their studies may sharply alter our concepts of what parenting is all about.

By Glenn Collins

For in the baby lies the future of the world:

Mother must hold the baby close so that the baby knows that it is his world;

Father must take him to the highest hill so that he can see what his world is like. —MAYAN INDIAN PROVERB

The tiny white room on the ground floor of the John Enders Research Building is only a block away from the quiet green quadrangle of the Harvard Medical School. The plate on the door reads "RESEARCH LABORATORY I," and the room is a jumble of television cameras, tape decks and video cables. There are, however, some unexpected items: boxes of Kimbies are stacked under a table; atop the table, a container of Wet Ones Moist Towelettes, and next to the Wet Ones, propped in a sturdy aluminum seat, a 96-day-old baby named Eddie. He is intently studying James, his father, standing before him. Two television cameras are capturing these moments on half-inch magnetic tape.

"Bet you're glad to see me!" says James, smiling. "Were you good with Mommy? You know, I missed you, all day. . . ." As he talks to his son, he taps him, tickles him, and smiles, his eyebrows moving in a language of their own. Eddie arches forward in his jumpsuit, kicks his feet in their little red socks, coos and giggles. After exactly two minutes, James leaves the room. The cameras observe Eddie for another 30 seconds, and then they are turned off.

Although a pediatrician on the staff of the Child Development Unit of Children's Hospital Medical Center in Boston will later play back the tape, James and young Eddie aren't patients: They are participants in one aspect of current research on fathering. Investigators will play back the tape at one-seventh speed, and will conduct a microanalysis of the facial expressions of James and Eddie, recording their vocalizations and their body move-

Glenn Collins is an editor of this Magazine.

ments on matching graphs. This information will be used to create a sawtooth chart that plots a father's typical interaction with a child; its signature is distinctive, different from the characteristic pattern of a mother's interaction.

☐

Near Princeton, N.J., a father holds open the front door and waits with his wife and two children as a team of researchers hefts a videotape camera and assorted television equipment into his house. It is just about dinnertime, and the investigators, from the Infant Laboratory at the Educational Testing Service, ask the parents where they normally eat their evening meal. "Tonight, in the dining room," says the father. The E.T.S. technicians start the camera running, and leave the house. Tentatively, the parents call their children to dinner. The 3-year-old waves at the television eye. The 6-year-old sticks out his tongue. The father seems a bit unnerved. They start eating, and, before any of them might have expected, they forget about the technological presence in the dining room.

After dinner, the E.T.S. researchers return for their equipment; later they analyze the behavior they view on the tape.

After studying 50 families, the E.T.S. investigators can generalize about what they have seen. Fathers talk more to their sons than they do to their daughters. Children talk less to fathers than to mothers, or to each other. And fathers, in their dinner behavior, tend to ask questions.

☐

In a laboratory at the University of Wisconsin, a father sits in an easy chair four feet away from a television monitor. He is about to see something unpleasant, but he doesn't know it. Electrodes from an eight-channel polygraph recorder — a lie-detector — have been attached to his index and middle fingers, and a rubber cuff has been inflated on his left bicep; his heart rate, skin conductance and blood pressure are being monitored. Soon the television screen glows with a six-minute videotape of a 5-month-old baby boy. The infant looks around gravely and

makes a sound; then he squirms, and soon he begins to cry. Loudly. Insistently. Interminably. Even though the baby on the screen isn't his, the father feels ever more uncomfortable under the assault; he moves, tenses, his heart rate rises and his blood pressure soars. Later, after testing 148 subjects, the investigators are able to report that there is no physiological difference between the reaction of a father or a mother to the sight of a squalling baby. To both, it is equally distressing.

R esearch examining what fathers do and how they do it has been booming in recent years. Not all of it employs computer analyses and electronic bric-a-brac. Much of it involves nothing more complicated than placing a trained observer in a room with a father who is playing with, or talking about, his child.

The impact of all this father-watching is beginning to be felt in courts of law, in hospitals and in universities, which face the task of redirecting the training of a new generation of doctors, pediatricians, psychotherapists, health-care professionals and teachers. "Our whole society has had the notion that a biological bond between mother and child made fathers less able, less interested and less important than mothers in caring for children," says James A. Levine, a Wellesley College researcher. "Courts have based decisions on that notion, therapists have treated patients on the basis of it, and men and women have made life choices because of it."

In fact, 44 percent of the mothers of children under the age of 6 in this country are working, only 24 percent of existing families are traditional nuclear families, and the "two-paycheck" marriage is the norm for nearly half of all two-parent families in America. In the changing society reflected by these statistics, the new knowledge about fathering has important implications for how children will be raised and educated, and will *(Continued on Page 74)*

Father and daughter enjoy a warm moment during their videotaped play session.

FATHERING

Continued

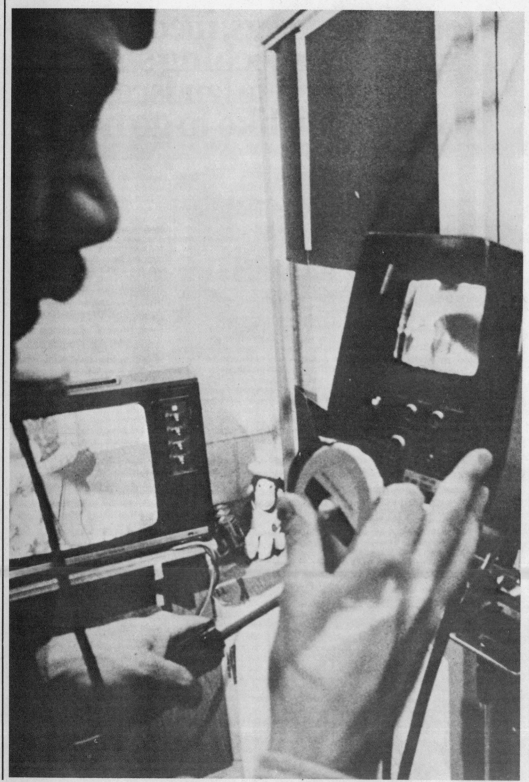

Behind a one-way window, Dr. Michael Yogman, one of the new fathering researchers, trains his videocamera on Kevin Nugent playing with his daughter, Aoife.

help to shape the kind of nation we inhabit in the 1980's.

☐

Fathers haven't always been a fashionable research subject for social scientists. "When I started out 17 years ago, there just wasn't much data," says Henry Biller, professor of psychology at the University of Rhode Island and a pioneering researcher in the field. "The recent increase in data collection on fathers is amazing. We have something of a revolution in thinking among those involved in early childhood development."

In past decades, researchers focused on the father as a role model, or studied him inferentially: by examining the impact of his *absence*, in families where the father had died, divorced or gone to war. Fathers have also long had a place in psychoanalytic theory, becoming important in the Oedipal stage, when the son competes with him for the mother.

Social-development theorists viewed the mother-infant relationship as unique, vastly more important than subsequent relationships; it was even termed the prototype for all close relationships. In 1958 and then again in 1969, John Bowlby, the British psychiatrist, published his elegant and influential theories of attachment, a word that is usually defined by behaviorists as the preference for, or desire to be close to, a specific person. "But the real synonym for 'attachment' is 'love,'" says Dr. Michael Lewis, a developmental psychologist at the Educational Testing Service. Bowlby, drawing on the animal-study work of ethologists and the parental-deprivation observations of cognitive psychologists, suggested that there was an evolutionary advantage to a unique bond between mother and infant; he reasoned that this bond was an imperative of the very growth and development of the species.

Subsequently, many researchers investigated the mother-child interaction, revealing the nature of the infant's early relationship with its caretaker. However, fathers weren't even present in most of the studies. "A major reason that fathers were ignored was that fathers were inaccessible," says E. Mavis Hetherington, a University of Virginia psychology professor who has studied family-related questions for 25 years. "To observe fathers you have to work at night and on the weekends, and not many researchers like to do that."

Studies of humans and of nonhuman primates began to suggest that infants had strong attachments to persons who had little to do with their caretaking and physical gratification; nor were these relationships necessarily derived from the child's bond with its mother. In a classic study, primatologist Harry Harlow demonstrated that the attachment process was not limited to a feeding context. Investigators also showed that the actual amount of time an infant and his mother spent together was a poor predictor of the success of their relationship. Consequently, a child's tie to its mother continues to be viewed as crucially important; however, its exclusivity and uniqueness have been challenged. The new research emphasizes the complexity of an infant's social world.

Researchers have now identified some of the ways in which fathers are important to children. Henry Biller sums up the findings: "The presence and availability of fathers to kids is critical to their knowledge of social reality, their ability to relate to male figures, to their self-concepts, their acceptance of their own sexuality, their feeling of security. Fathers are important in the first years of life, and important throughout a child's development." Frank A. Pedersen of the National Institute of Child Health and Human Development has also demonstrated that mothers can perform better in their parenting roles when fathers provide emotional support.

Researchers from a number of disciplines using ingenious new methods now suggest that the father-infant relationship is not what we thought it was; that, for example, there are few significant differences in the way children attach to fathers and to mothers; that fathers can be as protective, giving and stimulating as mothers; that men have at least the potential to be as good at taking care of children as women are; and that the characteristic interplay of father and infant, when scrutinized minutely, is distinctive in many fascinating ways. The new fathering research offers fresh insights about the "distant" father and about fathers' roles across disparate cultures; it reveals that fathers have been ignored in research and in medical practice in curious and interesting ways; and it offers a synthesis of the relationship between fathers, children, families and society.

James Herzog, M.D., a psychiatrist who teaches at Harvard, says this of the new findings: "We're in what I call the post-competency phase now. We don't need to prove that fathers 'can do it, too.' The question now is, what is the specific role of the male parent, and what is the difference between being a father and being a mother?"

☐

In 1970 a Harvard Ph.D. candidate named Milton Kotelchuck began a study of fathers that created a stir when it was presented in Philadelphia at the 1973 meeting of the Society for Research in Child Development. Kotelchuck, now director of health statistics and research for the Commonwealth of Massachusetts, had set up a classic "separation-protest" situation — a test of attachment — in studying the reactions of 144 infants when their fathers and mothers walked out of a playroom and left them with a stranger. Previous studies had observed the effects of a mother's departure on her child; Kotelchuck was able to determine that infants were just as upset when a father left them.

In four other studies, Kotelchuck and his associates found few significant differences in the way the infants attached to fathers and to mothers. They demonstrated that, in fact, children have extended social worlds and can attach equally well to siblings, peers and other figures.

Michael E. Lamb, research scientist at the University of Michigan's Center for Human Growth and Development, has carried on his investigations — including the crying-baby experiment described earlier — at Michigan, the University of Wisconsin at Madison, and at Yale, and was the editor of an influential 1976 anthology, "The Role of the Father in Child Development." His first key study of attachment appeared in 1975; in it, 7- and 8-month-old boys and girls and their parents were viewed in the home setting. An observer dictated a detailed account of the behavior he saw into a tape recorder. That narrative was then analyzed by applying 10 measures of attachment and affiliation: whether the baby "Smiles," "Vocalizes," "Looks," "Laughs," "Approaches," "Is in proximity," "Reaches to," "Touches," "Seeks to be held" or "Fusses to." Lamb and his co-workers found that no preferences were evident for one parent over the other among these infants, at the age when they should, according to Bowlby's theory, be forming their first attachments.

Lamb and his colleagues reported that when mothers held their infants, it was primarily for things like changing, feeding or bathing; fathers mostly held their children to play with them, and initiated a greater number of physical and idiosyncratic games than mothers did. This paternal play tended to be boisterous and physically stimulating. Furthermore, boys were held longer than girls by their fathers; fathers start showing a preference for boys at one year of age and this preference increases thereafter.

Currently, Lamb and his associates are studying 100 families from the time of pregnancy until their children attain the age of 18 months. The sample includes families where there are working wives; also represented are a few fathers who are primarily responsible for infant care. In Sweden, for the past six months, they have been observing role-sharing fathers and both mothers and fathers who have primary responsibility for child care.

If Michael Lamb has tended to focus on the child in his work, Ross D. Parke, professor of psychology at the University of Illinois at Champaign-Urbana, has centered his research on fathers themselves. In a 1972 study that is a classic in the literature of fathering, Parke and his colleagues haunted a hospital maternity ward in Madison, Wis., and observed the behavior of both middle-class and lower-class parents of newborn babies. They found, most strikingly, that fathers and mothers differed little in how much they interacted with their children. Fathers touched, looked at, talked to, rocked and kissed their children as much as their mothers did. The study suggested that they were as protective, giving and stimulating as the mothers were — even when the fathers were alone with their babies.

In later work, Parke and his collaborators measured the amount of milk that was left over in a baby's bottle after feeding time; infants consumed virtually the same amount of milk whether fathers or mothers did the feeding. They found that fathers were equally competent in correctly reading subtle changes in infants' behavior and acting on them; fathers reacted to such infant distress signals as spitting up, sneezing and coughing just as quickly and appropriately as mothers did. Parke asserted that men had at least the potential to be as good at caretaking as women. However, fathers tended to leave child care to their wives when both parents were present.

In the last three years, Parke, Douglas Sawin of the University of Texas and their collaborators have conducted two major studies involving 120 families. They observed family interactions, and used high-speed electronic "event recorders" with 10-button keyboards and solid-state memories to tap out four-digit codes that recorded behaviors as they saw them. Although there are very few differences in *quality* between mothers' and fathers' interactions with their children, one observed disparity is that fathers are more likely to touch and vocalize to first-born sons than to daughters, or to later-born children.

□

Eddie and James, whose close encounter began this article, were participants in a continuing investigation of children's early learning abilities and communication patterns at the Child Development Unit of Children's Hospital Medical Center in Boston. Originally, father-infant and mother-infant pairs were videotaped periodically during the first six months of babies' lives. The unit's newer work involves the father-mother-infant triad.

In Laboratory I, where young Eddie became something of an intramural television celebrity among social scientists, the research continues. Infants are placed in an alcove created by a blue-flowered curtain, and are taped with father or mother. These laboratory situations, though artificial, place the maximum communicative demand on the parent and child, the researchers say; they bring out the kinds of intense play situations that normally occur only during brief periods during an ordinary day.

Two trained observers play back the videotapes of these sessions and perform a "microbehavioral analysis" of the interaction of both the parent and the baby. The researchers assign numerical scores that rate such facial expressions as frowns, pouts and smiles; sounds like gurgles or coos; motions of hands and feet, and even eye movements. Ultimately, the observers note clusters of these behaviors and chart them during each second of elapsed time over the entire interaction.

Graphs of fathers' and mothers' behavior show distinctive patterns. In all of the families studied by the Child Development Unit, the chart of the mother's interaction is more modulated, enveloping, secure

and controlled. The dialogue with the father is more playful, exciting and physical. Father displays more rapid shifts from the peaks of involvement to the valleys of minimal attention.

There are other characteristic differences: Mothers play more verbal games with infants, so-called "turn-taking" dialogues that are composed of bursts of talking or cooing that last four to eight seconds, and are interrupted by three-to four-second pauses. Fathers tend to play more physical games with infants; they touch their babies in rhythmic tapping patterns or circular motions.

To provide conceptual models for the way babies interact with adults, the Boston researchers have employed the theories of cybernetics, the discipline that studies the control and regulation of communication processes in animals and machines. Researchers have broken with the traditional lexicon of rat psychology, and talk about the "interlocking feedback of mutually regulated systems" and "homeostatic balances between attention and nonattention." The baby, in its reciprocal interaction with an adult, modifies its behavior in response to the feedback it is receiving. Infants, they say, seem to display periods of rapt attention followed by recovery intervals, in an internally regulated cycle that maintains the balance of the infant's heart, lung and other physiological systems.

"It's important to say that father doesn't offer some qualitatively better kind of stimulation; it's just different," says T. Berry Brazelton, M.D., director of the Child Development Unit, a pioneer in the study of family interactions. "Mother has more of a tendency to teach the baby about inner control, and about how to keep the homeostatic system going; she then builds her stimulation on top of that system in a very smooth, regulated sort of way. The father adds a different dimension, a sort of play dimension, an excitement dimension, teaching the baby about some of the ups and downs — and also teaching the baby another very important thing: how to get *back* in control."

There are also interesting similarities in infants' relationships with both parents, says Michael Yogman, M.D., the pediatrician who videotaped James and Eddie and who has specialized in the study of fathers at the Child Development Unit since 1974. "With both parents," he says, "we see that behavior is mutu-

ally regulated and reciprocal, that there is a meshing of behaviors."

Dr. Brazelton says that "there's no question that a father is essential to children's development. Our work shows that babies have this very rich characteristic model of reaction to at least three different people — to father, to mother and to strangers. It shows me that the baby is looking for richness, that he's looking for at least two different interactants to learn about the world." For Dr. Brazelton, to whom Mayan Indians told the saying that preceded this article, its poetry is exceedingly descriptive.

"It seems to me," says Dr. Brazelton, "that the baby very carefully sets separate tracks for each of the two parents — which, to me, means that the baby wants different kinds of people as parents for his own needs. Perhaps the baby is bringing out differences that are critical to him as well as to them."

The Boston researchers plan to explore the later development of the paternal and maternal dialogues with children. They also hope to refine their procedures to the point where they may be useful as a diagnostic tool for practitioners.

□

Fathers are being studied from other perspectives. Although psychiatric clinicians, those who see patients, had always noted that the father played an important role in the psychological development of children, as late as 1973 the psychoanalytic literature bemoaned the lack of theorizing about the father's role during the first two years of life.

Building on Margaret Mahler's ideas on the successive stages of an infant's "psychological birth," psychoanalyst Ernst Abelin and others focused on the father's role in helping infants separate from mothers.

Some behavioral psychologists can't take the efforts of the psychoanalytic theorists very seriously, since the data for such work are often derived from the study of a single patient who may be going through the process of becoming a father, or coping with the difficulties of parenthood. "It's better to observe what's going on," says Alison Clarke-Stewart, a University of Chicago psychologist. "It's not distorted by retrospective recollection or the perceptions of the person who's being studied."

Psychoanalysts reply that the observational method is

limited. "How people behave is highly determined by their fantasies, conflicts and unconscious processes," says Dr. Herzog. "These are the causes of the behavior that others observe. I have nothing against documenting this behavior, but we need to look at the inner life, too."

Part of that inner life is a well-documented clinical phenomenon, the so-called "womb-envy" — the envy of women's capacity to give birth — among some expectant fathers and even among male children. Perhaps a societal counterpart of this is the "couvade" phenomenon that anthropologists have noted in many cultures, where men undergo elaborate rites of passage paralleling their wives' pregnancies and birth-giving.

"We know that the time of pregnancy and becoming a father is extremely important to men, a crucial and stressful time," says Alan R. Gurwitt, M.D., a psychiatrist, analyst and associate clinical professor at the Yale Child Study Center. "Yet the astounding thing in this society is that the father has come to be a subject of ridicule — there is no end to the cartoon and movie stereotypes portraying the expectant father, and fathers in general, as bumbling fools."

He says there is still a tendency to ignore fathers on the part of obstetricians, pediatricians, nurses and even child psychiatrists. "This failure to involve fathers even in the treatment of their children runs very deeply," says John Munder Ross, a Manhattan psychotherapist and clinical assistant professor at Downstate Medical Center, who is coediting an anthology of the new psychoanalytic views of fathering. "It may have to do with the relations of clinical workers to their own fathers. There seems to be an awful lot of stereotyping of fathers as 'absent and ineffectual,' or 'tyrannical and sadistic.' "

To the psychoanalysts, the process that is fathering continues. "The middle-aged father frequently finds himself in a painful situation," says Stanley Cath, M.D., a psychoanalyst and associate clinical professor at Tufts Medical School. "His adolescent children may be rebellious and challenging to him; he himself may be trying to separate from his own father, who may be aged or dying; and the grandfather himself may be looking for support" as he faces the debilitation of old age. "Of course, a man can be the father to his children, and also the father to his parents," says Dr. Cath. "We rediscover the father, and the definition of fathering, throughout our lifespan.''

□

As a social phenomenon, the evolution of fathering in man and various primate precursors is a matter of sheer conjecture. Paleontology provides little data on social interaction. Some cultural historians have tried to make inferences from the study of recent ''primitive'' societies, by which they mean complex societies that have not received the blessings of technology.

Margaret Mead's famous 1930 study of the Manus people of New Guinea reported that, at the age of a year, children were given from the mother's care into the father's. He would play with the baby, feed it, bathe it, and take it to bed with him at night.

Fathers in the Thonga tribe in South Africa, observed during the last century, were ritually prevented from having almost anything to do with infants until the babies were 3 months old. However, fathers in the Lesu culture of Melanesia commonly took care of babies while their wives were busy cooking or gardening. And among the !Kung bushmen in northwestern Botswana today, fathers have a great deal of contact with children, holding and fondling even young infants.

In analyses of all known cul-

Mary Maxwell West and Melvin J. Konner at Harvard University found that social and cultural conditions are related to the level of involvement of fathers with their children, and suggest that there is the potential among males for caring for their young if other conditions encourage it. West and Konner found that fathers observed in cultures with monogamous nuclear families were generally involved parents. So were fathers in ''gathering'' societies — the form of society that existed during 98 percent of human history. They suggest that distant fathering is associated with warrior cultures (''hunting'' societies) and with societies where men's agricultural or military activities take precedence.

Of course, the political and economic equivalents of warfare exist in modern industrial cultures, and it can be debated how much they affect males' involvement in fathering. There is conjecture that the tradition of the Roman paterfamilias had some influence on current patterns, as well as the Christian concept of the Old Testmament God. The few attempts at compiling histories of fathering show the Industrial Revolution to be a major disrupter of family life as it existed when many fathers were tradesmen or farmers working in the presence of their children.

The cross-cultural evidence shows clearly that the father

ers lactate as soon as a baby cries. But mothers also have the experience of carrying the baby for nine months, and if the business of attachment comes from sensitivity to being tuned into a baby, mothers have the advantage.''

''But there is a crucial distinction to be made here,'' says Milton Kotelchuck. ''Yes, pregnancy and lactation can make it easier for a mother to attach to a child. But the essential thing is that infants don't know that they are supposed to relate more to the mother than to the father.''

''It is my speculation — and I want to emphasize that word,'' says Michael Lamb, ''that we will find that biological differences are very small, and that they are exaggerated and magnified by the rituals and the roles that societies build around those distinctions. But are these differences genetic? My answer is 'Yes, but' — where the *but* is more important than the *yes*.

''Aside from the question of genetics,'' he says, ''there is good evidence to believe that mothers and fathers can be equally effective as parents. They just have different styles. Perhaps it's really not fathering or mothering — it's parenting.''

□

One researcher attempting to synthesize the relationship between child, family and society is Michael Lewis, director and senior scientist at the Institute for the Study of Exceptional Children at the Infant Laboratory of the Educational Testing Service. (It was his investigators who conducted the videotaped observations of Princeton fathers at dinner.) Lewis holds that different people — mothers, fathers, peers, siblings, grandparents, uncles, aunts and other relatives — serve the child's needs in different cultures in different ways: ''I am saying that a father's role is cultural and historic rather than biological and evolutionary.''

''There's no good data on any of this,'' Lewis says, ''but my impression is that, to an extent in the general culture, fathers are defining their functions in new ways — the 'new fathering' we hear about.'' He adds that ''we haven't assessed the basic question of values here yet, and that's what we need to do. If the cultural matrix is changing, is it assisting the values of our culture?''

To an extent, society has legitimized the needs of parenting men. ''In a sense, fathers have come out of the closet,'' says Mavis Hetherington. ''They feel more comfortable about being parents, and are more actively fighting for their rights.'' Recent revolutionary changes in the way society views men are now treated by the media as commonplace: men's

improved position in child-custody cases or men's right to single-parent adoptions in most states.

Nevertheless, it is James Levine's hunch that women are more aware of the issue of fatherhood than men are. ''I think it is becoming more of a question for women as more of them are working outside the home. Women make demands on men to parent in a way that fits in with their new concepts of how they will live their lives,'' says Levine, a research associate at the Wellesley College Center for Research on Women who wrote an influential 1976 book on male parenting options, ''Who Will Raise the Children?''

However, Levine says, ''the biggest push for change is coming from the economic pressures — the necessity for both parents to work. I think the bottom line in all of this is the economic situation of women.''

Michael Lamb believes that ''it's a depressingly small number'' of fathers who take on a large share of all that is involved in bringing up a child. He does not view the recent research about fathers' abilities as a new panacea. ''But,'' he says, ''I think we must realize that, in general, the average male won't be better than the average female as a caretaker. Yes, babies can attach to father. But that isn't to say that they won't be closest to the primary caretaker, which is usually mother.''

Says Levine, ''Where we really miss the boat is when we say the male role is changing, and cite as evidence the fact that men are changng diapers, bottle feeding, etcetera.'' The truly important part, Levine feels, involves a man's sense of emotional responsibility: ''It's not just the taking *care* of kids, but it's who carries around that inner *sense* of caring, that extra dimension of emotional connection.''

It is possible that there will be competition in parenting. ''At this point,'' says Dr. Brazelton, ''everyone is goading men on to do more, but the second that men get good at it, and really enjoy nurturing, it may cause problems that'll have to be faced. Fathers who are taking an equally nurturant role may threaten some mothers.''

For Levine, looking ahead, the most interesting question is, ''research for what?'' It seems to him that the next step, theoretically and practically, is to give some guidance to medical and mental-health practitioners: ''The most interesting area for research has to do with total family interaction, the family systems perspective.''

Virtually all of the father-watchers are wary, however, of being prescriptive — of saying that fathers should parent in a specific, more ''nurturant'' way.

''The crucial impact of the new research,'' says Douglas Sawin, ''should be that a father's role ought to be an optional choice — and that, with a little support and training and education, they can be primary parents — but only if they want to be. For they have the basic competence and warmth and nurturance abilities. Whether they implement them or not is their decision.'' ■

'We don't need to prove that fathers "can do it, too." The question now is, what is the role of the male parent, and what is the difference between being a father and being a mother?'

---

tures, anthropologists have suggested that, in about two-thirds of societies, wives and children accord the paterfamilias deference, that husbands exert authority over their wives and that most cultures trace descent through the father's line. In nonindustrial cultures, these analyses suggest, fathers generally play a small role in relating to young children. In other words, the similarities of men's roles outweigh the fascinating differences that may exist.

Male figures — though not necessarily fathers — are involved in child care in most cultures. Just how involved is another question. Applying a measurement called the Barry and Paxson Father-Infant Proximity Scale, researchers

has been many things in many societies; it suggests that, if the culture allows, fathering can be whatever fathers want to make of it.

□

Is there any answer to the question posed earlier by Dr. James Herzog: Is fathering the same as mothering? And, if not, is one parenting style superior to the other?

''You don't want to imply from these studies that people are interchangeable,'' says Alan Sroufe, a University of Minnesota child-development professor who is doing studies of attachment there. ''Sure,'' he says, ''an infant can attach to a woman or a man. But women have natural advantages in parenting. It's not just nursing — for example, moth-

DEATH IN THE FAMILY

Marsha and Gary, like many couples, learned that defects in their genes doomed their first child. Now medical advances help prevent such tragedy.

Marsha with her second child, Joshua. At right, the two are with the father of the family, Gary, and baby, David. Both Joshua and David were born without their parents'

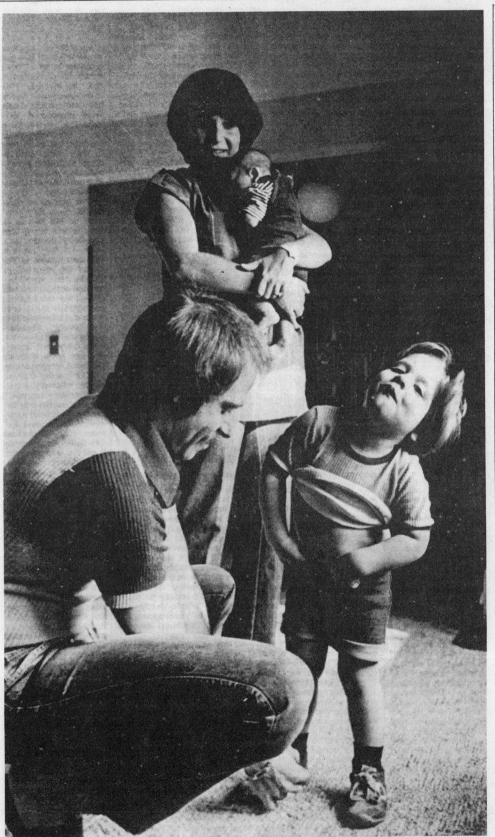

defective Tay-Sachs genes that afflicted their older sister, Jennifer, with a fatal disease of the central nervous system.

By William Stockton

How much despair can a young couple endure when they receive their baby's death sentence? When their intense feeling of anger and bitterness surfaces, at whom can they direct it? Their doctors? The rabbi and the elders of their synagogue? Their parents? Themselves? The cruelest truth of all comes later: If each one had married someone else, they would have been spared.

Such agonizing questions are facing more and more young couples these days because of steady advances in medical genetics. This once-arcane field is moving rapidly forward through research developments and a growing awareness among physicians and their patients that the birth of many genetically defective children can be prevented. But with such prevention comes inevitable suffering. People who once might have accepted the birth of an abnormal child as something over which they could have no control now often must confront frightening truths and difficult decisions.

Dr. Harold M. Nitowsky, a geneticist at the Albert Einstein College of Medicine of Yeshiva University in New York City who is an expert in Tay-Sachs disease, the "Jewish" disease, begins this poignant story that illustrates how parents and doctors must try to cope with these new dilemmas. Dr. Nitowsky is a quiet, seemingly shy physician who is regarded with awe by some of the couples who come to his office. He sits behind his desk, hands folded on top, the intensity of his feelings occasionally peeking through his quiet manner.

"Marsha's and Gary's case goes back to Oct. 6, 1975. I remember it vividly," he begins. "It was a Monday. The preceding day we were involved in a screening program for Tay-Sachs disease at Central Synagogue on Lexington Avenue in Manhattan. We have screened approximately 10,000 people in the New York metropolitan area and we have had screenings in Albany, Syracuse, Stony Brook and generally covered much of the state. Tay-Sachs carriers — people who have the Tay-Sachs gene — are 10 times more prevalent among the Jewish population who have their ancestral origins in central and eastern Europe. These are the Ashkenazy Jews. Among couples of this background the disease is present in about one in every 2,500 births. Among the non-Jewish population, it is much lower, perhaps one in every 250,000 births. So

William Stockton is director of science news of The Times; this article is excerpted from "Altered Destinies," published this month by Doubleday.

we wanted to attempt mass screening in this Jewish population. And that's what we were doing at that synagogue on that Sunday.

Tay-Sachs has an early onset and leads to a rapid deterioration of the central nervous system. Tay-Sachs children lose contact with their environment and show regression in whatever capabilities they develop in terms of motor development. They eventually have to be tube-fed. They become spastic and have convulsions. The end of their disease is a vegetative state. The cause of death is usually an infection or pneumonia. They become debilitated because it is very hard to maintain their nutrition. These children have been known to live up to 5 years of age, but many die at 2 or 3. However, Tay-Sachs is a genetic disease whose carriers can be easily screened.

"We received this call from a young man. He had heard about the screening and called the synagogue and asked to speak to one of the doctors. He was very, very distraught. He was crying, obviously terribly upset. He said his daughter had just been diagnosed as having Tay-Sachs. He asked if we could see her as soon as possible, so we made arrangements.

"Their names were Marsha and Gary. They had been married for five years. This was their first child, born 11 months earlier. She was apparently doing very well and even recognizing faces, rolling over and smiling and so on. But she was never able to sit up unsupported. At about 4 or 5 months of age she was noted to have a startled reaction to loud noise, which is an early clinical sign.

"Marsha and Gary began expressing some concern to their pediatrician and he allayed their fears, saying things seemed to be going along O.K. Of course, you never know what patients' perceptions are. But during the course of the next several months they noted that the pediatrician often was checking the baby's eyes with an ophthalmoscope. During that time the baby began showing less attentiveness. She rolled over less frequently and became less responsive. They really became quite concerned. They were eventually referred to an ophthalmologist because the pediatrician must have spotted something in the fundus of the eye. He probably saw the cherry-red spot. The ophthalmologist found the spot. On the basis of their Jewish ancestry and the infant's medical history, he made a presumptive diagnosis of Tay-Sachs disease.

"It's ironic that they are residents of Queens and live in a neighborhood where we had a Tay-Sachs carrier-testing program 18 months earlier.

They never heard of the testing. So this shows you how when you mount what you think is a very good education program, you still are not reaching all the people, particularly many of the young people whom you need to reach most. We had announcements on the radio and in the local press, but they weren't attuned. People hear what they want to hear. So in a sense this could have been prevented if we had gotten to them then. We could have detected the fact they were both carriers of the Tay-Sachs gene with a simple blood test."

□

Marsha and Gary are a handsome young couple who live in a single-family house they've remodeled on a small plot a few blocks from the Utopia Parkway. It is a neighborhood of tree-shaded streets and old single- and double-family houses. Gary's family business manufactures cardboard boxes in a plant nearby on Long Island. He works there with his father and a brother. Marsha is a financial officer with an air-freight company. I spoke with them in their home and quickly realized the intensity of the emotion that talking about their ordeal unleashed. At first, they both talked at once, constantly interrupting one another, their words pouring out almost uncontrollably. As we talked, they seemed to become calmer, as if the opportunity to tell their story to an outsider was a cathartic experience, a form of therapy.

It was a hot summer day and Marsha wore shorts and sandals, sitting cross-legged on the couch in their den. Gary sat at the other end of the couch, sipping a beer; he had come in from sunbathing in the backyard. "I honestly had never heard of Tay-Sachs until Jennifer was 8 months old," Marsha said. "We were in Florida with friends. One night a friend of my girlfriend came over and the topic came up. This girl was talking about this disease because she had received a telegram saying that her sister had been found to be a carrier. Therefore, her sister was telling her to go and be tested. She had gone to a doctor and he couldn't even tell her where to go for the testing. She was griping about how incompetent these doctors were. I said, 'What are you talking about?' She said, 'Tay-Sachs disease.'

"We didn't then make the connection to Jennifer at all, but we did feel something was wrong. I think Gary blocked it out more than I. From the time she was about 8½ months, I became convinced that something was wrong. My pediatrician would say that she was slow but would catch up. This must have been bothering me when I was in Florida because I made an appointment to take Jennifer to my girlfriend's pediatrician. He examined her and said,

'Are you happy with your daughter's progress?' I told him no. He did a quick test and said her motor skills were at the 4-month-old level. He said it could mean nothing, but if I was going to be in Florida for a while he could arrange a neurological examination. But we were coming back to New York, so he said I should speak to my pediatrician in New York. When I did, my New York pediatrician just said all the doctors in Florida are neurotic. They test for everything."

Gary picked up the story: "We were going to the same pediatrician who had cared for me as a child. I had a lot of confidence in him. I felt I knew him well. But we were new parents. We didn't know the progression that a baby makes as it grows. We did realize she was very slow. When she was 11 months old her eyes started to cross. Marsha was still complaining that

something was wrong and I was getting worried, too. Marsha was ill so I took Jennifer to the pediatrician. I told him our concerns, and he suggested we get her eyes checked. He sent us to the ophthalmologist."

Marsha interrupted: "You see, I was working full time. I had someone in the house during the day. But I was ready to quit my job. I felt that I was failing Jenny by working. I thought I didn't give Jenny enough time to teach her things. I felt guilty. I was convincing myself that I was at fault. But Gary said I shouldn't quit. He said, 'You can't teach her to crawl and sit up.'

"Gary was very confident up to this point because the pediatrician was a doctor who had been his doctor as a boy. Gary felt he was really qualified. But the ophthalmologist did a quick eye exam and said, 'I see a grave situation. I think you should go back and see your pediatrician.'

"Well, my God, I was about to explode. I cornered him and said, 'What do you mean, a grave situation?' I wasn't leaving until he told me. So he said he was pretty sure it was

Tay-Sachs, a 90 percent chance. We left his office and went right back to the pediatrician's office, and he just looked at us as if he couldn't do anything. And then we knew that he had known all along. He looked at Jenny there in his office, and she was smiling and moving around and he said, 'I wasn't sure because she was responding up to a certain point, more than the usual Tay-Sachs baby.' He made it clear that basically he didn't want to handle Jenny any more. He told us to contact the National Tay-Sachs Association."

"My parents had been concerned, too," Gary said. "They felt Jenny wasn't progressing either. They wanted to know the results of the eye examination as soon as we knew. They had gone away for the weekend. I called them and they drove back home and came to our house on Sunday. We were just sitting around the den feeling terribly depressed, and my dad suddenly said he remembered seeing a

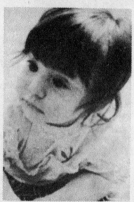

'The ophthalmologist saw the cherry-red spot. We went back to the pediatrician, and he looked as if he couldn't do anything, and then we knew he had known all along.'

notice about a testing program in some synagogue. He found out where the testing was and put me on the phone. I was a wreck by that time, not making much sense. That's when we met Dr. Nitowsky and Sandy Silverman and the others. They are fantastic people."

□

Sandy Silverman joined Dr. Nitowsky's staff as a genetics counselor in 1972. "Tay-Sachs disease is a particularly brutal one," she told me. "I've been in the room when couples have had that terrible death sentence pronounced for their child. It's horrible for everyone — the person giving the sentence and those receiving it. Very often, even though other physicians are fairly sure, they will push off on the geneticist the problem of telling the family.

"I've learned to watch the couple when they are told. They do one of two things. They either move together, physically reach for each other. Or they move apart. One man jumped up and ran out of the room. Gary and Marsha came together. They reached out and touched each other. I saw this happen again and again during their ordeal. When one was down the other was strong. Marsha was strong during the early period while Gary was nearly going to pieces. Later, when Marsha was down, Gary was strong.

"The first day they came to us Dr. Nitowsky drew blood for laboratory analysis. When he examined the eyes he could see the cherry-red spot. However, there are some other diseases that masquerade as Tay-Sachs. Dr. Nitowsky laid it out and was not optimistic. He said if the blood tests revealed that it was Tay-Sachs they were fortunate in one respect. There is a prenatal test for Tay-Sachs. That meant Marsha could have this test during the second trimester of each pregnancy and have an abortion if the baby was affected. They would be able to have normal children. When a couple finishes with Dr. Nitowsky and come out of his office, I kind of hover over them. They can't get rid of the emotions they're feeling. So Gary and Marsha came into my office and sat down and cried. And we talked. And after that Marsha just started calling me. Gary was really having almost a physical breakdown, and there was a time when she called me every day.

"Like so many people in this situation, they needed a neutral party to talk to. Friends or relatives don't understand the problem and they often talk in reassuring platitudes of hope when there is no hope. That can be more cruel in the end. Marsha needed a neutral party, someone who knew that her child was going to die.

"There is something you must understand about a recessive genetic disease. It's subtle. But eventually it occurs to most couples like Marsha and Gary. If their baby has a recessive condition, then that means each of them could have a normal baby with somebody else, with almost anybody else. When this realization comes, it challenges the relationship. I've seen marriages that became very shaky. I've seen couples separate. There is enormous guilt. It's a difficult test."

Each of us has from six to eight recessive deleterious genes. Most people can live a lifetime and never know they exist. They are revealed only if we marry someone carrying the same recessive gene and one of the children receives a double dose. This is how it works:

Rod-shaped chromosomes exist in similar, or homologous, attached pairs in the nuclei of human cells. Residing in these chromosomes are the genes that convey hereditary traits; in fact, each trait is controlled by a double gene or a double-gene sequence — one gene residing in one chromosome and the other in the other chromosome at the corresponding spot. In the initial formation of eggs and sperm,

the chromosome pairs split apart. One sperm or egg gets the genes from one chromosome, another sperm or egg gets the genes from the other chromosome.

A recessive gene, the Tay-Sachs gene, for example, exists at a certain spot on a chromosome. Its partner at the same spot on the other chromosome of the pair is normal. Thus, in someone who carries the Tay-Sachs gene, half of the sperm or eggs have the normal gene, half have the Tay-Sachs gene. At the moment of fertilization, a sperm and egg come together and the newly created individual has a full complement of genetic material — half from the sperm and half from the egg. The genetic makeup of this new person can reflect several possibilities.

There is a 25 percent chance that a normal egg and a normal sperm will come together, creating an individual who has no Tay-Sachs gene and need never worry about his or her children being afflicted. There is a 50 percent chance that an abnormal egg or sperm will fertilize a normal egg or sperm, creating an individual with one Tay-Sachs gene. This person won't be affected with the disease, but will carry the gene. If such a person marries somebody without the Tay-Sachs gene, their children will not be born with the disease. But if the spouse also carries the gene, there is a risk.

Finally, when two Tay-Sachs carriers marry, there is a 25 percent chance that a sperm with the gene will fertilize an egg with the gene, giving the baby the double dose, and causing the disease to be present. That's what happened to Jennifer.

Most people are blithely unaware of their genetic heritage. Because of the nature of recessive disorders, a family can go for generations without knowing about the gene's presence. In the daily course of their work, geneticists often see the result of people's ignorance of their genetic heritage. Geneticists recognized early that one means to deal with this might be to identify members of a particular population at risk because of some genetic malady and attempt to screen them before they have children — to uncover the hidden deleterious genes and at least apprise a person of the risk so he or she will not produce blindly.

Tay-Sachs affects people of Jewish descent whose forebears came from central and eastern Europe. A simple blood test can reveal whether such individuals carry the gene. Another genetic disease found in a specific population is sickle-cell anemia. An estimated two million blacks carry the sickle-cell gene, and a simple blood test can reveal

its presence. Although there are no tests for the cystic-fibrosis gene, an estimated 10 million whites carry it; when a blood test is developed — and it no doubt will be — a large segment of the population will benefit. Many states now require blood tests of newborn babies for a half dozen or more genetic diseases.

Nevertheless, the idea of screening various ethnic groups for a genetic defect, however laudable the intentions, brings chills to many civil libertarians. They worry that the results can be used for more than simply apprising an individual of the presence of a deleterious gene. Beyond such questions, there is also doubt about the efficacy of mass screenings — the cost, the ability to reach target populations and availability of medical manpower to carry out the screening and then properly counsel those in whom an unwanted gene has been discovered.

"Screening for Tay-Sachs has not received as much criticism as some other kinds of screening programs, notably for carriers of the sickle-cell trait among the black population," Dr. Nitowsky said. "In that case, I think there was inadequate education of the target population. People who were revealed to be carriers of the trait misinterpreted the information and thought they had the disease. A certain stigma came to be attached to being revealed as a carrier — again mistakenly. These people couldn't get insurance. They were refused jobs. . . . We geneticists are attuned to biological differences, but we must pay attention to cultural and social differences, too. . . . No one expects to be a carrier. It can't happen. So when someone is identified as a carrier, it can be quite disturbing. People need counseling and explanations. This approach generally works in adults, but I object strongly to mass screening of children, high-school students, for example. They simply aren't prepared. . . . Education is the answer in the end — education of the health-care providers and education of the public."

□

After learning that they are carriers for a recessive genetic condition, and that their daughter had received a double dose of the gene and would eventually die from it, Marsha and Gary felt an intense anger burning within them, building up to the exploding point. Gary in particular searched for someone to blame. He turned first to the pediatrician, the man who had been his doctor as a child.

"I called our pediatrician about two weeks later, because I felt I was close with him. I said, 'Did you know?' He told me he thought he had seen it some months before. 'I didn't see any reason to tell you any sooner until she displayed more evidence of the

disease.' He said he wanted to let us love her and enjoy her. He probably thought we wouldn't believe him anyway. Maybe we wouldn't have. I asked him when we should bring Jenny in next. He told us to call the Tay-Sachs Association. He said not to waste our money bringing her back to him. He certainly washed his hands of us.

"I had and still have a lot of resentment about what happened to us, because I feel it could have been prevented. I was just angry that we weren't given the option of knowing. Somewhere along the line someone should have told us. But no one wants to take the responsibility.

"I've heard some doctors say that informing Jewish couples about Tay-Sachs isn't their concern because it's a religious thing. We had a very progressive rabbi then. Actually he was my father's rabbi. We drove over to his house the day we found out and he was wonderful. Later, we had other meetings with him and at one meeting I asked him why a rabbi will marry a couple without first telling them about Tay-Sachs and the tests, and where to get them and what the risk is. He said there is enough pressure on a couple before getting married and he didn't feel that it was the proper thing to do. This was an attitude he said a lot of his fellow rabbis had. He said that it is a medical problem.

"We later became involved with another rabbi who is quite progressive. He started some seminars to discuss Tay-Sachs. Had doctors come as speakers and so on. But the attendance was poor. People don't want to know. They just don't want to know.

"I have a lot of guilt about not knowing about this disease. I have a lot of resentment about not being told. You might ask whether my parents were aware of it. Why didn't they say something? I don't think they knew about Tay-Sachs. To tell you the truth, I have never sat down with them to talk about it. I don't want them to feel I have any resentment against them."

□

There is a dichotomy of emotions swirling about the parents of a Tay-Sachs child. On one level is the guilt about what they have done to their child, the remorse and the desire to atone. But on another level is the desire to replace their defective child with a genetically whole child and a subconscious wish that the Tay-Sachs child die.

Within six months of Jennifer's diagnosis, Marsha became pregnant. At 16 weeks' gestation she went to Einstein for an amniocentesis. Dr. Nitowsky stood beside her and held her hand while the obstetrician performed the amniotic tap.

It takes three weeks for the fetus's cells in the amniotic fluid to grow out in tissue culture so that the tests for the presence of an enzyme called hex A can be performed. The enzyme levels reveal whether the fetus is normal, a carrier of Tay-Sachs or affected with the disease. But the Einstein doctors have developed a preliminary test that can be performed on the amniotic fluid the same day it is removed from the uterus.

The preliminary test indicated that Marsha's baby would be normal, and the formal results three weeks later bore that out. Marsha's and Gary's son — the amniocentesis had also revealed its sex — would not carry the Tay-Sachs genes at all. He need never worry about the disease striking his children.

Marsha's pregnancy was uncomplicated. Jennifer had been delivered by Caesarean section and Marsha was scheduled to have her second baby the same way. Ironically, Joshua was born on Jennifer's second birthday. He was normal and blood tests revealed that the amniocentesis was correct — he carried no Tay-Sachs genes.

"For about two weeks, Marsha and Gary just suspended their emotions," Sandy Silverman recalls. "They felt disloyal to Jennifer by loving this new baby. A Tay-Sachs couple must work out their feelings. They experience difficult emotional gymnastics. You know that on one level they are wishing their affected child would die. But another part of them wants the child to keep fighting. Marsha and Gary had these feelings. We talked about it.

"There is one word you can't say to a Tay-Sachs couple. I could never bring myself to utter it to Marsha and Gary. The word is replacement. Replacement of the Tay-Sachs child with a new, normal child. Marsha and Gary wanted a girl so badly and Josh was not a girl. They were unhappy about it but didn't understand why. . . . It's very difficult for a Tay-Sachs couple to admit to such feelings. They don't want to be disloyal to their ill child.

"Each couple must work things out for themselves. Each couple copes differently. Marsha and Gary went far beyond the call of duty. They kept Jennifer at home long after most couples would have given up. After an initial period of grief, Tay-Sachs parents begin to think of institutionalization because their child will soon reach a vegetative state. But not Marsha and Gary.

"They had a special nurse who lived with the child. When the nurse was gone they cared for her. One of them had to be with her all the time. If one went out, the other stayed home. They stopped doing things as a couple. They had Jenny on a suction machine because she could choke. They learned how to feed her through a nasogastric tube and did that. They used to take her onto their bed at night and

play with her long after there was any response. But they imagined response. They imagined that she still recognized them long after she stopped.

"We would see them periodically and advise them against keeping Jennifer at home, but Gary wouldn't hear of it. Finally, Marsha began to crack. She called one day and said she was near the breaking point. She was ready to institutionalize Jenny, but Gary said no. It was about time for them to bring the child in for an examination by Dr. Nitowsky. He examined her and told them she had to be institutionalized. Gary finally agreed. It was as if Dr. Nitowsky released him from part of his guilt.

"When Joshua was 1 year old, which was also Jenny's third birthday, Marsha told me something that reduced me to tears when I hung up the telephone. She told me how they had split the day so that they could spend part of it with Jenny at the hospital celebrating and the other part at home with Josh celebrating. At first you think what a lack of being in touch with reality that shows. But what we know is that a couple must work it out, assuage their guilt, make their sacrifice, cope as far as they can. . . . All the ritual Marsha and Gary have gone through I think in the long run will free them when Jennifer dies. It will free them to love Josh even more than they do now, to be able to have another child, to be able to be all right."

□

"We had a special nurse who lived with us," Marsha said. "She lived in Jenny's room and stayed with us from Sunday night to Saturday morning. So Gary and I took care of her Saturdays and Sundays. We would sleep downstairs with her. We did that for two years because I had a fear that she would choke and I wouldn't be in the room. You just couldn't leave her alone. We practically lived in her room. During the day we would bring Josh in and play with him on the floor.

"The only problem we ever really had was fevers, which we learned was usual for Tay-Sachs children. . . . She just lay there with her eyes open, no movement, no brain activity to speak of. As for the daily routine, she was being fed by a nasogastric tube. The tube had to be inserted three times a week. We went to North Shore Hospital and learned how to insert it so we could tube-feed her. But Gary was much better at it than I was. He was better even than the doctors in the hospital. Three evenings a week we took it out and she would sleep overnight without it. She was getting phenobarbital as a tranquilizer and we would take the tube out after we gave her the phenobarb. Then Gary would insert it the next morning.

"With a Tay-Sachs child all you can do is make them comfortable. They require little, if anything. Her care was never difficult at all, other than the constant moving her so she wouldn't lie in one spot and get bedsores, and the feeding. She had to be moved every hour. We had a suction machine in the house. She did have trouble with her saliva and choking. So we had to constantly suction her if she needed it."

"The rough time was in the night," Gary said. "Later on, when she began to have problems with the tube, we slept in the room with her. We took her out of the crib and brought her into the bed we had in her room. That way we could hear her immediately if anything was wrong — if she began to choke. On week-nights we had the nurse in her room. But we had an intercom in our bedroom upstairs and we heard every sound in Jenny's room. Marsha was always very worried that she might not hear Jenny choking. Marsha learned to listen for any change in the sounds coming from Jenny's room and she would wake up instantly. The nights were hard.

"Once Josh was born it became very difficult, particularly for Marsha. I was still all wrapped up in Jenny. I tried hard to love the new baby, but, frankly, it was difficult at first. As far as I was concerned Jenny was going to stay at home until I was told it wasn't right anymore."

"There came a point where I felt it wasn't a healthy family situation anymore," Marsha interrupted. "It had been over a year since we really did anything or were out of the house together. Any activity, such as family holidays, either Gary would go or I would go. One of us stayed home. Jenny had a very low resistance to infection, so we didn't have any children of our friends in the house for more than a year, and the reactions of our friends is another whole story. It was difficult for them to be with us. We were down about what had happened. We were jealous of their healthy kids. I remember seeing a child who was five months younger than Jenny doing things Jenny never did. Not that I wanted any harm to come to that child, but it was so hard to admit that here is this child doing things my daughter never did. I was jealous. Finally, everything started bothering me. . . . It wasn't fair to Josh. He was cooped up at home with us — tied to Jenny's room, too. It wasn't healthy for any of us.

"Yet the time I felt guilty the most was when we decided to put her in the hospital. I knew we had done more than most parents would do, but I felt guilty. How can a mother do something like that? . . . Still, she was better off."

□

A few months after Jennifer was hospitalized, Marsha be-

came pregnant for the third time. Late in 1977 she went to Einstein for an amniocentesis.

Unlike the first one, this amniotic tap was difficult. The obstetrician drew blood on the first attempt and had to keep trying until he retrieved some clear fluid. It was a long and anxiety-provoking session. Marsha left the treatment room badly shaken.

But the preliminary results the next day indicated that the child was unaffected. Marsha and Gary celebrated quietly. The third pregnancy was particularly important to them, because Marsha had been told she could tolerate no more than three Caesarean sections. This child probably was their last. Originally, they had wanted a large family.

The couple was unprepared for the call Dr. Nitowsky put in to Marsha at work three weeks later. He wanted her and Gary to come to his office as soon as possible. He said he had some bad news. The amniocentesis results indicated the baby Marsha was carrying would be affected with Tay-Sachs. A member of Einstein's obstetrics faculty examined Marsha and suggested a method of aborting the fetus vaginally, thus giving her a chance for another pregnancy. Shortly thereafter, Marsha successfully underwent the abortion, which was for her a devastating experience. She entered a period of depression and despair that rivaled the emotional extremes that had plagued Gary two years before when he learned his daughter would slowly deteriorate and die. Sandy Silverman saw the young couple reach for each other again — Gary strong now, holding up his faltering wife.

□

When we began talking in Marsha's and Gary's den, it was early afternoon. Josh had learned to walk and was into everything. First Marsha and then Gary would jump up and rescue him from some impending disaster as he tugged at a lamp cord or pulled at a bookcase. The affection they held for their son was obvious. They were the epitome of doting parents. But it was easy to imagine their conflicting feelings generated by this handsome boy and Jennifer, who at that moment lay in a hospital a few miles away, nearly 4 years old but unsmiling, unseeing, unfeeling and unthinking. Were they resigned? Were they still bound by their guilt and secretly resenting her? Or were they merely pulling away, anticipating her inevitable death? She already had lived longer than most Tay-Sachs children.

"She's a very strong girl. It's amazing," Marsha said, a curious note of pride in her voice. "Congestion, which interferes with their breathing and eventually leads to pneumonia, is the main problem that all the children in the ward have. But Jenny seems to be in very good shape compared to even the younger children, who are constantly choking and congesting. She's able to spit up the saliva and phlegm."

"We visit her several times a week," Gary said. "I go several times at night. Marsha makes it whenever she can. And then we go together on the weekends. My time there is maybe 10 minutes or 15 minutes. There's nothing to do. I hold her and I cuddle her a little bit. I'll kiss her. I'll talk to her, but of course I'm really talking to myself because she doesn't comprehend anything I'm saying. Sometimes I'll try to lift her. She's uncomfortable when being lifted. She has these seizures. On the weekends, we both bathe her and I do lift her up and hold her for just a couple of minutes."

"Bathing her gives you something to do," Marsha added. "I enjoy it. Somehow I feel better after I bathe her and fix her hair. She just looks so . . . pretty, so terrific."

After Jenny turned 4 and Josh 2, I spoke to Gary and Marsha again. I asked how they had marked the two birthdays.

"We went to the hospital," Gary said. "We usually go on Sunday morning and the birthday happened to be on Sunday this year. But there was no celebration. In our own way we said, 'Happy Birthday, Jenny,' and brought cards from the family and friends who do still remember. You buy her a new nightgown, which is really all she can use."

"When I visit her, I'm usually fine and under control,"

Marsha interrupted. "The only time I have a problem is on her birthday. I usually spend a half hour or 45 minutes on my visits. But on her birthday I cut it short. It's a difficult time. When I was in labor with Josh and realized that he was going to be born on the same date as Jenny, I wasn't too thrilled about it. But everyone said it was a good omen and there's a reason for it. In a way now I'm kind of glad. I come home from the hospital and I'm forced to be happy for Josh. It kind of balances out."

I mentioned Sandy Silverman's comment about how Tay-Sachs parents either come together or move apart when they learn the news and her memory of Gary and Marsha physically reaching for one another.

"We have become incredibly

close," Marsha said.

"It forces you to either talk about it or hate each other for it," Gary said.

They confided that Marsha was pregnant again. At the proper time they would go to Einstein for another amniocentesis. From the hesitant way they discussed it, I knew they were trying not to think about the 25 percent chance that the baby growing in Marsha's womb had received a Tay-Sachs gene from each of them. But six months later, a normal healthy boy, David, was born. Shortly after that, on July 14, 1979, the life of little Jennifer came to an end. "She was a fighter," Sandy Silverman said. "But Marsha's and Gary's ordeal is over." ∎

August 12, 1979

Mother Teresa, Receiving Nobel, Assails Abortion

Associated Press

Mother Teresa of Calcutta holding award she received in Stockholm.

OSLO, Dec. 10 (AP) — Mother Teresa, Calcutta's Saint of the Gutters, accepted the $192,000 Nobel Peace Prize today in the name of the poor, the sick and the unwanted children. She also denounced abortion as "the greatest destroyer of peace."

The 69-year-old Roman Catholic nun, a native of Yugoslavia, said she would use the money, along with $70,000 from her Norwegian People's Prize, to build homes for lepers in her worldwide Missionaries of Charity organization. She asked that the traditional Peace Prize dinner be canceled so the $6,000 allocated for it could be used for her work.

Prof. John Sanness, chairman of the Norwegian Nobel Committee, presented Mother Teresa with a check, a Nobel gold medal and a diploma in the afternoon ceremony in the great frescoed hall of the University of Oslo. Praising her three decades of selfless service to the poor, he urged that rich nations, in assisting poorer nations, emulate Mother Teresa's spirit and respect for individual human dignity.

Bridging Gulf From Rich to Poor

"Mother Teresa has personally succeeded in bridging the gulf that exists between the rich nations and the poor nations," he said. "Her view of the dignity of man has built a bridge. All aid given by the rich countries must be given in the spirit of Mother Teresa."

Mother Teresa told the audience: "Our

poor people are great people, a very lovable people. They don't need our pity and sympathy. They need our understanding love and they need our respect. We need to tell the poor that they are somebody to us, that they, too, have been created by the same loving hand of God, to love and be loved."

Her tiny figure clad in the traditional sari of India, she stood beneath biblical murals, beamed her radiant smile and led her listeners in St. Francis of Assisi's prayer for peace.

Though she usually steers clear of controversy, Mother Teresa chose to press her longstanding opposition to abortion. "To me the nations with legalized abortion are the poorest nations," she said. "The greatest destroyer of peace today is the crime against the innocent unborn child."

At her mission in Calcutta, where she started helping the poorest of the poor in 1946, Mother Teresa and those who work with her try to place unwanted children through adoption. They also teach a version of the traditional church-approved rhythm method of birth control.

The nun, the sixth woman to win the coveted award since its establishment in 1901, is to give a formal acceptance speech tomorrow. Since her arrival Saturday she has insisted on maintaining the fast-paced austerity of her life in Calcutta.

December 11, 1979

CHAPTER 7

Crime
and
Justice

The U.S. Supreme Court, under Chief Justice Warren Burger, ruled that judges have the power to bar the public and press from pretrial proceedings.
United Press International

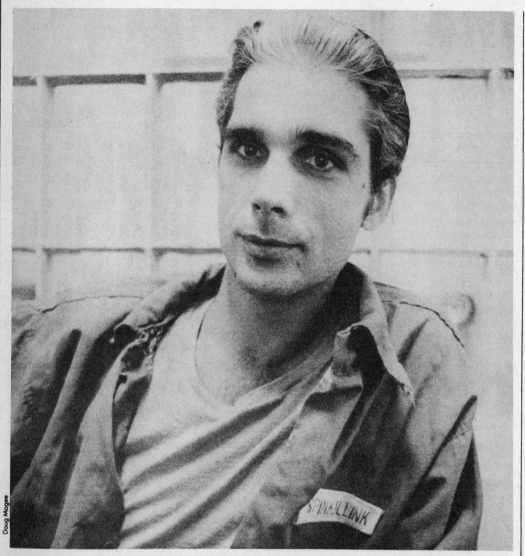

A guard tests Florida's electric chair.

John Spenkelink, a convicted murderer in Florida State Prison, may be the first person to be executed in Florida in 14 years.

WILL HE BE THE FIRST?

This month, capital punishment may become legal again — and 40 people on death row will be slated to die in 1979.

By Peter Ross Range

Peter Ross Range is a freelance writer based in Washington.

The electric chair at Florida State Prison is an unimposing machine. Standing empty, unconnected to the heavy black cables snaking behind it from an innocuous white power box, it appears almost antiseptically benign. Its broad, oaken flatness — it was built in 1924 from a single oak felled by in-mates — reminds one of those ageless, flat-backed armchairs that used to grace Y.M.C.A. lobbies before they went to carpeting and high prices. The beveled, wooden headrest, where the condemned gets the jolt that kills him, looks somewhat like that on an antique barber's chair. In the clean, fresh-painted brightness of this linoleum-tiled, first-floor room, it is hard to imagine the chair as an instrument of extermination.

But kill it will. Within this year — per-haps this month — John A. Spenkelink, 30, a work camp escapee who killed his traveling companion, an ex-convict, in what he describes as a domestic quar-rel over money, faces the probability that he will be shocked to death by 2,500 volts of direct current passed through his body eight different times at vary-ing amperages over a two-minute peri-od. If the death sentence is carried out, Spenkelink's dark hair with a swatch of premature gray in front will have been shaved, the better to establish electri-cal contact through the electrocution helmet. If the switch is thrown, his lightly tattooed body, lean and muscu-lar from five years of doing push-ups and jumping jacks in a 6-by-9-foot cell, will strain and jerk against seven new thick leather straps holding him in the chair; his scalp may begin to emit smoke before the two minutes are up.

The 12 state-appointed witnesses, seated in dainty, scallop-backed wooden chairs, face to face with Spenkelink, will not notice the noxious smell of burning human skin; they will be separated by an odorproof double window of thick plate glass. Yet, as if to give reality to their witness, the sounds of this grisly business — attaching of straps, reading of sentence and last words of the condemned, activation of power, the flipping of three switches be-hind the executioner's partition and the struggles of John Spenkelink — will be piped into the viewing room by inter-com. After two minutes, then, John Spenkelink will be dead at the end of a classically disheveled life; the State of Florida, in the name of the people, will have its revenge; and the floodgates of execution in the world's most de-veloped country will be reopened.

□

What has been in recent years a more or less theoretical debate over capital punishment will become a matter of corporeal urgency if Spenkelink's ex-ecution is scheduled. His avenues of ap-peal have been nearly exhausted. No-body knows how the United States Su-preme Court will decide Spenkelink's case, but if it turns down his current petition, it will short-circuit the process for many who have already seen their convictions and sentences upheld in the lower courts. Spenkelink's execution will sound the death knell for dozens of the nearly 500 men and five women now incarcerated under death sentences in 25 states. As many as 40 persons could be institutionally killed in 1979 — which may well be remembered as The Year of the Execution.

Death-penalty abolitionists already have plan- *(Continued)* ed overnight vigils and demonstrations on the greensward flanking the narrow highway outside Florida State Prison, located near the village of Starke in north central Florida, and officials there have agreed on the space to be used. Since the South, now as before, remains the chief bastion of capital punishment in America (84 percent of those on death row today are in 11 Southern prisons), Atlanta will this spring be the venue of a major national rally of those who oppose state executions.

It has been 12 years since a de facto moratorium was declared on the death penalty in America. A population of 600 persons accumulated on the death rows of the nation between 1967 and 1972, a period during which the states were reluctant to execute while capital-punishment laws were under fresh attack in the Federal courts. In 1972, the Supreme Court ruled the death penalty to be unconstitutional because it was arbitrarily applied, hence, "cruel and unusual." All condemned persons had their sentences automatically commuted to life imprisonment. Since then, 34 states have enacted new death-penalty statutes. Yet the only execution in America since 1967 was that of the suicidal, double-murderer Gary Gilmore, who qualified for the firing squad in Utah in 1977 by deliberately withdrawing his appeals.

With the present rate of sentencing to death at roughly two per week — especially in Southern states — there could be more than 100 executions per year in the 1980's. The annual figure could, in fact, equal that reached in the heyday of capital punishment in the postlynching era: 199 in 1935. "It promises to be a bloody decade," laments Henry Schwarzchild, director of the National Coalition Against the Death Penalty.

The death penalty, however, has ardent advocates. Ernest van den Haag, a conservative New York psychologist and sociologist, defends capital punishment, among other reasons, on the basis of retribution: "The motives for the death penalty may indeed include vengeance.... Legal vengeance solidifies social solidarity against lawbreakers and probably is the only alternative to the disruptive private revenge of those who feel harmed."

The rate of execution peaked at the height of the Depression and many students of capital punishment think economic woes and the public desire for

harsh penalties go hand in hand. "Inflation wears on middle-class minds with economic interests to protect," remarks Carol Palmer, a capital-punishment researcher at the NAACP Legal Defense and Educational Fund. "When times are tight, people think more about crime and they start looking for panaceas. The motivation to enact the death penalty always rises with inflation."

Our own inflationary period, some would say, is just such a time of vengeance. "This state is in a bloodthirsty mood," says one Georgia attorney who is fighting such a tenuous battle for the commutation of his client's sentence from death to life that he refuses to let his name be used. Indeed, the Chief Justice of the Georgia Supreme Court, H. E. Nichols, has been barnstorming the state, speaking out in favor of speedy executions of present death-row inmates in the interest of judicial efficiency. "I think it is a deterrent, especially to the more serious crimes," says Justice Nichols. "If they know they're going to get punished, brother, they're going to think twice about it."

Georgia, with 417 executions since it began keeping records in the 1920's, has put more people to death than any other state. Today, with 74 men and one woman on death row, Georgia ranks third in the nation behind Florida (119) and Texas (105). Georgia's death chamber and its massive, 43-year-old electric chair in the fifth-floor cupola atop the main building of the state prison in Reidsville have received a fresh coat of white paint. The three oversized black switches behind the simple, plywood executioner's partition have been checked — and they worked. The warden, Charles Balkcom — his father was warden before him — is a soft-spoken man who says he is ready. "I favor the death penalty. We'll carry out the law. But I don't intend to watch it. I have enough other problems as it is."

□

The remarkable thing about America's impending return to capital punishment on a significant scale is how few people realize it is about to happen. Most Americans seem to think the moratorium of the past 12 years, unofficial or not, is here to stay. While the new death-penalty statutes had to be approved by the state legislatures in full public view (in Oregon and California last November, they were approved by popular ballot),

many Americans seem unprepared for the reality of executions themselves. "There is a paradox, a schizophrenia about the matter," says Prof. Hugo Adam Bedau of Tufts University, editor of the definitive book "The Death Penalty in America." Death-penalty abolitionists expect a certain degree of public revulsion once the institutional killing resumes, but no one knows whether this will have significant political impact.

There is a sense of *déjà vu* about this curious current state of affairs. In 1972, a Supreme Court that was still dominated, 5-to-4, by pre-Nixon appointees declared existing capital-punishment laws to be unconstitutional because they violated the Eighth Amendment. They were so arbitrarily applied, mostly to the poor and the black, explained Justice Potter Stewart, that they were "cruel and unusual in the same way that being struck by lightning is cruel and unusual." This was a historic decision that was heralded in The New York Times with a rare banner headline. Four years later, however, the Court upheld newly written death statutes in Georgia, Florida and Texas — an equally historic decision which made banner headlines nowhere. Yet today, as a result of that decision, the death-row population of the country is quietly but quickly reaching the size it had attained when all sentences were commuted to life imprisonment in 1972.

The clock would seem to have been turned back. Since the responsibility for criminal executions was taken from the county courthouses (the hanging of a black man in Owensboro, Ky. in 1936 is thought to have been the last public execution under county jurisdiction in the United States), the national trend has been away from execution. It was around the turn of the century that most of the states made official killing a state-supervised affair, behind closed prison gates with only a small, carefully chosen audience. This coincided with the widespread adoption of the electric chair or gas chamber, as opposed to hanging, as the method of execution. Subsequently, the rate moved fitfully but steadily downward: from 124 executions in 1940 to two in 1967. The last man executed before the current moratorium began was Luis José Monge, gassed in Colorado on June 2, 1967, for the murder of his wife and three children. Public opinion was in agreement with the demise of the

death penalty: A 1965 Harris poll showed that while 38 percent of the American population favored capital punishment, 47 percent opposed it.

When the Supreme Court struck down capital punishment in its momentous decision in Furman v. Georgia in 1972, death-penalty opponents assumed they had finally won a battle that reformers had lost for two centuries; the United States, it seemed, had joined the three dozen other major countries, including all Western nations except France and Franco's Spain, that had abolished capital punishment. But that judgment was premature.

In 1972, the country was on the cusp of what has now been recognized as the conservative turn of the 1970's. Fears of crime in the streets were rising and, indeed, were being fanned by the Nixon Administration. The still unsettled war in Vietnam, the bitter national trauma of the previous year's conviction of Lieut. William L. Calley of My Lai infamy and the fresh memories of the killings at Kent State and Jackson State — all this contributed to a sharp political polarization that was further exacerbated by the Watergate break-in, which occurred just 12 days before the Supreme Court decision. It should have been no surprise that the largely conservative state legislatures rushed to rewrite their capital-punishment statutes.

The South led the way. Attempting to meet the Supreme Court's objections, Florida called its legislature into special session before the calendar year was out, passing a new, more carefully written death-penalty statute on Dec. 8, 1972. Georgia, Mississippi, Tennessee, Virginia and Arkansas — as well as Idaho, Indiana and Oklahoma — reenacted capital punishment by April of the following year. Being in favor of the death penalty became valuable political currency in legislative election campaigns and, by last year, became a crucial issue in gubernatorial races in several states. By now, 34 states have readopted the death penalty, and similar legislation is pending in three others. (Some states, such as Michigan, Minnesota, Wisconsin, Alaska, Iowa, South Dakota, North Dakota and Hawaii pride themselves on a consistent history of abolitionism.)

A CBS/New York Times poll taken in August 1977 showed that 73 percent of the population now favored capital punishment. A Harris poll taken a few months before showed 67

percent in favor, and the racial breakdown of the respondents was especially interesting: 72 percent of the whites favored capital punishment (22 percent opposed), while only 40 percent of the blacks were in favor (48 percent opposed). White America was now clearly unprepared to relegate capital punishment to the past. The only other industrialized Western nation still commonly using capital punishment is South Africa; France retains the guillotine but uses it only about once a year, and only for the most heinous crimes. In its new Constitution, adopted last December, Spain, released from the Franco era, abolished capital punishment except for treason.

□

The legal basis for America's return to state executions lay in the wording of the 1972 Furman decision. The five-man majority on the Supreme Court did not, as Chief Justice Warren Burger pointedly remarked in his dissenting opinion, rule "that capital punishment is per se violative of the Eighth Amendment; nor has it ruled that the punishment is barred for a particular class or classes of crimes." Instead, the decision simply struck down existing statutes for violations, *as then applied* — that is, primarily against blacks and the poor. Burger further pointed the way for rewritten statutes by stating in his dissent that "significant statutory changes will have to be made. ... Real change could clearly be brought about if legislatures provide mandatory death sentences...."

This clearly opened the door for the comeback of capital punishment, and this time its supporters meant to avoid the pre-Furman pitfalls of "arbitrariness" that had been struck down in 1972. A number of states wrote laws designed to apply equally to everyone by making the death penalty mandatory for certain crimes, such as killing a peace officer or murder by a life-term prisoner. This posed a problem for juries in capital cases in that they had no choice regarding punishment: Conviction meant death and the only lesser penalty was outright acquittal. Hence, some states, such as Ohio and North Carolina, found themselves sentencing criminals to death in unprecedented numbers. Recognizing the dangerous rigidity of these laws, the Supreme Court, in Lockett v. Ohio and Woodson v. North Carolina, struck down mandatory death-

'More than the race of the murderer, it is the race of the victim which determines whether the guilty party will receive the death penalty.... The taking of a black life ... is one-tenth as likely to be punished by death as the taking of a white one.'

penalty statutes in a number of states in 1976.

However, in the same 1976 rulings, the Court upheld a different kind of statute written in other states, notably Florida, Texas and Georgia (Gregg v. Georgia). Their laws contained no mandatory death penalties, but rather outlined what is called "guided discretion." This meant that while the law contained a specific list of capital murders, it also gave the juries discretion to decide on life imprisonment or death, according to a suggested list of "aggravating and mitigating circumstances." This decision was to be reached in a second trial immediately following a guilty verdict in the main trial itself. Known as a bifurcated trial, this system, primarily as written in the Florida law, has become the model for other states.

Opponents of capital punishment regard the bifurcated trial as just a more sophisticated masking of the same old system of arbitrariness and caprice in the matter of who gets death and who gets life. In fact, even at the present rate of sentencing, only a tiny percentage of murderers is sentenced to death. And there seems to be no consistent pattern dividing the one group from the other. One woman in Georgia is on death row for a murder that her husband committed; she was convicted of being a co-conspirator, which she vehemently denies. In another Georgia conspiracy case, however, two men recently received life imprisonment for a clearly proved "contract" killing. There are child-killers serving life, while, in Florida, for instance, one teen-aged boy received a death sentence for the sale of heroin that led to the purchaser's accidental death by overdose. The Supreme Court, meanwhile, has denied all attempts to attack the death penalty on the basis of continuing arbitrariness. "The Court just doesn't want to hear about capital punishment," complains Atlanta lawyer Millard Farmer, director of the Team Defense Project, which successfully defended the Dawson Five in south Georgia last year. "They just threw the whole issue back to the states and said, 'Do what you will.' "

☐

In their attempts to force Americans to deal with what they consider the hor-

ror of institutional killing, liberals have adopted the unlikely position of opposing any trends toward making executions "more humane." While reform-minded groups originally supported the move from the disfiguring and brutalizing method of hanging to the modern and allegedly more humane techniques of electrocution and gas chambers, the new abolitionists see such changes today as inimical to their cause. Thus, their opposition to newly enacted statutes in Texas, Oklahoma and Idaho that call for execution by injection into the condemned person's arm of a lethal dose of sodium thiopental. Aside from the overtones of Nazi-style euthanasia or "putting a dog to sleep," abolitionists see this new technique as yet another way of masking the central issue of whether a modern society should take a life for a life (rape and kidnapping no longer being capital crimes, as a result of the decision in Coker v. Georgia, 1977).

"There is no such thing as a humane way of killing," observes Clinton P. Duffy, the retired longtime warden of California's San Quentin prison, who had to carry out 90 executions. "I'm against the death penalty because it is wrong to kill. It was wrong to have the first murder and it's wrong for the state to premeditate another murder. Besides, it's a privilege of the poor. I don't know of a wealthy person ever executed in the United States."

The racial implications of capital punishment have taken a subtle and ironic turn. A decisive majority of the 600 persons on death row when the 1972 decision was issued were indeed black (at a time when rape was still a capital crime, and when whites almost never received a death sentence for rape). Today, 54 percent of those on death row are white, 41 percent black. But Northeastern University sociologists William Bowers and Glenn Pierce have unearthed startling data which indicate that while whites and blacks may receive the death penalty in more or less equal numbers today, more than the race of the murderer, it is the race of the victim which determines whether or not the guilty party will receive the death penalty. Abolitionist Alabama lawyer Morris Dees once pointed out to Bowers and Pierce that he always knew whether to prepare for a capital trial in homicide cases by finding out if the victim was white — regardless of the race of the killer. If the victim was black, no matter what the race of the offender,

the prosecutor would probably not request the death penalty.

Bowers and Pierce undertook a detailed study of homicides in Florida over the five-year period of 1973-77. He then correlated the race of the victim with the race of the offender and came up with the table below.

This research shows that the 72 white men on Florida's death row (by the beginning of 1978) had *all* killed other whites. Not one of the 111 whites who killed blacks received a death sentence. It also shows that 92 percent of the men on death row — white or black — had killed whites, although, in fact, an almost identical number of murders of blacks had occurred (2,432 offenders arrested for killing whites versus 2,431 offenders arrested for killing blacks). The taking of a black life, even by another black, was one-tenth as likely to be punished by death as the taking of a white one. And yet, a black who took a white life was five times as likely to receive the death penalty as a white doing the same thing. John Spenkelink's lawyers used this line of reasoning in their petition to have his sentence overturned, only to have it rejected by the Fifth Circuit Court of Appeals; it now rests with the Supreme Court. The man Spenkelink killed was, like himself, white.

Death-penalty opponents claim that the obvious bias demonstrated by the race-of-the-victim argument is but one example of the continued arbitrary application of the death penalty — the very practice that the Furman decision was supposed to strike down. There are more invidious problems, such as the obvious tendency for people of means who commit homicides to retain skilled counsel to plea-bargain for a lesser charge or manage to obtain a life sentence even when their clients are convicted of first-degree murder. Another example of arbitrariness sanctioned by law is "prosecutorial discretion," the decision of one man, the local district attorney, as to what charge to bring for any given homicide.

Spenkelink's crime is a case in point. He had been traveling for some months with a man with a much longer and more serious criminal record than his own. His companion, testified Spenkelink, carried a pistol and occasionally threatened him with it. He once forced Spenkelink to have homosexual relations with him. During a brief motel stop in Tallahassee, Fla., Spenkelink discovered that all of his money was missing from his suitcase. He argued and fought with his companion. Later, he testified, he returned to the motel to pick up his clothes, tried to recover his money and leave. He brought the pistol in from the car. His companion attacked him, both men went for the gun and Spenkelink shot the other man. Since the body was found in the bed with two gunshot wounds from behind, the prosecution argued that

Spenkelink had premeditated the killing.

First-degree murder? Second-degree? Manslaughter? Justifiable homicide in self-defense? That question must be answered by the prosecutor in making his decision as to what charge to bring before the grand jury. It is argued that even in another part of Florida, Spenkelink, like several hundred other men serving life or long terms at the state prison, would have been charged with something less than first-degree murder. Indeed, the Bowers-Pierce study showed a distinct geographic break between conservative north Florida and more liberal south Florida. For example, the killer of a white in the Florida Panhandle is 24 times more likely to receive the death penalty than the killer of a black in Miami. Tallahassee is in the Panhandle.

This very circumstance prompted Florida Supreme Court Justice Richard W. Ervin to dissent from his court's affirmation on appeal of Spenkelink's death sentence: "Truly characterized, the sentencing to death here is an example of the exercise of local arbitrary discretion. The two actors in the homicide are underprivileged drifters. Their surnames, Spenkelink and Szymankiewicz, were foreign and strange to the Tallahassee area. They have no family roots or business connections here. All of the ingredients were present for the exercise of invidious parochial discrimination...."

☐

The argument that the return to the death penalty merely reflects a vengeful turn in public sentiment is supported by the age-old deterrent debate. First, there are virtually no reputable data to prove that capital punishment is a deterrent to murder. But second, just as people once opposed the death penalty but now favor it, the 1977 CBS/New York Times poll showed that 61 percent believed in its deterrent value. Third and most interesting, when the 1977 Harris poll asked the people being polled whether they would support capital punishment *even if* they were shown that it had no deterrent effect, whites still favored it by 49 percent (39 percent against). This leads directly to the conclusion that it is the retributive aspect of capital punishment that has come on strongest in the past 10 years. Deterrent value or no, people want a death penalty.

Despite the intensity of these battles between the abolitionists and the lawmakers (and prosecutors) who favor the death penalty, most of the country seems to have been blissfully oblivious to the controversy. The capital-punishment debate has never been raised to the level of a national issue which can be fully investigated with national resources. A Senate committee last year held hearings which probably helped kill a movement to re-enact a Federal death penalty, but no sweeping report was issued. Instead, each state operates with limited data, local political pressures and expediency on an issue that promises once again to become a serious moral dilemma.

Last year Alabama attorney Morris

FLORIDA, 1973-77			
Victim/Offender Race	Estimated Number of Offenders	Persons Sentenced to Death	Probability of Death Penalty
B kills W	286	48	.168
W kills W	2,146	72	.034
B kills B	2,320	11	.005
W kills B	111	0	.000

Dees capitalized on his standing with Jimmy Carter — Dees was the President's chief fund-raiser during the 1976 primaries — to urge him to depoliticize the death-penalty debate. President Carter, who signed the Georgia death-penalty statute and personally favors capital punishment for such crimes as killing a peace officer, listened over a private lunch in the Rose Garden at the White House while Dees made his proposal. Following the model of the British Royal Commission on Capital Punishment, whose 1952 report led to the eventual abolition of capital punishment in Great Britain in 1965, Dees suggested that Carter establish a National Commission on Capital Punishment to prepare a thorough report on the subject. "This would take the political pressure off the governors, the legislatures and the President himself," insists Dees, who obviously hopes such a commission would recommend abolition.

So far nothing has been heard from President Carter. In the meantime, the death penalty moves daily closer to being not just a statute but a reality. John Spenkelink spends his time in R wing reading Edgar Cayce and the Bible, helping illiterate death-row inmates write and read letters, and smoking his one pack of Benson & Hedges per day. His girlfriend from Jacksonville comes to visit every weekend and her children call him "Dad." Says Spenkelink: "I can see a bright future someday. I've got things to do when I get out of here. I shouldn't die, but the state wants to make sure I burn." ■

March 11, 1979

JUSTICES, 5-4, LIMIT COURTROOM ACCESS BY PRESS AND PUBLIC

JUDGE'S PRETRIAL BAN UPHELD

Supreme Court Appears to Extend Its Analysis in New York Case to Curb on Trials as Well

By LINDA GREENHOUSE

Special to The New York Times

WASHINGTON, July 2 — The Supreme Court, ruling today that "members of the public have no constitutional right under the Sixth and Fourteenth Amendments to attend criminal trials," granted trial judges broad discretion to order courtrooms closed to the press and public.

The 5-to-4 decision affirmed a ruling of the New York State Court of Appeals, which had upheld the power of a trial judge to close a pretrial hearing if he found a "reasonable probability" that pretrial publicity would hurt the defendant's chance for a fair trial.

Today's majority opinions appear to go considerably further. Although sometimes limiting discussion to pretrial proceedings, the majority at several key points extends the analysis to actual trials, thereby suggesting that they as well can be closed upon agreement by the defendant, the judge and the prosecution.

No Reason Required

Also, there is no requirement in Associate Justice Potter Stewart's majority opinion that a judge, before barring the public from the courtroom, make a preliminary finding that an open proceeding will be harmful to the defendant.

Indeed, as Associate Justice William H. Rehnquist observed in a concurring opinion: "It necessarily follows that if the parties agree on a closed proceeding, the trial court is not required by the Sixth Amendment to advance any reason whatsoever for declining to open a pretrial hearing or trial to the public."

The Sixth Amendment provides, "In all criminal prosecutions, the accused shall enjoy the right to a speedy and public trial." The 14th Amendment made this provision binding on the states.

The decision aroused immediate strong criticism from both the press and the legal profession.

In addition to Justices Stewart and Rehnquist, the majority opinion was joined by Chief Justice Warren E. Burger and Associate Justices Lewis F. Powell Jr. and John Paul Stevens.

Justice Powell also filed a separate concurring opinion in which he said that before closing a courtroom the trial judge should be required to balance the defendant's interests against the First Amendment rights of the press to report on the proceedings. The four other members of the majority explicitly declined to rule on whether there is a First Amendment right to attend trials.

Justice Powell said that a judge, when presented with a request to close a hearing, should first decide "whether there are alternative means reasonably available by which the fairness of the trial might be preserved without interfering substantially with the public's interest in prompt access to information concerning the administration of justice." The judge in this case had properly considered the issue before closing the hearing, Justice Powell said.

Associate Justice Harry A. Blackmun wrote a dissenting opinion and took the unusual step of reading excerpts from the bench this morning. The 44-page dissent, joined by Associate Justices William J. Brennan Jr., Byron R. White and Thurgood Marshall, accused the majority of surrendering "to the temptation to overstate and overcolor the actual nature" of the case and producing a "wooden approach" that they said "is without support either in legal history or in the intendment of the Sixth Amendment."

No 'Sensational Journalism'

The dissenters said that the pretrial publicity in the case, Gannett v. DePasquale (No. 77-1301), which involved the disappearance and apparent murder of a former policeman in a Rochester, N.Y., suburb, had been "placid, routine, and innocuous," containing "nothing that a fair-minded person could describe as sensational journalism."

There was no publicity for 90 days before the start of a hearing on whether the two defendants' confessions should be admitted at trial. The trial judge granted the defense attorneys' request to close the hearing, a decision that was eventually upheld by the state's highest court.

The majority opinion today held that the Sixth Amendment's guarantee of a right to a public trial belongs only to the defendant himself. "There is not the slightest suggestion" in earlier cases, Justice Stewart wrote, "that there is any correlative rights in members of the public to insist upon a public trial." The framers of the Constitution, he said, did not intend "to create a constitutional right in strangers to attend a pretrial proceeding."

The dissenters argued that the right to a public trial, dating back through several centuries of English common law that preceded the drafting of the Constitution, protects the interest of the general public in the integrity of the criminal justice system as well as a defendant's personal interest in a fair trial.

"Open trials," Justice Blackmun wrote, "enable the public to scrutinize the performance of police and prosecutors in the conduct of public judicial business."

He continued: "Secret hearings — though they be scrupulously fair — are suspect by nature." Justice Blackmun observed that in New York, as in most other states, some 90 percent of all criminal cases are disposed of before trial, leaving pretrial proceedings as "the only opportunity the public has to learn about police and prosecutorial conduct, and about allegations that those responsible to the public for the enforcement of laws themselves are breaking it."

Rulings in Earlier Conflicts

The public's interest in open trials, he concluded, is so "fundamental" that it cannot be waived by the individual parties to a trial. A judge should require a defendant who requests a closed hearing to show that he would be harmed by open proceedings, that less drastic alternatives would not adequately protect him and that "there is a substantial probability that closure will be effective in protecting against the perceived harm."

A majority of the Supreme Court had in fact adopted such an approach in earlier cases involving conflicts between the rights of the defendant to a fair trial and of the press to report on the proceedings. The earlier cases had held that while closing the courtroom was a step a judge might have to take as a last resort, First Amendment interests require that less drastic alternatives first be considered and found to be inadequate.

That such an approach did not command a majority today will undoubtedly lend support to the view of press lawyers and others who believe that the Court's commitment to the First Amendment has eroded over the last few years.

Justice Blackmun today quoted a dissenting opinion from a 1965 Supreme Court decision that held that the presence of television cameras had deprived Billie Sol Estes of a fair trial. "The suggestion that there are limits upon the public's right to know what goes on in courts causes me deep concern," the dissenting Justice wrote in that case. The dissenter was Justice Stewart, the author of today's majority opinion.

July 3, 1979

Ruling Draws Criticism From Lawyers and Press

By WARREN WEAVER Jr.
Special to The New York Times

WASHINGTON, July 2 — Today's Supreme Court decision broadening judges' power to close criminal courtrooms to the press and public aroused strong criticism today from both the press and the legal profession.

Generally, authorities on civil rights and press freedom argued that the high court had gone too far in the name of curbing prejudicial pretrial publicity, failing to consider lesser remedies that would not have limited so severely public access to the criminal justice system.

Allen Neuharth, chairman of the American Newspaper Publishers Association, called the ruling "another chilling demonstration that the majority of the Burger court is determined to unmake the Constitution."

He said that the decision established the judiciary as a private club "which can shut the door and conduct public business in private when it chooses."

Mr. Neuharth is also chairman of the board of the Gannett newspapers, whose papers in Rochester, N.Y., had brought today's case challenging judicial secrecy.

The American Civil Liberties Union charged that the law established by the Supreme Court majority today "would have let the Nixon Justice Department accept a guilty plea from the Watergate burglars behind closed doors" and "erected an iron curtain between the criminal process and the inquiring press."

Professional organizations of editors and journalists uniformly condemned the decision and expressed considerable fear of its consequences on criminal court coverage.

Charles W. Bailey, chairman of the Freedom of Information Committee of the American Society of Newspaper Editors, said that it was "very unfortunate" that the Justices had authorized courtroom closings "on anything other than a last resort basis."

"The Supreme Court said that judges can close their courtrooms pretty much when they want to," Mr. Bailey, who is editor of The Minneapolis Tribune, said. "I think it's self-evident that it is a bad way to handle the public's business."

John Finnegan, who heads the Freedom of Information Committee of the Associated Press Managing Editors Association, predicted that the ruling would invite pretrial courtroom closings all over the country, "which is certainly not in the public interest." Mr. Finnegan is executive editor of The St. Paul Pioneer Press and Dispatch.

'Against Public's Interest'

Robert D.G. Lewis, chairman of Sigma Delta Chi's Freedom of Information Committee, said that the decision "goes against the public's interest in open courts and may undermine confidence in the judicial system." He is a member of the Washington bureau of the Newhouse newspapers.

The Reporters Committee for Freedom of the Press emphasized that local trial judges, to whom the Supreme Court was giving sweeping new powers, "are frequently part of the partisan political process" and could be influenced by that fact.

"The overall impact of this decision will be to deprive the public of timely and critical information about the criminal justice process, bearing in mind that 89 percent of all criminal cases in this country are settled during these pretrial proceedings," Jack C. Landau, director of the committee, said.

July 3, 1979

JUVENILE JUSTICE: A PLEA FOR REFORM

By Irving R. Kaufman

On July 27, 14-year-old Luis Bonilla sat impassively in the starkly modern courtroom of the Bronx Supreme Court, waiting for the jury to return. He was accused of shooting and killing a teen-ager who had resisted Luis's attempt to steal his portable radio; at the time of his crime Luis was 13 years old. If the jury convicted him on charges of second-degree murder, his sentence would be mandatory life imprisonment, with no chance of parole for at least five years.

That Luis had been tried in an adult criminal court at all was extraordinary, for criminal suspects below the age of 16 had long been under the exclusive jurisdiction of the family court. Even for the most serious offenses, that court could impose at most five years of "restrictive placement," only the first 18 months of which would be spent behind bars. But two months before the fatal Bonilla shooting last October, New York's new juvenile-offender law took effect. It requires that 13-year-olds accused of second-degree murder (as well as 14- and 15-year-olds indicted for a variety of offenses, including burglary and assault) be treated as adults in the eyes of the law and tried in the adult court system. (To protect young convicts from premature contact with adult offenders, however, the new law still requires that a convicted youth be

Irving R. Kaufman, chief judge of the U.S. Court of Appeals for the Second Circuit, is chairman of the Joint Commission on Juvenile Justice of the American Bar Association and the Institute of Judicial Administration.

134

confined in juvenile facilities until he or she is old enough for transfer to an adult prison.)

Spurred by a series of brutal and highly publicized killings in the New York City subways, the State Legislature, in an extraordinary session in July 1978, enacted this far-reaching transformation of New York's criminal law within a few hours after the bill's submission. Luis Bonilla was the first 13-year-old tried under the new law, but though convicted, he was not sentenced as an adult. When the jury finally did return, after 15 hours of deliberations, it found him guilty of manslaughter — an offense not within the ambit of the new law for 13-year-olds. Accordingly, he was returned to family court for sentencing. The process had come full circle. The New York Family Court does not place manslaughter in the highest severity category, so the maximum sentence available — which Luis received — was three years, of which only the first will be in secure detention.

The case of Luis Bonilla illustrates only one of the problems facing those who have attempted to reform laws dealing with juveniles. Eighty percent of the 328 youths arrested under the new law during its first seven months were charged with theft, rather than with the more vicious crimes that were the principal target of the lawmakers. Over 60 percent of the youths initially charged as adults were returned to the family court — but not before some had undergone such experiences as pretrial detention in the filthy, windowless holding pens used by the Kings County Criminal Court in Brooklyn to house adult defendants.

All this, critics assert, demonstrates that the new law was drafted too broadly, sweeping within its purview many youths who *(Continued)*

An 1882 drawing of a 3-year-old arraigned in a New York court for "inaugurating a reign of terror."

should not be exposed to the adult criminal-justice system. Most states have long provided for the transfer to adult courts of juveniles who have committed particularly serious crimes. But in virtually all these states the family court determines whether transfer is necessary, thus minimizing potentially harmful pretrial contact with adult criminals.

The new law's defenders, however, point out that cases of neglected, abused, runaway or disobedient children, as well as child-support controversies, make up the bulk of the family court's docket. This court, they contend, was established to deal with youthful criminal mischief and petty thefts, not the brutal, heinous crimes that the public fears today. They also emphasize that many dangerous offenders are treated too leniently, simply because the few available secure juvenile detention facilities are already overcrowded by runaways and other nondangerous offenders. Removing youths accused of serious crimes from the family-court system, these policy makers concluded, would result in stiffer penalties, and thus serve as a more powerful deterrent of juvenile crime. But perhaps because juries are unwilling, except in compelling cases, to condemn youthful criminals to adult prisons, the "tough" new law has produced fewer and shorter detentions than the juvenile-offender law that preceded it.

New York's unsatisfactory experience with its new juvenile-offender law is illustrative of the age-old dilemma in this field. It is a fact of life that juveniles, no less than adults, are capable of killing and destruction, and must be punished for such deeds. Yet society is unwilling to abandon young delinquents to the often hopeless fate that awaits adult criminals.

On one issue, however, both sides in this often bitter debate are in agreement: the present system of juvenile justice is plagued by serious problems, problems that run too deep to be rectified by last-minute legislative fiat. Recognizing

this fact, the American Bar Association, in cooperation with the Institute of Judicial Administration, a research organization, has sponsored a comprehensive nine-year study, now nearing completion, of our entire juvenile-justice system. This project's goal has been to re-evaluate and produce a detailed set of guidelines, or "standards," that may serve as the basis for reform of every aspect of the child's manifold interactions with the law.

For each of 22 distinct topics, an expert, termed "reporter," was appointed to draft proposed standards and commentary. The work of the reporters was then reviewed by drafting committees drawn from the full spectrum of professionals in the field. Many were lawyers; the roster included three past presidents of the American Bar Association. The project also consulted judges, scholars, psychiatrists, psychologists, social workers, educators and experts in corrections and police work. This process insured that the standards would be well-integrated and internally consistent. In addition, they would reflect the varying perceptions of those involved in different aspects of the juvenile-justice system.

The standards, now being considered by those concerned with juvenile-justice reform, are not an exercise in theoretical armchair sociology. Rather, they draw on the best of existing, tested statutory reforms. Covering such areas as police handling of juvenile problems, youth service agencies, pretrial proceedings, juvenile records and information systems, and rights of minors, the first 17 volumes of the standards were approved by the American Bar Association this February. Five more volumes will be submitted to the A.B.A. at a later date.

The standards represent the latest chapter in the history of troubled children in America, a history that reflects some of the meanest and the most noble aspects of the American character.

Seventeenth-century London was beset by every type of

problem child now encountered by our family courts. Young criminals terrorized the middle classes, and were tried and punished in the same manner as adults. "Rogues and vagabonds," as well as "idle and needy" children who had run away from their parents and masters, were brought to the courts for punishment; orphans and neglected or abandoned children were bound out or cared for at public expense. But the Londoners had a "dispositional alternative" (in the jargon of the day's juvenile justice) that is unavailable now: they shipped many of these children to the Virginia colonies as indentured servants. The Dutch, for their part, pursued a similar policy in New Amsterdam.

The colonial policy toward children and their families was marked by a stern Puritanical view of the responsibilities of the child to his parents and of the parents to the state. The dead hand of this view continues to mold today's policy. The colonial government lent its force to the absolute rule of the parents, permitting up to 10 stripes at the public whipping post for incorrigible children who displayed a "rebellious carriage" toward their elders.

The colonies also expected much of parents. Virginia provided in 1748 that unfit parents who were either too poor to raise children properly, or who failed to bring them up "in honest courses," could lose custody of them forever.

The New England colonies (and later New York as well) provided for dependent persons, including young abandoned children, by a reverse auction known as vendue. The state offered subsidies of ever-decreasing amounts to prospective caretakers, with the lowest bidder winning. Apart from a cursory check to ascertain whether the child was being grossly abused, the child's best interests were not even considered.

By the 1820's, this reality gained broad recognition as the nation's increasing prosperity alleviated the financial pressures that had made the vendue acceptable. The

search for a more humanitarian approach was galvanized by a dilemma in governmental policy toward juvenile delinquency, which was already beginning to emerge as a heated issue with the massive influx of European immigrants. In the eyes of the criminal law, youths reached maturity at age 14, and suspects as young as 6 were legally considered as adults if the state could show that they knew right from wrong. Thus, several cases are recorded of 12- and 13-year-olds tried for murder, and of 7- and 8-year-olds locked up in adult prisons. The sympathy of judges and jurors toward the young mitigated the harshness of this common-law rule; even in 1820 all were aware that brutality and instruction in "the most artful methods of perpetrating crime" awaited the youth incarcerated with adults. But if acquitted, as the New York Society for the Reform of Juvenile Delinquents noted in an 1826 report, "they were returned destitute, to the same haunts of vice from which they had been taken, more emboldened to the commission of crime, by their escape from present punishment."

The New York reform society therefore recommended the establishment of specialized institutions for children. These homes would provide a refuge not only for young criminals but also for those who had been dealt with by the vendue — the homeless, the neglected and the vagrant — and children beyond their parents' control or whose parents were considered "unfit."

The institutional movement spread rapidly, at first under private auspices and later supported by public funds. These early "child savers" deemed several years of stern, regimented discipline essential to reform. They therefore thought it necessary that institutions and the state be given extensive authority over these children, even over parental protests. In 1838, the seminal case of In re Crouse established this power.

Mary Ann Crouse was consigned to the Philadelphia House of Refuge on her moth-

er's assertion that she was "incorrigible." The Pennsylvania Supreme Court rejected her father's petition to recover custody, defending its usurpation of the rights of "unworthy" parents by stating, "She in fact has been snatched from a course which must have ended in confirmed depravity, and, not only is the restraint of her person not unlawful, but it would be an act of extreme cruelty to release her from it."

With these noble-sounding words, the court planted in American jurisprudence a dangerous and alien seed that flourished most fully in the juvenile-court system founded 60 years later. A central principle in Anglo-American law is that crimes must be charged and proved precisely. One is convicted not of "being a robber" but of robbing John Jones at 2 P.M. on the corner of Fifth and Main. A law that confines a child in an institution because a parent is "unworthy" or because the child is destined for "depravity" presupposes that a court can divine the most salutary remedy for an individual by peering through some magic window into the soul. A homogenous community like Puritan New England may tolerate such official moral judgments, but in the diverse society dedicated to free thought that the United States has become, such institutional presumptuousness appears pernicious.

These considerations did not impress those active in the institutional movement, nor reformers who attempted other means of grappling with the problems of youth. The Children's Benevolent Society, for example, like its predecessors in London 150 years before, sought to "drain the city" of its delinquent children and ship them westward to "the best of all asylums, the farmer's home." And as the frontier began to close after the Civil War, yet another new phenomenon — the reform-school movement — swept the North and the West.

Two principles were central to the reform school ideal. First, youths could be properly treated as criminals for offenses peculiar to their status

as minors: for example, truancy, running away, incorrigibility. Second, because the mission of the reform school was a benign, rehabilitative one, the normal procedural protections of the adversary criminal trial — the rights of confronting witnesses, of prior disclosure of specific charges, of counsel and of trial by jury — could be abrogated with no constitutional qualms.

Initially, these principles suffered a major setback. In People v. Turner, the Illinois Supreme Court rejected a statute that mandated the confinement of children under 16 "growing up in mendicancy, ignorance, idleness or vice." "Vice is a very comprehensive term," wrote the court. "Acts, wholly innocent in the estimation of many good men, would, according to the code of ethics of others, show fearful depravity. What is the standard to be? What extent of enlightenment, what amount of industry, what degree of virtue, will save from the threatened imprisonment? The principle of the absorption of the child in, and its complete subjection to the despotism of the State, is wholly inadmissible in the modern civilized world."

The Turner decision represented a subtle wisdom that would not become widely evident for nearly a century. The Illinois child savers ignored its warning and lobbied to create a new institution that would further their interventionist, paternalistic goals within constitutional limitations. Finally, in 1899, at the prompting of the Chicago Women's Club, Illinois founded the nation's first juvenile court. Its jurisdiction encompassed all children in trouble, including those found living "with any vicious or disreputable person." This idea of a separate court for juveniles, able to address and cure the problems of youth with a special fatherly concern, quickly spread to every major city in the Union.

The juvenile court was the culmination of a growing belief that children were institutionalized far too frequently, both in degrading, unhealthy almshouses and in reform schools, which were already perceived as dehumanizing in their regimentation and brutal in their operation. Foster care and adoption were the practices preferred by the juvenile court for children with unsatisfactory homes, and for the first time home relief was offered on a broad basis to "worthy" mothers of dependent children. Juvenile probation officers were another innovation. These men and women — at first, unpaid volunteers — were to supply the court with investigations of the young offender's background. More importantly, they would supervise the child living at home, thus providing an alternative to punishment in an institution.

"Juvenile courts," noted Thomas D. Eliot, a critic of the system, in 1914, "were to fill every gap in the child-caring system. However, if this be granted, no line can be drawn short of a court administering all the children's charities. . . . Some courts actually state this as their ideal; they are all things to all men." The juvenile courts had, indeed, set themselves an ambitious, perhaps impossible, task. As a leading juvenile-court judge, Paul Alexander, noted in 1944, the judge must apply "not only man-made law, but the moral law and the laws of social science, psychology, psychiatry and the general laws of human nature." But the states never gave the juvenile courts the resources to carry out their mission. As late as 1967, one-third of the juvenile courts in the country did not have social workers or probation officers available to them, and 83 percent lacked psychologists or psychiatrists.

Confronted with massive waves of runaways in the 30's and the 40's, as well as a steady increase in youth crime, the system steadily broke down. Insufficiently staffed to follow its proclaimed purposes, the juvenile court -- despite its origins as an alternative to institutionalization — with increasing frequency referred delinquents to an ever-growing network of misleadingly and euphemistically named "reform" and "training" schools. By 1966, 51,000 children were living in 292 institutions for delinquents.

Discipline, rather than rehabilitation or education, preoccupied these prison-like facilities. As late as 1949, the punishment for possession of tobacco at the Indiana State School for delinquent boys was 10 strokes on the bare back with a leather paddle; for impudence

and vulgarity, 15 strokes. At the Hudson School in New York State in the 1950's, rebellious girls could be locked in solitary confinement for up to 81 days and fed only milk and bread for two of their daily meals. And reports emerged in the 1960's from one Massachusetts training school of children who were punished by "staff dunking youngsters' heads in toilets, dousing them with pails of water, forcing them to march around with a pail over their heads, and using one 12-year-old boy as a mop and dragging him through urine in the boys' washroom."

With the heightened social consciousness of the 1960's, many of these aspects of the juvenile-justice system began to appear archaic, if not cruel. Recognizing at last the essentially punitive nature of reform schools, Justice Abe Fortas, writing for the Supreme Court in 1967, noted that "however euphemistic the title, a 'receiving home' or an 'industrial school' for juveniles is an institution of confinement." Before making a finding of delinquency, therefore, juvenile courts were required to observe the "essentials of due process and fair treatment."

The 1960's also witnessed a second, more massive deinstitutionalization movement. It took its most striking form in Massachusetts, where the entire training-school system was closed and replaced with a network of small-scale, community-based residences.

There was, therefore, little disagreement in 1967 when the President's Commission on Law Enforcement and the Administration of Justice wrote: "The great hopes originally held for the juvenile court have not been fulfilled. It has not succeeded significantly in rehabilitating delinquent youth, in reducing or even stemming the tide of juvenile criminality, or in bringing justice and compassion to the child offender."

□

The Juvenile Justice Standards Project of the American Bar Association and the Institute of Judicial Administration thus convened in a day of growing disenchantment with the institutions used to deal with problems of youth. At that time, it was clear that a substantial number of adult criminals had started as juvenile offenders and invariably returned to lives of crime. The depth of dissatisfaction, however, provided the project with an ideal opportunity to achieve its ambitious goal: to structure a new juvenile-justice system that by heeding the lessons of the past would better serve us in the future. After nearly a decade of research and discussion, the project members con-

cluded that the juvenile court must abandon the "moralistic" model that had guided it since the turn of the century. Some members supported this view because they believed courts could not meaningfully pass upon the "worthiness" or "depravity" of those who came before them. Others argued that such judgments have no place at all in modern society. All agreed that the juvenile courts — like all other courts — should be tightly bound by the rule of law. Acts that would be crimes if committed by adults should be charged, proved and punished in a similar manner. Senator Edward Kennedy, chairman of the Senate Judiciary Committee, has forcefully decried the cruel ironies of the present system: "If juveniles want to get locked up, they should skip school, run away from home or be deemed 'a problem.' If they want to avoid jail, they are better off committing a robbery or burglary." Children whose actions do not amount to adult crimes — the runaways, truants and incorrigibles — should be dealt with outside the judicial system. The term "delinquency," therefore, with its attendant stigmatization and possibility of incarceration, should be reserved exclusively for juveniles who have committed crimes.

A few examples will illustrate the basic philosophy that undergirds the Juvenile Justice Standards Project:

Parental Abuse and Neglect. Under the standards, children may not be removed from their homes because of alleged parental misconduct without a finding that specific physical harm, clinically recognized psychiatric disturbance, or sexual abuse has occurred or is imminently threatened. This standard is intended to reduce sharply the widely deplored extent of state intrusion into family life that the child savers thought proper. Although state intervention is clearly called for when a child is, for example, battered by a parent, physical abuse was reported in only 10 percent of the 300,000 neglect petitions filed annually. More often, the petition for juvenile-court intervention to remove the child from the home is based on such complaints as the tendency of a single mother to bring home overnight male visitors. Separation of child from mother in such cases, which once may have appeared imperative, has become distasteful in a society that has learned respect for the dignity of all and has come to recognize the necessity of keeping hands off the parent-child relationship except in particularly dangerous situations.

Experts agree that unnecessary intervention often harms rather than helps the child, shattering parental bonds. The wiser course, the standards conclude, is for the state to leave the child with the family and supply voluntary social services to assist the family unit in the home.

Youth Crime. In cases of juvenile delinquency, the standards frankly recognize that although reform schools may have rehabilitative goals, they are dominated by their punitive aspect. Clichés about saving a child from a life of crime do not alter the stark reality — and necessity — of punishment. For the protection of society, it is necessary that legislatures increase many of the maximum prison sentences for juveniles who have committed serious crimes. And, under carefully defined circumstances, juvenile courts must be permitted to transfer the most hardened young criminals to the adult system for the severest punishments allowed by law. But in either court, juveniles accused of crime should have the right to the same adult procedural protections that guard against conviction of the innocent. Consequently, the standards accord accused juveniles, beyond the minimal due process that the Supreme Court has held is constitutionally re-

quired, the rights to counsel and to a six-person jury. The juvenile-court judge must remain free to act leniently in an appropriate case. The standards require no minimum sentences, but where the judge deems a heavier determinate sentence necessary, no parole commission should be permitted to release the offender prematurely. Under the existing system, as Senator Kennedy recently said, "certainty of punishment is a joke."

Noncriminal Misbehavior. The standards permit incarceration only for conduct forbidden by the adult criminal law. Large segments of juvenile court jurisdiction would therefore be removed, for adults cannot be convicted for disobeying their parents, truancy or running away. In recent years, some states have begun to treat youngsters who have committed such acts, but no crimes, as "Persons [or Children] in Need of Supervision" (PINS or CHINS), rather than as delinquents. The change, however, is one in name only. PINS children continue to be confined alongside young criminals, an experience destined to do more harm than good. In fact, through PINS petitions, juvenile courts have intervened in family dis-

putes over household chores, abusive language and undesirable boyfriends. The juvenile courts, in these cases, have allowed themselves to be used as birch rods by angry parents disgusted with, but unable to control, their adolescent children's behavior.

The project concluded that all services rendered to PINS children should be provided on a strictly voluntary basis, and preferably outside the judicial system. For example, in recent years, concerned adults — many of them former runaways — have established "crash pads" and runaway houses where young people who have left home can live for a time, safe from the dangers of the street. At first, these shelters were often hostile both to parents and authorities, but today, with the help of government funding and state certification, they have evolved into an important adjunct to the social-services system. Experience has shown that after a runaway child spends a few days in such a shelter, an amicable reunion with parents usually ensues. The standards endorse this system of licensed shelters and recommend that it be augmented substantially.

For other troubled children

who have not broken the law, the standards prescribe voluntary psychiatric or other medical care, and educational, vocational and legal counseling appropriate to the needs of the particular child. Where a child chooses to avail himself of such services, the chances of success will be enhanced by the fact that he has chosen this course voluntarily. The key, then, is to make these services adequate, well-known and easily accessible.

□

The long history of justice and injustice to our nation's children counsels that the juvenile court focus its power, its wisdom and its compassion on the gravest problems that confront it: the twin evils of violence by and against the young. Courts must mete out fair and even-handed justice to young criminals. No longer should they try to serve as surrogate parents in attempting to control youthful mischief. As the legislatures of our states join in this effort and draw upon the standards' recommendations, the age-old promise of justice for our nation's youth may at last be fulfilled. ■

October 14, 1979

The Consequences Of an Open Door

By BABETTE ROSMOND

IT was a warm June day. The typewriter made sporadic little bursts of sound as I sat working at my desk. Our apartment is in an enclave of narrow brownstones in east midtown New York, with windows fac-

Babette Rosmond is at work, behind a locked door, on her 12th novel. Her most recent book is "Monarch."

ing a communal garden. No one can be admitted downstairs without being buzzed in, or unless that heavy guard door is inadvertently left open for a moment or two.

Once inside the downstairs hall, an intruder would have to climb three flights of stairs before facing the locked door of our two-story apartment. But on this day I had neglected to lock our door when I'd brought the mail up, around noon, and downstairs in the hall there were workmen going in and out

on a special job.

My desk is near the window. I stretched, looked at my watch — noting that it was about 3:30 — and walked idly to the open door of my room leading to the entrance hall. There followed an inconceivable chain of events:

¶Someone was in the house, quickly maneuvering himself behind me.

¶Someone tall, skinny, with sneakers.

¶Someone carrying a white shopping bag that bore the words "Outpatient Clothing."

¶Someone grabbing me and thrusting a soaking mask under my nose.

It knocked me out almost at once. Later I was grateful. It prevented me from knowing at the time that someone had set out to kill me.

My husband came home about 6:30. Usually at that time, I'm in the kitchen, on the floor above the entrance hall. When I didn't answer his greeting, he decided I'd gone to visit my mother, who'd been ill. After a while, he began to worry, and went around the house looking for me in the way people do when they're half-serious, half-deprecatory about a search. It made him open the hall closet, where winter clothing is kept. The light was on, and there, in the back of the closet, beneath a hastily-flung pile of coats, I lay on the floor, with a hospital mask tied around my nose and mouth. Under me was a plastic bag, which still clung to the back of my head but cleared my face.

He didn't know if I was alive or dead, just that I couldn't hear or respond. He tried to lift me out, but discovered that I was dead weight — even at 95 pounds. Somehow, he managed to carry me to the bed and, still not knowing, certainly not believing it was all happening, in a state of half-shock, he called the police and an ambulance.

They arrived promptly. By this time it was nearly 8 o'clock and I was beginning to come out of it. And my first thought was irrational: I felt that I was about 19 years old and incredibly drunk.

When I could open my eyes, and saw several police officers and my husband, I apologized. "Oh, I'm *so* drunk, I'm afraid I'm going to be sick." They all looked so sympathetic, remaining silent, that I began to understand that my high didn't come from alcohol and that something extraordinary had happened.

• • •

They began their questions in the ambulance. Did I see the intruder's face? No. Only wrists, the sense of deft movement, the sneakers, and then the overweening, burning sweetness of the anesthetic.

Still in the ambulance, I began to awaken fully, with a frightening sense of time and memory lost. Another pair of policemen, young and polite, asked me more questions and told me some of their colleagues had stayed behind in the house to investigate clues and search for possible fingerprints.

As I was put into a wheelchair at the hospital, I recall urging my husband not to leave for a *second*. My blood count was taken, to be sure no residue of anesthesia remained. Then a detective introduced himself and began the questions again. He asked if I had any enemies, if there was any person in the building itself who might hold a real or imagined grudge against me. No...no...once before, in another apartment, there had been a robbery, but never before had I been assaulted.

The only signs of physical injury were on my left side. From shoulder to ankle, I was beginning to turn shades of blue. I had not been raped, nor had a knife been used. The bruises weren't enough to warrant a hospital stay, so I was allowed to go home. This time we didn't need the ambulance; the police and my husband and I drove back to join the three or four other investigators in the house, who had found massive evidence of my attacker's ineptitude.

He had broken the needle arm of the stereo, and had flung two records down the inside stairwell. On top of the records was a crushed beer can, out of which stale beer was floating. In the mess were two crumpled $20 bills. It was my household-emergency cash, and my assailant had forgotten to take it with him. He hadn't been very deft with the surgical mask, either, and his heart wasn't in the securing of the plastic bag.

One of the policemen stayed in the bedroom with me while my husband went upstairs with the others. I was rather cheerful. He took my hand and said, "You know, you haven't realized yet how close you came. When it hits you, don't be too frightened. Just be grateful that this guy was so strung out he didn't have his head together."

Perico Pastor

Then he began chatting aimlessly, and I was thankful simply to listen. At one point, I wandered out to see the fingerprint men pouring black powder on white surfaces and white powder on dark surfaces, and inconsequently noted that there'd be a lot of cleaning up to do.

Whatever latent prints were found weren't very helpful. Toward midnight the police were finishing up, and my nice young friend, who had led me back into the bedroom, said, "Now when you think about this, don't let yourself be scared, because it's *over*. It won't happen to you again. It was one chance in a billion."

"Do you think you'll find him?" I asked.

He smiled and shook his head. "Nope. If I were to tell you the number of times this happens every day around here. . ."

"But this was attempted murder," I broke in, as though I were losing star billing and being put down among common robberies.

He nodded. "That's what I mean. If every one of these meaningless attacks got into the papers, there wouldn't be room to print anything else."

I gasped. "But that's . . ." There wasn't anything to be said. It was a fact of life I'd been dimly aware of, but it hadn't penetrated.

When my children, now grown, were young, we lived in a suburb and the kitchen door was left open all day. It has been difficult for me to remember to lock the door every time, lock the door every time, every time.

• • •

I think the message has penetrated now. The bruises took a few weeks to heal, but I couldn't help feeling grateful to my assailant for having somehow obtained enough anesthetic to knock me out at once. How had he gotten it? The shopping bag I saw so fleetingly indicated he might have been a patient in a drug program; perhaps a swap was made, some heroin for some chloroform, or whatever he used. I'll never know.

What I do know is the ever-growing need for watchfulness. It may not be the most relaxing way to live, but it is a fact of life, transcending almost everything else in the city.

When the police left, I thanked them heartily; they wished me well, and one of them stuck his head in the door to say farewell, adding, "By the way, be careful about breathing that fingerprint powder. It's a carcinogen."

I nodded. Life and death were so closely meshed that night, one more fact-to-be-filed didn't seem really frightening. I was one of the lucky ones.

October 17, 1979

MURDERER IN CASINO EXECUTED IN NEVADA

Prisoner, Rejecting Offers of Aid, Told Officials of 18 Slayings

By WALLACE TURNER
Special to The New York Times

CARSON CITY, Nev., Oct. 22 — Jesse Walter Bishop was put to death in the gas chamber early today for a murder he committed two years ago. It was the first execution in Nevada in 18 years, and only the third in the nation in the last 12.

Mr. Bishop, a 46-year-old drug addict who told a doctor in prison that his occupation was "armed robber," said in court of his pending execution, "I don't want to fight it. I don't want to drag my family through it." Another time he said, "Commute me or execute me. Don't drag it out."

Justices Thurgood Marshall and William J. Brennan Jr. of the United States Supreme Court had urged in vain that the High Court delay the execution, which they called "state-administered suicide."

After he was pronounced dead, officials in Las Vegas disclosed that Mr. Bishop had told them of committing 18 murders for hire. Charles Lee, a Clark County homicide detective, said the taped interview had not been transcribed for checking, but he tended to believe the story. He said no names or locations were given because the condemned man "didn't want to go out a snitch." Officials here were less ready to credit the story.

Mr. Bishop's crime was committed on the night of Dec. 20, 1977, 11 months after Gary Gilmore, 46, another armed robber who committed murder, was shot to death by an execution squad at the Utah State Prison, resuming deaths by capital punishment that had been halted by court decisions in the late 1960's. Mr. Bishop was the second person to be executed this year. John A. Spenkelink was put to death in Florida's electric chair on May 25.

Mr. Bishop went into a small casino in the El Morocco Hotel on the Las Vegas Strip and told the cashier he was a robber. She saw no gun, and fell to the floor screaming for help. Larry Thompson, a casino employee, grappled with the robber and was joined by David Ballard, 22, a Volkswagen mechanic from Baltimore, who was on his honeymoon.

Mr. Bishop pulled out a gun and shot Mr. Thompson in the stomach. He recovered. Mr. Ballard tried to run and was shot in the back. He died some days later without regaining consciousness. He had been married three hours before the shooting.

After the casino robbery, from which he got $278, Mr. Bishop went through two days of further armed robberies, taking cars, kidnapping their drivers and aimlessly running. He was arrested as he slept beneath a trailer house at Boulder City, Nev.

"I was using narcotics," he told officers who arrested him.

Could Have Delayed Death

Mr. Bishop could have delayed his execution last night in his last hours if he had accepted an invitation from Charles L. Wolff Jr., the prison director, who asked him if he was willing to cooperate with attorneys in filing an appeal.

Mr. Wolff said the offer was refused. Mr. Bishop had resisted all offers from public defenders and other lawyers to help him. At his arraignment he pleaded guilty to first-degree murder.

Mr. Bishop was born in Glasgow, Ky., and grew up in Los Angeles. He said his first armed robbery was committed at the age of 15. He had four felony convictions for drug offenses and armed robbery when he committed the crime for which he received the death sentence.

His first adult arrest was in 1951 when, as a paratrooper in Korea, he was arrested for possession of heroin. He served time at Fort Leavenworth, Kan., and received a dishonorable discharge. He later served time in California and in Federal prisons, and was on parole the night he killed Mr. Ballard.

He is survived by his mother, two brothers, a sister and two children of two marriages that ended in divorce.

October 23, 1979

Nation's Crime Rate Is Up Again, With Smaller Cities Leading Rise

By JOHN HERBERS
Special to The New York Times

ALBUQUERQUE, N.M. Oct. 27 — Increases in every kind of major crime have contradicted some experts' predictions of a year ago that the national crime rate would begin to decline.

Statistics for the first six months of this year show an overall rise of 9 percent in serious crimes. There were considerable increases in every category reported to the Federal Bureau of Investigation — murder, rape, robbery, assault, larceny, burglary and automobile theft.

The rise in crime has been especially pronounced in cities with populations of up to half a million and in small towns and rural areas, suggesting a widening pervasiveness of both violence and crime against property.

The statistics, of course, are preliminary and subject to challenge. But many national and regional authorities on crime say that they expect a real and substantial increase this year.

Violence Now Called Endemic

In cities such as Albuquerque, which has a metropolitan population of about 400,000 and a crime increase of 20 percent, the rise is easily documented and measured. A few days ago the courts in Bernalillo County, of which Albuquerque is the county seat, scheduled more time for criminal cases and less for civil ones, a direct result of the increase in crime.

Some authorities had expected a national reduction in the crime rate, extending into the 1980's, because of the decline in the number of young people in the 15-to-24 age group who historically commit the most crimes. The increase that has come instead suggests that the surge of violence and defiance of the law that began in the 1960's has become an indelible part of the social fabric and that the causes are much more complex than was once believed.

There is evidence, too, that much of the violent crime is impulsive and is abetted by the large number of weapons that citizens have accumulated over the last two decades, many of them purchased as a protection against crime.

"There are now 100 million guns in the hands of the civilian population," said Donald Lundy, a professor of law and social psychology at the Stanford University Law School. "This is an appalling situation that exists nowhere else in the world."

The expenditure of many billions of dollars on new anticrime weapons and techniques, investment of additional billions in private protection forces, years of speeches against crime by political figures at all levels, the publication of thousands of studies on the causes and prevention of crime and a toughening of public attitudes and laws against criminals — none of this seems to have had much effect.

Patrick V. Murphy, the former New York City Police Commissioner who is president of the Police Foundation in Washington, said crime had become so much a part of life in the United States that the national commitment to its eradication had lessened substantially in recent years.

Growth of Neighborhood Patrols

The current effort in improvement is at the community level. Neighborhood groups, some with their own patrols, have sprung up in inner-city, suburban and rural communities in every region of the country.

In one neighborhood of small, neat homes of working-class people in Albu-

querque, where an anticrime drive has been under way, a quarrel after a bout of gambling and drinking spilled into the streets from one of the houses last Thursday.

A 28-year-old woman identified as Ethel Louise Clements drew a gun and began striking another woman with it. The gun fell to the ground in the scuffle. Someone in the crowd, witnesses said, picked up the weapon and fired several shots into Mrs. Clements's chest.

"This is a bucket of blood," the Rev. Herschel Patterson, pastor of a nearby Church of God and Christ, said after the slaying. "People shouldn't have to live with this in their community." There have been two other murders and numerous assaults on the same street this year.

A few years ago, a police official said, a brawl such as the one in which Mrs. Clements died would have ended with pulled hair and cuts and bruises.

Turbulent Changes Blamed

The spreading crime in the growing, prosperous cities of the South and Southwest is increasingly committed by whites as well as blacks and Hispanic-Americans, women as well as men and the middle-class young as well as the poor. As a result, some authorities have concluded that the high incidence of crime may be as closely related to the turbulent social and demographic changes of recent years — the breakup of families, the transient population and severing of ties with old institutions — as to the heavy concentration of poor, unemployed members of minority groups in the large cities, the usual explanation for high crime.

Charles B. Silberman, author of a recent book, "Criminal Violence, Criminal Justice," which was written with the research support of the Ford Foundation, says that, rather than being a phenomenon of just the last two decades, the high crime rate may be a return to the norm in this country.

The National Commission on the Causes and Prevention of Violence, established in 1968, concluded after an exhaustive study that, for most of its history, the United States had been a violent nation to the extent that outlaws, gunfighters, murderers, swindlers and robbers were made folk heros in some periods.

Mr. Silberman, who quoted some results of the study, wrote that it was not until the 1930's that crime and violence began to decline. The Depression and World War II gave people a sense of community and sharing of interests. This period of relative stability, with a low crime rate and little fear of crime, extended into the 1960's. Then began the surge that continued into the 70's. In 1978, the murder rate alone per 100,000 people was three times higher than it was in 1938.

Breakdown of Some Institutions

The causes cited then included not only the turbulence of the times, the urban riots, the Vietnam War and the heavy concentrations of poor and unemployed blacks in the central cities, but also the fact that the post-World War II baby boom was sending millions of people in the so-called crime-committing age into the society at a time when institutions that traditionally helped keep order — the family, schools and churches — were in decline.

When the crime rate continued at a high level after the Vietnam War and urban riots ended, the failure of the society to assimilate much of the black population, as it had done over the years with other members of minority groups, was widely regarded as one of the main roots of crime.

The crime rise, coming after a generation of relative stability, had a profound effect on American society. Middle-class people fled the central cities, only to have crime follow them a few years later.

Albuquerque: Frontier of New Turbulence

Special to The New York Times

ALBUQUERQUE, N.M., Oct. 27 — With its sprawling subdivisions, clean new public buildings and aspens and adobes that sparkle in the autumn sun, Albuquerque is not a city where one would expect to find a high incidence of crime.

A few years ago, when the Federal Bureau of Investigation released statistics showing that Albuquerque led the nation in violent crimes per capita, city officials raised a howl of protest; in fact, the latest statistics list it as 24th in the nation.

The statistics show that, while 18 persons were murdered in the first eight months of 1978, 32 were killed in the same period this year. In the first six months of this year, the number of robberies rose to 370 from 295 in the comparable period last year, assaults to 762

from 708, burglary to 3,512 from 2,810 and larceny to 7,310 from 6,259.

Assistant Police Chief Louis Powell, acknowledging the accuracy of the statistics, said in an interview that although the city had grown, its citizens had not permitted it to expand the police force because they had rejected tax increases.

A growing difficulty is drug traffic, he said, adding that it was spread through all levels of the society. Heroin has been replaced by cocaine and pills of all kinds as a chief cause of concern.

Bizarre Crimes Committed

Whatever the causes, there have recently been bizarre crimes by persons of disturbed minds — rapes accompanied by torture, along with ordinary muggings and murders.

In many cities, burglaries and robberies have been attributed to a large

black population in which unemployment among teen-agers frequently approaches 50 percent. But blacks make up less that 5 percent of Albuquerque's population.

The city does have a large Hispanic population, but in some of the barrios where there is extreme poverty there is also a low crime rate because of family and community ties.

It is the unattached, those who abandon old moorings, who seem to commit many of the crimes. According to statistics provided by the state in a recent year, there were 52 divorces for every 100 marriages in New Mexico.

The transiency of the population is also evident in the public schools, where there is an annual turnover of 32 percent. Albuquerque is at the forefront of the nation's social turbulence; in the past, crime has frequently accompanied rapid change.

Much of the society lived in fear, and the fear had a political effect that included new opposition to social programs and demands for harsher criminal penalties.

Then, in 1977, the nation's overall crime rate declined by 7 percent. In some of the old neighborhoods in the Eastern cities, the number of murders, robberies and burglaries was down substantially and there was a public impression of less crime. Young middle-class white couples began returning from the suburbs and refurbishing old houses in neighborhoods that their parents grew up in but abandoned out of fear.

Last year the decline appeared to be continuing at about the same rate as the decline in the number of people between the ages of 15 and 24.

Professor Predicted Decline

Dr. Marvin E. Wolfgang, professor of sociology and law at the University of Pennsylvania, told a Congressional subcommittee at the time, "We have found that the rates of crimes of violence are likely to decline in the late 70's and continue to decline in the 80's, just on the basis of change in age composition alone."

Dr. Wolfgang, who is one of the most highly regarded scholars on the subject, said in a telephone interview last week that he still believed the crime rate would level off, at least for a few years, but then would probably rise again. In any event, the consensus is that the high level of crime has become so pervasive that it is here to stay for the foreseeable future.

The uniform crime reports issued by the Federal Bureau of Investigation have long been suspect because the statistics are subject to political manipulation or to unreliable reporting methods because they are compiled by local police officials throughout the country. Recently, however, the reports have come into better repute because the F.B.I. has been more careful in checking the figures and because surveys conducted among thousands of crime victims around the country show that, with the exception of murders, many more crimes are committed than are reported to the police.

Complete figures for 1978, released this week, showed an overall rise in the crime rate of 1.3 percent over the previous year. Murder, with 19,555 victims, increased 2.3 percent. Forty-nine percent of the murders were committed with pistols, and 14 percent with rifles or shotguns.

Rape was up 6.5 percent with 67,131 cases reported. Robbery increased by 3 percent with 417,038 incidents. Aggravated assaults rose 7.8 percent with 558,102 cases. Burglary rose 1.7 percent, with 3,104,496 cases reported. Larceny increased by 1.3 percent with 5,963,401 cases. And motor vehicle theft rose 2.4 percent with 991,611 thefts reported.

In all, the number of people who were victims of crime last year, just on the basis of cases reported to the police, exceeded the population of New York City.

Nevertheless, 1978 was a reasonably stable year. But this year the crime rate began to rise dramatically. In the first six months, the number of murder cases was up 9 percent, rape 11 percent, robbery 15 percent, aggravated assault 11 percent, burglary 7 percent, larceny 10 percent and motor vehicle theft 13 percent.

In the older cities of the Northeast and Middle West, the rise was smaller than the national average. New York, for example, showed less than a 5 percent overall increase for the first six months of this year, as against the same period for 1978.

It was the large rise in crime in the growing cities of the Sun Belt — much larger than the growth in population — and the nature of many of the causes of violence that indicated the close relationship between crime and turbulence in the society to the analysts.

Dr. Lundy said in a telephone interview that, in addition to this societal instability, the closing of mental health facilities in California and many other states turned many people with severe mental problems loose on the society. He added that some mass murders of recent years fell into this category — Herbert Mullen, for example, who killed 13 persons in Santa Cruz.

"More of the mentally ill, rather than being confined to mental institutions, are being crowded into the already crowded criminal justice system," Dr. Lundy said. "And this is having an impact on the whole area of crime."

October 28, 1979

Galante Mob Said to Have A New Boss, Now in Prison

By LEONARD BUDER

Philip Rastelli, now in Federal prison in Lewisburg, Pa., has emerged as the new leader of the crime organization that was headed by the late Carmine Galante, police intelligence officials say.

Mr. Rastelli, who is 61 years old, began serving a 10-year term for extortion in 1977, but could be eligible for parole in 1981. He recently lost a move to have his sentence reduced. Authorities believe that even behind bars he is capable of exercising leadership over a criminal empire whose tentacles reach into many states and Canada.

"Being in jail doesn't stop a boss from being boss," said a New York City detective who is an expert on organized crime.

Some Family Activities

Police intelligence sources say that late in 1973 and early in 1974 Mr. Rastelli was acting boss of the organization known to the police and Federal authorities as the Bonanno crime "family," after its first leader, Joseph Bonanno. Mr. Rastelli was thought to have relinquished the top spot to Mr. Galante when the latter was released from prison. He was also known to have served on the three-member council that ruled the family early in the 1970's in an effort to end internal strife.

Law enforcement sources say the Bonanno organization's illegal activities include narcotics, gambling, labor racketeering, loansharking, truck hijacking and extortion. According to a 1978 internal report of the New York City Police Department, the organization "controls the large majority of the heroin shipped through Canada into the United States."

The July 12 murder of Mr. Galante remains unsolved, but the police think he was killed because he got too greedy —

"he was probably demanding more than his normal cut," one detective said — and was encroaching on the operations of other organized crime groups.

There was speculation immediately after the slaying that Nicholas Marangella, identified by law enforcement sources as the organization's "underboss," would

move up to become boss.

Intelligence officers who regularly monitor the activities of organized crime figures now think that Mr. Rastelli was selected for the top position even before Mr. Galante was killed. They also believe Mr. Rastelli knew about the intended "hit" before it took place and that he gave his approval.

According to a confidential police report, Mr. Rastelli had a stream of visitors from the Bonanno organization — before and after the Galante murder — while he was temporarily at the Federal Correctional Center in Manhattan last summer.

Some of His Visitors

Mr. Marangella, the underboss, visited him on July 11, the day before the killing. Steve Cannone, who is listed in police files as the counselor of the organization, was a frequent visitor.

Other reputed Bonanno members who visited Mr. Rastelli at the Manhattan facility were Philip Giaccone, Frank Lupo and Armand Pollastrino, identified by police intelligence sources as "captains" in the organization; William Roddine and Philip Cimmino, both described as "associates," and Joseph Messino, identified as a "soldier."

Mr. Rastelli is reported to have been visited in the Lewisburg prison by Mr. Giaccone and Mr. Messino, who the police believe are carrying oral messages in and out of the prison.

In Trouble at the Age of 8

Mr. Rastelli, who weighs 190 pounds and is 5 feet 8 inches tall, was born in Brooklyn on Jan, 13, 1918. He first attracted the attention of the police at the age of 8 when he was charged with delinquency. In 1950 he was arrested for assault and robbery and sentenced to five to 10 years, but it is not clear if he actually served time. In January 1953, was questioned in connection with the shooting of Michael Russo. In December 1954 Mr. Russo was shot again, this time fatally. Mr. Rastelli turned himself in to the police, was tried in that killing and was acquitted.

In 1962, according to police files, Mr. Rastelli's estranged wife, Connie, told the Brooklyn District Attorney's office that she and her husband had killed Mr. Russo. A month later — two days after she reportedly told the Federal Bureau of Investigation that her husband was an international narcotics dealer — she was found shot to death in the hallway of the building she lived in.

Early in 1971, when he reputedly was a captain in the Bonanno family, he served 30 days for evasion and refusal to answer questions by a Nassau County grand jury. In December 1972 he and four confederates were convicted of loan-sharking.

While appealing that conviction he was indicted in March 1975 on Federal charges of operating an extortion and protection racket involving mobile lunch wagons. After his conviction in April 1976 he was sentenced to 10 years following the completion of the state sentence in 1977.

Reputed Leaders Of the Bonanno Organization
(According to law enforcement officials)

Joseph Bonanno
1930-64
Deposed

Frank LaBruzzo
1964-65
"Invalidated"

Gaspare DiGregorio
1965-66
Died

Paul Sciacca
1966-69
Retired

Frank Mari
1969
Disappeared

"Family Ruling Council"
1969-72

Natale Evola
1972-73
Died

Philip Rastelli
1973-74
Stepped aside

Carmine Galante
1974-79
Killed

Philip Rastelli
1979-

Mr. Bonanno was deposed by the so-called "National Commission" of organized crime, which later "invalidated" Mr. LaBruzzo's claim to leadership. Mr. Rastelli served as the "acting leader" from 1973 to 1974.

Growth in Rural Regions Brings Rapid Crime Rise

By JOHN HERBERS
Special to The New York Times

HILLSBORO, Mo., Nov. 2 — For the past few years, crime in rural areas and outward suburbs across the nation has been rising faster than in the cities. Jefferson County — 667 square miles of open space, villages, small houses and mobile homes south of St. Louis — is part of the pattern.

The county lies in the foothills of the Ozarks. Except for the new shopping centers and houses, mostly in the northern part, it has retained the appearance of rural calm. Distance predominates.

In their persistent efforts to escape the cities, Americans have opened up this countryside and others to some of the same forces that historically brought crime to the core of large urban areas — a transient restless population, an influx of poor immigrants from other regions and impersonal relationships that make it easier for a person to steal.

Employers have moved out to the ring of suburbs around St. Louis. Many of those who work there live on the hills and in the hollows of places such as Jefferson County and get to their jobs quickly on the interstate highways.

At the same time, middle-class residents of the suburbs who moved from the city years ago are now building full-time homes on farms or in rural subdivisions.

In St. Charles County, another largely rural area, north of St. Louis, the police a few days ago performed a "sting" operation that resulted in arrests of drug, burglary and car theft suspects. The "sting," a technique of planting undercover agents in crime rings, has been used mostly in the cities.

"We've got a very bad drug problem here," said George Williams, a longtime county resident. "I'm glad to see the police doing something." The authorities say they have infiltrated a high-level drug syndicate that was selling LSD, cocaine, "angel dust" and marijuana in St. Charles County.

In Jefferson County, burglaries have risen so much that residents of towns and subdivisions have, with the cooperation of Sheriff Walter (Buck) Beurger, set up citizen patrols, a phenomenon that has been spreading in the cities and suburbs for several years.

Adjustment to Statistics

For 1978, Jefferson County reported six murders, five rapes, 26 robberies, 13 aggravated assaults, 729 burglaries, 504 larcenies and 254 motor vehicles thefts. Sheriff Beurger said burglaries would run susbstantially higher this year.

Although the statistics show crime to be rising in rural areas, many authorities believe the rise is actually higher than re-

Crime in Rural America
(Number of crimes per 100,000 population

Source: F.B I.

2,012 in 1977

423 in 1960

The New York Times / Nov. 4, 1979

ported because of the tendency of people in rural areas not to report crimes and of sheriffs to minimize the seriousness of those reported.

The increase of crime in rural areas and distant suburbs, according to a wide range of authorities, is one reason national crime rates have begun to soar again after a period of decline in the mid-1970's. According to the uniform crime reports released last week by the Federal Bureau of Investigation, serious crime rose 7 percent in rural areas and 10 percent in suburban areas in the first six months of this year, as compared with the same period last year.

Those classifications, however, were somewhat misleading. Vast areas of rural countryside such as Jefferson County are classified as suburban because many of those who live there work in the metropolitan area.

Big Crime Rise in 20 Years

In any event, rural crime as reported to the police has increased more than five times its rate two decades ago, while urban crime has quadrupled.

Statistics collected by the F.B.I. for the past 20 years show that rural crime is now roughly equal to urban crime rates of 1967, said Howard Phillips of Ohio State University.

Between Sheriff Beurger's office here and the Gateway Arch in St. Louis are a series of concentric circles that tell the story of a restless population.

The first encompasses a small downtown of hotels, convention and sports centers, and the high-rise financial center.

The ring beyond that delineates vast open spaces leveled by two decades of urban disorder, public housing and a few remaining old residential and commercial buildings. That area has had a net loss of 350,000 people in two decades.

Work Places in Suburbs

The third ring encompasses some middle-class neighborhoods remaining in the city but a large number of older suburbs in St. Louis County that have for several years sustained a variety of urban ills, including crime. Within the fourth ring are the newer and more exclusive suburbs, factories and office buildings that make up the backbone of the area's economy. Beyond this are the rural areas.

The diameter of the outer circle, which defines the St. Louis Standard Metropolitan Statistical Area, is 80 miles. Within this area live 2.5 million people. Difficulties usually associated with urban areas have now reached them all.

In the process, some new lessons are being learned about crime. The high incidence of crime generally in the past has been attributed to large concentrations of the poor, who tend to be blacks and other minorities. While the crime rate continues high in those areas it is becoming high also in areas far removed from large black populations. And in rural areas most of the perpetrators are people who live nearby.

In St. Louis, the black middle class has been moving either back to the South or migrating to the suburbs to the north and west. Whites have been moving south and west. Thus Jefferson County is largely white, yet its crime rate is about the same as counties nearer the poor black population.

November 4, 1979

A DECADE OF CONSTITUTIONAL REVISION

By Sidney Zion

t seems safe to say that, until fairly recently, most Americans believed they possessed certain rights that were so fundamental no governmental authority would dare challenge them, much less take them away. They went without saying, these beliefs — they were the "givens" of our American heritage:

●That all people who were accused of crimes were entitled to trials by juries of 12; that in every case the trial judge was required to instruct the jury that the accused was presumed innocent until proved guilty; and that no jury could convict an accused person unless its verdict was unanimous.

●That the courts belonged to the people, and that the public and the press alike had the constitutional right to enter the halls of justice of our cities, our counties, our states and our nation.

●That indigent defendants who were charged with nonpetty crimes had the right to free legal counsel.

●That bank and telephone records were private, and could not be seized by government agents without a person's knowledge and consent.

●That the police could not legally ransack newsrooms, and that reporters could not be jailed for protecting their confidential sources of information.

●That a person's good name and reputation were sacred, and were secured by the Constitution against wrongful assaults by public officials.

But in every instance these beliefs, these "givens" of our heritage, were

Sidney Zion, a regular contributor to this Magazine, has written extensively on the law during the past 15 years. He is a former legal reporter for The Times. He is also a member of the Bar in New York and New Jersey, and was an assistant United States Attorney.

wrong. We do not have the rights we thought we had. Indeed, we are told we did not have them in the first place. And we are told this by the men with the power to make their opinions prevail: the justices of the United States Supreme Court.

A decade ago, the Supreme Court would have been the last place where

these rights were called into question. Under Chief Justice Earl Warren, the Court protected and expanded civil liberties to a degree that had never before been approached. "The essential scheme of our Bill of Rights was to take the Government off the backs of people," Justice William O. Douglas wrote in 1972. In large measure, the Warren

The Supreme Court: Has the Bill of Rights been diluted by the Burger years?

Evelyn Hofer

Court enforced that purpose.

But the Court that Earl Warren left 10 years ago last June was markedly different from the one that his successor, Chief Justice Warren E. Burger, called to order on the opening of its new term last month. Gone, along with Justice Warren, were Justices Douglas, Hugo Black, John Harlan and Abe Fortas. In their place were four men appointed by President Nixon: Justices Burger, Harry A. Blackmun, Lewis F. Powell Jr., William H. Rehnquist; and one Justice, John Paul Stevens, appointed by President Gerald Ford. Remaining from the Warren Court are Justices William J. Brennan Jr., Thurgood Marshall, Potter Stewart and Byron R. White. Of the latter, only Justices Brennan and Marshall were consistently part of the libertarian march that characterized the Warren era. That march has been largely turned back by the Burger Court. The broad-scale revision may be summarized this way:

●The criminal-justice revolution forged by the Warren Court has been virtually dismantled through a series of decisions that have sharply limited the rights of suspects — in the streets, in interrogation rooms, in police lineups and in the courts. This has been accomplished by narrow interpretations of the Warren Court's landmark rulings, rather than by outright reversals.

●Conflicts between an individual and the Government have been resolved mainly against the individual. For a Court that was heralded as "conservative," as a counterbalance against the power of government, this is ironic. So, too, is the fact that the Burger Court has proved to be as "activist" as the Warren Court.

●Conflicts between an individual and the press have been resolved mainly against the press. This includes conflicts between the criminally accused (even the criminally convicted) and the press. Thus, an apparent paradox: the Burger Court, far less concerned about the rights of suspects than the Warren Court was, has been far more concerned about their rights when the press has been on the other side.

●In response to all of this, a small but importarnt trend has been emerging in the state courts, where — instead of relying on the Supreme Court — judges are looking to their state constitutions to protect individual rights.

☐

Despite this wide-scale revision, the Burger Court has been pictured by many in recent years as nonideological and unpredictable. This was not the case in the earlier years of the Burger Court, when it was generally viewed as

a "conservative" bench bent on substantially altering the constitutional jurisprudence of the Warren Court. As we shall see, the Burger Court has changed little from its early days; its stance has been essentially consistent. But the perception of it by some of those in the news media has changed considerably.

Chief Justice Burger has probably had much to do with this development. In a rare news conference three years ago, Justice Burger astonished civil libertarians by stating that there had been "no significant changes in the Court's attitude toward the rights of criminal defendants." In the ensuing years, Chief Justice Burger — who declined to be interviewed for this article — has repeated this theme, and more and more it has been picked up by various law professors around the country. These professors have found a receptive audience largely because none of the Warren Court's landmark rulings have actually been overruled by the Burger Court. However, legal scholars know that a case may be as effectively undermined by interpretation as by reversal. That some of them choose not to point this out is not so surprising as it may seem. Few law professors make a habit of criticizing the Supreme Court in the mass media. They need the Court for various reasons: they wish to have their articles cited by the Court, they vie to place clerks with the Justices, and some even argue cases before the Court.

Moreover, the professors generally resent what they consider to be "simplistic" reporting of legal matters. By nature they prefer complexity and the examination of small distinctions to anything resembling a sweeping overview of the work of the Supreme Court. Indeed, some of them deny that there is even such a thing as the Burger Court.

These legal scholars point to cases where, for example, President Nixon's appointees have not voted as a bloc. How, they ask, could there be a "Burger Court" when sometimes Justice Blackmun, or sometimes Justice Powell, do not go along with Justices Burger and Rehnquist, even in criminal cases, even in press cases? It is altogether too complex to be labeled, they say.

But this analysis is in itself simplistic. As legal scholars know, Justices White and Stewart had dissented from most of the Warren Court's criminal-justice decisions, and they vote with the majority in most of the Burger Court criminal-justice rulings. Thus, it makes no difference that, occasionally, one or another appointee of Mr. Nixon breaks away from the majority — so

long as, in the end, there is a majority to eviscerate the Warren Court holdings.

Those who say that there is really no such thing as a "Burger Court" are hard pressed to explain why Justice Brennan, a bulwark of the Warren Court, and Justice Marshall, a late Warren Court arrival who upheld the traditions of that Court, now nearly always find themselves dissenters in civil-liberties cases.

In the few instances when they are in the majority, the Court is almost invariably involved in striking down police procedures that are so flagrant — like the random stopping and searching of automobiles — that they would probably not have been attempted by the police, and surely not defended by prosecutors, without the earlier encouragement of the Burger Court. It is mainly in these kinds of cases that the justices appointed by President Nixon split; that is, where Justices Burger and Rehnquist take pro-police positions that are considered too extreme for Justices Blackmun and/or Powell, not to mention Justices White, Stewart and Stevens.

Still, the Burger Court is looked upon as a "moderate" one by many, and in some quarters even as a "progressive" force.

Meanwhile, the libertarian critics of the Court have generally been relegated to the law reviews or to the small journals of opinion. Criminal suspects have never had much of a constituency. The media, which obviously has a voice, more often than not refuses to challenge the Court's press rulings on the grounds that to do so would be "self-serving."

☐

But there is a Burger Court, and nobody knows this better than the prosecutors and the police. They feel they have a friend in the Burger Court, no ambiguity about it, and they express their appreciation in the most eloquent, effective way they can: by never blaming the Supreme Court for fostering crime.

In the 60's, hardly a day went by that some police chief or district attorney did not blast the Warren Court for "coddling criminals" and encouraging "crime waves." Decisions that many believed breathed life into the Fourth Amendment's proscriptions against unreasonable searches and seizures, the Fifth Amendment's guarantee of trial by jury and protection against self-incrimination, the Sixth Amendment's pledge of assistance of counsel, and the Eighth Amendment's prohibition against cruel and unusual punishment,

Yoichi R. Okamoto

The members of the Supreme Court: From left to right, Justices Stevens, Powell, Blackmun, Rehnquist, Marshall, Brennan, Burger, Stewart and White.

were often bitterly denounced by law-enforcement people. More often than not, these attacks were joined in by some of the press and the public, which provided a kind of rhythm section to an omnipresent chorus of policemen and prosecutors.

Although the Burger Court has not stemmed the rate of crime, its rulings have served to keep the police and the people off the backs of the justices. This is as it should be. Whatever one may say about the Burger Court, one may not say that it is responsible for the rising crime rate.

Nor was it fair to pin that responsibility on the Warren Court, a point underscored, albeit inadvertently, by the record of the Burger Court. For if that record teaches anything, it teaches that the Supreme Court can have no appreciable effect on crime.

Yet the Burger Court has continued to eviscerate the holdings of the Warren majority, as if to vindicate Mr. Nixon's 1968 campaign oratory that the Warren Court had "tended to weaken the peace forces as against the criminal forces in this country."

On the other hand, the Burger Court has continued to press for school desegregation and has attempted, most legal scholars believe, to strike a fair balance in the so-called affirmative action, reverse discrimination cases. This article does not study the affirmative action cases, or the sex-discrimination cases, because in these areas the Burger Court was writing on a relatively clean slate, and there was little question of revising the Warren Court rulings.

This piece surveys the two landmark areas that span both the Warren and Burger Courts — criminal justice and freedom of the press, which are at the core of the civil liberties guaranteed by the Bill of Rights.

Prof. Yale Kamisar of the University of Michigan Law School, one of the preeminent scholars in the field of criminal procedure, observed recently: "Reading the criminal law decisions of the Burger Court is like watching an old movie run backward." Here are some of those movies — involving the Burger Court's interpretations of the Fourth, Fifth and Sixth Amendments — run forward.

The 5th and 6th Amendments:
CONFESSIONS

The Warren Court delivered its landmark ruling on confessions in 1966, in *Miranda v. Arizona.* The Court held that when police have a suspect in custody, they may not interrogate him without first warning him of a number of things: that he has a right to remain silent, that anything he says can be used against him in a court of law, that he has the right to the presence of an attorney, and that if he cannot afford an attorney, one will be appointed for him prior to any questioning if he so desires.

The purpose of the ruling was to secure for suspects, in a meaningful way, the Fifth Amendment's protection against self-incrimination and the Sixth Amendment's guarantee of assistance of counsel — both of which, previously, had been held by the Warren Court to

apply to the states through the 14th Amendment (" . . . nor shall any State deprive any person of life, liberty or property, without due process of law").

Before the *Miranda* decision, no admonitions to a suspect were necessary. Confessions, or other statements damaging to a defendant, were admitted into evidence at the trial if they were found to be "voluntary." And the courts had held that the determination of voluntariness depended on the "totality of the circumstances" surrounding the taking of the statement.

Under this standard, however, statements elicited by virtually every method short of the third degree were held to be "voluntary."

The effect was the establishment of a double standard: one for the interrogation room of the police station, where the defendant's rights were often honored in the breach; another for the courtroom, where his rights were scrupulously honored. However, once a case got to court it no longer really mattered, because once a person confessed, it made little difference how properly he was treated in court, for the confession sealed his fate.

In the *Miranda* case, the Warren Court sought to narrow, if not entirely abolish, this double standard. A suspect could still be questioned, without the presence of a lawyer, but only if he waived his rights, "voluntarily, knowingly and intelligently."

The decision created an uproar among law-enforcement officials, in judicial circles and in much of the press. High-ranking police officials across the country, arguing that some 75 percent of reported crimes were solved by confessions, threw up their hands en masse.

But various studies, including one by the Los Angeles District Attorney's office, showed that the importance of confessions had been grossly exaggerated. (Most cases are not solved at all; but the vast majority of the cases that *are* solved employ means other than confessions — by catching the culprit red-handed, by relying on information from eyewitnesses, or by employing ordinary detective work.) Although confessions "cement" a case, they proved necessary in less than 10 percent of the cases studied by the District Attorney's office.

That the *Miranda* warnings themselves have had a minimal impact on police effectiveness is perhaps best indicated by the fact that the police outrage over the decision had largely died down by the time the Burger Court first considered the implications of the case in 1971.

That year, in *Harris v. New York,* the

147

The criminal-justice revolution forged by the Warren Court has been virtually dismantled by the Burger Court through a series of decisions that have sharply limited the rights of suspects — in the streets, in interrogation rooms, in police lineups and in the courts. This has been accomplished by narrow interpretations of the Warren Court's rulings, rather than by outright reversals.'

Burger Court ruled that if statements have been obtained from a suspect in violation of the *Miranda* safeguards, those statements can be used to "impeach his credibility" when the suspect testifies at trial. Thus, if the defendant takes the stand and says anything that contradicts what he told the police, the prosecution can bring to the attention of a jury an otherwise inadmissible statement.

Of course, if he does not testify, the statement cannot be used. But as every lawyer knows, a defendant who does not take the stand in his own behalf is far more likely to be convicted than one who is willing to explain his alleged actions. By offering this Hobson's choice to a person who has been unconstitutionally questioned by the police, the Burger Court has undercut the fundamental purpose of *Miranda*. Or, as Justice Brennan wrote in dissent in the *Harris* case: "The Court today tells the police that they may freely interrogate an accused [person] incommunicado and without counsel, and know that although any statement they obtain in violation of *Miranda* cannot be used on the State's direct case, it may be introduced if the defendant has the temerity to testify in his own defense."

The Court was also telling something to state and Federal judges. As Professor Kamisar wrote at the time: "After [the] *Harris* [case], a lower court judge unhappy with *Miranda* has cause to believe that almost no emasculating interpretation of *Miranda* may be too outrageous."

Before the Burger Court decided the *Harris* case, 23 courts faced the precise issue of "impeachment," and 20 of them held that statements obtained by the police in violation of the *Miranda* rules could *not* be used against a defendant who took the stand in his own defense. The main reason the courts cited: The Warren Court, in the *Miranda* case itself, had specifically said that such statements could not be used for impeachment purposes. Therefore, the argument that *Miranda* could be so easily circumvented was considered too outrageous, no matter what the

judges might think of the *Miranda* decision.

It was not too outrageous for Chief Justice Burger. Writing for a 5-4 majority in the *Harris* case, he said that *Miranda's* discussion of the issue "was not at all necessary to the Court's holding, and cannot be regarded as controlling." That is to say, according to Chief Justice Burger, *Miranda* could have been written without reference to the possible use of a statement for impeachment purposes.

Generally speaking, the same thing could be said of all precedents — since, technically, all that is "necessary" to a decision is what was directly before the Court. But *Miranda* was specifically structured to canvass a wide range of problems that were not directly raised by the case, in order to "give concrete constitutional guidelines for law enforcement agencies and courts to follow."

In any event, the Burger Court has never held a single item of evidence inadmissible on the authority of the *Miranda* case. Furthermore, Geoffrey R. Stone, associate professor of law at the University of Chicago, pointed out in a definitive article in the Supreme Court Review:

"Despite the relative frequency and complexity of these decisions, neither Justices White or Stewart, both of whom dissented in *Miranda*, nor any of the four justices appointed by Richard Nixon, has found it necessary to cast even a single vote to exclude evidence because of a violation of *Miranda*."

The Burger Court's decisions in the dozen-plus cases it has taken up — not to say the scores of lower-court rulings it has left standing — indicate a desire to discount the *Miranda* holding and return to the old "voluntariness" test for confessions.

Indeed, five years ago in *Michigan v. Tucker*, the Court appeared to have done just that. Speaking for the majority, Justice Rehnquist took the position that the *Miranda* warnings were not themselves "rights protected by the Constitution," but were merely "prophylactic rules" designed to "pro-

vide practical reinforcement" for the constitutional privilege against compelled self-incrimination.

Therefore, Justice Rehnquist said, the failure of police to give the full *Miranda* warnings does not violate the Fifth Amendment's self-incrimination clause; in order for such a violation to occur, the ensuing admissions must be involuntary, "as that term has been defined in the decisions of this Court."

As Professor Stone has observed, Justice Rehnquist's reading of *Miranda* is an "outright rejection" of its core premises. *Miranda* was anchored squarely in the Fifth Amendment. Had it not been, the Supreme Court would have been powerless to reverse *Miranda's* conviction. The High Court has no supervisory powers over state police practices; it can only strike down procedures that violate some Federal constitutional guarantee.

Justice Rehnquist would strip *Miranda* of its constitutional basis, and leave the case in an analytical vacuum. Had he stopped there, *Miranda* would have been interred. But, for whatever reason, Justice Rehnquist ultimately found a narrower ground to hold against the defendant without overruling *Miranda*.

Since then, the Court has continued to dismantle the case in a piecemeal way, and most observers believe the Court will always stop short of a direct reversal. In any event, the case has already been so confined by the Court to its basic facts that, some have said, now only Ernesto Miranda himself could take advantage of it. And Ernesto Miranda himself is dead.

The 4th Amendment: SEARCHES AND SEIZURES

What of the individual's protection against the unreasonable incursions of Government? Before the *Miranda* case came along, the Warren Court's most controversial criminal-procedure ruling was *Mapp v. Ohio*, decided in 1961. There, the Court required state judges to bar from trials any evidence that had been seized in violation of the Fourth Amendment's proscriptions against unreasonable searches and seizures. Previously — in 1914 — the Court had applied this sanction, known as the "exclusionary rule," to Federal courts. Now, through the Due Process Clause of the 14th Amendment, the exclusionary rule was made applicable to the states.

The reasons for this were clear and simple. If a court could not sanction a search or seizure before the event — because, for example, the police lacked sufficient cause to make the search, or were unable to describe the items they sought with the particularity required by the Fourth Amendment — then a court could not, or at least should not, affirm or sanction the search or seizure after the event. To do otherwise — to permit into evidence items unconstitutionally seized by the police — would violate the imperative of judicial integrity by making the courts partners in police lawlessness. The Court quoted the famous remark of Justice Louis Brandeis:

"Our Government is the potent, the omnipresent teacher. For good or ill, it teaches the whole people by its example. If the Government becomes a lawbreaker, it breeds contempt for law; it invites every man to become a law unto himself; it invites anarchy."

Without the exclusionary rule, the Warren majority concluded, the Fourth Amendment would "remain an empty promise," for the privacy rights protected by it would be "revocable at the whim of any police officer who, in the name of law enforcement itself, chooses to suspend its enjoyment."

The *Mapp* case was met by a firestorm of protest comparable to the one that later greeted *Miranda.* Enforcement officials — ignoring the fact that the Federal Bureau of Investigation had operated effectively throughout its existence under the dictates of the exclusionary rule, and that 26 states had imposed the rule on themselves — cried out that they could not protect the citizenry with the new handcuffs that had been clamped on them by the Warren Court. Paraphrasing a line out of an old New York Court of Appeals opinion by Benjamin Cardozo, the police asked rhetorically: "Should the criminal go free because the constable has blundered?"

It was hardly a matter of "blundering," however. The police systematically ignored the Fourth Amendment in those states that had no exclusionary rule, a point implicitly conceded in the outraged reactions of the police, as well as in the broad-scale efforts undertaken after *Mapp* to "retrain" officers in their understanding of the law of search and seizure. Sometimes the concession that the Fourth Amendment had been ignored was explicit. At a post-*Mapp* training session in New York City, Leonard Reisman, then the Deputy Police Commissioner in charge of legal matters, said, "The *Mapp* case was a shock to us. We had to reorganize our thinking, frankly. Before this, nobody bothered to take out search warrants. Although the Constitution requires [search] warrants in most cases, the

The members of the 1967 Supreme Court, pictured here from left: Justices Harlan, Black, Warren, Fortas, Stewart, Douglas, White, Brennan and Marshall.

George Tames/The New York Times

Supreme Court had ruled that evidence obtained without a warrant — illegally, if you will — was admissible in state courts. So the feeling was: Why bother?"

Mapp was the Warren Court's answer. But its effort to rescue the Fourth Amendment from its steerage-class status has been gutted by the Burger Court — despite the lack of evidence that *Mapp* has substantially curbed the efforts of the police, and against the clear knowledge that state-court judges seldom grant motions to suppress evidence that has been gathered in alleged violation of the Fourth Amendment.

Last spring, in response to a request by the Senate Judiciary Committee, the Comptroller General of the United States produced a study showing that evidence was suppressed on Fourth Amendment grounds only in about 1 percent of Federal cases; motions to suppress were made in only 10.5 percent of the cases studied. Nonetheless, many legal experts predict that the Burger Court will soon emasculate the exclusionary rule by requiring defendants to prove that police officers did not act in "good faith" when conducting otherwise unconstitutional searches and seizures.

In the meantime, the Burger Court

has conducted a substantial watering down of the Fourth Amendment itself; and so, it has often managed to get around the exclusionary rule without purging it of all significance.

This has been accomplished by sharp limitations on the meaning of "probable cause," the constitutional standard upon which arrests, searches and seizures are permitted; by a dramatic expansion of the concept of "consent searches," wherein suspects "voluntarily" consent to searches that would otherwise be illegal; by narrowing the scope of the exclusionary rule through holding that it does not protect grandjury witnesses; by depriving defendants of the right to test the legality of searches in Federal habeas corpus proceedings once the state courts have ruled against them, and by holding that some things are outside the purview of the Fourth Amendment — that they may be seized by Government agents without a warrant, without probable cause and without the knowledge and consent of the individual.

Here are examples of some of the above; the last concept will be dealt with first, because of its potential for "Big Brother" abuse:

Outside the Fourth Amendment: In 1976,

the Burger Court ruled that the Government can subpoena from banks a person's checks, deposit slips and financial statements without regard to the Fourth Amendment. The reason: a depositor has no "legitimate expectation of privacy" in his accounts because, by dealing through a bank employee, he has "revealed his affairs to another" and thus has "assumed the risk" that they "will be conveyed by that person to the Government." And this, "even if the information is revealed on the assumption that it will be used only for a limited purpose, and [that] the confidence placed in the third party will not be betrayed."

Since it is next to impossible to survive in today's world without having a bank account, the Court leaves a person with no choice but to "waive" his privacy — unless he decides to deal only in cash, and to keep his money under the pillow.

Moreover, as Justice Douglas once observed: "In a sense, a person is defined by the checks he writes. By examining them, the agents get to know his doctors, lawyers, creditors, political allies, social connections, religious affiliation, the papers and magazines he reads, and so on ad infinitum."

If bank records do not provide the Government with all it needs to know about a citizen, the names of the people he calls on the telephone may help to bridge the gap. Last term, the Burger Court held that a phone company's installation and use, at police request, of a "pen register" to record the numbers dialed from a telephone at a suspect's home did not constitute a "search" within the meaning of the Fourth Amendment. (The "pen register" is a device that records the numbers dialed from a particular phone; the original instruments, now obsolete, used a pen to mark coded dots on tickertape paper.) This, the Court said, was because the pen-register device does not record conversations, but only makes a record of the numbers dialed from a given phone, and a record of the time the number was dialed.

As in the bank-record case, the Court ruled that a person has no "legitimate expectation of privacy" in the numbers he or she dials, for the person has "voluntarily conveyed to the [phone company some] information that it had facilities for recording, and that it was free to record." Thus, the person "assumed the risk that the company would reveal to police the numbers he dialed."

The dissenters argued — there were two dissenting opinions, one by Justice Stewart (joined by Justice Brennan), the other by Justice Marshall (also joined by Justice Brennan) — that it could not be said that the defendant voluntarily turned over any information to "third parties." As a practical matter, a person had no alternative if he or she wanted to use the phone. Wrote Justice Marshall: "Unless a person is prepared to forgo use of what for many has become a personal or professional necessity, he cannot help but accept the risk of surveillance." The majority's "assumption of risk" analysis, Justice Marshall said, is therefore out of place. The question instead should be: what risks should a person "be forced to assume in a free and open society."

Ernesto Miranda, right, during a 1967 Phoenix trial on kidnap charges. His first conviction was overturned by the Warren Court in its 1966 landmark ruling.

fused to review the ruling.

While such a refusal — known as a denial of certiorari — does not put the Supreme Court's imprimatur on a decision, there is little reason to believe that the Court will, in the future, put a stop to this kind of surveillance, given its reasoning in the pen-register case, and its general view that the press stands in no better position than any other citizen, the First Amendment notwithstanding.

The implications for freedom of the press are serious, to say the least. By checking long-distance numbers — and in the "pen-register" case the Court rejected any distinction between long-dis-

Journalists must assume the risk of Government surveillance when using telephones quite as much as those persons engaged in criminal activity. In 1974, the Reporters Committee for Freedom of the Press sued the American Telephone and Telegraph Company, contending that both the First and Fourth Amendments require the phone company to give newsmen prior notice before turning over their long-distance telephone billing records to law-enforcement officials. This modest demand was rejected by the Circuit Court, and, in 1979, the Burger Court re-

tance and local calls, so both can be seized — Government agents are in a good position to discover confidential sources of information. The Burger Court, as we'll see in Part II of this article, has effectively held that the First Amendment does not give reporters a privilege to protect these sources. So the seizure of phone records provides another — and less politically sensitive — route for Government to uncover "leaks" and to otherwise induce a chilling effect on a robust, investigatory press. It also allows a kind of end run around the *(Continued)* state "shield laws" that are designed to protect sources; it does this by permitting agents to search out the leakers without giving notice to the reporter, thus preventing the reporter from protecting the sources with the shield law.

Consent Searches: The easiest, most propitious way for the police to avoid the myriad problems presented by the Fourth Amendment is to obtain the consent of a suspect to a search of his person or his premises. Once consent is given, the search is legal, and any contraband it turns up will be admitted into evidence, even if there was no probable cause to make the search. This is because the suspect (or any other person on the suspect's property who can give consent for the suspect, in what is known as "third-party consent") is deemed to have waived his right to privacy.

But the question the courts must decide in such cases is whether the alleged consent was that of a free and intelligent choice, or whether it was coerced; or, to put it in nicer words, whether it was a "true consent" or a "peaceful submission to authority."

In 1973, the Burger Court dealt a crippling blow to the nature of consent. In *Schneckloth v. Bustamonte*, the Court ruled 6-3 that a person can consent to an otherwise unconstitutional search — even though he didn't know, or wasn't told, he had the right to refuse the search.

The facts were simple. A police officer stopped a car in Sunnyvale, Calif., because one headlight and the license plate light were burned out. There were six men in the car, and they were asked to get out. As soon as they did, two other policemen appeared on the scene. The officer who stopped the car asked one of the men if he could search it. The answer: "Sure, go ahead." The search produced three checks that had been stolen from a car wash; they were wadded up under the left rear seat. The admission of these checks into evidence resulted in Bustamonte's conviction. Although he was in the car at the time, he was not the man who "con-

sented" to the search. This was a "third-party consent," but it is not what makes the case important.

The crucial point was the Court's holding that the police need not tell a person that he has the right to say "no," when that person is asked to consent to a search that would otherwise violate his Fourth Amendment rights. Consent, the Court majority said, "cannot be taken literally to mean a 'knowing choice.' " It is enough that the consent was "voluntary," i.e., free of coercion. The prosecution therefore need not show that the person made an "intelligent waiver" of his rights, only that the police didn't force him to waive them.

In dissent, Justice Brennan declared: "It wholly escapes me how our citizens can meaningfully be said to have waived something as precious as a constitutional guarantee without ever being aware of its existence."

But the majority opinion makes it clear that the Burger Court does not consider the Fourth Amendment a "precious" guarantee. The opinion agrees that to establish the waiver of a constitutional guarantee — according to the dictates of a 41-year-old Supreme Court decision — the state must prove "an intentional relinquishment or abandonment of a known right or privilege." But the majority said that this doctrine was only meant to protect a defendant's right to a fair trial:

"There is a vast difference," the Court said, "between those rights that protect a fair criminal trial and the rights guaranteed under the Fourth Amendment."

Like the old movie running backward, the Burger Court thus went a long way toward again demoting the Fourth Amendment to its steerage-class status.

"The holding today," Justice Marshall wrote in separate dissent, "confines the protection of the Fourth Amendment against searches conducted without probable cause to the sophisticated, the knowledgeable, and, I might add, the few."

"In the final analysis, the Court now sanctions a game of blindman's bluff, in which the police always have the upper hand, for the sake of nothing more than the convenience of the police."

Ironically, the *Bustamonte* case might not even have reached the Supreme Court today, for Bustamonte's bid — to suppress the evidence the police had found against him — had been denied by the California state courts. His conviction was subsequently reversed by the Federal Court of Appeals on a writ of habeas corpus, a judicial procedure for testing the legality of a person's detention. The Burger Court reinstated the conviction, but, three years later, it barred Federal habeas corpus relief for prisoners who had previously been afforded an opportunity for "full and fair litigation" of their Fourth Amendment claims in state courts. The practical impact appears to be that prisoners can no longer rely on lower Federal courts to overturn their convictions on Fourth Amendment grounds.

This habeas corpus ruling was a startling decision even for the Burger Court, for it orphaned the Fourth Amendment, making it the only provision of the Bill of Rights (so far) which may not be vindicated by habeas corpus — the Great Writ, so-called because it is considered to be the single most important safeguard of personal liberty known to Anglo-American law.

"This denigration of constitutional guarantees and constitutionally mandated procedures," Justice Brennan wrote in a long, bitter dissent, "must appall citizens taught to expect judicial respect and sup-

port for their constitutional rights."

What was the Court's reason for this "denigration"? The same reason for its antipathy to the Fourth Amendment and to the exclusionary rule that enforces it and keeps it from being an "empty promise." When police make an illegal search and find stolen goods or guns or drugs, they nearly always have the guilty party. Shall the criminal go free because the constable has blundered — or even *plundered* — his rights?

But the fundamental purpose of the Fourth Amendment — of the Bill of Rights, in general — was to protect the guilty as well as the innocent.

As Justice Brennan added in his dissent: "Even if [the] punishment of the 'guilty' were society's highest value (and procedural safeguards [were] denigrated to this end) in a Constitution that a majority of the Members of this Court would prefer, that is not the ordering of priorities under the Constitution forged by the Framers, and this Court's sworn duty is to uphold that Constitution and not to frame its own."

The 6th Amendment: THE RIGHT TO COUNSEL

The Burger Court has emphasized, in Fourth Amendment cases, the "vast difference" between (1) a person's rights that are protected by the prohibition against unreasonable searches and seizures and (2) "those rights that protect a fair criminal trial." One would think, then, that when the very question of guilt or innocence is involved, the Court would be especially concerned

> Although the Burger Court has not stemmed the rate of crime, its rulings have served to keep the police and the people off the backs of the justices. This is as it should be.

with a person's right to a fair trial. It might be expected that the Court would lend a sympathetic ear to a defendant's allegation that the police nabbed, and the jury convicted, the "wrong man."

It hasn't turned out that way, even in the area that legal experts universally deem the most suspect of all: eyewitness identification. In 1967, the Warren Court, noting that "the annals of criminal law are rife with instances of mistaken identification," ruled in *United States v. Wade* that an arrested suspect has a constitutional right to have his lawyer present when he is paraded in a police lineup before possible eyewitnesses. The major reason: to protect him from the suggestive techniques that are often employed by police and Federal agents (for example, when the accused may be the only black person, or the only tall or short person, in the lineup). Such suggestion, once accomplished, may be irretrievably devastating.

"It is a matter of common experience," the Court said, "that once a witness has picked out the accused at the lineup, he is not likely to go back on his word later on, so that in practice the issue of identity may . . . for all practical purposes be determined there and then, before the trial."

Unless the lawyer for the suspect is present at the lineup, he can neither guard against improper suggestion, nor even know what really happened there. At a trial, this, in turn, deprives the accused of "that right of cross-examination which is an essential safeguard to his right to confront the witnesses against him." The lawyer must conduct the cross-examination in the dark, so to speak, making the assistance of counsel, guaranteed by the Sixth Amendment, an empty right.

In 1972, the Burger Court, in *Kirby v. Illinois*, had its first opportunity to interpret the Warren Court's *Wade* case. In the *Kirby* case, the lineup took place before the defendant was indicted. In *Wade*, the lineup took place *after* the in-

dictment. The Burger Court chose to make the fact of the indictment the controlling distinction — the fact that determined the court's decision — and therefore ruled that *Kirby* was not entitled to counsel at his lineup.

Justice Brennan was in a peculiarly good position to say whether *Wade's* indictment had anything to do with the decision in that case, since he wrote the opinion for the Warren Court. He said the fact of the indictment was "completely irrelevant," and that "even a cursory perusal" of the *Wade* decision "reveals that nothing at all turned upon that particular circumstance." But now, in 1972, Justice Brennan was writing in dissent.

It is instructive to compare the Burger Court's use of distinctions—those facts that are crucial to a ruling — in the *Kirby* case with its treatment of the *Miranda* case. In the Burger Court's *Harris v. New York* decision, statements taken illegally from a suspect were allowed in to "impeach his credibility" when he took the witness stand. The Burger Court said that while the *Miranda* decision had barred the use of such statements, it was not a "controlling" precedent because "it was not at all necessary to the Court's holding."

Whatever one may say of that viewpoint, one can only wonder at a Court that would, one year later, create a constitutional distinction out of a mere description of a defendant's status — i.e., whether he was in a lineup before or after his indictment. And one wonders at a Court that, in creating such a constitutional distinction, has signaled the police that there is an easy way to circumvent *Wade*: by holding lineups before filing formal charges.

In subsequent eyewitness identification cases, the Burger Court has moved step-by-step toward what Justice Brennan calls "the complete evisceration of the fundamental constitutional principles established . . . in *United States v. Wade*."

This development, says Professor Kamisar, is "in some

ways more depressing than anything else the Burger Court has done in the criminal-procedure area." Why? "Because unlike *Mapp* and *Miranda*, which furthered societal values not usually — certainly not always — related to guilt or innocence, the Warren Court's 1967 lineup cases were explicitly designed to protect the innocent from wrongful conviction. What is more important than that? And where is the countervailing balance? The defense lawyer's presence in the interrogation room may well cut off police questioning altogether, but the defense lawyer's presence at a lineup will not — and cannot — eliminate lineups, only discourage the holding of *unfair* ones. How does that harm effective law enforcement? In fact, it helps it; if the wrong person is convicted, the system has failed and the real culprit is still at large. Even if one is convinced that the Warren Court substantially weakened the 'peace forces' — I'm not, but some justices evidently are — the Burger Court's retreat from the 1967 lineup cases is not responsive to that need."

☐

In the lineup cases, as in some of the confession cases and the search-and-seizure cases, the Burger Court has reached out to reverse the actions of lower courts that have read the Bill of Rights more liberally than is to the High Court's liking.

This is the opposite of the situation that prevailed during the halcyon days of the Warren Court, when that Court continually patrolled the state and Federal courts, which not only gave niggardly interpretations to its rulings, but often fought the Court in the press.

During the past few years, however, a trend has developed. In response to the Burger Court's reluctance to afford what they consider proper protections to the criminally accused, a number of state courts have dusted off long-ignored state bills of rights. Since the United States Supreme Court may only set minimum standards of jus-

tice, the states are free to grant their citizens more extensive rights. In cases in which they wish to afford such rights, in order to foreclose reversal by the Burger Court, the state courts need only say that they have based their rulings on state law, rather than on the Federal Bill of Rights.

This development, which has reportedly annoyed and occasionally frustrated the Burger majority, is an event in the law. The trend is in its infancy, and has been taken up by only a handful of state courts — most notably those in California, Michigan and Pennsylvania. In the large majority of states, the judges seem only too happy to go along with the Burger Court and to accept its "signals," and go even further. But the new movement the other way is not without significance, and it has been encouraged by Justices Brennan and Marshall, dissenters in the tradition of Justices Holmes and Brandeis, and of Justices Black and Douglas.

In 1977, Justice Brennan published an article in the Harvard Law Review "saluting" those state courts that chose to use their own bills of rights to vindicate liberties no longer recognized by his Supreme Court brethren.

☐

At age 60, John McNulty, the great Irish-American journalist and author, wrote to his old boon companion, James Thurber: "Dear Jimmy, I think that maybe threescore years and 10 is subject to change without notice."

It turned out that way for Mr. McNulty, who died a few days later. In the law, it is supposed to go the other way: the older the precedent, the less subject it is to change without notice. But in the matter of the Burger Court, this comfortable maxim has not held true.

President Nixon's appointees to the Court were heralded by him as apostles of "judicial restraint," men who would adhere to precedent, "strict constructionists" who would not allow their political, social and economic views to influence their decisions. These themes dominated the Senate confirmation hearings, at which each of the four swore his dedication to such principles.

Here, though, are a few examples of how these principles have been practiced:

The Presumption of Innocence: In 1895, the Supreme Court, tracing the venerable history of the presumption of innocence from Deuteronomy through Roman law, English common law, and the common law of the United States, wrote: "The principle that there is a presumption of innocence in favor of the accused is the undoubted law, axiomatic and elementary, and its enforcement lies at the foundation of the administration of our criminal law."

On May 14, 1979, the Burger Court said that the presumption of innocence has "no application to a determination of the rights of a pretrial detainee during confinement before his trial has even begun." In so holding, the Court reversed two lower Federal courts in New York, which had granted relief to inmates awaiting trial while housed at the Metropolitan Correctional Center. The inmates complained that they had been treated as convicts rather than as persons presumed innocent until proved guilty. They were subjected to body-cavity searches, following visits from friends and relatives; they were forced to "double-bunk" in rooms built for single occupancy; they were prohibited from receiving books unless the books were mailed by the publishers or by book clubs or bookstores; they were forbidden to receive food and personal items from outside the institution.

The lower courts enjoined these practices as unconstitutional. They relied primarily on the presumption of innocence as the source of an inmate's right to be free from these sorts of conditions. In a lengthy opinion by Justice Rehnquist, the Supreme Court lifted the injunctions of the lower courts, stating, among other things, that the pre-

sumption of innocence provides "no support" for the relief that was granted to the prisoners by the lower courts.

"Without question," Justice Rehnquist wrote, "the presumption of innocence plays an important role in our criminal justice system." But that role, he said, is confined to the trial itself.

One week later, on May 21, 1979, the Burger Court held that, at the trial itself, a judge could refuse a defendant's request to instruct a jury that he was presumed to be innocent until proved guilty beyond a reasonable doubt. So, now a jury need not be told that a defendant starts a trial with a clean slate, a right that was considered fundamental even in biblical days.

"No principle is more firmly established in our system of criminal justice than the presumption of innocence that is accorded to the defendant in every criminal trial," wrote Justice Stewart. But now, this principle was relegated to a dissent, and Justice Stewart was joined by only Justices Marshall and Brennan.

Trial by Jury: In 1952, Justice Felix Frankfurter, in delineating for the Court those provisions of the Bill of Rights that have a "rigid meaning," as opposed to those without a "fixed technical content," wrote: "No changes or chances can alter the content of the verbal symbol of 'jury' — a body of 12 men who must reach a unanimous conclusion if the verdict is to go against the defendant."

By 1970, the Burger Court, saying it was unable "to divine precisely what the word 'jury' imported to the Framers," ruled that six-person juries were constitutional in criminal cases. Two years later, the Court held that a 12-man jury need not reach a unanimous verdict to convict a defendant, upholding votes of 11-1, 10-2 and 9-3.

Indigent Defendants: Last term, the Burger Court ruled that an indigent defendant charged with a crime carrying a possible one-year prison sentence was not entitled to free legal counsel as long as the

judge did not sentence him to jail.

This was a real surprise to the dissenters who had assumed — as did most lawyers and even most states — that at least when a person had a right to a jury trial he had a right to free counsel. Since the Supreme Court had already held that any crime punishable by more than six months in prison carried with it a right to trial by jury, it was natural to expect that it also required a lawyer — especially in view of the Court's earlier holding that the right to counsel occupies a higher constitutional status than the jury-trial right.

But Justice Rehnquist, writing for a 5-man majority, said no, arguing in part that such a rule would economically burden the states.

In dissent, Justice Brennan termed the ruling "intolerable," noting that the crime that was involved — theft — carried a "moral stigma" indicating "moral depravity" and was therefore by no means petty, whether or not a prison sentence was applied. As to the economic burden argument of the majority, Brennan pointed out that 33 states provided for counsel in such cases and that, in any event, the argument was "both irrelevant and speculative."

Judicial Immunity: This is the one area in which the Burger Court has managed to divine an absolute right for a class of people — namely judges — despite the fact that the Constitution nowhere makes any provisions about judicial immunity.

In 1978, the Court ruled on a case involving a judge who signed an order to sterilize a 15-year-old girl — without a hearing, and merely at the request of the girl's mother, who said she was "somewhat retarded" and had been staying out overnight with "older youth or young men." The Court held that the judge was immune from a subsequent lawsuit by the girl — who was now a young woman — and her husband.

The girl was attending a public school at the time of the sterilization — despite the "retarded" appellation — and

was told that she was going to the hospital to have her appendix removed. Instead, the doctors, acting in accord with the judicial order, performed a tubal ligation.

Two years later, the girl was married. Her inability to become pregnant led her to discover that she had been sterilized. She and her husband sued the judge. The Federal District Court in Indiana dismissed the action on grounds of judicial immunity, but the Circuit Court reversed, holding that the judge had forfeited his immunity due to "his failure to comply with elementary principles of procedural due process." The Burger Court reversed, by a 5-3 vote (Justice Brennan took no part).

Citing an 1872 ruling by the Supreme Court, the Burger majority held: "A judge will not be deprived of immunity because the action he took was in error, was done maliciously, or was in excess of his authority; rather, he will be subject to liability only when he has acted in the 'clear absence of all jurisdiction.' " The Court found jurisdiction in the sweeping language of the Indiana judicial code, which conferred jurisdiction "at all cases at law or in equity."

In sharp dissent, Justice Stewart (joined by Justices Marshall and Powell) wrote: "A judge is not free, like a loose cannon, to inflict indiscriminate damage whenever he announces that he is acting in his judicial capacity."

What gave a judge this freedom, according to the Burger majority, was the doctrine of judicial immunity. Where did that come from? From the Supreme Court itself, surely not from the Constitution.

It is quite remarkable, to say the least, that the same justices who disparage the exclusionary rule of the Fourth Amendment as "merely judge-made," think differently when judges are defendants in civil-law suits. But as it is said: "Where you stand often depends on where you sit."

The Presidency

A boyish Edward Kennedy with Robert and John, after the Democratic Convention in July 1960. Below, the Senator today at Hyannis Port.

THE KENNEDY MYSTIQUE

He's ahead in the Presidential polls, sparring with Jimmy Carter. But as his boosters' enthusiasm mounts, the key question about Edward Kennedy is his character. *The first of a two-part series.*

By Anne Taylor Fleming

This is a story about Edward M. Kennedy. It is a story about his family, mainly about the men in that family and the men who came to spend their lives with them, who came to envy them their brotherhood, who came to want the love they had for each other, a love they expressed shyly, as men seem to, as Irish men particularly seem to, in banter and bravado, their fierce tenderness camouflaged in semigenial competition. There are women in this story, too, often beautiful women, some to whom these men were married, some to whom they were not. Their names and faces are well known. But they are not central to the story. They are widows, victims, night times. The women who matter are the mother, at 88 still touchingly vain and rectitudinous, a woman capable of quieting her remaining son with a simple "Teddy," and the three sisters who can, better than most, banter with their baby brother as their brothers did. "Whatever contributions the Kennedys have made," Ted Kennedy has said, "are very much tied into the incredible importance and power of that force in our lives, the family."

As the final child in that family of nine, Ted Kennedy seemed young forever, harmlessly young, then, in the days after his brother Robert's death, defiantly and then recklessly young, baby-faced even in his grief. It is hard to know exactly when he stopped being young that way. There is no longer a hidden apology in his charm. He seems adjusted to his appetites; they are not calm. He works hard, laughs hard, eats hard. He likes a drink but not when he is alone; he likes a cigar even when he's alone. His large handsome face is permanently flushed, his not-quite-Ivy League suits tight across his broad back. He wears insubstantial shoes that make him look a bit top-heavy. "Ted," a friend said, "was always big trying to be little; Bobby was always little trying to be big."

For the fourth time in the 11 years since Robert Kennedy's death, Ted Kennedy must again decide if he wants to try to be President. Since the day of that death on June 6, 1968, many people in this country have been tenacious in their affection for him. He has been hated, too, obviously; there is always that. But every Presidential election year, many voters have expressed their readiness to back up their affection with their votes. He is a politician. He has liked that affection, toyed with it, invited it even as he has tried to shake it. All the way along he has seemed genuinely ambivalent about the Presidency, saying no to it three times, using his excesses, his recklessness to disabuse a public that holds fast to a loyalty that matches the Kennedys' own. He seems to half want the job, or think he should, because all of his brothers before him wanted it. But at the same time he has behaved in odd ways, bad

Anne Taylor Fleming is a regular contributor to this Magazine.

The Congressional Kennedy: In his Senatorial power base, he is now chairman of the Judiciary Committee; here, he presides at a hearing last April.

The Senator talks to an aide in his Washington office last October. Kennedy runs a staff of 90 as his father ran the family: they compete for his attention.

The Senator, buttressed by charts and statistics, discusses tax reform, one of the favorite Kennedy issues along with health insurance and gun control.

ways, sad ways, seeming subconsciously somehow to want to walk away. The voters just won't let him. Again he is at the top of the polls.

"Ted is like an athlete who's always in condition, always doing his calisthenics," said a close political friend. "If there's an opening in the line he'll go through. It's hard to keep from doing something if it's in you and the situation. He lives in four-year cycles. He's always wondering if this year he'll have to take the test."

Kennedy keeps insisting that this, like the ones past, is not the year, not his year. His friends say he is happier than he has been in 10 years. He says he is happy, mostly, that his work is good and his children are well. He is, after a long journey, taken seriously by serious people; it feels good. He is head of the Judiciary Committee, a force in the Senate, using his leverage as the perennial Presidential pretender. If he can convince himself that his work there is important enough, as important in its way as being President, then maybe he can walk away and stop living a life pegged to a possibility.

"There has always been a dichotomy in Ted," said longtime friend John Tunney. "We both came from families that demanded success. He is a very enthusiastic, warm, fun-loving person who at the same time has a strong sense of duty. If he runs, it will be a denial of one part of his personality; and yet if he doesn't run it will be a denial of the other part." For Kennedy, it will be a most solitary decision. Friends like Tunney, especially old friends, will say nothing to him about whether he should run. His sisters will say nothing. "Because of, uh, what, uh, happened to my brothers," Kennedy said, "nobody close to you will advise you."

Meanwhile he plays the game better than anyone; he's had a lot more practice. "Carter, he's my candidate," he says, grinning, and in the next breath inquires of certain autograph seekers whether they're from primary states. He is bantering with Carter, a lot, so much so that some Democratic Party members are accusing him of dividing the party, some say impishly, some say malevolently, and there is a modest backlash of sympathy for Carter. The polls indicate that it is Kennedy's nomination if he wants it. (The most recent Gallup Poll shows 58 percent of Democrats for Kennedy, 31 percent for Carter, and 11 percent undecided.) The prevailing analysis is that Kennedy will not get in unless Jerry Brown does, Brown all groomed and gloomy and so unlike Kennedy who, at day's end, is usually rumpled but hopeful, still hopeful. If Brown can show Carter to be wobbly, then Kennedy, the scenario goes, will make his long-awaited move. If he does wait until the New Hampshire primary next February to see whether Brown damages Carter, Kennedy will have missed the filing dates for almost a third of the other primaries. Some think that a fatal delay. Others say that for a Kennedy it just won't matter, that even in the primaries he misses, there will be "un-

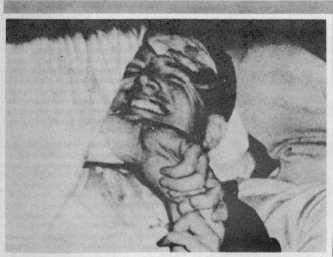

The tragic Kennedy: in agony on a stretcher after his near-fatal 1964 air crash.

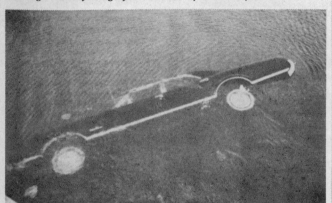

The Kennedy car in eight feet of water off the Chappaquiddick bridge in 1969.

In 1973 with 12-year-old Edward Jr.: bone cancer forced the amputation of a leg.

committed'' delegates who would wait for him, and that an announcement as late as midyear, next year, would create such excitement that he could not be stopped.

If Kennedy does challenge Carter, a bland incumbent with an appealing symbiotic marriage and convincing piety, Kennedy will look both politically and personally flagrant, exaggerated, liberal. But, except for his commitment to national health insurance, his politics are not that dissimilar to Carter's and not flagrantly liberal. That would leave Kennedy to answer once more for his personal life, an exhumation of which he cannot welcome. There is Chappaquiddick and there is his marriage and they are all tangled up. His wife is living away from him in Boston, trying hard to stop drinking, loving him still, friends say, probably unable to be beside him if he runs. There are those who will have sympathy for Kennedy for this and others who will not, others who will say — if he does run, if her sorrow becomes more visible — that what happened to the girl he married is his fault. There are people who will say that. His marriage is not something Kennedy will want to talk about. He is, beneath his garrulousness, a shy man; when people come up to him on the street, he often looks down and pets his tie. He rarely completes a sentence, speaking still in the deflective shorthand of his childhood dinner-table conversations when there were so many others it was hard to finish anything. He is, even with younger people, deferential. After most of a lifetime of being the youngest brother, one does not stop being that just because the others are gone.

□

On the early August day of 1944 when the first of his brothers died a hero, Ted Kennedy was 12 years old. The family was together in Hyannis Port that Sunday afternoon when two priests came with news of Joe Jr.'s death at age 29 in a plane crash over the English Channel. He died his father's favorite. "My father took the two men upstairs," Eunice Kennedy Shriver said. "A little while later he came down and said, 'We've lost Joe.'" Then he went back upstairs to his room and locked the door. "It was the first thing that he valued tremendously that he ever lost," Eunice Shriver said.

John Kennedy then became the oldest. With two years between them, he and Joe Jr. had battled through childhood. Joe Jr. was big and healthy and hot-tempered, a scholar and an athlete, while Jack was frail and not particularly studious. At Choate prep school, Joe Jr. was a football star and Jack was a cheerleader. Years later, Jack Kennedy would say, "Joe was the star of our family. He did everything better than the rest of us."

On the day he heard of his brother's death, John Kennedy, who had returned from the war skinnier than ever but a hero, took the rest of the family sailing. "I remember racing that day afterwards," Eunice Shriver said. "Jack said, 'Let's go for a sail — that's

what Joe would have wanted us to do.'" Then Jack put together a book about his brother for his family, a collection of memorial essays, a tribute which his father could never bear to read. Then he went on to become the first Roman Catholic President of the United States, what Joe Jr. had told his father he meant to be.

Teddy's childhood was not really that of his brothers because he was seven years younger than his closest brother, Bobby. They would be close friends much later, the closest of friends only after John Kennedy's death. In the early family pictures, Teddy is a round little boy in knickers with little-boy bangs while the other boys are already in long pants and ties, the girls in proper dresses. "In the touch football games," Eunice said, "the girls had to be able to cover the boys, too. Covering Bobby wasn't easy. We all wanted to cover Teddy because he was so little and pudgy."

Teddy did have a surrogate brother, a cousin, his mother's sister's child who came to stay. Joey Gargan was two years older than Teddy, a big tenderhearted pal who became a big puppy dog of a man, remembered not for that childhood he shared with Teddy — shifting beds whenever the real brothers came home — but for a party he arranged years later on Chappaquiddick Island. Gargan loved that childhood he shared and the father that went with it. "Just the way he looked at you over his glasses — that was it, man," Gargan said. "No one I've ever seen in my life could control a situation more clearly and distinctly than he could. At the same time, he was the kindest, most considerate, most understanding man."

So there was discipline and in it, apparently, love. Such is the testimony of the children, the only thing in the world Joe Kennedy loved, making a lot of money for them in not so nice ways, the money not for himself but for them so that they could tell him, he later said, to go to hell, knowing, of course, that there was no way they would do that. With an Irish chip on his shoulder, a serious chip, a memory of being a saloonkeeper's son, an upstart Irish Catholic kid in Boston, a town of old monied WASP's, he had ultimately moved his children into two big beach homes facing the sea — a compound at Hyannis Port and a big stucco Spanish house in Palm Beach — almost defiantly inelegant places where they became a world unto themselves. For Joe Kennedy, to the outside a strangely sour man, an isolationist, a lonesome capitalist, a man who had not been able to join the best clubs at Harvard, a man who had been employed and then abandoned by President Franklin Roosevelt — a lifetime of adding chips on chips — to this man, this father, his children were not only his family but his place and finally his business and finally his revenge. And they loved him without revision.

□

A day in 1962 after his father had had a stroke, Robert Kennedy and a friend

Scene from a family album: The Kennedys at leisure in Hyannis Port, in the fall of 1978. From right: Edward; his mother, Rose; his wife, Joan; and sister Pat.

were visiting him in Hyannis Port. Joe Kennedy, 73, sat in a wheelchair at the far end of the room; he could not speak. Although there was no supporting medical evidence, all of his children believed he could still understand them; they telephoned him weekly, sometimes daily, as they had always done. That day one of Robert Kennedy's children came into the room after a boat race. "How did you do?" Robert Kennedy asked his child, ruffling his hair, touching always, needing to — in that way like Ted and not like Jack. "I came in fourth," his child said. Turning to his friend, Kennedy said very quietly, "That's the difference between them — my children — and us; if we didn't win we were sent to bed without dinner." He said it without rancor.

The Kennedy children married late, mostly, bringing their spouses back home, into the family, into the family business. "It was like the Dallas Cowboys; you were not going to beat them," said Peter Lawford, divorced now from Pat Kennedy. "I always felt that her love for her father took prece-

dence in a funny way over her love for me. The more I saw it in our daily life, the more I understood it. She worshiped him; in the long pull that was stronger."

"The girls still think he was the most marvelous father on the face of the earth," said an old family friend. "Rose Kennedy saw that they got their shots and she held the family together, but deep down she didn't matter the way he did, even to the sisters."

If rumors that Joe Kennedy was having romances reached his wife or children, certainly they do not say. Friends of the brothers would kid them when they were grown about being chips off the old block; light stuff, locker-room stuff, so they must have heard the innuendoes. Perhaps the father let them know that it was O.K., grown up, all those women, that they didn't matter, the way the woman one married did. "Marriage is a contract for life," Joe Kennedy would say in front of his children. It was just different — there were women to marry and women to lie down with, and the women his sons would do

that with — later when they were Senators, and one a President, and finally legends — those women were lying down with history and everybody involved knew that. When Ted Kennedy was being talked of as a possible President before 1972, before Chappaquiddick, one of his sisters said, "The women won't matter anyway; look at F.D.R."

Rose Kennedy was happy with Joe Kennedy; that is what she has told her children. Ted Kennedy says that his mother, Rose, balanced his father's discipline with love. Eunice Shriver says that they were the same kind of people. "Mother was equally competitive about everything," she said. "It wasn't that you wanted to destroy anyone; you just wanted to do better than they could. I was 24 before I knew I didn't have to win something every day."

Joey Gargan said of Mrs. Kennedy: "She was the most disciplined person I ever met. She organized the house terribly well. Everybody knew what he was supposed to do. Our lunches were a horror show; Teddy and I were cross-

examined on the news of the day and on our catechism." She went to mass twice a morning, at 7 and 8, carrying rolls — a millionaire's wife — to eat between services. On the day she heard that her third son, Robert, had been killed, she walked, immaculately groomed as always, Catholic to the core, up and down the Hyannis Port driveway, bouncing a tennis ball.

Being the youngest, Ted missed the heart of the competition; he developed less of an instinct for the jugular than any of his brothers. Looking back, it seems that the love of the father fell most easily on the oldest son, of whom the most was expected, and on the youngest, of whom the least was expected, leaving the two in between to earn it more, which they did. "Jack," a childhood friend of all three brothers said, "was as tough as anyone who ever came down the pike. He was the most like the father of any of them. He was disciplined in what he wanted to do, a creature of self-will, brilliant and hard but also understanding. Bobby was very religious, disciplined also, but

Kennedy the crowd-pleaser, on his swing through New York City last month.

much more emotional. Jack was never moved by that, by heart. Ted was a combination of his father and his older brothers but he also got some of the softer qualities of his mother. It was Bobby who always wanted all the Kennedys to play on the same team."

These men were not close until they were adults, through their work, and then, as Ted said, "we were still kind of private from each other." Because of their childhood, they had a susceptibility to the notion of heroism (Ted, at bottom, was a little less susceptible), which is, after all, a solitary business. They were men who, for all their closeness, could talk most personally to women — not necessarily the ones to whom they were married. They were men, all of them, with a real, almost exaggerated tenderness for children, wanting as many of their own as possible.

What was expected of Ted, early and always, was laughter. "Being that far down in the family you have to be likable," said Milton Gwirtzman, also the youngest of many children, who for years was Ted's main speech-writer. "You're the weakest, so you just have to be liked. Youngest children understand Teddy better."

His brothers threw him in the water and threw him in the air and dared him to jump off high places and he seemed not to mind. He laughed, and laughing, made them laugh. It was what he did; it was his place in that family to make everyone laugh, to make those brothers laugh. "I always remember him being good-natured," Eunice Shriver said, "more than the others. Joe Jr.'s humor was more sardonic — he'd kid you about being fat. But Teddy was much more sensitive. He'd try to bring everyone together by making them laugh."

It came easily to him. He had, his mother said, a sunny disposition, right from the beginning. This is a family that embraces the clichés about itself,

about its members so that Teddy, the baby, being told he had a sunny nature, had it, elaborated on it, as later, being told by the brothers that he was the best natural politician of the three, he believed them — and believing them was, by which time there was just the tiniest bit of shrewdness in his affability.

□

When John Kennedy was running for office and Robert Kennedy was running interference for him and they were as close as two essentially private men can be, it was Ted who made John Kennedy smile. "He was the shining light," Peter Lawford said. "He'd come home from school for the weekend. The old man wasn't too well and everyone was treading softly and when Teddy arrived it was like the sun coming out; Jack's face just beamed. Every New Year's Eve we all had to go to this fancy party in Palm Beach. It was a crashing bore but a ritual. The host had a display room full of priceless china birds. Teddy, who was around 20, said that at the stroke of midnight he was going to stand on his head in the bird room. And he did. Always afterward we'd say to one another, 'Hey, do you remember the night Teddy stood on his head in the bird room?' "

Later, in John Kennedy's Presidential campaign when Ted was 28 and politically green, he was assigned to the Western states. It was a nonassignment of sorts because they didn't expect to carry those states and didn't, losing 10 out of 13 to Nixon; and Ted — his humor still impulsive, nervy, prankish, not smart and self-deprecating like John Kennedy's or Robert's, as it has become in their absence — actually went off a 180-foot ski jump for the first time in his life. Then in Montana he rode a bucking bronco, for votes he said, for fun possibly, for his brother really. He got back what he wanted — a long-distance laugh. "Jack called me shortly

after that," Ted recalled, "to say that he had just talked with the state chairman in Wyoming who told him I was coming there . . . They were arranging to have a sharpshooter shoot a cigarette out of my mouth at 20 yards. Jack said, casually, that he didn't think I had to go through that if I didn't want to, but . . ."

Later still, after everything, his closest sister, Jean Smith, can say: "When I see him he's always up. I have never seen him down, sad yes, but in all the years, never down, never depressed."

"It's just that Irish thing," a boyhood friend said. "When you want to cry, you laugh, like at a wake. There were times when there were tears in his eyes or he was a little hysterical in response to sorrow . . . but Kennedys never cry."

Ted Kennedy's school days were distinguished by one thing: movement. His father was the Ambassador to Britain so there was quite a bit of early shifting and Teddy was sent to 10 different schools before he was 14. He says simply that the moving around made him more genial. He did not make real friends until he entered Harvard in 1950, and then they were good-time pals for a lifetime, men like Claude Hooten, now a real-estate developer in Houston, a sentimental man who wears cashmere pullovers and loafers without socks and shoots big game. "We were typical jocks," Hooten said, "which at Harvard meant you were identified as stupid. But it didn't matter. The war was over; there were no riots. It was inconceivable that we would march on the dean's office. It was a happy, happy time. Ted was the kind of guy who, if we were going on a double date, would buy two corsages, one for my date and one for his."

Then in the spring of his first year, Ted, as a lark, Hooten said, cheated on a Spanish exam, allowing someone else to take it for him. Kennedy's official version, released to the national press when he first ran for office, was that he was having difficulty in the course and was apprehensive enough to let someone else take the test for him; so, trying not to fail his father in one way, he failed him in another. He was expelled, went home, then went in the Army, coming back to Harvard in 1953, "a different guy," Hooten said. "You couldn't shake him from the books after that — it was, how far can you go before you really screw up. It wasn't just his father he worried about, or his mother; I think Ted worried about disappointing everyone."

John Kennedy said that his father cracked down on Ted at a crucial time. "If it hadn't been for that," he said, "Teddy might just be a playboy today."

□

Back at Harvard, head down, studying harder than so many of the people around him — which he would always have to do — he managed also to play football, catching a few winning touchdowns, recouping his father's affection and outshining for once two of his broth-

ers, Joe Jr. and Robert, who had both played at Harvard. "Bobby was a more defensive type," his freshman football coach was once quoted as saying. "He didn't have the size that makes for a blocker, but boy, was he mean. . . . The outstanding thing about Ted was that he'd do everything you asked him to do. If you gave him a job to do, he'd do it, exactly as you asked it to be done."

Robert Kennedy had gone through Milton Academy and then Harvard, picking up, in both places, genuine golden boys like David Hackett and Kenneth O'Donnell, football stars, gifted, graceful boys whom he later employed, keeping them beside him for life, forever, until he was the hero. David Hackett now runs the Robert F. Kennedy Memorial Foundation. John Kennedy's football hero was even bigger, in Harvard's football hall of fame, Torbert Macdonald, the roommate who went with him to war, then worked in all his campaigns. On the night of John Kennedy's 1952 Senate victory celebration, Macdonald was giving Kennedy a massage when a friend walked into the headquarters. "I said, 'For Chrissakes, Torbie, is that what you do for him; you're not a masseur.' But that's what happened to them, all of them, the athletes first and then the scholars."

There was no meanness in what happened to these men; it just happened. They became Jack's men or Bobby's men or Teddy's men and they worked in all the campaigns and were taken care of between, and if they crossed over and worked for a different brother than their own, it didn't change anything. So that when Kenny O'Donnell went to work for John Kennedy in the White House, he was still Bobby's man and everyone knew it.

It was a little different with Ted. His grip, or his magic, was less firm. At Harvard, he did befriend a sort of star, fullback-scholar John Culver, a big man like himself with a big laugh and a matching temper, and Culver did then work for him as an aide for a few years. But Culver got out from under, became a Senator from Iowa and is now Ted Kennedy's closest colleague. They walk the halls together, outsized equals, bumping into people and making each other laugh. "All Ted needed back then," Hooten said, "were his brothers. Jack really, Jack was his hero. He didn't need any others."

After Harvard, Ted went to the University of Virginia Law School and made another lifetime pal, another big man who became almost a brother, John Tunney, from whom also later became a Senator. From the first they were inseparable, playful, Tunney more earnest than Kennedy, sometimes his straight man, always, always devoted. About Ted at law school Tunney said: "He just wanted to get through; it required a lot of discipline which did not come easy to Ted at that time. He always thought both of his brothers were extremely intelligent. But he knew he'd have his time. He is the most fatalistic person I know; he knew all his life that he'd eventually be in politics." To which Kennedy *(Continued)*

KENNEDY

Continued

says: "I knew by college that I'd probably run for something. I was pushed, then driven."

□

In 1958, the year before he graduated from law school, Ted Kennedy entered politics and married; and the way he did both seems somehow so obvious. John Kennedy was running for election to the Senate and Ted Kennedy was simply given part of his campaign to run; so was his brother-in-law Stephen Smith — married for two years to Jean Kennedy — a sharp, carefully wry man, fit forever, who would later manage the Kennedy money and various Kennedy campaigns and various Kennedy embarrassments, more vigilant about the Kennedy legend because he had married into it. Ted, large, easy-to-grin, moved around the state with Smith and a couple of his brother's Irishmen in attendance, meeting working people, real working people, for the first time at plant gates and lunch lines; and he was good at it, and he knew it. People meeting him said he was sweeter than his brothers, "a guy who got down closer to the level of the grass," was the way one put it. "To say old Joe groomed one and not the others is pure nonsense; they were all groomed."

That same year Ted Kennedy married the perfect wife for the life he was going to live, or so it seemed. Later, it is hard to remember beginnings. "She was blue-eyed, blond and beautiful and she thought that's what Ted wanted," a friend of Joan Kennedy's said. "He wanted to love her, very much," a friend of his said. "He still does. He's a sentimental jerk; when he comes to Boston he takes her out to dinner."

In between are two whole lives together and three children and who knows what happiness but everyone knows what sadness. People who knew Ted Kennedy then, when he was 26, say he had no intention of marrying until he was 35, that he was too full of energy, but that it happened, that she was the ideal wife, sweet, religious, had done well in school — Rose Kennedy called and checked — and he was obviously going to run sometime soon for some office and needed a wife, as pretty as possible, and she loved him. Not that he calculated all that, that way; he just married. "We all did that at that point," a friend of his said.

He took her home, to them, and he pleased them with her thriftiness. He called her Joansie. He taught her to play tennis as earlier, courting, he had taught her to ski. She was drawn to Jackie Kennedy, the other outsider, Ethel Kennedy, everyone said, being more Kennedy than the Kennedys. "It is hard for anyone to remember Jackie Kennedy then," said a family friend, "but she was a lovely, bubbly little girl, though she was always tougher than Joan. At Hyannis Port, she kept her own schedule; I could always tell you where Jackie was at what hour."

Soon Ted and Joan had their own house in the compound, and children he could bounce. One day, passing Ted with baby daughter, Kara, on his shoulders, John Kennedy quipped: "Relax Ted, you can put her down now. There aren't any photographers around."

By then it was 1961 and Ted Kennedy was at last running, for the Senate seat John Kennedy abandoned when he became President. After the 1960 Presidential campaign, when he had ski-jumped and ridden broncos and roamed around the West chasing minimal votes with Claude Hooten and trying to get his brothers back East on the phone, Ted Kennedy thought about staying in California. "He liked the freedom," Hooten said. "He liked the idea of making it on his own bootstraps. He thought his place in the world might be somewhere else. But he went back. His father said, 'Ted, you've got a base here, family, friends. Why go off someplace and prove yourself for nothing? You're silly.'"

So he went home to Massachusetts to prove himself. He ran for his brother's vacated seat, and his brothers in Washington were not happy. John Kennedy was just President and had just appointed Robert as his Attorney General and they were already sensitive to charges about a Kennedy dynasty. They were also afraid — and these kinds of fears about one another they did not voice directly, it was all elliptical, careful, but their aides voiced them — that Ted would be perceived as less than substantial by the Cambridge intellectuals to whom the Kennedys, all Kennedys, were always trying to prove they were serious. That is exactly what happened. At some function, a Harvard classmate of Ted's came up to him and said, "You're an insult to our class, running for this." And Ted said, "Well, that certainly is a point of view."

But none of that mattered, neither the objections of the brothers nor those of the academics. He had his father. That was all he needed. "The others were taken care of," Kennedy said, "so he was starting over with me." It was, he said, their best time together. "He knew his father was for him 1,000 percent," an aide said. "So if his brothers weren't, it was O.K."

□

Then in December of 1961, when Ted hadn't even announced yet because he was not yet 30, and wouldn't be until the next February, his father had a stroke and could no longer talk, so Ted had to go on without him. His brothers kept out of it, kept a low profile, though his campaign manager, Gerard Doherty, did go to Washington to be interrogated by both of them before he took over. Kennedy assembled a staff, overseen by Smith, his own mix of Irishmen — Doherty and Joey Gargan and small, bright-eyed Eddie Martin, his press secretary, who said "youse guys," and

driver Jack Crimmins, and Harvard men — John Culver and speech-writer Milton Gwirtzman, who had previously worked for John and Robert — his first very own staff who would later be so proud, long after they had left him, when people said Ted was the best Senator of the three brothers as if it were somehow a personal victory. "You partially wanted to close your eyes and think you were the one, you were the candidate," Gwirtzman said, "though you knew you never could be."

In the race, the cheating came up and he survived it. Then there were debates — Robert Kennedy and Theodore Sorensen, John Kennedy's right arm, slipped up from Washington to put Ted through his paces — and in one of them, Ted's opponent, State Attorney General Edward McCormack, lashed out at him with the accusation "If his name was Edward Moore ... the candidacy would be a joke. . . ." — and the reversal of sympathy assured his primary victory. In the general election he ran against a Lodge and the President came home to Boston to vote for his brother. "The President's people wanted to make a big deal out of it," Eddie Martin said, twinkling. "They wanted him to land in the Common in a helicopter, walk across the Common, cast his vote on Beacon Hill, visit his brother's headquarters, then go to Locke-Ober's for a bowl of his favorite lobster stew. We wanted him to land at Logan, cast his vote and leave immediately." And that's what he did. He later said, "I was allowed to come to Boston for 20 minutes."

Ted beat his Lodge, called his father, then went to join his brothers. And for the next 11 months there were three Kennedys in Washington.

For a moment, they were men together, similar and dissimilar, men who thought of themselves — another one of those familial clichés — as men of action. "For all of them," a family friend said, "boredom was a very powerful dynamic; they were always moving, all of them, to stave off loneliness, melancholy, introspection, whatever, playing games, chasing broads, to use their words, skimming books. After he was gone, Jack's secretary, Evelyn Lincoln, said, 'He was always in so much of a hurry, even in relaxation. Bobby was the most introspective, but even he would get up at 4 in the morning and ride a horse. He couldn't sit down and write an

article.'"

That restlessness charged John Kennedy's Presidency, that and his inherited cold-war view of the world and his own dogged sense of history as heroics. He inherited a war and expanded it, and Robert Kennedy was beside him. Whatever the ultimate political assessment of the Kennedys, the shaping of their political instincts began, as always, in the family. The family was too much with them; nobody ever moved them as they moved each other. They had a genetic preference for each other's company and the money to support that preference. That was the importance of Joe Kennedy's money, not the fact that his sons could win elections with it. That was secondary. The theory holds that when Robert Kennedy lost John Kennedy, he was released to care for someone else, he no longer had to repel his brother's enemies so he became softer and more liberal and that when Ted Kennedy, going in softer, the baby, lost Robert Kennedy, he became softer still and more liberal so that in a family of late bloomers and very late liberals, Ted Kennedy is the tenderest and most liberal of his brothers by virtue of survival.

On the late November day of 1963 when his second brother died a hero, Ted Kennedy was 31. He lost one brother and got another one back. He would not be as close to Robert Kennedy as Robert Kennedy had been close to John; it just wouldn't happen. "Bobby and Teddy had a closeness that came from the genes, not really the heart," said a friend of both. "They were terribly close but not instinctively close, if that makes sense. Bobby was hanging around with Schlesinger and McNamara and Teddy with Tunney and Hooten."

In 1964 Ted Kennedy was in a near-fatal airplane crash; he lay on his broken back and wrote a small book about his family. His wife campaigned for his re-election and his staff said she was terrific, and he won big. In the days on the road without him, she had begun to drink and, a very close family friend says, after that, could not stop.

Robert Kennedy ran that year for the Senate from New York, surrounding himself with smart Manhattan lawyers — not for him the down-home Irish-Harvard mix of John Kennedy's staff and Ted's — and when they got to the Senate they laughed at Ted. "They thought him a jerk," one of Ted's aides said.

"Everybody knew they thought that. But Bobby never did. He knew the blood."

The brothers did everything together. They called each other childhood names, Robbie and Eddie. They sat next to each other at hearings, jesting, jabbing each other with all that stiff-upper-lip inverted Irish affection, competing as they had learned to do as boys. If Robert Kennedy made a speech about something, Ted would say to his staff the next day, "Why didn't we have that?" He had his turns. A bill came up before the health subcommittee on which they both sat, and Robert Kennedy, bored by detail, openly bored from the beginning, always about to be President, waded in and couldn't get out. The Republicans came after him requesting an elucidation of a certain paragraph and with his eyes he implored his brother for help. Ted sat. His brother finally scribbled him a note, saying, "Goddamn it, why don't you help me out?" And Ted scratched on the bottom, "You're not in real trouble yet." Robert Kennedy got in deeper and Ted, who'd done his homework, who was beginning always to do his homework, bailed him out. "He didn't want Bobby badly embarrassed," an aide said, "but he wanted to sting him a little bit. He was saying, you're just passing through; I'm the senior Senator here."

But Robert was the premature patriarch. Ted always checked things with him, selecting for himself adjunct issues, not major ones, leaving those for his brother. When Robert Kennedy began speaking against the war in Vietnam — never apologizing for his late brother's escalation there, hoping, believing maybe, that John Kennedy was turning against the war before he died, and, anyway, Kennedys do not revise one another — as Robert turned, Ted went, too, selecting for himself a small, manageable cause, that of the Vietnamese refugees. And he has stayed with it. Similarly, his involvement with health care began with his 1966 sponsorship of neighborhood health centers, a modest beginning, and he stayed with that, with health, until he knew it and felt it in more than a paternalistic way.

In 1965, David Burke, a lean, smart policeman's son, joined Ted's staff and turned it around. With Kennedy he was custodial, protective, prodding and funny, even sometimes about the possibility of Kennedy's being shot. Burke would say to him if they were going to some big function that he was going to carry a large placard with an arrow on it that said, "It's him, not me." Only Burke could say that. He reined him in sometimes, pushed him at others, helping him to hire other smart young men. It was during Burke's stay that Kennedy became known as a staff-supported Senator and he is still sensitive to that charge long after it could be leveled. Burke was too new in 1965 to keep Kennedy from nominating Frank Morrissey, an old Irish pol/pal of his father's, for a Federal judgeship. His father wanted it done and his brother didn't do it so he did it. He knew that everyone was saying that Morrissey was unqualified for the position but he nominated him anyway — with President Johnson's clever approval — Johnson keenly aware, as Carter is now, that the heart of the country still belonged to the Kennedys and probably would in his lifetime no matter what he did — then had to withdraw the nomination on the Senate floor. It was Ted Kennedy who was embarrassed; it was Robert Kennedy who handled the generalship of the defeat. "Back then," an aide said, "big things went to Bobby."

Robert Kennedy's 1968 campaign for the Presidency was 11 weeks long. That's all. The people who were part of it, the people who worked for him, and many of the people who covered him, those people simply loved him, loved his barbed sorrow, some of them so much so that to talk to them about Robert Kennedy in front of their wives is somehow embarrassing. Robert Kennedy had been reluctant to run. A little bit of his reluctance centered around Ted; he relied on Ted but not to the same extent he remembered John Kennedy as relying on him, and he didn't see who would be his Bobby. He was reluctant but then Eugene McCarthy, whose cerebrations and cynicism irritated him, won in the New Hampshire primary and showed Johnson to be vulnerable. The people around Robert Kennedy pressed him. Aide Adam Walinsky, for one, kept writing him notes encouraging him to get in and when Ethel Kennedy saw him, she'd say, "Adam, terrific kid, keep it up," and he did, and Robert Kennedy finally got in and Johnson got out. Ted had been against it. He felt that enough had been done for a while. Then the campaign started, and Ted and Stephen Smith were given matching corner offices in the headquarters, Robert Kennedy smiling when he saw the assignments, unwilling to name a campaign manager, smiling at the smartness of one of his people, and he never did name Ted campaign manager and it didn't matter because the whole thing was over practically before it began.

□

On the early June morning of 1968 when his third brother died a hero, Ted Kennedy was 36. He did what he had to. He gave the eulogy at St. Patrick's Cathedral; more than half of it, the heart of it, simply quotations from the speeches of Robert Kennedy. Then he walked the funeral train, ever polite, ever the politician, shaking hands, thanking people with Robert Kennedy's oldest son, Joseph, in tow, in training, shaking hands with people who could not quite forgive him for not being his brother, or for outliving him. For a long time Ted Kennedy had trouble forgiving himself.

He spent a lot of the summer on his boat at the Cape. People came and went, friends, women, one after the other. He could not share his mourning with his wife. What solace he took, one friend said, was from "the freneticism of booze and sex. Afterward, there were times of great guilt," the friend said. "He was so raw; there was no need for banter; it was all understood. He would go through the necessary performances and then a lot of times it would be — goddamn it, it's me now, it's me for the bullet, it's me for the brass ring."

On Aug. 21, in Worcester, Mass., his first speech since his brother's death, he spoke strongly against the war not as Robert Kennedy had but in simple economic terms. It was tagged the "fallen standard" speech, and it led to a flurry at the '68 convention, where Hubert Humphrey was trying to hold on, hold out against the memory of Robert Kennedy. Mayor Daley of Chicago pressured Ted, and Steve Smith went to Chicago to look around. Ted liked the momentary game; it gave life back. But he knew there was something a little distasteful about it and he had given Humphrey his word, so he backed off.

Then it was fall again and Congress started, and he began the transition from being a staff-supported Senator to being a self-initiator. Nights he would turn up at Robert Kennedy's home, Hickory Hill, at 9:30, 10 o'clock, just to touch base, not always sober, according to friends — all those children now somehow his children. Other nights he organized weekly seminars at his house, as his brothers had done, brain-pickings, and he held himself together, plunging into a fight with Russell Long of Louisiana for the Senate whip position, happy for a moment in combat, happier in victory. But it would not hold. The mourning was not done. In April 1969 he went to Alaska, a trip his brother Robert, as a member of the Indian education subcommittee, had promised to make. He made a speech in Michigan on the way out and sat around in the bar afterward with a couple of reporters, jovial, talking about his weight problem. "It's the sauce, boys," he said. Then it was Alaska for a week, one village after the other, deploring the plight of the Eskimo, meaning it as he means those things, moving hard. On the plane ride back, he began to drink, seriously, and then he got silly, not mean, just loud and silly, barging around, throwing pillows and laughing his large open laugh. The reporters aboard did what they thought they should do, among them Hays Gorey of Time, who informed his editors of what had happened and they said, "Well, if it happens again, but this is the first time, so let's not bother with it." John Lindsay of Newsweek sent a memo to New York, and his editors also decided against running anything. Sylvia Wright of Life simply sent a memo to her editors saying: "He's living by his gut; something bad is going to happen." And it did.

On the night of July 18, 1969, Ted Kennedy left a party and drove a car off a bridge into the water at Chappaquiddick Island and he got out of the car and the woman with him didn't, and there are some who will never forgive him for that. Then he did not report the accident for 10 hours, and there are some who will never forgive him for that. That silence is at the heart of the hostility for Kennedy, in the hearts of some people. It is a silence Kennedy — in shock then, he said in those first hours after — has maintained, through an inquest, through the years, through the investigations of numerous journalists [including Robert Sherrill's thorough 1974 examination of the event in these pages, "Chappaquiddick Plus 5"] and will continue to maintain, as presumably so will the other men at that party, motivated not by money but by fraternity. The unanswered questions, the conflicting stories, the tacit cover-up — in these there are innuendoes that hang on, which in his silence Kennedy must live with as, he said, he lives with what happened to that young woman that night. He has called his actions there "irrational and indefensible and inexplicable."

Joey Gargan, who organized the party, draws maps of the island on cocktail napkins to explain how it happened because it was the worst thing that ever happened to him. He knew of the accident that night and also didn't report it. There was talk that he and his friend, lawyer Paul Markham — the two men Kennedy called on that night for help — might be charged with failure to report it, but other members of the Massachusetts bar argued that reporting the accident without Kennedy's approval would have been a breach of the lawyer-client relationship. For Joey Gargan, it would have been a breach of heart. There was no way he would do that, no way because of who he was and who the Kennedys were and the way he loved Ted Kennedy.

Everybody came to Hyannis Port: Jack's people and Bobby's people and Teddy's people, Sorensen and Schlesinger, McNamara and speech-writer Richard Goodwin; Burke Marshall as well as Culver, who was not happy with the way things were handled, left after a few days and didn't talk to Kennedy for some months; and Gwirtzman and Tunney and Hooten, and Smith, everybody. And, with Smith clearly in charge, they caucused for nearly a week to decide what Kennedy should say. He later said he felt like an object, a nuclear missile in some crisis; "they talked about me as if I weren't there," he said.

There were two schools of thought. The younger men, Teddy's men, with David Burke included and Goodwin a crossover from the older group, favored a simple, flat statement, no muddy waters and mea culpas. Goodwin drafted such a statement, or at least notes, then left. The older men, with Smith firmly in control, Sorensen closely behind and Gwirtzman, their crossover from the younger group, favored the speech that Kennedy finally gave on television Friday night, a week after the accident. They became known as the send-in-your-box-top group because the speech included an emotional supplicating plea to the Massachusetts voters for their support; 100,000 of them responded. At one point it was suggested that as part of the speech he should say, "I will never follow in the path of my brothers; I will never seek the Presidency." But some of his sisters said, no, don't include that. And they didn't.

Whatever the speech — some think it saved his political life while others think it soured it — Chappaquiddick would come up again and again. As one reporter says, "If he runs this year, it's the first question I'll have to ask him, the first question we'll all have to ask him: 'Given the way you behaved at Chappaquiddick, how do you think you would behave as a President in a crisis?' " Kennedy knows that; he knows it will come up, always; he knew that then. Worse, those who were there said that anyone at that party, any one of those men would have taken the rap for him and that possibility must have been stunning. Worse still, he knew, a friend said, that Mary Jo Kopechne did not even really like him, that she thought he was a playboy, that it was Bobby she worked for, Bobby she worshiped; so what he had done, in his mourning for Bobby, he had done to one of Bobby's girls. ■

June 17, 1979

Since Chappaquiddick, Edward Kennedy has publicly emerged as a Senator who is taken seriously by serious people. His Presidential decision, however, will be a personal and family matter. *Second of a two-part series.*

KENNEDY: TIME OF DECISION

The Senator and the President — sparring partners, of late — spoke at a February Judiciary Committee press conference.

By Anne Taylor Fleming

The youngest of the nine children of Joseph and Rose Kennedy, Edward Kennedy lost the last of his three brothers on June 6, 1968. His father was still alive, but not able to speak because of a stroke. So on that June day, at the age of 36, Kennedy became not only a possible President but also the head of a large family emptied of its men, and he was not ready. His mourning for his last brother was a time of turmoil, ending somewhere near midnight in an overturned car, in a pond off Chappaquiddick Island. It was an accident which he did not report for 10 hours and which some still think stands between him and the Presidency — that and a troubled marriage, a wife living away from him trying not to drink. Yet all the polls indicate that Kennedy could beat anyone, Democrat or Republican, that many can forgive him for everything, but something stops him from running. The question is not whether he will or will not ultimately run. The question for now is: What makes Ted Kennedy not run?

Chappaquiddick was, all his friends agree, a turning point for Edward Kennedy. He stopped being frenetic in the way he had been those summers. "He had been through a lot before," one friend said, "personal trials, the plane crash, his brothers' deaths, his father's stroke. But in a funny way, it was the only tragedy totally of his own making, and he felt that. I think it quieted him, it killed all that, the excesses. There are two sides to Teddy, the public and the private. The implication to the public side after Bobby's death was: Now you've inherited the scepter, the mantle, you'd better get your act together. But between that death and Chappaquiddick, the freneticism had intensified with the pressure. And after Chappaquiddick he just was never that frantic again."

He stopped, too, feeling guilty, said another friend, for not having felt as strongly about Robert Kennedy's death as Robert Kennedy himself had felt about John's. Sorrow did not come as easily to him as it did to Robert Kennedy, and for that he was truly sorry.

There was another turning point that summer. A month after Chappaquiddick, Joan Kennedy miscarried for the third time. There would be no more children. "After that I just thought we'd tried hard enough," she said.

Ted Kennedy went back to the Senate. The day he got back, Senate majority leader Mike Mansfield came over and put his arms around him and said, "I'm glad you're back." Claude Hooten, his Harvard pal, sitting in the gallery, had tears in his eyes. It was not until 1971 that Kennedy began to emerge from the shadow of Chappaquiddick, too late to hold on to the Democratic whip seat that he then held, which was taken by Robert Byrd of West Virginia. Ted talked about quitting. "What do you think about family in politics?" he asked Fred Dutton, a Kennedy political overview man, in his typically elliptical way. "I mean, don't you think there is something more meaningful?"

Kennedy stayed and his chief aide David Burke left, knowing that Kennedy couldn't, or wouldn't, be a Presidential candidate in 1972. There is always that chance that if he doesn't go, his best staff people will, and those are not small losses because after days and nights and asides together, it is hard to start with someone new. George McGovern did call Kennedy in the summer of that year at the Cape and asked him to be his running mate. McGovern said it was for the good of the country. Kennedy went out and walked on the beach for 45 minutes. Hooten, who was with him, said: "I just wanted to beat McGovern up; it was so cruel to come at Ted that way, with that line about the good of the country." Kennedy said no, and there are those who think he should have said yes, that he should have run then, with McGovern, and that Chappaquiddick would have come up big and been over with.

With Burke gone, Kennedy brought Eddie Martin back to be his administrative aide, the friend with whom he could banter and semireminisce — Kennedys don't reminisce much about good times or bad — in his family way. "I always viewed Eddie Martin as a piece of the Senator's personality," another aide said. "All Kennedys have their Eddie Martins. They could shoot the bull together, laugh about people back home. He couldn't do that, that way, with the rest of the staff." He hired, too, Paul Kirk, a very savvy, very blue-eyed young man, of whom it is said on the staff, as the Irish expression goes, "he wouldn't tell you if your coat was on fire." He hired, too, Carey Parker, an unassuming man, a Rhodes scholar and a tax specialist; and, for a while, Kennedy was working more or less with contemporaries. He slowly began to speak out on big things, on tax reform and on campaign finance reform and on health, always health, and on gun control. His attempt to ban the shipment of long guns through the mails had begun just three weeks before his second brother's death, and he was gone during the summer after — missing the debate, there for the vote on Sept. 18 — but somehow not there, distant, uninvolved. The legislation passed. He has, in the years since, introduced bills and various amendments to bills to require the registration of all handguns and the licensing of all gun owners. None of them have passed and his staff is readying a new gun-control bill that Kennedy will introduce in about a month.

In health care, it is the same story, a long one with few victories to tell of, and for a man from money, his heart seems in this, too. He introduced the first big national health insurance bill, S.3, in 1971, then again in 1973. When it was obvious it would not pass, he joined Wilbur Mills in the House to sponsor a

Anne Taylor Fleming is a regular contributor to this Magazine.

163

modified plan which included, as his original had not, a $150 deductible clause and made the insurance companies intermediaries, and he lost his labor-union support and was accused of waffling. "There is a ball game going on up there," he said to a downtown Washington group just before the Kennedy-Mills bill was introduced. "And I'm going to be part of that ball game. I don't want to be outside the park listening to the cheers and yells." So there was that, too, the political instinct in there beside the heart, always.

The Kennedy-Mills bill did not pass, and Kennedy reintroduced S.3 again in 1975. It still did not pass, and he has just now, in 1979, unveiled a new plan, much more a man for the long haul than either of his brothers. The most quietly radical legislation he ever introduced, in 1974, a small thing but somehow so radical, would have required all medical students to serve in underserved places, for two years after graduation. It was radical because the shifting of people that way in this country is radical. That did not pass, either.

□

In November 1973, Kennedy's oldest son, Edward Kennedy Jr., then 12, had a leg removed because of cancer. In the months he was in the hospital, his father stayed with him, often sleeping there. A little less than a year later, he called a press conference in Boston and announced that he would not be a candidate for the Presidency in 1976; his family problems would preclude it. His oldest son was not all well; his youngest, Patrick, had severe asthma; his daughter, Kara, was unhappy and running away from home, and his wife, who was O.K. when her son was very sick, nursing him, collapsed afterward, and was sad or whatever else it is that makes people drink. Kennedy himself gave up drinking for a year. "He wanted to clear his conscience," a friend said. "He always felt so guilty about having her into the environment he did. He said she ought to have married a New York banker. He felt terrible about Chappaquiddick but not guilty; he thought he'd done all he could. But he kept feeling guilty for years about Joan."

Eddie Martin left in 1977; Paul Kirk left in 1978. Carey Parker stayed to anchor the staff and a new, a third generation, came in, men younger than Kennedy, so he was finally the father there, too. He runs his current staff of 90 much as his father ran the family, everybody competing and jockeying for the Senator's approval and affection. One morning he had scheduled a health subcommittee hearing and a Judiciary subcommittee hearing at the same time. He called his health people. "How many TV cameras have you got me?" he asked. They said six. Then he called his Judiciary people and said, "They have six cameras over there. How many do you have?" They had eight. So he called the health people back and said, "Sorry, they have two more, I guess I'll go to theirs."

His main young men — his 26-year-old administrative aide, Richard Burke; Dr. Larry Horowitz, 34, his

chief health adviser; Kenneth Feinberg, 32, counsel to the Judiciary Committee, a former prosecutor who is the architect of his crime bill S. 1437, and his press secretary, former reporter Thomas Southwick, 30, and his political operative, Carl Wagner, 33 — these young men want to please him the way he and his brothers wanted to please

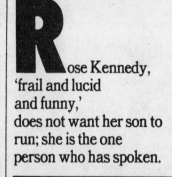

Rose Kennedy, 'frail and lucid and funny,' does not want her son to run; she is the one person who has spoken.

their father, and he lets them try. They bring statistics to him like presents, learning early to understand his half sentences. At staff meetings he sits, the patriarch, slumped, smoking a cigar, peering over his glasses at them, mostly listening, "sucking the information out of them," as one adviser said, "as if they were tulips. They work very hard; there is always the possibility that they might someday be Metternichs or Kissingers."

This is a typical Kennedy beginning to a staff meeting, this one on energy: "But there are, you know ... I talked with someone at the White House ... we have to set up a scenario on, you know, a variety of efforts to impact [his current favorite verb] energy, to see what the hell we can ... we need to take a look at Scandinavia ... I mean, I don't know what they're doing there but we ought to find ... and the Japanese ... I don't know what ... well, what do the rest of you think?" Someone knows

about Scandinavia already, in detail, and tells him, quietly, and then someone else talks about power plants and someone else talks about coal and the meeting goes on for half an hour and everyone leaves knowing exactly what is to be done, what is expected. "Work it out," Kennedy says. It is his favorite phrase. And they do.

They carry with them the conviction that working for a Kennedy, their Kennedy, is like working for no one else on earth. "It's a whole life working for him," Eddie Martin said. "If you left the office before he left, you always felt guilty." That conviction and that guilt makes the staff productive and eager and there is nothing cool and coy about them; they have no pretenses about being laid back, just as "the boss" doesn't. It is a crucial difference between Kennedy and Carter, the kinds of staffs their temperaments have invited — Carter's all low-key, unvivid, and Kennedy's eager, strong in their office chauvinism.

"Kennedy's staff is far more aggressive than anyone else's," said someone who worked for him, "far less sensitive to the ego needs of other staffs and other senators. There was a health hearing in West Virginia and they didn't contact the home senators. They worked for Teddy and he views himself as a national figure. The policies are his; they just give him the mechanisms. Once you're his staff, you're his staff. If you leave, you just always miss him, you miss his office."

In this world there is not much room for women. He hires them but rarely for key positions; some of them leave quite embittered. Kennedy does support women's issues like the equal rights amendment and the right to Medicaid-financed abortions — he is not for abortion per se, but as it is legal, he supports Medicaid coverage so poor women are not denied access — but staff women, professional women who work for him, make him nervous, make him look away. Irish and Catholic and chivalrous, an inveterate door-opener, he simply does not know how to be in charge of them. He cannot banter with them, and he has to be able to banter.

So around him are bright young men. Nights on the road, after long days of shaking hands and touring slums, he buys them beer — often borrowing their money to buy them beer — $5 bills, he likes fives — he talks about hockey scores and makes them laugh, referring to himself in the third person. Questioned on a home trip to Massachusetts by someone at a speech about the drinking age at a time when the state was about to raise it, Kennedy, not wanting to get into it, mumbled something about people not drinking until they reached the age of maturity. Back in the car, pressed on what age that might be, he quipped: "I don't know. I haven't reached it yet." Later he said, "Well, what did you think of the Senator's answer to that question? ..." They knew what question and what answer and they laughed. Some of them make him laugh, but it is not quite a give and take. Too many years and too many reference points and too many Holiday Inns lie between. They know

that if he runs for President, they are not quite the first string; David Burke would probably come back and Paul Kirk and maybe some of the old-timers, some of Jack's men or Bobby's men, though they have made new lives now, as much as possible.

Washington is still full of men who belonged to Bobby — John Kennedy was so long ago, now — men who have moved from cubicle to cubicle, from office to office carrying their Kennedy pictures. One feels them in that town like an undertow. "With the passage of time," said Edwin Guthman, a former press secretary to Robert Kennedy, "the sense of loss just gets worse rather than better. Bobby was the kind of man other men would follow into combat feeling that they could not die."

"Ted Kennedy is a much more conventional figure than his brothers," said a man who worked for them. "There were things about them that were larger than life; they had a literary dimension to them. But I think Teddy is finally no longer intimidated by their memory. Virtually anyone close to Bob would be willing to do anything for Ted, not out of blind loyalty, but out of a strong feeling that he has the capacity. He knows his limitations. There is clearly a level of consciousness that was not there 10 years ago. Pain, drop by drop, does produce a deeper human being."

Of himself Kennedy says: "I like to think I have, over the process of life experiences, developed or refined my own values. I think we Kennedys are sort of constant learners. As life moves forward, our range of interests grows. Am I calmer? I suppose so. I suppose I am. I suppose ... well, that's all."

"Remember," Fred Dutton said, "he's been No. 1 a lot longer than Bob was No. 1. And the father was still alive when his brothers were, so Ted's really the only man since Joe Kennedy to be the head of that family. And that means a lot."

Eunice Shriver, Kennedy's sister, who is sometimes referred to as the head of the family because she is terribly energetic and older, and because Ted Kennedy would not and could not displease her, said of her brother: "He's had a hard time because he's had to do it all on his own. President Kennedy had my father and Joe Jr.; Bobby had all three. But Teddy's been without all of them for a long time."

□

Kennedy has no sense — or none that he speaks of — that his political heritage was anything less than liberal. To questions about the Vietnam War or the Bay of Pigs, he says simply that we must learn from our mistakes, must stay in touch with the opposition leaders of other countries so we don't find ourselves on the wrong side. His father's politics and the way he made his money are history to him. "I'm a believer in the here and now," he says time and again. Once, when Robert Kennedy was running for the Presidency, he had a lunch with an impassioned young woman who worked with Cesar Chavez. "Your father owned Cutty Sark," she said accusingly. "He did not," Kennedy said. "Well then, he owned Schenley's," she said. "I don't know," he said calmly, "but whatever my father owned he doesn't own it anymore." He was not angry. The man who had arranged the lunch said, in the four years he had worked for Kennedy, that that was the only time he heard him speak of his father in that way.

If similarly challenged about his father, Ted Kennedy, one imagines, would have been no less directly defensive, perhaps a bit more flustered. His politics now, in 1979, are a mix of left and right, of social consciousness and old-fashioned capitalism. On one side is his devotion to the nonderegulation of oil and national health insurance. Last month, he unveiled another rendition, combed as he is in the mornings, in the room where his brothers both announced their Presidential candidacies. Carter and Kennedy agree on the need for national health insurance and disagree on the way to effect it, both insisting on exaggerating the differences between their plans as if it were a mini-Presidential referendum. Kennedy's new bill is a comprehensive, $40-billion birth-to-death plan. It is a testimony to his political savvy that he has again included the insurance companies as intermediaries while holding, this time, the support of the unions. Carter's plan, unveiled earlier this month, is quiet by comparison, a $24-billion phase-in plan, pegged to inflation, which would provide catastrophic coverage for medical expenses above $2,500. Kennedy says that's not good enough, that he would "vote no" on Carter's plan. Kennedy disagrees with Carter, too, about the deregulation of oil, favoring, as in health, a controlled market. This, for lack of a better description, is Kennedy's left political flank.

On the other side, mitigating against his big liberal image, are his crime bill and his various deregulation bills. Introduced by Kennedy in early 1976, the crime bill, S. 1437, got through the Judiciary Committee and passed in the Senate. But hard-core liberals jumped on the bill because it seemed to broaden the definition of crime, of conspiracy, and it died in the House committee. It has been modified and Kennedy will reintroduce it in a few weeks. He is also against the deregulation of oil, because he says the oil market is not a free market going in, so it must be controlled. But he favors the reinjection of competition into other industries, such as trucking and the airlines, and he quotes endless figures (his memory for them is prodigious) as to why these moves make financial sense. So there is not a consistent party line; it is sophisticated piecemeal politics, no binding philosophical overview. It is smart politics, Presidential politics, precluding Carter from painting Kennedy as simply an out-of-step big spender, a man going one way as the country lists the other. Kennedy has tailored his liberal tenets as much as possible to the careful, if slightly ungenerous, mood of the country.

Unlike Carter, he is a master of the elegant political gesture. The night before he was going to vote against an arms sale to their countries, he called both the Egyptian and Saudi Arabian ambassadors to explain why. When he went to the funeral of Pope John Paul I, in Rome, dark-suited again, he stopped by the place from which Aldo Moro was kidnapped. Some would say that these are simply gestures, that they do not matter, that Kennedy, if he is still at all weak somewhere, is weak in the area of foreign affairs. His speeches about world matters do tend to be a bit rambling and rhetorical. As a shadow President, he must always have a position on Carter's positions, which he does, praising Carter for his efforts on strategic arms limitation, and for his human rights program, differing from him on the treatment of Taiwan, and voting against Carter's proposed Middle East arms sale.

Kennedy feels bolder on domestic ground. About what goes on at home, about the future of the country, Kennedy can talk easily, in full voice, to an audience of one or 1,000. He talks about there being no magic wand; about the increasing power of special interests and of single-issue constituencies; about how the American people need to be challenged; that they have not, Proposition 13 aside, lost their central compassion; that they need to be challenged the way his brothers challenged them — not mentioning their names, just "my brothers" — knowing that the myth of them grows with his obliqueness. There is shrewdness there, too, in that obliqueness, along with the sadness. There is something politically provocative about the aloneness of this Kennedy, a feeling that if he wants the Presidency, the American people almost somehow owe it to him because of what happened to his brothers.

But does he want it? Kennedy knows he is at the top of the polls; the latest New York Times/CBS News Poll on June 10 showed that 73 percent of Democrats regarded Kennedy favorably, compared with only 43 percent for Carter. "These are very good odds for anyone else," said a sometime adviser, "but there is a part of all Kennedys that wants to be crowned, not elected." Kennedy says he's encouraged by the response of the American people, but he keeps insisting that Carter is his candidate. Kennedy says he met privately with the President on March 21 to indicate his "tentative support" of Carter for the Democratic nomination, yet Kennedy was quoted in The Boston Globe early this month saying that he would think "terribly seriously" about running for President in 1980 if Jimmy Carter withdrew from the race.

Kennedy lights his cigars from matchbooks engraved with a picture of the White House on one side and Jimmy Carter's name on the other. He grins when he does it. But, on the other hand, he is sparring with Carter more and more, about oil, about health, grinning when he does that, too. And Carter is beginning to scrap; he told several Democratic Representatives at a White House dinner that he would "whip" Kennedy if the Senator opposed him for the nomination. "It's been kind of fun," Kennedy said of the sparring. "I suppose there'll be a certain amount more." If it's a game, it might go on too long. Kennedy likes the sparring and maybe it's for real, for final real, and maybe it isn't and maybe, just maybe, he himself doesn't know. He hasn't categorically, absolutely said no, and he knows he hasn't said no, not really. So it's spinning now, and he's letting it spin, the adulation and the longing, and if he's not careful maybe this time he won't stop it in time and he'll have to run, the unions and the blacks and the Carter defectors will somehow make him run. And if he doesn't do it for them this time, there is a chance that having waited for him for so long, they will finally take their love away and give it to someone else.

Walking beside him on a prematurely hot spring day through a Washington park, it is hard to imagine he will run, whatever the odds. The heat makes his sweet cologne very strong; he blushes when asked its name. On the way back to his office he stops in midpark to deliver a fervid, full-voiced defense of one part of his crime bill — as if a full audience had magically appeared — earnest like the kid in class who is pleased to know all the answers. Then he drops back into his customary trot, tilting forward. There is buoyancy, and beneath it what a friend called the knickers syndrome, the baby-boy syndrome, and maybe one doesn't ever stop being that, not totally, not from that family. That part of him, the boyish part, the soft part, does not seem to be heading for the White House. Just everything tells you when you're beside him that it is so. But then there is all the bantering with Carter and the constant visibility and the morning man, in the chambers, all smooth and smart and commanding, his brothers' brother who somehow feels he has to do what they did.

Carter is currently worried that that might be so. "To the extent that Hamilton [Jordan] reflects the President," said someone close to Carter, "it must be Kennedy he's afraid of. Carter has the same desire to be loved; he's just less skillful at getting it. He doesn't engender the same kind of love that Kennedy does — but then he also doesn't engender the same kind of hate."

There is that hate, hate for all Kennedys, hate for this Kennedy, and in it envy and the suspicion, still nurtured by some, that he is the lightweight made heavy only by sorrow. With Teddy, the baby, there was always the fear that he hadn't grown up, that he wouldn't, that the cheating and the women and Chappaquiddick were at the heart of him, of his character, and that the other side, the work he did and the father he was, were not somehow the real part. For a long time it was chic to dismiss this Kennedy, the last, as the least; that has turned now, and if John Kennedy is remembered as the coldest Kennedy, and Robert Kennedy as the toughest, then Ted Kennedy, it seems, is to be remembered as the shrewdest. There is always the bandwagon, the attempt to second-guess history, and the current catch-all judgment of Kennedy is that he is a good Senator, the best of the brothers, and there is to some extent a quiet repentance among the reverse snobs who held him once in contempt. "There isn't a new Ted Kennedy," said someone who has known him for a long time, "just as there wasn't a new Nixon. There isn't a new anybody. But since Chappaquiddick, he has grown and matured. Even those who don't agree with him mostly agree with that. His capacity to absorb information was always evident, and now he has a sureness of hand and judgment. I don't worry about his qualifications to be President, I just worry that people will expect too much of him, too much magic."

Kennedy's days are very long. A typical one begins at 7:30 or 8, when he is briefed at home by one of his aides. He then makes the 25-minute drive to his office at the Senate; sits through two or three sometimes enlivening, often tedious hours of health or Judiciary Committee hearings; has lunch with somebody; has staff meetings; goes across the street to the Senate to vote; is interviewed; shakes hands and sees all the people who can't get to Carter and some who can. Through it all, he is mostly good-humored, kind, rumpling as the day goes on. When he does get angry at someone on his staff, he just goes, flapping his hands in his pockets, glaring over his glasses as his father did, saying, "Jeeesus Christ, how could you do that to me; how could you brief me like that. You've got to do your homework. I can't go out there like that. I'm a leader."

Those on Kennedy's staff, though younger than he, are protective of him, without ever saying so. They try to keep the hate mail away from him; he gets about 2,500 letters a day (10 full of hate, two serious enough to go to the F.B.I. each week), and then there are phone calls, and the numbers

165

of letters and calls go up the minute he goes up in the polls. And on the anniversaries of the assassinations of his brothers, his staff people are quiet because he is quiet, and they don't schedule much.

Nights, he goes home to his big comfortable house in McLean, Va., often driving himself, a little fast, in his slightly beat-up 1971 Le Mans convertible. He lives there with his two sons. Joan Kennedy is in Boston going to school and going to Alcoholics Anonymous meetings. He sees her maybe once a month. His daughter, Kara, is in college near her mother. His oldest son will be going to college next year so Kennedy will be alone with Patrick, who is 15.

He is a man of small ceremonies. When he gets home, he lights the candles beside the fireplace in the den. He gives monosyllabic tours of the house. Everywhere there are pictures of his brothers and of his wife and framed family letters, even in the bathroom. He drinks daiquiris, which he drank with his brothers, and always refers to John Kennedy as "The President."

"I don't consider myself a lonely person," he said. "A lot of it is what you permit yourself. Those kinds of emotions were possible to me many times. It is important to learn, to rediscover things that are important, like taking Patrick sailing. I'm very close to my family, my children, to the cousins and to a very few friends, very few. I enjoy life. I'm challenged by it. I enjoy competition — everyone always wants to beat the Senator. I've had my moments of pause involving personal disappointments. Joan is better, facing up to her problems, so that's hopeful. It's not stopping Jerry Brown that ... I mean, it doesn't have anything to do with that; it has to do with the impact on the people you know, the people one loves, and that family thing was so, I have ... with my children. I'm very close to them."

Whether Ted Kennedy stays married is both a personal and political decision. Can a man, and a Roman Catholic man, run for the Presidency half-married, or divorced, or remarried? Can Kennedy get divorced and remarry in time for 1984 when, certain friends say, he'd rather run anyway, because more people would have forgotten Chappaquiddick and because, if Carter holds on, he will not then be running against an incumbent, a political boldness for which his temperament is not well suited?

These are political questions and they obscure matters of the heart. If there is something left to salvage in the marriage, no one knows; Joan Kennedy says she doesn't know. She says she is proud of herself for moving to Boston, for being on her own. After everything, he was not an easy man to leave. But she did. And, sitting in that house in McLean full of pictures of her when she was young and pretty and unbruised, and of her brothers-in-law when they were young and handsome and in the White House, it is hard to imagine how she could come back. People who see her now say she is still wounded, still fighting hard not to drink, loving him still, and that she is unable to be by his side if he runs; that he cannot run with her trying to mend, cannot divorce either — too Catholic, too traditional, too much his mother's son.

There is always his mother. He visits her at least once a month in Palm Beach or in Hyannis Port, taking his children often. In the winters, she lives in the big house in Florida with one maid. She is frail and lucid and funny and demanding. She sits at one end of the long dining table and her son sits at the other. There is roast beef and wine and afterward, Jell-O and chocolate-chip ice cream. He interrogates his own children and their cousins about their schoolwork as he remembers his father doing with him and his brothers and sisters. He is funny about it but it is just something he has to do. After dinner, his mother's Palm Beach friends come in for a slide show of his trip to China, and he talks to them in his serious senatorial voice. They leave and his kids go off to some disco, and his mother goes to bed early and he goes to bed. The next day he gets up and he takes her to church and they are pleased with each other. She does not want her son to run; she is the one person who has spoken and people say that if she has spoken, he will not run. "The temptation to be the one to kill the third Kennedy brother is just too great," she said, cutting through everything else. "It's ironic, the polls indicate that he could be President ... but then there is that."

"All I want, if someone's going to blow my head off," Kennedy once said to a friend, "I just want one swing at him first; I don't want to get it from behind."

Today he says: "The idea of death doesn't bother me. I love life. I've had a full life. I don't mind about my own children. I think they're all right now. It's the others, my brothers' children; I'm sort of an anchor in their lives."

And then he adds, quietly, "I would be much happier if Carter were successful." ■

June 24, 1979

REAGAN: THE 1980 MODEL

He hasn't announced yet, but he's ahead in the polls and his team is busy repackaging his image and message.

By Adam Clymer

Run, will you please, so I got someone decent to vote for," implored a burly Chicago cop on duty at O'Hare International. He had been watching passengers board a plane when he spotted a handsome, black-haired man with a tanned, deeply lined face, and felt he had to speak to him. A few minutes later, two aircraft mechanics, whose union is as loyal in its way to the Democrats as is the Chicago police force, offered Ronald Reagan the same kind of greeting.

They were cheerful and important reminders of an important political fact — that Mr. Reagan has considerable appeal in the rank and file of groups that have little use for conservative Republicans. In fact, the January New York Times/CBS News Poll — the most thorough recent public poll measuring candidates and their supporters — showed that about as many independents and Democrats as Republicans want him as the Republican nominee.

Neither the policeman nor the mechanics, nor the more stereotypical Reagan supporters among upper-class Republicans, still consumed with such issues as the Panama Canal, need worry about whether he is running. For, after a full life, when another man might be content to ride a horse to the high ground of his 688-acre ranch north of Santa Barbara and enjoy the view of an Exxon drilling rig, Ronald Reagan is setting out on his third campaign for the Republican Presidential nomination, and his first truly well-prepared campaign. This time he may win. A late June Gallup poll gave him an edge over President Carter for the first time — 49 percent for Mr. Reagan as opposed to 45 percent for President Carter.

Indeed, repackaging Ronald Reagan is the key growth industry in American politics today. In Washington and Los Angeles, experienced politicians are plotting the finances, the branch offices and franchise distributorships, the sales pitches and the promotional tours of the *new* candidate Reagan.

But while the design of the new model is generally known, the trim and accessories are yet to be revealed. Mr. Reagan is waiting for the opportune moment to announce his specific policy positions, explaining, "You have to ration your ammunition." On SALT II, for example, he is only "suspicious," not yet a declared opponent. He says he is studying the Senate debate and his briefing papers. For inflation, he finds a "sole" villain, the Federal Government, with its excessive spending and ever-expanding money supply. As for energy, he would not create new Government agencies but would end Government controls on private industry and turn it loose to produce all the energy the country needs.

He dismisses President Carter's July 15 energy speech, noting that Mr. Carter has been in charge of the very Government that he criticized for two and a half years.

Adam Clymer covers national politics for The Times.

Michael Evans/Gamma-Liaison

Ronald Reagan is all smiles amid admiring Iowa women who call themselves "Dutch's Dollies" (after his college nickname, Dutch).

And when news of Cabinet resignations reached him, Mr. Reagan saw them as only "further indications of confusion and indecision." He said, "There really isn't any crisis in the country. There's just a crisis in the White House."

Beyond broad statements and occasional partisan jabs, there is a basic concern in the 1980 Reagan strategy that the candidate must sound, above all, responsible. His head of sales, campaign committee chairman Paul D. Laxalt, the Republican Senator from Nevada, summarized the pitch: "You're not talking about a right-wing nut with horns growing out of his ears. You're talking about a responsible conservative." In fact, as Governor of California, Mr. Reagan did not operate as a knee-jerk conservative. His 1976 Presidential campaign, however, was given to sudden, strange and basically right-wing declarations. For example, Mr. Reagan urged slashing $90 billion from the Federal budget, and then spent several weeks explaining himself. He also advocated sending American troops to Rhodesia to support the white regime, selling the Tennessee Valley Authority and scrapping the strategic concept of limited war.

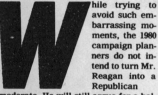

hile trying to avoid such embarrassing moments, the 1980 campaign planners do not intend to turn Mr. Reagan into a Republican moderate. He will still argue for a balanced Federal budget, increased reliance on state and local governments, the elimination of regulations that bother business, and a tougher stance in the world, whether the antagonist be the Soviet Union or the OPEC nations.

What the Reagan planners want is a candidate who can explain, anticipate and avoid self-inflicted wounds, a candidate whose statistics will not be so often subject to challenge, a candidate with more than a slogan for a complicated problem.

It is not that the Reagan planners undervalue these slogans. Mr. Reagan's ability to communicate, according to Mr. Laxalt, was a key reason why he should be President: "He could get on the tube and cut through all the bull on something like energy." What Mr. Reagan's supporters call a talent for synthesis, or blunt summary, his critics describe as oversimplification, as in his summary of the Panama Canal issue in 1976: "We built it. We paid for it. It's ours, and we're going to keep it."

Or consider his current shorthand for the national malaise about this country's standing in the world: "I'm beginning to wonder if the symbol of the United States pretty soon isn't going to be an ambassador with a flag under his arm climbing into the escape helicopter." This kind of summarizing is a singular political talent, and it gets through to voters.

☐

Mr. Reagan comes into this campaign with solid Republican popularity.

In the June Times/CBS Poll, 66 percent of the Republicans said they had a favorable impression of him. Even the Democrats were 34 percent favorable, 46 percent unfavorable — a ratio comparable to Senator Edward M. Kennedy's standing among Republicans. That popularity was steady across the country, even in the once unfriendly Northeast.

In the January survey, Mr. Reagan's supporters strongly favored cuts in Federal spending, convinced that such cuts would surely curb inflation; they were also heavily pro-Taiwan, but not especially anti-Peking. They were a group as in tune with a candidate as can be found in American politics, and thus make up a base of support that such foes as Senator Howard H. Baker Jr., George Bush or John B. Connally will find difficult to shake.

This Reagan group, unimpressed with spending for schools or cities but very much pro-Pentagon, is not a collection of right-wing grouches. They are older and more likely to call themselves conservative than the rest of the population, but not very much so. Like the public as a whole, but unlike the fairly substantial group that somewhat forlornly wants former President Gerald R. Ford for 1980, the Reagan supporters credit President Carter with exceptional honesty and integrity. But Mr. Reagan is also well regarded in this area, which is President Carter's only really strong suit at the moment.

In fact, integrity and loyalty are the first things campaign chairman Laxalt brings up when asked about Mr. Reagan's qualifications for the Presidency. He cites the Californian's refusal to consider the suggestion by Senator Richard S. Schweiker — named, in an imaginative but unsuccessful gamble, as Mr. Reagan's potential 1976 running mate — that the Senator withdraw because he seemed to be costing Mr. Reagan delegates. He remembered Mr. Reagan's saying, in 1976, "We came in here together, Dick, and we're going out of Kansas City together."

According to campaign strategist John P. Sears, Mr. Reagan's knack for leaving things to others is another major asset. This sets the former California Governor apart from the compulsiveness of the other Presidential figures Mr. Sears has worked with (such as Richard Nixon), who "want to do everything themselves." To others who have worked in various Reagan campaigns, this tendency can seem lazy, especially when Mr. Reagan seems to tolerate personal backbiting among campaign aides. If he is not lazy, then he's "the ultimate, laid-back Californian," in the words of former Reagan aide David A. Keene, still an admirer but now running George Bush's campaign.

Mr. Reagan has the intense self-discipline of many successful office seekers, though he doesn't talk about it much or make a display of it. Despite a developed taste for good wine, he drinks it only at dinner parties. Ice cream and even his beloved jellybeans are avoided as he works at holding his weight down. Along with horseback riding, a daily

routine of exercise helps keep his stomach muscles flat.

This discipline has made him give up reading novels. Now it's all briefing papers and articles, but there are still some gaps in his studies. "Maybe he doesn't spend much time thinking about slums in Oakland," conceded Mr. Sears, but he said Mr. Reagan knew enough to rely on those who do.

☐

Technically, Mr. Reagan says he isn't running — yet. He only has an "exploratory committee," so that he can continue making speeches to trade groups for fees as high as $5,000, writing his newspaper column, which King Features sells to more than 200 newspapers, and, especially, making radio broadcasts without being threatened by equal-time rules.

Sometime this summer, that committee will complete its "explorations" and is expected to report back that there is a demand for Mr. Reagan out there. He is expected to announce his candidacy in the fall. Until then, the committee is raising money, setting up state organizations, taking polls and planning advertising.

In a sense, it is carrying forward, but in high gear, the holding operation maintained since 1976 by Citizens for the Republic, a political-action committee Mr. Reagan set up with leftover Presidential campaign funds. Since 1976, this committee has helped Republican candidates with money, polls, advice and Mr. Reagan's stumping. It has also built up a conservative fund-raising list and kept Mr. Reagan in the public eye.

☐

The people who hear and see Ronald Reagan trust him. A wealthy businessman came up at a receiving line and declared, "I believe in you, and I believe in America very much. My dad came over when he was 18, and we made it."

Mr. Reagan obviously enjoys this kind of attention. His conservative view of the country and the world is plainly sincere. Still, with the pleasures of the ranch, well earned after his many years as sportscaster, actor and two-term Governor of the nation's largest state, and after a good run at the Republican Presidential nomination in 1976 — why is he about to go through the ordeal once again?

He smiled as the plane passed through a storm over Indiana and replied, "Well, obviously there's a side to any of us, there certainly is a side to me that would find [retirement] very attractive. I love that ranch. It is my personal Shangri-La. I love doing the work that is required there. But there's a sense of obligation. If I am in a position, due to the circumstances of my experience as Governor of California, in which I can be of help with the problems that are confronting us, then I have to do that. Wasn't it Vandenberg, Senator Vandenberg some years ago, who said, 'Once having shouldered responsibility, it is not easy to put down'?"

In all ways but one, the 1980 Reagan campaign has an easier task than the

1976 effort. This time Mr. Reagan is not challenging an incumbent of his own party, where insurgency is not exactly a tradition. Mr. Reagan now has time to put together a campaign, while four years ago his supporters were spending their energies urging him to run, not planning how to win. He now has a tested team of campaign operatives. And he starts off ahead in the polls. So it is not implausible for his staff to feel that the nomination is Mr. Reagan's unless he (or his staff) somehow blows it. Mr. Reagan is now waiting, letting other candidates try to make it appear he is fading. He is running what is essentially an incumbent's campaign.

But there is one new problem: age. Born Feb. 6, 1911, Mr. Reagan would be older, if inaugurated in 1981, than any other incoming President. After four months in office, Mr. Reagan would be older than Dwight D. Eisenhower was at the end of his second term.

His competitors for the Republican nomination are profiting, a bit, on the age issue. They leave it unmentioned, and just try to look vigorous and youthful. It is most obvious in the otherwise troubled campaign of Representative Philip M. Crane. One erstwhile Reagan backer in New Hampshire, state legislator Irene Shepard, put the reason for backing the 48-year-old Crane bluntly: "Reagan is a lot older than I think he ought to be."

Mr. Reagan insists that his age is no handicap. "The world has changed," he says. "The advances that have been made in every form of health care are such that I don't think you go by numbers anymore with regard to age. It's an individual and his capacity, and I feel fine."

He conceded that a poll, taken last year for Citizens for the Republic, showed that the issue mattered to about 5 percent of the voters when asked about Mr. Reagan. Those numbers may be growing slightly, and the June Times/CBS Poll showed that three-fifths of the people said they would be less likely to vote for someone who would be over 70 while in office; the poll, however, didn't show how much less likely. Mr. Reagan's pollster, Richard B. Wirthlin, president of Decision Making Information in Santa Ana, Calif., called the age numbers not especially significant.

But Mr. Sears, the chief Reagan strategist, conceded age was a problem. Mr. Sears observed that Mr. Reagan's age, like John B. Connally's problem with his indictment (despite the acquittal) for bribery in the milk-fund case, was "a convenient excuse for people not to support you. Often, you know, you can get people to support you if they can't think of a reason not to."

The campaign approach to this problem is to meet the issue head-on, talk about it and hope that people will get bored with it. When Mr. Reagan came to the Capitol to visit with a group of senators last January, Mr. Laxalt raised the issue before anyone else had

Reagan reviews papers at the ranch.

the chance to. Mr. Reagan told the senators he had just had a fine report from his doctor, and some friendly lawmakers chimed in that Russian and Chinese leaders were older than he was.

Another way of dealing with the problem will be to expose Mr. Reagan more through television and less through personal campaigning in 1980. Mr. Reagan, who has done well on television since his days on the General Electric Theater, largely ignored the medium at the beginning of 1976 because of a fear, never borne out, that he would be thought of as just a movie actor. But that March, when he gave his basic speech on television all over North Carolina, it brought him his first primary victory in the 1976 race.

As Mr. Reagan put it, "In New Hampshire, for example, I would do as many as a dozen towns in a day. I like that kind of personal campaigning, and yet at the end of a day you had to say to yourself, 'Well, in all of these 12 or 14 hours I have really spoken to only 1,000 people at most.'" He said television news coverage of such campaigning was little real help — "maybe 30 seconds is just your arrival, and maybe shaking hands with people or an opening remark or something." But by making speeches on television, a candidate has "the same thing that way back in an earlier day was called the political stump, where the country was small enough that a candidate could go out and get up on a stump or a platform or a wagon and address the people, the voters."

□

There will have to be some 12-town days, just to prove that Mr. Reagan isn't too old to handle them. But, however the message is transmitted, there is determination to have more than slogans this time. Mr. Laxalt, who made a reputation for himself last year in Washington for his thoughtful, careful opposition to the Panama Canal treaties, observed that Mr. Reagan is "saying essentially the same things he said for Barry Goldwater in 1964. I'm amazed that he gets away with it."

In 1976 Mr. Reagan usually got away with it, although the Ford campaign attacked his $90 billion plan to cut the Federal budget, and occasionally re-

porters would point out that while he said (as an argument against national health insurance) that the United States had more doctors "in proportion to population than any other nation on earth," the fact was that this country was 14th. Mr. Reagan's talk about a Chicago welfare queen or reduction in the welfare caseload in California did not check out, either.

In this race, however, there is a determination to use most of 1979 to plan positions, so Mr. Reagan's 1980 campaign statements do not invite unnecessary challenges. Several aides speak of having more time for briefing the candidate, and explain that having their campaign headquarters in Los Angeles, near his home, rather than in Washington will make it easier to keep the campaign on track.

The 1980 campaign," Mr. Laxalt said, "will have more issue development in far more detail." To that end, Martin Anderson, a senior fellow on leave from the Hoover Institution at Stanford, moved to Los Angeles in March, to put the research operation together. A veteran of both the 1968 Nixon campaign, with its emphasis on large-scale issues and thorough research, and a 1976 Reagan issues operation so small that it could have been carried around in a shoebox, Mr. Anderson wants to do more, this time, than explain the candidate's statements. "Things have gotten to the point," he said in Los Angeles, "where any serious Presidential candidate has got to have a pretty serious, formulated position on dozens of issues."

Mr. Reagan is a bit defensive about the suggestion that some of his past ideas were half-baked. To this day he insists that the proposal to cut $90 billion in Federal spending was a net plus for his cam-

paign. He does concede that he may have been too specific about it: "It was unnecessary for me to price that out. I wish I hadn't done it." But he still thinks it a good idea to eliminate a wide variety of Federal programs in such areas as education and health, and in his appearances around the country he continues to revive the idea, but without numbers. He also emphasizes an idea that he developed in 1976, amending the tax laws to specify that some percentage of Federal taxes be left in the state where they are collected, to pay for state takeovers of some terminated Federal programs. His idea went over well last month in places from West Orange, N.J., to Roanoke, Va.

Mr. Reagan agrees that he needs more help from experts this time. He cites Social Security as an area where he needs the help of economists to map out a program that would avoid ever higher payroll

taxes, but would "not strike fear into the senior citizens dependent on it."

He says he wants to develop an urban program that would leave taxes where they are collected. He also speculates about developing a national health-insurance plan that would use tax credits, or perhaps tax deductions, to help every American buy some kind of standard private insurance policy — a post-1976 concession to the need for better health coverage.

This latter idea, if anything ever comes of it, would fit in with Mr. Sears's notion of disproving the "legend that conservatives are people who don't care about people." The other main problem for a candidate like Mr. Reagan, as Mr. Sears sees it, is that while a liberal must prove that he is concerned about national defense, a conservative "has to explain that he's a peaceful man."

Intentionally or not, the fact is that Mr. Reagan sounds much less aggressive on foreign affairs now than he did in 1976. His main foreign-policy speech of that year warned that "the enemy at the gates is lean and combat-ready." These days, however, his basic foreign-policy speech emphasizes the relative weakness of the American economy and Presidential indecision about defense priorities. It is a calm, reasoned, and even dull speech. In press conferences, however, his ad-lib comments still ring with alarm, and in private conversation he contends, about defense, "I'm frightened."

□

Even if the Reagan campaign is not yet producing rafts of position papers, it is obviously active in all other areas. Fund-raising, for example, is the chore of Lyn Nofziger, the rumpled punster and long-time Reagan aide who ran Citizens for the Republic until he joined the staff of the exploratory committee last March. He said not long ago that the Citizens' group mailing list was "the second-best conservative money list around," inferior only to that compiled by Richard A. Viguerie, the postmaster general of the far right, who is mailing for Representative Crane. Mr. Nofziger's work is one of the key ways for the Reagan campaign to show progress and back up the impression that it is ahead, perhaps by raising, in 1979, the whole $9 million that would entitle it to full Federal matching money in 1980. One other task is lining up — and announcing — prominent

supporters. That was the point of the Senate luncheon where Mr. Laxalt brought up the age issue. It was the point of Mr. Reagan's campaigning around the country last fall. Mr. Sears said in February that he wanted to "bank" some names of prominent supporters for later unveiling. He seems to have done a bit of this already, with the June appointment of Drew Lewis as his state chairman in Pennsylvania. Mr. Lewis thwarted Mr. Reagan in the Keystone State in 1976 by holding the delegation for President Ford when Mr. Reagan picked Senator Schweiker as his running mate.

This salesmanship has another purpose. Even if Mr. Reagan's appeals cannot turn a Senator Jacob K. Javits, for example, into a supporter, the personal effort and the impression that the repackaged Mr. Reagan might be a winner could make Mr. Javits or others decide it would be futile to oppose him actively.

But ultimately it will be the campaign marketplace that will determine whether he gets nominated. Voters will be measuring him against the competition. Despite Mr. Bush's friends in New Hampshire, Representative Crane's early organizational strength in Florida, Mr. Connally's oratory and new allegiances among businessmen, and Senator Baker's strong potential in the polls and in his role as minority leader during the SALT II debate, most of the managers in those camps acknowledge privately that the nomination is Reagan's to lose.

What Ronald Reagan's campaign team has to sell is a politician who believes that there are simpler answers to most problems than the country has been given to assume, and that most problems require less, not more, from a Government that, in Mr. Reagan's words, "invents miracle cures for which there are no known diseases."

No matter how much research backs up that approach, how much money is raised, how young the candidate looks or what skillful politicians join his committees, there is only one basic question for Republicans in 1980. It is whether they think the repackaged Reagan offers appealing but simplistic solutions with little depth, or whether they believe, with him, in the vision of a hardier, more self-reliant, and simpler America for the 1980's, and want to try to make it happen. ■

CARTER AGONISTES

A struggle rends the soul of
Jimmy Carter, the author says,
and the President 'is unsure
of his faith in himself.'

PRESIDENTIAL PASSAGE: *The physical transformation of Mr. Carter is apparent in these pictures from two press conferences — above, on Nov. 30, 1977, and right,*

By Eugene Kennedy

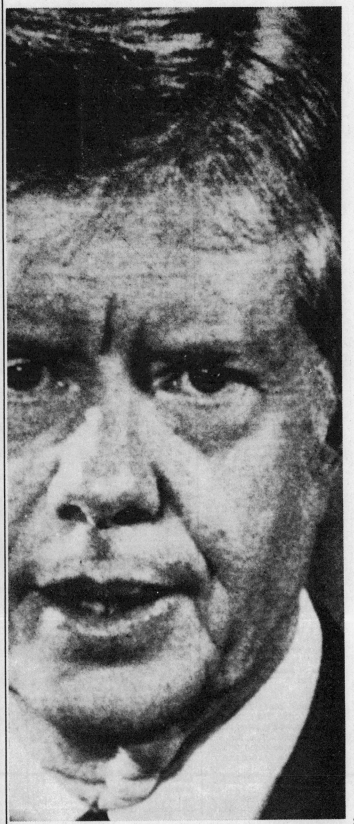

on July 25, when he faced questions about Cabinet firings and his energy speech.

Jimmy Carter isn't so much new as he is different, and that is what worries most of us. Ever since he skipped his Hawaiian vacation and went fishing, apparently concluding that the people he had once courted as decent and good were not so virtuous after all, the President has stunned Americans by lighting down now here and now there in their consciousness like a distressed angel, passing judgment on us all, turning half his Cabinet into salt, making the barren trees of bureaucracy bloom with questionnaires, and speaking solemnly not of blood and sweat but of oil and sin. Apparently the President feels that he has lost something, and that we can find it and give it back to him. Keeping his promise never to tell a lie, Mr. Carter has revealed the bittersweet truth of his own sadness, not just in action, not just in words, but in tones and gestures as well as in the pressure of expectation which he has forced upon us ever since he went up into the mountain.

He has said the malaise is ours, in the tones of a man who knows that it is his own. The peacemaker has made his own uneasiness of soul manifest in the conflict which blew like a wind of uncertain prophecy through the structure and themes of his energy address and in the subsequent Presidential manifestations, some imperial and some evangelical, but each one perplexing. If Americans remain puzzled at their own two-sided reactions to their President — to their agreement with him and confusion about him, to their relief and impatience — they merely reflect the way people always feel when someone delivers more than one message at the same time. Americans may be unsure of their faith in the country, but part of the reason is that Jimmy Carter is clearly unsure of his faith in himself. The nature of that loss of certainty involves a personal crisis in the soul of the President, a crisis that we learn more about every time he appears before us.

The President cannot praise the Norman-Rockwell-painting goodness of the people, and then list the devils of laziness and selfishness that consume them, nor portray government as isolated and then place it above the free market, choosing government as the agency of solution for the energy problem, spending $142 billion to reduce America's dependence on oil imports. The President cannot make a claim to new leadership, and then insist that the populace instruct him in the exercise of that leadership, without generating bewilderment and dismay, the dark-veiled cousins of the ambivalence that haunts Jimmy Carter's templelike soul.

The President has presented himself as the parson who, having studied hard and said his prayers, is scandalized to discover sin in the congregation and, depressed by this realization, mounts the pulpit to blame the people for the

Eugene Kennedy is a professor of psychology at Loyola University, Chicago.

miscarried innocence of his own calling; *they* are to blame for the fact that he has never understood them. Such externalization of personal crisis only makes it clear that he needs them, as, sometimes, a professor emotionally needs his students: the President needs them both to accept the burden of sin and to heal his own bruises.

President Carter had a vision of remembered light and peace, a clearing in which salvation seemed at hand for the good and decent people whose only requirement was a virtuous leader. But the vision has clouded over, and now all will be well if he can only find his way back there, to that place of retreat where a man might be born again.

Congregations put up with a lot from their preachers, but they will not be taxed with sin when the problem is really whether to buy a new boiler for the church or not. They cannot understand, even though they are given to kindly interpretations, a man who makes a moral issue out of a basic economic problem. Nor do they expect that he will place on them the burden not only of paying his salary, but of telling him how to do his job.

Mr. Carter is right; he cannot dissemble. Even in his Sunday night speech, from the moment that the seal of the Presidency gave way to his image, the President was telling us what he was really like — a man who had dutifully assumed his pose, his hair parted in the manner of his brother engineer Herbert Hoover.

Mr. Carter is the parson who must have his picture taken but whose heart isn't in it. Still he does not want to lose his job, despite the problems of being pastor to a sinful congregation, and, flaw in his saintliness, he likes the extra expense money too. Thus the surprisingly slow and unsmiling delivery during his energy speech, a man who was unhappy, going through the motions, confessing that it was all beyond him in a voice that lacked enthusiasm or passion — a voice of curious detachment and weariness that only an act of the will could raise a decibel above melancholy. This is not a convinced believer but a man who, with a certainty he can barely remember, once braced himself against the great wheel of history, only to find that he could not make it budge.

With Mr. Carter one feels invisible presences and absences — of Rosalynn, pitting her will against his hesitation and reluctance, pushing and pulling him out of the trap of his own moods. Or of Brother Billy and Bert Lance, no longer alter egos for the President preoccupied with good and evil, no longer there to afford him a whiff of the indulgence he habitually denied himself. And one senses, not far away, the Georgians who have managed the recent media events, heightening the suspense in order to get the attention of the country — first, with 10 days that did not shake the world, in a doggedly run mountain kingdom that was surely not Camelot; next with Cabinet changes managed with the grace of an execution in a medieval town — these Georgians are surrogate Carter sons who must have wondered at the darkness that in-

(Continued on Page 20)

Carter Agonistes

Continued

vaded their leader's soul.

We have made passage together through the summertime passion of Jimmy Carter, and we have winced together at the anger that was the underside of the President's own sadness, anger that radiated out subtly but surely to all of us. The President's recent prime-time press conference revealed again Mr. Carter's studied effort to climb Mount Rushmore as he underscored the hard-edged words, styling himself a maker of "aggressive" and "competent" decisions. At times he appeared more like a youth reciting memorized words in an oratorical contest, determined to sound as strong as Patrick Henry — but sure in advance that the audience was against him, and angry that the judges were not carried away by his delivery. In his earlier energy speech, somber and humorless, he quoted the remarks of a balanced ticket of Americans, "This from a Southern Governor," "This from a young woman in Pennsylvania," "And this from a young Chicano," "And this from a religious leader." And so on, a black woman mayor, a labor leader — the President had taken notes, which he read with a mixture of rue and curious inner savoring, the tragically satisfied smile of a man who knew someday his parachute would not open despite all his precautions, and who, in our last fleeting glimpse of him as he passed out the plane door, laid the blame for it on all of us.

This portion of the President's speech was as uncomfortable a moment as we have ever shared, for we were all in touch with the President's own troubles, and we could feel the raw edges of his own conflict as he said, ". . . After listening to the American people, I have been reminded again that all the legislatures in the world can't fix what's wrong with America."

Anyone who has truly listened to America knows that its citizens are as sick of self-accusation as they are of war, and they are rightfully wary of the masked hostility in a leader who devotes the first half of a major speech to a doleful repetition of the played-out thesis that we are all guilty for everything that has ever happened, that we have splashed the moon with blood as well as with the boot prints of astronauts, that we collectively gave birth to the howling Manson, to the fiery god of napalm, and to every kind of self-gorging sin, original or not, in the archives of wickedness. It all seems new to the President who, while vaguely complaining about our lack of faith, shows that he has listened more to outmoded social critics than to the people, that he has nourished his disillusionment, not by hearing the voices of Americans who struggle and fall and get up to try to lead good lives anyway, but to the echoes of his own profoundly troubled spirit.

Mr. Carter is mad at us, projecting onto us whatever problems he is experiencing with faith in his own judgment. President Carter may be like the individual, not infrequently observed, who works obsessively all his life, rising through the ranks by a relentless pursuit of duty, only to find when he achieves the office of his ambition that it is different from what he expected, that he cannot fill the position by using the same tactics he used to get there in the first place, that it is too much for him.

In this context, and in the time since the Camp David meetings, Mr. Carter has forced himself to act like a decisive man, and has infected Americans with his own gloom at the very moment they anticipated some optimistic inspiration. The President has spoken words that seemed to be so carefully prepared, but he also invested them with the downbeat music of his own depressed attitude. For now his own immobilized spirit has been laid bare, as he almost plaintively asks the people to show him how to lead them. The enormous passivity of the man, as revealed in the last section of his speech, distresses Americans because it pulls at them, manipulating them like an ethnic mother in a situation comedy who battles for self-esteem and dominance through making other people feel guilty, through making them feel responsible for her slightest twitch or indisposition.

Georgia was never like that, an observer of the new Jimmy Carter might muse, only to add that Georgia must always have been like that. The resignations handed in by the White House staff within a few days, and the distribution of rating forms to be filled out on the professional and personal qualities of government workers, and the turmoil that has followed, is all of a piece with what we experienced during the energy address: a President adjusting to his own difficulties by projecting the problems in all directions around him, by blaming others for the house of woe that he alone inhabits, thus getting at least temporary relief for his abiding anxiety.

The reaction of the money markets, cruelly unsentimental, has also been instructive for those trying to fathom their continuing uncertainty about how to understand this President. The rise of gold and the fall of the dollar have occurred not because of any external event in American experience; these drastic fluctuations have occurred because of an internal event, because of what is going on inside Jimmy Carter, and because narrow-eyed traders in currency make their choices by reading people, as well as by reading charts and graphs. Their judgment, unclouded by mercy, is that they cannot believe in a man who has so much trouble believing in himself.

It has not helped to have so much of Mr. Carter's activity shrouded in secrecy, so much going on behind closed doors and across fields of distance that have become the ironic symbols of the man who wants to spend his "open Presidency" closer to the people. Nothing has heightened America's uneasiness more than the still photographs of the President that were shown on television during his domestic summit, grainy evidence that the man, who was hostage to his own spirit, still existed.

Ordinary persons pick up disturbing messages from such phenomena, but they are almost infinitely patient and, in the long run, fair in their evaluation of the mute and elusive cues by which public figures reveal themselves. But they do not miss them, and they will not accept much more of what they have seen and heard recently from their President, without concluding that the problem is more his than theirs.

□

What is Jimmy Carter trying to tell us? And, equally important, what are his closest aides, especially his chief of staff Hamilton Jordan — whose glance, like that of Bill Sikes in "Oliver Twist," could cause a pet dog to cower — doing to help him and the country? Jimmy Carter is not hiding anything; he has told us publicly and repeatedly that he is more than uncomfortable as a person and as a President. Respect for him, whom everyone judges to be a man of fundamental goodness, and for his office, which Americans still revere, dictates that the truth about his crisis of confidence be neither hidden nor distorted.

This is not a moment for the amateur theorizing of the psychohistorians, nor for clinical guesses made at a distance, even by the most sensitive professionals. Ordinary people suspect that the President has undergone some change in his personality but they rightfully mistrust the easy analyses that describe Carter's conflict as between his pastor's heart and his engineer's mind, between a "good" Jimmy and a "bad" Jimmy, or in a dozen other scenarios, which, like the first plays that showed the influence of Freud, are long on simplistic explanation but short on wise instruction. This is too important a situation to be devoured as delicious speculation, and what we need is the kind of truth the President has always promised us. It is time to take the President seriously as a man and as an officeholder and, as we would with a troubled family member or a close friend, offer support and sympathy along with a demand that he do something about the internal conflicts that he has so publicly advertised.

The Presidency has traditionally been an office of crisis. The difficulties that cannot be resolved at lower levels rise finally to the Oval Office; they are the "buck" that is supposed to stop there. President Carter has had a buck-passing Congress as well as an energy problem so complex that it defies any quick solution. The people understand that, and they can live with such a state of affairs, but they cannot live comfortably with a leader who is reluctant, troubled, and unsure of how to use his power. Genuine leaders do not talk about the nature of leadership, they do not muse on the art of its exercise. Neither do they listen to polls, nor calibrate their responses to the urgings of public-relations men.

True leaders act by surveying the alternatives, and finally choosing one — something which they decide, more by instinct than any other way, is the right thing to do. So Franklin Roosevelt closed the banks, and did dozens of other things to stir the depressed spirit and economy of the country. Many of these moves proved later to be ineffective or wrongheaded, but no one doubts that he rallied the people, that he acted boldly and without regret at a moment of historic challenge. Harry Truman made numerous decisions — including dropping the atom bomb and taking over the railroads — decisions whose merit has been debated for years; but he is recalled as a vigorous and responsible leader. Even Gerald Ford made his own decision to pardon former President Nixon because he felt it was the right thing to do.

Such men were not enslaved by their own personal devils nor, in the final analysis, to the assembled advice of their aides. Indeed, they frequently dumbfounded them, by following their own judgment, drawing together in their decisions the lines that had been set out in their souls over dozens of years of political and executive experience. Genuine leaders act, and they do so because they understand their people, they understand that the people will allow them to make mistakes but they will not tolerate indecisiveness, endless meditation, or an attempt to lay on them a style of moralizing as self-serving and inappropriate as that in a Somerset Maugham minister.

Thomas Aquinas once posed a question about the choice between a proud man and a pusillanimous one. Take the proud man, he advised, because you will be sure that he will at least do something. In the same way, Americans, who know that they are neither the best nor the worst of the world's people, do not want the burden of leading their leader; they are, on the other hand, ready to place their faith in him and to follow him, even at the price of self-denial and sacrifice. Mr. Carter and his aides have shown contempt for the people because they do not really believe in them and, under the guise of being honest, they have revealed the truth about their own troubled selves. Americans sense when things do not fit together, and they hear the President better than he hears them. They know that before he can lay claim to being their leader, he must solve his own problem of self-confidence by facing and healing the conflict that is in him rather than in them. ∎

August 5, 1979

CONFLICT OVER THE CABINET

As Jimmy Carter discovered, battles between Presidents and their Cabinets are inevitable. And, the author adds, desirable.

By Thomas E. Cronin

"Was it worth it to you to cause some destabilization of the dollar, some demoralization of the Federal Government, spreading doubt throughout the land, in order to repudiate much of your Cabinet?" That question, posed by a reporter at President Carter's news conference on July 25, offered a small sampling of the risks faced by Presidents who fire their Cabinet officers.

Andrew Johnson won the prize in that regard. His removal of Secretary of War Edwin M. Stanton led Congress to seek Johnson's impeachment. But even when the impact is less traumatic, a President knows that he is in political trouble.

Americans want their Presidents to be both tough, smart bosses and compassionate, broadminded leaders. A President who removes his own appointees admits that he was less than smart about hiring them to begin with. He also raises doubts about whether he possesses the patience and self-confidence to live with a diversity of views in his Cabinet and to blend them into an effective team. Moreover, the erstwhile Cabinet officer becomes an overnight celebrity, and his potentially damaging comments on the Administration he has been kicked out of are eagerly solicited by the news media.

Thus, even though Jimmy Carter was chiding his Cabinet as far back as the spring of 1978 about its failure to cooperate sufficiently with the White House, he held off any more forceful action. But in the end, as have many Presidents before him, Mr. Carter opted for an open break. The tension between the

Thomas E. Cronin is a professor of political science at the University of Delaware and the author of "The State of the Presidency."

LINCOLN President and Cabinet at the time of the Emancipation Proclamation, in a detail from a painting by Carpenter. Secretary of War Stanton, far left, was removed by Lincoln's successor, Andrew Johnson.

The New York Times

ROOSEVELT On the White House lawn, F.D.R. and Cabinet at what news services called an "impromptu" meeting in the spring of 1936. Beside the President is the Navy Secretary, Claude A. Swanson.

President and a handful of strong-minded men had become too severe, the desire for a Cabinet of unquestioned loyalty had grown too great.

John Adams fired his Secretary of State and forced his Secretary of War to resign. Harry Truman arranged the "resignations" of several of his Cabinet, and so did Richard Nixon. Even Gerald Ford, that most mild-mannered of men, shuffled the Cabinet in the middle of his short Presidency. And Jimmy Carter, who had insisted upon weekly Cabinet meetings, who had bucked the trend of recent Presidencies and publicly committed himself to greater Cabinet involvement in his Administration, found himself following suit.

Such fireworks in the relationship between President and Cabinet tend to lead to public uproar, in the course of which a fundamental fact is often obscured: More often than not, the tension between the two power centers has been a positive element in our history. If that tension is removed — as by the substitution of Cabinet officers who avoid challenging the President — the nation is the poorer for it.

□

"A good Cabinet," wrote the late Harold J. Laski, of the London School of Economics, ought to be a place "where the large outlines of policy can be hammered out . . . where the essential strategy is decided upon, where the President knows he will hear, both in affirmation and in doubt, even in negation, most of what can be said about the direction he proposes to follow." In other words, a body somewhat akin to the British concept of a cabinet.

But the Founding Fathers decided to leave things flexible. Which is to say, they wrote a Constitution that pretty much ignored the issue. As opposed to the British, or parliamentary, version, the American Cabinet meets only at the request of the President and exercises only such authority as he gives it. Votes are not taken unless a President asks for one, and even then — as Abe Lincoln took some pains to point out — the only vote that counts is that of the President.

Traditionally, there are two species of Cabinet in the United States, the mythical and the real. The mythical comes closest to being realized during the postelection euphoria, as one of the first manifestations of the Presidential honeymoon. Thus, Harry Truman proclaimed the Cabinet to be "not merely a collection of executives administering different Government functions. It is a body whose combined judgment the President uses to formulate the fundamental policies of the Administration." But in practice, as one of Mr. Truman's biographers points out, "many important decisions were made by ad hoc groups."

One can actually chart the deterioration of White House-Cabinet relations over the period of an Administration. During the first year, the newly staffed Executive branch, busy recasting the Federal policy agenda, seems to bubble over with new possibilities, daring ideas and imminent breakthroughs. The new Cabinet chieftains are installed and welcomed in a ritual round

The New York Times

COOLIDGE In 1923, Harding's death put Coolidge in the White House; he is shown here with his inherited Cabinet.

of White House ceremonies. The President is ready and eager to lend them his ear. A member of the Kennedy Cabinet once recalled the early days of that Administration and the President's assurance to his Cabinet that they would meet frequently and that individual Cabinet officers should telephone him or Vice President Lyndon Johnson about anything of importance; when in doubt, they should "err on the side of referring too much" on policy matters.

But all too soon, domestic crises and international developments begin to monopolize the Presidential schedule. He has less time for personal contacts with Cabinet officers, and they in turn become wary of being rebuffed and inclined to save their calls for critical issues only. Meanwhile, the President's priorities have become clearer, and his program is becoming fixed; budget ceilings force Cabinet officers to rein in their dreams. It becomes the task of the White House staff to "handle" the Cabinet.

Now the Cabinet has reached what will be its typical condition for the balance of the Administration. Meetings of the Cabinet as a whole are few and far between, and then they are scheduled more for their symbolic value, as public indications of collegial activity, than in hope of any substantive debate. The Presidential interest in his Cabinet officers and their views has reached its nadir.

John Kennedy, for example, at this stage of his Administration, was little available to Cabinet officers, for all his proclaimed good intentions. One of his aides told me about a Cabinet member who persistently called about getting an appointment with the President: "So finally, about the 43d time — after I had told him over and over again that this wasn't the type of problem the President wanted to discuss with Cabinet members — I finally relented and

scheduled an appointment. Immediately after the Secretary had completed his appointment and left, Kennedy stormed into my office and chewed me out for letting him in!" The President's language, according to the aide, was most emphatic.

A similar impatience was expressed by Richard Nixon in his memoirs about the Cabinet meetings of the Eisenhower years: "Most of them were unnecessary and boring." On the other hand, the Cabinet meetings of the Nixon-Ford years got nothing but low grades from Elliot L. Richardson, who wrote that they "ordinarily focused on bland common denominators like the economic outlook, displays of budgetary breakdowns, or reviews of the status of administrative proposals." He added that, "as a special treat, Vice President Spiro T. Agnew would occasionally give us a travelogue."

As time passes, Cabinet members grow bitter about being left out of White House deliberations and decisions. A former Secretary of Commerce once made a joke of his infrequent White House contact, saying that the President "should have warned me that I was being appointed to a secret mission." For most, though, it is no joking matter. Nixon's first Secretary of the Interior, Walter J. Hickel, publicly protested he had had only two or three private meetings with the President in two years; he was fired for his pains. Most Cabinet officers keep their complaints to themselves.

As the ties between a President and his Cabinet officers become strained, the officers are exposed to ever greater pressure from their other constituencies. As Bradley Patterson Jr., a veteran Eisenhower White House aide, pointed out in a study, though the Cabinet Secretary is so closely bound to the President, he is also tied to the Congress: "His or her appointment is sub-

ject to Senate confirmation . . . every penny he or she expends must be appropriated by the Congress . . . every act is subject to oversight by one or more regular or special Congressional committees. . . ." Thus, the members of Congress interested in a Cabinet member's department must be listened to, their views taken into account when major moves are contemplated. Moreover, they constitute a power center that the Cabinet secretary can sometimes count on to run interference for him with the White House staff and the President.

A Cabinet officer has another natural constituency: the individuals and groups and corporations interested in the area of his department. The environmental activists and teachers' unions and tobacco companies, for example, who are among those who seek to influence what the Secretary of Health, Education and Welfare does. Inevitably, after a Cabinet officer is in the job for a year or more, he will be influenced by these special interests and by members of his staff pushing their views; inevitably, to some degree or other, this will bring him into conflict with the President.

When this happens, the White House typically complains that the Cabinet member has "gone native." (Richard Nixon once said of the independent Cabinet member that "rather than running the bureaucracy, the bureaucracy runs him.") Such Cabinet officers tend, for example, to seek extra money for their departments from the President's budget — which led one budget director to complain that Cabinet members are "vice presidents in charge of spending, and as such they are the natural enemies of the President."

Those who fall back on the "going native" explanation for the tension between President and Cabinet are, of course, greatly oversimplifying the phenomenon. A useful instance of the complexity involved is illustrated by the events that led to the resignation of President Ford's well-regarded Secretary of Labor, John T. Dunlop.

In the spring of 1975, Secretary Dunlop had talked the President into backing a long-time union proposal for legislation to broaden picketing rights in the construction industry. In return, union leaders pledged support for another part of the legislative package that Mr. Dunlop was putting together. It sought to create a labor-management panel to preside over the collective bargaining process in the construction industry.

But as the measure moved through Congress, President Ford found himself increasingly under attack by his own party's vocal right wing, at a time when Ronald Reagan was showing strength as a Ford challenger for the Republican Presidential nomination. Moreover, the construction industries, with whom Secretary Dunlop thought he had a deal, refused to support the legislation. Still, the bill was passed by both houses.

Mr. Ford's political advisers urged him to veto the measure. The President publicly acknowledged that he had promised to support the bill, but he ended up vetoing it. And Mr. Dunlop ended up leaving the Cabinet, though Mr. Ford urged him to remain.

What had happened? A White House aide offered this comment: "Dunlop got way out ahead of Ford. He sold Ford on a deal without the extensive debates and evaluations or other considerations that should have taken place on a measure like this. Dunlop really let the President down. He thought he had a deal, but in the end the contractors weren't on board." On the other hand, it could be argued that Mr. Dunlop was no more to blame than Mr. Ford, that he had been caught between his labor constituency and a politically vulnerable President.

Ironically, the view of Cabinet members as advocates, as having "gone native," often proves to be a self-fulfilling prophecy. Starting from that assumption, for example, Presidents tend to use their Cabinets to appease interest groups, to "give" top department posts to spokesmen for appropriate special interests. Moreover, as the President gives Cabinet members less and less of his time, they in turn are forced to use that precious time with him to push for their departments' needs first and let other, larger matters of Presidential interest go by the board.

What is a good Cabinet officer? A year ago, a Carter White House aide gave me this description:

"First, he should be clearly in charge of the department. Everyone in the department should know that he is in charge and know what he and the department are doing. Second, he should be very sensitive to the department's interest groups and have access to them and fully understand their views. But, third, he should also at the same time be able to distinguish the President's interests and political needs from the department's clientele interests. He should be able to say that this is what they want but it is or is not compatible with your interests. Fourth, the effective Cabinet officer is one who can work on most of his Congressional-relations problems without running to the President for help. Finally, he should be able to follow through on Presidential policies — that is, to see that they don't get watered down, or lost in the shuffle."

"Jimmy Carter," former Attorney General Griffin B. Bell commented as the President was midway through his midsummer Cabinet purge,

"had a thing about holding his Cabinet together." Indeed he did, and for a variety of reasons. Like other Presidents before him, he wanted to be viewed as a competent manager who could make the Cabinet process work. And he wanted to create a strong Cabinet — at least, one that was perceived as being strong — to emphasize the openness of his Administration and avoid any charges of an "Imperial Presidency" or "Palace Guard Government" left over from the Nixon era.

President Carter took the Cabinet so seriously that he held more than 60 Cabinet meetings during his first two years in office — something of a record in recent times. The full membership gathered Monday mornings at 9 o'clock for two-hour sessions. And the President pledged that the Cabinet would play an instrumental role in setting policy in his Administration. He announced that "there would never be an instance while I am President when the members of the White House staff dominate or act in a superior position to the members of our Cabinet." He allowed his Cabinet officers an unusually free hand in selecting their subordinates and in setting many of the initial priorities in their departmental policy areas. And Cabinet officers had unusually free access to the President during his first two years in office.

In fact, Jimmy Carter did not really use the Cabinet as a policy-setting collegium. The policy contribution of the Cabinet members came principally through their participation in task forces associated with the National Security Council, the domestic policy staff or other such groups. Some of these units were headed by a Cabinet officer, but others were led by a White House aide. The purpose: to coordinate policy planning and develop policy options, which were passed on to the President for his decision.

However, President Carter went far beyond most of his recent predecessors in emphasizing the Cabinet, and even some of the Cabinet members felt he went overboard. One said that the meetings "were tedious, boring and virtually a waste of time." Another said there were two or three times as many meetings as were really needed.

Most of the Cabinet meetings had no agendas. The President raised issues that were on his mind, then solicited the views of those around the table, both on the subjects he had raised and on other

matters they thought appropriate. Several meetings were criticized as nothing less than an adult version of a grade-school "show-and-tell" session.

Gradually, as the first months of his Administration passed, President Carter began to be disillusioned with his Cabinet. He was thoroughly upset, for example, when Secretary of the Treasury W. Michael Blumenthal spoke out favoring limited tax changes, notwithstanding the Administration's position in favor of more sweeping and comprehensive tax reform.

The White House staff began to put pressure on Mr. Carter to cut back the role he had given the Cabinet. One aide informed me that Carter was "naïve" about how the Government works: "He really believes all this stuff. He is an optimist and an idealist when it comes to working with people. He doesn't think people are evil or capable of disloyalty."

Late in 1977, another aide complained to me. "Frankly," he said, "the powers of the Cabinet secretaries of the first year of this Administration badly hurt us." He insisted that most of the Cabinet members had appointed aides whose loyalty was primarily to those who hired them and whose chief goal was to build up their bosses, even if it was at the expense of President Carter.

By the spring of 1978 the honeymoon was over. The President moved to strengthen the position of his own staff vis-à-vis the Cabinet. Hamilton Jordan, his top political aide, was designated as the de facto chief of staff. Mr. Carter also decried the countless instances when the Cabinet or top department officers spoke out in criticism of each other or of White House policy directives.

Meanwhile the Cabinet members themselves were upset. Secretary of Transportation Brock Adams complained that he was sometimes left out of energy discussions. Secretary Blumenthal was annoyed when the White House appointed Robert Strauss as the chief anti-inflation officer of the Carter Administration, an appointment that Mr. Blumenthal had opposed. Secretary of Health, Education and Welfare Joseph A. Califano Jr. found himself often in the position of being contradicted by the White House on tobacco policy.

A White House lawyer surveyed the first 18 months of Carter's "Cabinet administration" this way: "All of our problems are aggravated. . . .

You can't run the Government that way — from 10 different locations. . . . There needed to be a coordinating place. There had to be a legislative package prepared by the Administration. . . . It took us most of the first year to figure this out and to begin to iron it out. . . ."

In the summer of 1978, while conducting a series of White House staff interviews, I was informed time and again by frustrated White House staffers that the President had given his Cabinet too much freedom. "He should discipline some of them," one man said. "He should fire a few people, just to show them who's in charge." But it was a year before Mr. Carter took that advice.

When the Carter Administration is studied in years to come, the verdict will probably be that the Cabinet failure of his first two years was caused more by Jimmy Carter himself than by the members of his Cabinet. If he had been a stronger leader of his Cabinet, he might have molded them into a positive force in his Administration. If he had been more popular in the country as a whole, higher in the polls, he would have had greater respect from the Cabinet officers — and they would have tried fewer end runs around him in pursuit of their own, particular interests.

What does all this portend for the second-generation Carter Cabinet? He has clearly strengthened his White House staff and sent an unmistakable signal to the Cabinet officers, new and old, that loyalty and team play are what is wanted. In effect, he has told them that their advice is welcome, but that policy is made in the White House. He is demanding as well that, once policy is made, Cabinet officers must loyally and enthusiastically carry it out.

Carter's appointment of Hamilton Jordan as White House chief of staff has raised the specter of another Sherman Adams (Eisenhower) or H. R. Haldeman (Nixon) in command, not only of the White House but of the door to the President's office. President Carter contends that Mr. Jordan's power will be limited to the staff, insisting that "I will be the chief of the Cabinet." But veteran observers of the Presidency have their doubts.

The American Cabinet has had a strange and anomalous history, its position relative to the President, and the Presidential staff, varying often from one Administration to the next. Over recent decades, however, one can discern a

transfer of power from the more public offices of the Cabinet to the relatively hidden offices of the White House staff. And that trend presents the modern President with a major dilemma.

While most important decisions are made by the President and a small number of his personal aides, the basic responsibility for effecting these programs rests with the Cabinet officers and their departments. Thus, there has come to be an increasing separation between policy formulation and policy implementation. And with that has come a separation between the exercise of power and accountability. Presidential aides, accountable only to the President, may have little understanding of what can be effectively implemented through the departments. And since they are not responsible for program implementation, the aides have little incentive to devise an inherently workable policy. At the same time, Cabinet officers have less incentive to implement programs efficiently if they have no real stake in creating them.

Under such circumstances, a President is inevitably hard put to maintain civility, much less camaraderie, in his relationship with Cabinet officers. The pressures, present throughout the history of that relationship, have increased. But though this condition may be difficult on all sides, it should be recognized as being part of our system of Government.

Cabinet-White House conflict is really no more than extreme witness to an important and necessary process. The battle of ideas, the adversary proceedings between President and Cabinet are part of a creative tension that serves the nation well. As one former Cabinet officer has put it, "A good Cabinet member — one who isn't just filling some political niche — can be a very excellent corrective to the White House 'hothouse' staff, who are confined there and are virtually locked up 14 hours a day. The President needs to hear from his Cabinet."

Yes, the outbreak of battle between a President and his Cabinet can temporarily paralyze a Government. But the tension behind that kind of battle is a positive good. It can protect the President from two enemies that have wrought havoc in recent years — isolation and ossification. ∎

BAKER: THE COOL CANDIDATE

By Steven Rattner

A cold Feb. 12 in Springfield, Ill. The nation's highest-ranking elected Republican official is standing in Lincoln's Tomb patiently enduring endless tributes to his party's most famous leader. Finally, after all the dignitaries have been introduced and each has carried a plastic wreath to the front of the sleek marble catafalque, Howard Henry Baker Jr. steps to the microphone. Trying as he does to give every word the ring of instant history, he invokes the memory of Lincoln. "He was a great custodian of the heritage of his forefathers and we are his legatees," says Senator Baker in confident, honeyed tones.

The stress on the legacy is important to Baker, for if he were truly to open up, he would admit that he fancies himself in the Lincoln mold — thoughtful, measured, statesmanlike, forward-looking, bullish on America and practical but with a tinge of idealism. "The greatness of Lincoln was in his view of the future," Baker tells a luncheon audience. Lincoln would be unhappy "about the lack of national resolve." Lincoln would have "a clear view of our future, where we will go next in our experience with the perplexities of government."

How well Howard Baker has managed to live up to the self-image, to follow in the Lincoln tradition, has now become an important national question. The 53-year-old Tennessean — whom millions remember saying on television of Richard Nixon during the 1973 Watergate hearings: "What did he know and when did he know it?" — is

Steven Rattner is a reporter in the Washington bureau of The Times.

now trying to parlay that prominence into the Presidency, the elusive goal of at least nine Republicans at present. And he is in better shape than most; he runs second in nearly every poll to Ronald Reagan and was further strengthened by the decision of Gerald Ford not to run. Like Mr. Ford, Senator Baker is perceived as a moderate, and for that reason, he also holds promise of being able to challenge Jimmy Carter on the President's own Southern turf. For those reasons, he is the Republican whom White House aides and many other Democrats say they fear the most.

□

Senator Baker's character and his ambitions of climbing to lofty plateaus have, in fact, carried him far, already. They have earned him a reputation as a skilled centrist, one who extols the political system that allows him to thrive by building consensuses. And they have brought him respect — even in cynical Washington — as a decent and kind man, unfailingly polite, who prides himself on rarely losing his temper. His brown eyes twinkle, and his off-center grin breaks out readily, though not overbearingly. Like most politicians, he mixes well, remembers faces and names, yet, at the same time, he is an intensely private man who displays very little of his inner self even to close aides.

Howard Baker's qualities of statesmanship may be flawed, however, by what seems to many to be a certain emotionlessness, by a sense that he lacks a certain spark within himself. That image is enhanced by his apparent ambivalence, of seeming not quite to know what he wants, trying to be visionary without having a vision. "It's almost as if Baker wants to be President because everybody has told him to run but he doesn't quite know what he wants to do when he gets there," says one sympathetic onlooker.

And this lack of direction — perhaps it is, rather, a natural preference for trying to reconcile the opposing directions of others — can create difficulties for a political candidate who must at some point galvanize voters who stand strongly on one side of a question or another. Further, contradictions often appear in the Senator's own politics, such as in his desire simultaneously to run for President and work closely with Jimmy Carter. On energy, he announced in July that "this is not a time to nickel-and-dime" the President to death. At another point, he presented himself at the White House, offering to negotiate on SALT with the President and to furnish him with Republican head counts along the way. "There's a strong ambivalence in the way he handles being minority leader and a Presidential candidate," says one close observer. That image is enhanced by a tendency to speak of the process rather than the substance, and, at times, he dislikes being pressed on specifics. But he frequently extricates himself with his wry, folksy humor: "My father always told me never to speak more clearly than you think."

Although Mr. Baker is not pleased at the thought — perhaps because he realizes the adverse political implications — he and his candidacy will inevitably invite comparisons with the current occupant of the White House. On the most superficial level, both are Southerners, liberal for their backgrounds. Both are short in stature and sensitive about it. Both are certainly among the most intelligent candidates in their respective parties, with enough raw intelligence and faith in their own judgment to drive them to glibness. Both studied engineering. And perhaps because of that, both — despite attempts to be idealists — tend to be pragmatists, making essentially moderate decisions at the margin, after carefully weighing pros and cons. That's good public policy by the best academic doctrines but probably not the best politics, for like the current Administration, a Baker Presidency would most likely be judged passionless.

Why is Howard Baker running for President, anyway? "The Presidency is the only place in America where one man or one woman can have maximum leverage to change America, to formulate policies the way I want them, the way you want them," he says late one evening at home. "Teddy Roosevelt said that it's a bully pulpit and what he was talking about as I interpret it was that it's the only place where you can identify problems and state them to the country and persuade people to follow you."

"I think I have a clear vision of where I want to lead the country and I think I can state it more clearly than the others can," he offers a bit later. That vision, to Baker's thinking, is twofold: A need "to reiterate our belief in free enterprise, free market, free country." And a parallel requirement that "we signal that we intend to remain strong to protect our vital interests and that we're going to stop giving up a little at a time, whether it's in the Caribbean, whether it's in Africa, whether it's in the Middle East."

☐

Senator Baker has to some extent separated himself from issues that at present draw close public attention, by becoming closely identified with foreign policy. Strategic arms limitation, for example, is a complex, arcane, bewildering subject which, from the start, failed to excite public interest in anything but the most superficial way. The old adage that the political interests of the people begin at home was proved anew as energy shortages and an economic downturn relegated SALT to a clearly subordinate place. "If we don't get those gas lines straightened out, the Russians could be in the next county before anybody would worry about them," he remarked ruefully last summer. Increasingly, he has tried to broaden his profile.

Nonetheless, Senator Baker seemed to be creating a foreign-affairs image when he made headlines last winter at a Republican conference at Easton, Md. He called for an end to the bipartisan foreign policy of the post-World War II era of Senator Arthur Vandenberg, the powerful chairman of the Foreign Relations Committee who persuaded isolationist Republicans to support the Truman doctrine of extensive involvement in world affairs during the beginnings of the cold war.

"Some say it is vaguely dishonest to speak of *(Continued)*

Baker and his wife, Joy, campaigning recently in Concord, N.H.

John Earle/Alamo

foreign policy in a political context," he said later at a Boston press conference. "The statement that politics ends at the water's edge was right in Arthur Vandenberg's time. But I am right in our time. Foreign policy is too important not to be involved in politics." But since Easton, Senator Baker has not explained precisely what that meant. What specific changes would take place in the way the Senate deals with foreign policy? How would a political context make the SALT debate different from the Panama Canal treaties question a year earlier? Given the new task of making himself into a Presidential candidate, the Senator's call for a new partisanship in foreign policy may well suggest a concern with being labeled a middle-of-the-roader, particularly at a time when Ronald Reagan, an unequivocal conservative, is the G.O.P.'s leading Presidential contender. In any case, Senator Jacob K. Javits, Republican of New York, who calls Howard Baker a "very bright fellow," says the declaration for a new partisanship "hasn't changed a thing."

While Mr. Baker was recalling Arthur Vandenberg, he was also watching as SALT negotiations came to a head in Geneva and visualizing for himself the commanding role in the Senate debate that

would be his due as the minority leader and as a member of the Foreign Relations Committee.

As he conceived the prospect, SALT would be a replay of the Panama Canal treaties consideration, the great foreign-policy debate of 1978 in which the Tennessean emerged with a giant boost to his public image. The Baker formula was simple: strike a responsible pose at the outset, seek the counsel of experts on both sides and then make a decision of conscience. To be sure, in the case of Panama, he delayed and delayed and delayed in his decision making. But in the end, he bucked his party, voted for the treaty and, by all accounts, provided a crucial margin of victory to an Administration agonizing over each of the 67 votes it needed. Again, in a case like this, he makes use of his humor to blunt the attack of those angered by his stand: "I'm tired of doing the right thing," he said. By the time he began campaigning for President, his defense of his vote had shifted somewhat, however. "If there's going to be a standoff with Russia, it's not Panama where the real danger is," he told one New Hampshire audience.

As the SALT debate unfolded, Senator Baker took much the same pose. He discussed his philosophy of arms limitation but remained silent on the all-important question of which way he would vote. Finally, on June 27, he declared for the opposition. The declaration, whether one of conscience or not, probably did not enhance his statesman's image. However, to most strategists, it helped, for a time at least, his prospects for winning the Republican nomination. (For obvious reasons, he has been a leader in the thus-far unsuccessful fight to have the debate on the treaty on the Senate floor televised.)

To Senator Baker, the issue became the relative military strengths of the forces to be allowed the two countries. With his ordered, precise mind, he mastered the technicalities and what he saw as the technical defects of the proposed treaty. At the top of his list was the provision allowing the Soviet Union to keep 308 SS-18 missiles, mammoth weapons with firepower equal to the entire United States arsenal. "That is not equality," he insisted.

Maybe not, but as the debate moved into committee hearings, it was also clear that firepower limits were not the central issue. For while the treaty would provide only slight confinement for the Soviet defenses, it would provide even less for our own; virtually no American defense expenditure that could now be envisoned would be barred by the treaty. What emerged instead was that the treaty became a basis for a much broader debate over what America's own strategic posture was to be — a debate Senator Baker never really got into. Command of the opposition then fell mainly to a second group, led by Sam Nunn, Democrat of Georgia, who wanted to use the treaty really to win new commitments on American defense spending. If the treaty was approved without such commitments, Senator Nunn and his allies contended, then the Government would be lulled into a sense of false strategic security and might swing the fiscal ax on the defense budget.

Meanwhile, as the summer hearings unfolded, Senator Baker's strategy failed to score any points, and, indeed, a number of observers believed it contributed to the perception that the Administration's position was gaining. In July, he became all but invisible at the hearings. It is perhaps indicative of Howard Baker's reputation for switching sides that Senator Javits and others still believe he can be brought to support the treaty.

Whether the treaty ever gets through the Senate has become more problematic because of the Soviet troops in Cuba, but if fails it will probably not be because of Senator Baker. And on the Cuban matter, he struck a typically Bakeresque pose — noncommittal or statesmanlike, take your choice. The President's handling of the Cuban situation was "unacceptable," he pronounced , while demurring on

specific advice as to how the problem should be handled. "Other than listing the options, I think no President should telegraph in advance what he would do in that situation; and frankly, I don't think a candidate for President should say what he would do in a future confrontation with the Soviet Union."

□

With Presidential campaigns starting earlier and earlier, Senator Baker has had steadily to move up his private timetable for the race in every respect except for the formal announcement, which he had long set for Nov. 1. In February, he took his first brisk campaign - style canter through five key states.

In New Hampshire, a full year before the primary, he found the Republican state chairman, Congressman, two National Committeemen and a host of local party officials all already committed to other candidates — such as George Bush, who had begun campaigning there a full six months earlier. "We've found a great deal of the organization is committed but a lot isn't," said Senator Baker after that visit. "I can't lay claim to the power structure, but a lot of people have volunteered to help."

A realization of the intensity of the early campaigning pushed him ahead at a torrid pace. For one thing, he was forced to stop equivocating about whether he would run. Although he deferred the formal announcement in the hope of concluding a successful fight against SALT first, he virtually committed himself to a race in April: "I'm running hard, I have a good organization, and I plan to win," he said then. He has maintained a grueling schedule since, traveling nearly every week from Thursday afternoon until late Monday, often covering several thousand miles in the

course of a long weekend. During the Congressional recess in August, he visited 25 states, made 100 speeches and traveled some 30,000 miles.

Like those of most candidates, Senator Baker's campaign speeches have slipped into a comfortable patter, dwelling heavily on platitudes about America, foreign policy and the philosophy of politics. In keeping with his general political manner, he and his strategists have tried to garb him in a cloak of learned responsibility. "He will evolve a style where he will be able to be critical, but constructively so, and Presidential," says Senator Richard Lugar of Indiana, Mr. Baker's campaign chairman. "He will have partisan stints, critical but delivered in a way different from some backward haranguing politician."

Like Senator Baker himself, most of the speeches are subdued, unemotional. And like Senator Baker himself, the speeches walk a fine line between the thoughtful and the clichéd. "The Republic is in a new time; we're between eras," he begins for one New Hampshire audience. He declares politics to be honorable. "Party politics is, in a way, the fourth department of government," he tells another group. He exhorts his audience to talk about what it has to offer and then, moving to foreign policy, announces that "I do not suggest or intend to see that foreign policy is politicized, but we dare not deny it access to the political forum."

In his campaigning, his talk of domestic issues has been largely subordinate and his views on those matters — particularly economic questions — hew closely to conservative dogma. He extols the "free-market system" and believes a principal question is "how to reduce the size of the Federal Government." His general tenets include deregulation and unshackling the nation's productive abilities; his specific positions range from a call for a constitutional convention to mandate a balanced budget, to a moratorium on new Federal regulations, to more tax incentives for business.

Generally, Senator Baker's Republican audiences feel he makes good sense. But their measured reactions match their speaker's delivery. They listen carefully, applaud politely, are rarely roused, rarely offended. "I don't think people are looking for epic challenges on the right and left," Mr. Baker explained one morning over breakfast. "They are looking for a clearly stated and easily understood set of positions and ideas. My speaking technique doesn't lend itself to staccato applause. It lends itself to careful silence while the audience tries out my ideas and issues." He digresses as he pokes at a grapefruit, then adds, "Besides, I don't especially enjoy making speeches."

What he does enjoy are small groups in which his relaxed manner works nicely. But there is little indication that even in the smallest groups much of the real Howard Baker appears. A conversation with him leaves one with the sense that there is more to him than lofty ideas and country humor — but not more than a vague sense. "This is a man who doesn't show a great deal of his inner self," said one close associate. "He is a private, secretive man who does not like to be intruded on." He wants to go home at the end of the day and read or watch television — he watches a lot of television — or go down to his darkroom — he's an amateur photographer. "He is a public person only when he has to be. I don't think anyone knows very much about his private life." One Republican who recently saw him at the funeral of the mother of his wife, Joy, said that he looked like a man "who wanted publicly to show grief but was under such control that he couldn't let go."

☐

Senator Baker's attempts to put together an effective campaign staff have met with mixed results. One of the few overt Baker flaws has been a long-evident inability to recruit top-notch people to work for him; he has tended to rely on altogether amiable but undistinguished aides. William D. Rogers, a lawyer and former Assistant Secretary of State whom Senator Baker enlisted to argue the pro position during the Panama debate, describes Mr. Baker's Senate staff as "not in their own right people of wisdom, stature or toughness." Thus far in his campaign organizing, he has asembled a curious blend of mostly young and inexperienced aides, at the head of whom is Don Sundquist, a 43-year-old Tennessee resident whose previous political experience was limited to heading the national Young Republicans.

Although aides maintain that Senator Baker is changing, he has in the past disliked delegating responsibility, preferring to have decisions made by the person whose judgment he trusted most — himself.

If there is doubt about whether a politician can simultaneously hold office and run for President, the Baker experience only adds to the doubt. As he has accelerated his campaign, the amount of time he has spent at his Senate job in Washington has simultaneously decelerated. Anxious for the public exposure the job brings him, he has thus far declined to surrender, even temporarily, his post as minority leader, as he previously said he would do. But for all intents and purposes, his deputy, Senator Ted Stevens of Alaska, has become the functioning leader of the Republican contingent.

Even on his three or so days a week in Washington, much of Mr. Baker's time has often been consumed by campaign matters, by his talking with Senator Lugar and his issues chairman, Senator John C. Danforth of Missouri, or meeting with trade groups and possible contributors, or courting out-of-town politicians who might be persuaded to sign. But he has managed to wage some legislative battles, including the fight to authorize the completion of the nearly finished Tellico Dam. Senator Baker supports the dam — which will destroy the original habitat of the endangered snail darter, in riffles of the Little Tennessee River near Knoxville — despite a generally proenvironmental stance. He backs tighter regulations, for example, on strip mining, in Tennessee and elsewhere, but says the economic advantages of the Tellico Dam outweight the need to save the endangered snail darter. "I am

UPI

Baker in Moscow with Brezhnev and American Ambassador Toon.

Baker's Republican audiences feel he makes good sense. They listen carefully, applaud politely, but are rarely roused. 'People are looking for clearly stated, easily understood positions. . . . My technique doesn't lend itself to staccato applause. It lends itself to careful silence while the audience tries out my ideas.'

locked in mortal combat with the lowly snail darter for what now seems an eternity, and I am embarrassed to say that I have taken a sound thrashing," he said at one point, but added, "So far." And the "so far" was significant, because he recouped, as he can do, and went on to win the the Senate's approval.

As he raced between campaign and Senate, Howard Baker paused recently to endure a brief meeting in his Capitol office with a Tennessee delegation anxious for the appointment of a black judge. "There ought to be a black district judge in Memphis," he told the group. With his reverence for politics and political history, he delighted in stopping "right here, at this table where L.B.J. and Dirksen [Senator Baker's late father-in-law, Everett M. Dirksen, one of his predecessors as Senate Republican leader] hammered out the Civil Rights Act of 1964." Then he halted by the window overlooking the Mall. "This is the second best view in Washington," he told one of the guests. Then to the obvious next question, he smiled and said: "The one from the Oval Office."

☐

As tenure on Capitol Hill grows, legislators often find that somehow their commitment to home communities softens, not just in the sense of constituent service, but also in the sense of the place they regard as home. In the case of Senator Baker — again, much like President Carter — no such doubt seems present. Home is definitely Huntsville, Tenn. (pop. 387), where the millionaire Senator has a 10-acre estate — an elegant main house, a guest cottage originally built for the Dirksens and a variety of other facilities, including Scott County's first tennis court.

Huntsville is East Tennessee, a Republican stronghold that never really seceded from the Union during the Civil War. The region is a part of the Appalachian Mountain range, and its residents are as rugged as the terrain. Though somewhat set apart from their neighbors by perhaps broader ambitions and commensurate success, the Bakers have lived in the backwoods country since the 18th century; the family's political involvement is almost as lengthy. Howard Henry's grandfather was a judge; his grandmother, still alive at 100, was the first woman sheriff of the state.

Senator Baker is pleased to note frequently that the Second Congressional District, of which Huntsville is a part, is the most Republican district in the county, having been occupied continuously by the G.O.P. since it was formed in 1858. Howard Baker Sr. held the seat from 1951 until he died in 1964 and was succeeded by his second wife, Irene Bailey Baker, the Senator's stepmother.

Young Howard Henry, as he is known back home, watched much of his father's early political career, which included spectacularly unsuccessful races for Governor in 1938 and and United States Senator in 1940. Today, Senator Baker calls his father his most important in-

spiration. "My father had a profound impact on me in a way I don't think I could ever explain," he said simply in a recent talk.

While his father's political carrer was floundering, Howard Baker was sent at 14 to the McCallie School in Chattanooga, and later, World War II being underway, he entered the Navy's V-12 program in engineering at Tulane University in New Orleans. He finished up in time to do brief service on a PT boat in the Pacific. After the war, he returned to school, at

'The Presidency is the only place in America where one man or one woman can have maximum leverage to change America — to formulate policies the way I want them — the way you want them.'

the University of Tennessee, where he earned a law degree (being elected president of the student body in the process). To this day, Senator Baker maintains that he meant to continue his engineering studies — but that on registration day the line for law school was shorter.

In 1951, he met and married the charming, direct, occasionally unpredictable Joy Dirksen, whose father was just then embarking on his Senate career. Anxious to leave politics, Joy Baker has nonetheless remained close by her husband's side, campaigning with him as required, despite a personal distate for it. She never campaigns alone, however, or gives speeches, or even offers opinions to strangers on issues. She has preferred to remain in the background, serving a few charitable interests and their two children, Darek, 26, a banker, and Cynthia, 23, who is on leave from her job as a news producer to campaign for her father.

Her one involutionary moment in the public eye came in 1976; it was whispered during her husband's private campaign for the Vice-Presidential

nomination that she was an alcoholic. As it turned out, her father's death in 1969 and the strains of political life had, indeed, pushed her to alcohol but, calling on a reserve of toughness, she quit quickly. "People always ask about it," she said in an interview recently. "Frankly, I'm getting kind of tired of it; it's kind of like whipping a dead horse." Baker's attitude toward Joy is revealing of himself. "I cannot remember his ever calling her down," said one staff member who has traveled extensively with the Bakers. Instead, when Joy becomes a bit unwound or flustered by the pressures, Baker becomes self-consciously cool, affecting an easy manner and calling his wife "Miss Joy."

For their first dozen years together, the Bakers' work was confined almost exclusively to Knoxville, Scott County and environs as he earned local renown as an excellent criminal lawyer with the firm founded in 1885 by his grandfather. And in the best country-lawyer fashion, he dabbled in bank, coal and real-estate deals that made millions.

His entry into electoral politics did not occur until 1964, when, presented with the possibility of an easy race for his father's House seat, he chose instead a difficult race for the Senate made vacant by Estes Kefauver's death. The race was made still more difficult than it would have been for him anyway by the presence at the head of the ticket of Barry Goldwater, who proposed, among other things, selling the Tennessee Valley Authority.

Perhaps inexperience was to blame for his defeat, but the campaign was marked as well by a more extreme Baker, who warned that "our Government is headed toward state socialism." He proposed ending Government foreign-aid and education programs and said that he would have voted against the Civil Rights Act of 1964. "I was a young man in his first race, which was a tumultuous campaign," he recalled recently.

In 1966, he ran for a full term in the Kefauver seat, and, striking a more moderate pose, won support from a variety of traditionally Democratic groups, including 35 percent of the state's blacks, a very high percentage for a Republican. The result was the first popularly elected Republican Senator in the state's history and a new Baker theme — the need to broaden the underpopulated Republican Party by attracting blacks, who have continued generally to support him in unusual numbers for a Republican.

As a Senator — he has easily won re-election twice — he has talked a moderate line but walked a more conservative one. Ask Senator Baker what his proudest legislative achievements are and he will talk about things like his work on environmental issues — including the sensitive issue in Tennessee of the anti-stripmining measures — and revenue-sharing legislation, genuine achievements to be sure. And despite his 1964 statements, he has been largely supportive of efforts in the civil-rights area, particularly on behalf of open housing, where he was one of only three Southern Senators to vote for the 1968 legislation. He even voted for the equal-rights amendment.

A look at Senator Baker's overall voting record shows a man with an average rating in the past five years of 15 percent from the Americans for Democratic Action and a far more robust 67 percent from the Americans for Constitutional Action. Over the years, he has, for example, opposed busing and gun control, supported the antiballistic missile and the supersonic transport, opposed labor-supported common situs picketing and supported natural-gas deregulation and the breeder reactor.

Mr. Baker's Senate service has been made notable by his tenure as the ranking Republican on the Select Committee on Presidential Campaign Activities, better known as the Watergate Committee. Named to the committee by then minority leader Hugh Scott as punishment for having challenged Mr Scott for the leadership post, Senator Baker to all appearances escaped with his reputation greatly enhanced, despite extraordinarily difficult circumstances.

The Baker tendency toward weighty, serious statements served him well, leaving most Americans with the impression that he had nothing at heart but the interests of the country. His lawyerlike questioning kept him from alienating the Nixon supporters within his own party while allowing him to build his credibility with the majority that thought Nixon guilty. Senator Baker's instincts to stay close to the center stood him in good stead.

He was subsequently attacked by some who felt that he was maneuvering off-camera to protect the White House. However, Terry Lenzner, former deputy counsel to the committee, said he "never saw any attempt to undermine our investigation because it was a Republican President." For his part, Senator Baker says "nobody on that committee was more diligent in doing what I announced that I was going to do — that was, to follow the facts wherever they led. Remember, I started those hearings convinced Nixon was innocent. And it wasn't many days though until I realized that we had a quite different situation on our hands and I made that statement — that was going to be painful; it was going to be difficult but Republicans were going to follow the inquiry wherever it led us and let the chips fall where they may. And I think we did that with a vengeance."

Indeed, it was the Watergate service that provided Senator Baker with his chief platform from which to seek the Presidency. His Watergate image is what allows him to speak with credibility of the problems of the past — an allusion he never fails to make — and the hopes for the future. "Vietnam and Watergate are behind us and America's greatness is ahead of us," he tells one campaign audience. Perhaps most important, Watergate gave the studious, sometimes sardonic, occasionally ponderous Howard Baker national television exposure that has helped him to a high public recognition — which in turn allows him to avoid some of the town-by-town slogging to which lesser-knowns are fated. And it brings him 200 speaking invitations per week. "Aren't you Senator Baker?" a patron in a Springfield, Ill., motel coffee shop asked him one morning. "I watched you during Watergate and you're a much more handsome person than you were on television."

It is certainly generally true that Howard Baker's career in the Senate as minority leader has been most successful when he has forged compromises out of numerous factions. Fairness and reason have earned him the respect of his colleagues, several of whom are active in his campaign. "He's a very wise person," said John Chafee of Rhode Island, who is helping with New England. "I've seen him reach decisions on tough problems and I admire the way he goes about it." But Senator Baker's detractors contend that while he can often bring disparate Republicans together, he has difficulty leading his flock. "If you take the view that one function of the loyal opposition is to oppose, to take strong positions, then Baker's been unsuccessful," said one critic, who nonetheless calls the Senator a "complete gentleman."

(Pettiness is also a quality Senator Baker eschews; a few months ago, he called for the establishment of a special prosecutor to investigate President Carter's peanut business — as was later done — and the White House, in a moment of pique, pulled back an Air Force plane promised for a Baker trip to China. Senator Baker declined to make a public issue of the matter, preferring quietly to cancel his trip.)

□

For all of Senator Baker's ambivalences, he also possesses a deep reservoir of determination, a trait required of any serious Presidential candidate. He is no superman, and the strains of his current double life as minority leader and Presidential candidate often show in a tiredness in his eyes and around the corners of his mouth. But his easygoing manner belies an intense self-control, a tenacity that led him to run two unsuccessful races for minority leader and that would prevent any personal strain from forcing him from the race easily.

One evening, early in his campaign, he and his wife and a small entourage were in Brooklyn for a county Republican dinner in a garish catering hall. The heavily ethnic audience was as foreign to the Bakers as the setting. (The Senator said later that they felt as if they were on a set for the "The Godfather.") As he sat preparing in his careful fashion for his speech, an aide arrived from Washington, unexpected by Joy Baker.

"Who's minding the store?" asked Joy of no one in particular. The Senator looked up over his reading glasses. "This is the store, honey bunch," he said in his at-once-soft, at-once-hard way. ■

November 4, 1979

CONNALLY: COMING ON TOUGH

As the 1980 Presidential campaign heats up, the former Governor of Texas
hopes to tough-talk his way into the White House.

By Steven Brill

At 6 o'clock on a fall evening in New York, John B. Connally strides into a cocktail reception at the Board Room, a Park Avenue businessmen's club. He looks taller than he is, his 6-foot-2-inch frame enlarged by perfectly tailored suits and a magnetic self-confidence. All eyes turn to him.

The men in the Board Room, each of whom has paid or pledged $1,000 to the Connally Presidential campaign, are not accustomed to playing supporting roles: John Loeb, of Shearson Loeb Rhoades; John deButts, former chairman of A.T.&T.; William Spencer, president of Citibank; John Whitehead of Goldman, Sachs; I. W. Burnham of Drexel Burnham Lambert; Robert Belfer of Belco Petroleum; William Ward Foshay, the managing partner of the Wall Street law firm of Sullivan & Cromwell.

John Connally, the 62-year-old former Secretary of the Treasury and Governor of Texas, moves easily among them, breezily patting backs and acknowledging support, the businessman's rock star surrounded by his groupies.

In his introduction, Spencer of Citibank, the chairman of Connally's New York fund-raising drive, suggests the appeal of the big, cocky Texan. "I was over in Europe recently and people there were taking pity on me because of the country I come from," he said. "I don't like to be pitied. I want a country that's on top again."

John Connally does not inspire pity. His off-the-cuff speaking style is riveting, his posture erect, his cadence perfect. As he proceeds through a catalogue of gloom — from Communist advances to burgeoning bureaucracy to setbacks in international trade — he laces his points with data on trade rates, barrels of crude, tons of edible grain, all of which check out to the last digit when researched later.

Steven Brill is the editor of The American Lawyer magazine and author of "The Teamsters."

Connally, a dynamic speaker on the stump, talks to New Hampshire businessmen.

His finale reads like a civics lecture, but leaves his audience nodding in approval: "I'm not asking you to help *me* as a favor. It's *your* business. It's *your* country. It's *your* currency. It's *your* privileges, *your* freedom. It's *your* leadership — the leadership of *your* country, which reflects on every one of you." When he is finished, they move in on him. One partner in an investment-banking house sputters that he has a partner who knows so-and-so who served on a board with Connally, much the way a high-school kid would remind a politician that he'd shaken his hand at the last Memorial Day parade. "Oh, yes," Connally says absently, "he's a terrific guy. Nice to see you."

For Connally, the chemistry in the Board Room is just right. The man billed as this season's big-business favorite is not overwhelmed by the biggest of big businessmen. "There's an amazing amount of mediocrity among even the top, top businessmen. I know, I've seen them. Ninety percent are mediocre, pompous, narrow, stupid Neanderthals," he had said offhandedly three nights before during a breezy conversation aboard a rented jet flying from Texas to New Hampshire, entertaining his wife, Idanell, a campaign aide and a reporter with discourse on subjects ranging from ranching to private jets to Houston law firms, to campaign issues to the merits of his rivals ("They just put speeches in front of Reagan to read." "Bush sat on his butt in those appointed jobs").

☐

In 1976, a country shocked by Water-

gate elected an admitted antipolitics outsider who promised with a tear in his eye only that he would never lie to us. Four years later, a country browbeaten by OPEC and a falling dollar may well turn to the smart, tough Texan, the self-proclaimed wheeler-dealer who, he tells us, can throw national self-doubts to the winds and get things done.

It is an appeal that is not confined to places like the Board Room. In a New York Times/CBS poll taken last June, 66 percent of those responding admitted that they would be more likely to vote for "someone who would step on some toes and bend some rules to get things done."

John Connally is perceived as a doer. "I guess I'll vote for him," a Fort Worth cabdriver who voted for John Kennedy, Lyndon Johnson, Richard Nixon and Jimmy Carter said a week before Connally's New York trip. "He may be a crook, you know, with all that milk-fund stuff. But he knows how to operate. He's smart, and he won't take any bull from anybody. We've been so worried about what people around the world think of us, or about pleasing minorities or social planners or some ethics teacher in some college somewhere that we're going under. He'll stop that."

Front-runners for 'a nomination the fall before the election have been jinxed in recent years, and the 1980 Republican front-runner, Ronald Reagan, could easily suffer the fate of George Romney, Edmund S. Muskie and Henry M. Jackson. If that happens, Connally could emerge as the Republican to beat — an irresistible choice for those frustrated by a well-meaning but paralyzed Presidency, those who want a country that swaggers again the way this candidate does.

Behind the swagger is one of the fastest, most experienced minds in national politics. "Yes, he has a supreme self-confidence," says Julian Read, the Austin lobbyist and public-relations man and former Lyndon Johnson aide who is Connally's communications director. "But it's because he has no blind spots. Talk about business. He's been there. He's been in the board room [as a board member of Dr Pepper, Pan American World Airways, Greyhound, Falconbridge Nickel and Justin Industries, the nation's largest brick maker]. He's been to the banking syndicates for financing and he's been a banker [as a board member of First City Bancorporation of Texas and a part-owner of two smaller banks]. He's been a farmer and a rancher [he built from scratch a highly profitable 8,000-acre ranch in Floresville, Texas]. Talk about government. He's been a Governor [of Texas from 1963 to 1969]. Talk about law. He's been there. He runs one of the nation's largest law firms [285-lawyer Vinson & Elkins in Houston, where he is still a senior partner, drawing $465,000 a year]. He's been a Cabinet member [Secretary of the Treasury, 1971 to June 1972]. Talk about the military. He's been there [as a decorated World War II naval officer and as John Kennedy's Secretary of the Navy in 1961]. Talk about Congress. He's been on the Hill. Why, he answered and typed Congressional mail himself as Lyndon Johnson's Congressional secretary [in 1939]. So, you see, he's not awed by any of it.

"Now," Read continues, with a voice and mannerisms so like Connally's that it seems he has been totally consumed by the boss, "we're going into the World Series, and this guy's got it. Because you name it and he's been there."

According to Connally, that "World Series" will pit him against Senator Edward M. Kennedy. "It'll be a great election, a classic test between a liberal and conservative philosophy," he promises. It might also be a classic gutter fight. At a recent private session with a small group of businessmen, Connally was asked where he stood on nuclear power. "I'll tell you one thing," he shot back. "More people died at Chappaquiddick than at Three Mile Island."

Connally does not pull any punches. In 1968, he was one of the most virulent attackers of critics of the Vietnam War, frequently using words like "treason" and "appeasement" to characterize positions taken by Senators Robert Kennedy, George McGovern and Eugene McCarthy.

As a Democratic candidate for Governor in 1962, he branded his Republican opponent, Jack Cox, a "renegade, turncoat opportunist" because Cox had once been a Democrat but had switched parties. Earlier in the gubernatorial campaign, Connally had charged that his liberal Democratic opponent, Don Yarborough, was getting outside help from "the most vicious, dangerous undercover political campaign in Texas history"; that "the architect of this diabolical action is the left-wing radical Americans for Democratic Action" and that the plan "is being carried out with fanatical dedication by [an] elite corps of henchmen sent into Texas to divert this attack."

Yet for now, Connally is taking the high road, hitting away on major issues. His campaign statements reflect his self-assurance. They are scrubbed of the qualifiers that mark most political rhetoric. In a speech that has drawn the most flak so far, he said in October that the United States should push Israel to withdraw from all the territories gained in the 1967 war in return for an Arab guarantee of Israeli security and uninterrupted oil supplies to the West.

He is just as unambiguous about nuclear power: *(Continued on Page 181*

"My staff tells me to be careful of what I say about nuclear power up here," he told an Exchange Club luncheon in Manchester, N.H. "O.K. I'll be real careful of what I say. I'm for it, period."

He is also for the neutron bomb, the MX missile, the B-1 bomber. He wants a $50 billion to $100 billion tax cut in the next two years. He's against SALT. He'd "remove" the much-respected and much-feared Stanley Sporkin as Director of Enforcement at the Securities and Exchange Commission, because "he has clearly abused his authority." He's against attacking the oil companies, because "when you have a problem, you should work with the people who know how to solve it." He's against gun control. He's against expanded health insurance. He's for the Equal Rights Amendment, but he's against protecting gays from job discrimination. More generally, he likes the Burger Supreme Court's decisions limiting defendants' rights (but notes that this is "not a particularly outstanding court intellectually"). He thinks Nixon was "a particularly good President in foreign policy" and "not a bad" President on domestic policy, but he was "ill at ease with people." He wants to start "individual investment accounts" for small investors to roll over investments in stocks and bonds, with all tax payments deferred. And, above all, he is the champion of free enterprise and the enemy of the "parasitic, burgeoning government bureaucracy."

Yet with Connally there lurks the question of how much conviction there is behind the tough language. The candidate who says he'll fight government bureaucracy and set business free to restore the economy is the Governor who lectured the Texas Farm Bureau in November 1962 that it "must learn to live with governmental controls, like it or not" and the Treasury Secretary who initiated wage and price controls in 1971. The candidate who says that the Federal deficit must be eliminated is the Treasury Secretary who told 1,000 businessmen assembled by the United States Chamber of Commerce in January 1972 that there was "no alternative" to the Federal Government's $40 billion deficit, and that "for restoring vitality to this economy, you ought to be applauding [the deficit]."

Supporters insist that Connally's switch from the Democratic to the Republican Party in 1973 (at the height of the party's Watergate crisis, they

note) and his switches on issues only prove that he's a practical problem solver who takes on issues one by one, day by day. But a look behind the rhetoric of Connally's latest positions does not always reveal the thinking of a problem solver.

For example, a recent speech to the International Association of Chiefs of Police in Dallas was devoted to an attack on the civil-rights suit filed this past summer by the Justice Department under then Attorney General Griffin Bell against the Philadelphia Police Department, charging the department with systematically allowing brutality. Because he's so much better speaking extemporaneously, Connally gives speeches from a prepared text only when they are major pronouncements on key issues. This speech had a text, and his staff called it a major law-and-order policy statement.

Connally called the Philadelphia suit an "ideological inquisition." With a declaration that local officials do not "need to be dictated and lectured to by bureaucrats who profess to be the self-anointed judge and jury of all that transpires in American life," it was a real crowd-pleaser.

A week later, Connally was asked if he really thought Attorney General Bell was conducting an "ideological inquisition."

"I doubt that he has that much influence in that decision," he replied. "No Cabinet officer has any great amount of leeway in these decisions any more than a President does. These departments start preparing these papers and by the time it gets up to an attorney general his options are all gone." (When the Philadelphia suit was filed, Attorney General Bell went out of his way to stress his personal involvement in the decision to bring it.)

"Well, then, what about you? Would that suit have been filed under your Presidency?"

"I don't know. I don't know enough about it."

Realizing that he had just backed away from the thrust of his 30-minute speech, Connally paused and added, no, he didn't think he "would have indicted the Mayor and the city and everybody." (The Justice Department suit is a civil suit, not a criminal indictment.)

Another tough pronouncement, this one repeated almost everywhere, has to do with foreign trade. "I'd tell the Japanese," he says, "that unless they're prepared to open markets for more American products they'd better be prepared

to sit on the docks of Yokohama in their Toyotas watching their SONY televisions, because they aren't going to ship them here."

"That kind of thing sounds terrific," says one foreign-trade expert who worked in Connally's Treasury Department. "But you know if it were that easy, we'd have done it a long time ago." During Connally's tenure as Treasury Secretary (February 1971 to June 1972), when his abrasive style reportedly offended trade ministers from other countries, the American balance of trade deteriorated from a surplus of $2.6 billion in 1970 to a deficit of $2.2 billion in 1971, to a deficit of $6.7 billion in 1972.

Other than his brief stint as Treasury Secretary and a briefer (11 months) period as Secretary of the Navy, where he was known as a fast study but had little time to establish himself as much else, Connally's six years as Governor of Texas, beginning in 1963, offer the only preview of how he might perform in the White House. Though he played politics in areas such as awarding architectural contracts to campaign supporters, he ran an honest government. Generally, he appointed well-qualified people to top posts and hired more women and minorities for these jobs than any Texas Governor before him. He aggressively built the Texas university system and established the state as a magnet for tourism and new industrial development.

In other areas, however, he was less effective. His top priority, a reform of the state constitution and a reorganization of the government, never passed the Legislature. "He was resented by the legislators because he was so aloof, so he had trouble passing some things," says Waggoner Carr, a Connally ally who was State Attorney General during Connally's governorship. "He just didn't want to sit with them and talk, the way L.B.J. would have. His style is more to confront people. He'd get bored otherwise."

Attentiveness to administrative detail does not necessarily make a good President. But conservative voters in 1980 might also be disappointed by Governor Connally's record as a budget-cutter. A real leader can cut waste, Connally promises. "I'd just sit down with Congressional leaders and the heads of all the budget subcommittees and tell them they've got to do it, and they'd do it," he explains. In 1962, Connally said he would cut the Texas budget by 10 percent, but he offered no specifics. In the six years from his first budget to his last (covering the fiscal year ending in August 1969), Connally increased the state budget 63 percent. In the Governor's own office, expenditures increased 121 percent. Money spent on welfare and social programs, now a favorite Connally whipping boy, increased 88 percent. And, by his own account, he recommended tax increases "to every single session of the Legislature."

One old friend of Connally's, who is now a close adviser, explains these budget numbers by describing Connally as a Texas version of Nelson Rockefeller — a compulsive builder and doer. "John likes to do things," he says. "He likes to get his hands on everything. That's why you had wage and price controls at Treasury. That's why you had his university building program. That's the one thing consistent about him. If all those business people supporting him think he'll let them and the economy alone, they're going to be real surprised."

Critics charge that another consistency in Connally's career is his disregard for minorities and the poor. Even Lyndon Johnson is said to have frequently remarked that his old Congressional aide and protégé had no compassion for the poor, despite the poverty of his own early years. (A staple of the current Connally campaign speech is a description of his childhood as one of seven children, whose father "never finished the eighth grade. . . . We lived in a house on a dirt road without electricity or indoor plumbing. . . . We hauled our own corn into town on a wagon. We slaughtered our own hogs and made our own lye soap. We rendered our own lard. I studied my last year in high school by a kerosene lantern.")

On Aug. 19, 1962, Connally went out of his way to assure a Dallas newspaper that, though he favored better educational opportunities for blacks, he wasn't talking about "social mingling" of blacks and whites. But in that interview and elsewhere Connally supported equal rights and never succumbed to the open racism that marked much of Southern political rhetoric through the early 60's. On the other hand, when it comes to government activism to combat racism or poverty, Connally has always been a hard-liner — and one who seems to relish telling minorities, the poor, or unions that he's not going to help them. In the 1962 campaign he vowed to fight federally subsidized medical care for the aged and Federal aid to education.

As Governor, Connally tried to hamstring Federal voting rights and antipoverty efforts

in Texas, consistently resisting Justice Department efforts to monitor voting and often rejecting Federal antipoverty programs. In 1964, while running for re-election, he appeared at the Texas A.F.L.-C.I.O. convention simply to tell the unionists that he would promise them nothing.

In 1966, when thousands of Mexican-American migrant workers were near the end of a summer-long march from the border to the state capitol in Austin to protest their exemption from minimum-wage laws, Connally got into his car and drove to their gathering outside town to tell them in person that he didn't support their cause and that he wouldn't meet them. "That showed his total arrogance," says state A.F.L.-C.I.O. leader Harry Hubbard, who was then the union's legislative coordinator. "He enjoyed rubbing their noses in it."

"I got to feeling sorry for these poor Mexicans who'd walked from the valley . . . and thought I'd give them the courtesy of telling them myself that I didn't support them and wasn't going to meet with them," Connally maintains. "And, yes, I didn't walk or ride a burro. I drove a Lincoln Continental there. That was the Governor's car. . . . I'm not Jerry Brown."

"I think because John made it so well on his own, he doesn't respect people who need a union or a civil-rights group or a special poverty program to make it," one old business associate says. "You know, some people who come up from poverty never forget those they left behind. But others only have contempt for those who lost the race. It works both ways."

Connally would contend that his rise from poverty worked the right way for him. When asked how he would try to persuade someone sitting on a stoop in the South Bronx to

vote for him, he said confidently, "I'd go in there and explain to him that I understand his plight and his poverty better than any man in this race on either side."

Connally's plans for helping him don't show it. Perhaps it is a sign of our times, or of the party whose nomination he seeks, but the ghettos and the poor are issues Connally has clearly not prepared himself to address. The only remedy he offers for poverty is a vague plan for "reshaping the entire educational curriculum of the high schools. . . . We ought to start with a massive vocational-training program."

"Federally funded?"

"Well, it could be. . . . But there would be no strings attached."

If there would be no strings, how would he "reshape" the curriculum? "I think I can persuade the school people of this country to change the curriculum."

A bit later, he adds what, it seems, he really wanted to say all along. Asked if something like Proposition 13 helps the middle class at the expense of the poor, he raises his voice and says, "The middle class are *the* forgotten people of this country and I *am* going to speak for them."

"Aren't the poor forgotten?"

"No, they certainly are not. They've gotten every kind of program under the sun devised to help them. You've got to remember there are over 97 million working people in this country. They're the ones getting it in the neck."

Connally often reminds his audiences that most of those 97 million working people are paid by private business, that "business is the lifeblood of our country, the source of our greatness." The admiration is mutual. Connally's financial-disclosure reports show contributions from the business community unmatched by any

primary candidate in history. He bristles at press reports labeling him the candidate of *big* business, stressing that all kinds of business people, big and little, support him. He's right. With the election law now limiting contributions to $1,000 per person, there are no longer any fat cats to bankroll a campaign. Fund raising is a semi-grass-roots affair now, and Connally's success has been phenomenal.

In all, from the beginning of 1979, when he declared himself a candidate, up to Sept. 30, Connally has received a total of 6,512 contributions for $4.3 million, compared with 2,227 contributions of $2.8 million for Ronald Reagan and 4,020 contributions of $2.4 million for Carter. Such is the level of the Connally people's confidence that campaign manager Edward Mahe, a veteran Republican strategist, notes that "finance is our big disappointment so far. We'd planned to raise $6 million by now. But we're improving. We need to expand our base, and we are." In the first two quarters of 1979, 52 percent of the contributions were from Texans. In the third quarter, 40.5 percent were.

Political Action Committees (P.A.C.'s) formed by business groups are also contributing heavily to Connally. One of the most interesting contributions came from a P.A.C. called "Committee for Good Government." Actually, this is the P.A.C. of Temple Eastex Incorporated/Time Inc., which is a subsidiary of Time Inc. Temple Eastex chairman Arthur Temple, who owns the controlling interest in Time Inc., and his wife each personally contributed $1,000 to Connally, and Time Inc. chairman Andrew Heiskell weighed in with $200. Heiskell and other key Time executives also contributed to the P.A.C. (Heiskell gave $2,500), which in turn

gave Connally another $2,000, that P.A.C.'s largest contribution to any candidate.

Many who know Connally well say that his campaign's success in raising money is as much the result of his personality as it is of his political appeal. Sid W. Richardson, the multimillionaire oil man who hired Connally as an in-house lawyer and all-purpose staff man in 1952, was obsessed with money, and friends of Connally's say that Richardson gave the young University of Texas law graduate the same zest for the bottom line.

Connally seems to factor much of life into dollars and cents. Campaigning one morning at a New Hampshire shopping mall, he set out to buy something at most of the stores he visited — $11.50 in candy at The Fudge Factory; $96 for boots at Stride-Rite; $1.19 in vitamin C tablets at the drugstore, and so on. (His staff, usually sensitive to his image as the big-business candidate, didn't see to it that he had something smaller than a $100 bill to make the first purchase.)

When asked the night before that shopping spree-campaign stroll about three people who might work in his White House, Connally responded like an actuary. Yes, Richard Keeton, a top Texas litigator, might come to the White House staff. "He's made enough money to go into government." Then there was Frank van Court, a shrewd young lawyer who worked for Connally at Vinson & Elkins and left to run the American business ventures of Ghaith R. Pharaon, a Connally client who is one of the major Saudi Arabian investors in the United States. "No, Frank won't be ready," Connally said. "He hasn't made enough money yet." Asked 10 minutes later if Edward Bennett Williams might make a good Attorney General, he replied, "Oh, yes. He's made enough money to go into government."

Connally's own career has been a series of careful jumps from public life to private money-making and back to public life. Even while on the campaign trail he is still an active senior partner at Vinson & Elkins, where he has been a major business getter and is a highly regarded corporate counselor. He still meets with clients, including Saudi investor Pharaon, and often arranges for partnership meetings to be held on the rare days he is in Houston so that he can attend.

Skeptics view Connally's public-private-sector shuttle as a parlay of public positions into private profit. Connally argues that it's just the opposite — "I believe in public service" — that the way to be an independent public servant

The Connally's entertained Elizabeth Taylor Warner at a charity tennis fund-raiser in 1978.

is not to "need the job," financially or emotionally.

"John could have won again, and again, and again," says his wife (whom everyone calls Nellie), seated beside him on the campaign jet. While President Carter encourages Rosalynn to campaign on her own, John Connally insists that Nellie, whom he met when she was the 1938 Sweetheart at the University of Texas, travel with him. The two are extremely close and both are devoted to their three children and seven grandchildren. There's no denying that he and Nellie savor the good life his forays into the private sector have brought them and their family.

Connally is not as wealthy as many of his campaign benefactors. But with directors' fees and capital gains and dividend income of about $370,000 in 1978, ranch income of about $200,000, speaking fees of some $16,000, and a draw from the law firm of about $465,000 in 1978, he earned some $1,051,000 in 1978. His net worth is well over $3 million.

☐

Connally has made the system work. The free-enterprise system. The political system. And the criminal-justice system.

On July 29, 1974, Connally was indicted by a Watergate grand jury for having taken $10,000 from Associated Milk Producers Inc. (A.M.P.I.), the nation's largest dairy cooperative, and for then perjuring himself and obstructing justice in order to cover up the crime. On April 17, 1975, he was acquitted on the charge of

taking the money. The perjury and obstruction of justice charges were never tried; the Watergate prosecutors dropped them after the acquittal on the money-taking charges.

Were he a private citizen, Connally's acquittal would close the case. But since he is the first serious Presidential candidate in history whom prosecutors and grand jurors have believed to be a felon, the milk-fund trial will not be forgotten if Connally's campaign gains momentum. Questions about it will dog him every day much the way Chappaquiddick will hound Senator Kennedy, although Connally will be able to reply that, unlike Kennedy, he stood trial and was found not guilty by a jury.

The Government's case rested on testimony provided by Jake Jacobsen, a Texas lawyer and former White House legislative assistant to Lyndon Johnson, who was an old friend of Connally's.

In 1971, the year Connally was Treasury Secretary, Jacobsen was a Washington lobbyist for A.M.P.I. The dairy cooperative's overriding goal that year was to get the Nixon Administration to boost price supports on dairy products.

It was not disputed at the trial that A.M.P.I. and Jacobsen found a friend in John Connally. Connally's personal appointment book, admitted as evidence, showed that Jacobsen logged more than twice as many visits to Connally's office during the first half of 1971 than any other nongovernmental official. And a portion of the Oval Office tapes played at

the trial revealed that Connally had indeed successfully interceded with Nixon and Secretary of Agriculture Clifford M. Hardin so that on March 25, 1971, Hardin reversed a decision he had made 13 days earlier and raised the price-support level.

What was disputed was whether Connally had received $10,000 for his help.

By late 1973, Watergate prosecutors investigating Nixon-campaign finance practices had begun to look into the dairy donations — and Jake Jacobsen. Facing charges of bankruptcy fraud in Texas, and looking for a good plea bargain, Jacobsen agreed in early 1974 to tell the Watergate prosecutors what he said he knew about A.M.P.I. and John Connally. He told them, and later testified at Connally's trial:

● That on April 28, 1971, at a meeting in Connally's Treasury office — after Connally had helped A.M.P.I. with the price-support decision — Connally had remarked to Jacobsen that the A.M.P.I. people had been generous with contributions to politicians and that he would appreciate it if Jacobsen could get some A.M.P.I. money for him.

● That Jacobsen had then asked for and received $10,000 from A.M.P.I. officials to give to Connally. The A.M.P.I. officials testified, too, that Jacobsen had solicited and received $10,000, claiming he was going to give it to Connally.

● That on two occasions — May 14, 1971, and Sept. 24, 1971 — he had gone to Connally's office and given him $5,000 in cash.

● That on Oct. 26, 1973, Connally and Jacobsen, aware of the Watergate investigation, met in Austin and concocted the following to cover up the crime: Since the A.M.P.I. people had already told the Watergate investigators that they had given the money to Jacobsen, who said it was for Connally, Jacobsen would acknowledge that he himself had received it, but would claim that Connally had never solicited it and, in fact, had turned it down when Jacobsen had offered it. In short, Jacobsen, not Connally, had kept the money all along. At the trial, Connally's defense was that this is exactly what had happened.

● That on Oct. 29, 1973, Connally gave $10,000 back to Jacobsen at Connally's Houston law office, so that, in keeping with the cover-up story, Jacobsen would be able to produce it as the money he had taken from A.M.P.I. in 1971 but not given to Connally.

● That Connally soon realized that some of the bills he'd given back to Jacobsen had been printed too recently to have been the ones Jacobsen had received from A.M.P.I. in 1971, so, on Nov. 25, 1973, he gave Jacobsen another $10,000 containing older bills.

For all the publicity the Connally trial received, it may be one of the more inaccurately reported events in recent history. The news media repeatedly referred to it as a *bribery* trial. Connally was *not* charged with bribery. He was charged with taking an illegal gratuity. The difference is critical, and the failure of the prosecution to articulate it ef-

fectively may have helped Connally enormously.

A government official is guilty of bribery if he performs an official act in return for a payment that is agreed upon in advance. A government official is guilty of taking an illegal gratuity if he performs an official act, even for good reason, and *then* takes a payment as a thank-you. Yet the case was widely perceived as a bribery trial, and the fact that Connally was known to be a man of means, a man considered unlikely to sell himself for $10,000, made the charge seem implausible. But those were not the facts alleged. It was charged that Connally took a $10,000 gift for something he had already done. The jury, perhaps encouraged by the recital of Connally's business achievements when he testified and by the failure of the prosecutors to emphasize the gratuity charge, may not have appreciated the difference. One juror willing to discuss the trial says, "One of the things that swayed us was that he was a millionaire. Why would he take a $10,000 bribe and risk everything?"

Beyond that possible misunderstanding of the crime, there was the performance by Connally's defense lawyer, Edward Bennett Williams. The famed Washington litigator more than earned his $250,000 fee. His cross-examination of Jacobsen skillfully tore into the star witness's general credibility. He used character witnesses such as Billy Graham (one juror was heard saying, "Amen," when the evangelist finished) and Barbara Jordan (the jury had 10 blacks) and his summation clearly evoked the jurors' sympathy.

But his most important move actually came before the trial, when he persuaded the judge to sever the perjury and obstruction counts from the gratuity trial and let the gratuity trial proceed first. When Connally was acquitted on the gratuity counts, the prosecution, following normal procedures, dropped the related and lesser obstruction and perjury counts that theoretically would have been tried later, had he been convicted on the gratuity charges.

Yet the prosecution had the strongest evidence on those perjury and obstruction counts. According to the original indictment, Connally told a Watergate grand jury on Nov. 14, 1973, that it was "a long time ago" when he had last talked to Jacobsen about what he (Connally) claimed was only A.M.P.I.'s *offer*, through Jacobsen in 1971, to give him $10,000. "Have you discussed it with him recently, within the last three or four weeks?" he had been asked. "No," he testified. Adding, when pressed further, "I don't recall having any major discussions with

The same bullet that killed President Kennedy in Dallas Nov. 22, 1963, also struck and wounded Governor Connally (lower right).

him since last fall." It was eventually proved, and Connally later conceded in another grand-jury session in April 1974 and at his trial, that he had met with Jacobsen on Oct. 26 in an Austin, Tex., hotel room to discuss what the two would tell investigators about the A.M.P.I. money. The prosecutors charged that Connally lied at the Nov. 14 grand-jury session because he and Jacobsen had agreed to keep their Austin meeting a secret and he did not know Jacobsen was going to crack. The meeting, the prosecution charged, had been held to nail down details of the cover-up story about Connally's having turned down the money after Jacobsen offered it. This was one instance in which the jury would not have had to rely only on Jacobsen's word: a hotel waiter was found who testified to having served Connally and Jacobsen breakfast in Connally's room.

With Jacobsen talking to the prosecution and the waiter ready to corroborate him about that meeting in Austin, Connally, the prosecution charged, was caught in a lie and was, therefore, left with the lame excuse at the second grand-jury session and at his trial that he "misunderstood" the first grand-jury questions about when he had last met with Jacobsen.

Watergate prosecutor Frank Tuerkheimer did introduce the facts alleged in the perjury count as evidence in the gratuity trial. But had the gratuity trial included the perjury charge, the effect of that evidence would have been much stronger. "What happens in a situation like that," explains a criminal lawyer who has handled similar cases, "is that if you have a perjury count as part of the indictment, the jury *must* face it and consider it in its deliberation, because they have to render a verdict on it. So they're constantly reminded that it looks as if he lied. If it's just part of the evidence, they can forget it, or at least not necessarily face it squarely. That's why you put those counts in an indictment, and that's why Williams's getting it out was so important. That's where he won the case, I think. Or at least one of the places where he kept from losing it."

Williams's performance notwithstanding, a review of the 1,300-page trial record shakes one's confidence in Connally's indignant claims that the whole affair was a frame-up put together by a desperate plea bargainer and some overzealous prosecutors.

True, Jacobsen is a convicted felon. (His plea bargain with the Watergate prosecutors required him to plead guilty to one count of having given Connally the gratuities that Connally was then acquitted of having taken.) But his

Connally, Lyndon Johnson's Congressional secretary in 1939, makes a White House call in 1964.

account of various runs to and from safe-deposit boxes to pay off Connally and later to take the money back from Connally squared exactly with bank safe-deposit logbooks and with Treasury records of when Connally met with him. For example, the records show that he emptied his Washington, D.C., safe-deposit box 74 minutes before meeting with Connally on Sept. 24, 1971, for the purpose, he said, of giving Connally the second payment. What makes this "paper trail," as prosecutors call it, especially impressive is that all those records would have to conform exactly or Jacobsen would be caught in a lie. Yet when Jacobsen first told his story to the prosecution, he hadn't reviewed those records and didn't know what they would say; he simply told the prosecution, confidently, that the records would prove his story. Jacobsen's travel and telephone records also checked out impeccably.

What emerges, then, from the Connally trial record is, first, that he was accused of a lesser crime than people perceived. Politicians often take "gratuities" in the form of campaign contributions. Second, it seems clear that, guilty or innocent, Connally benefited from excellent criminal defense work. He was also the beneficiary of his own testimony. His voice ringing loud and clear, under questioning by Williams and cross-examination by Tuerkheimer, he projected the perfect mix of polite respect for the court, self-assurance and a not-too-heavy touch of indignation at being brought to the dock by a plea-bargaining convict.

Four years later, he is all indignation, telling anyone who asks, "I think everyone now agrees I got a bum rap. I was framed. Sure, I'll talk about the trial. The jury said I was innocent."

The jury having made its judgment, the next decision is for the voters. Connally aide Julian Read is confident that

Connally's high negative rating in the polls, derived in part from the milk-fund trial, will soften as the campaign goes beyond the 1979 phase of raising money and meeting party leaders and reaches out to voters. (And, in fact, a recent New York Times/CBS poll may bear him out. The question was asked whether or not a bribery charge against a candidate would make any difference and 48 percent of those responding said they would not hold such a charge against a candidate, provided he was found not guilty.) Mahe also asserts that those negatives that remain are "the ones who would never be for him anyway."

Although the campaign is not nearly as well organized in key primary states as those of Reagan, George Bush or Philip Crane, campaign manager Mahe claims, "In three or four weeks we'll have city and town chairmen in every place in New Hampshire," adding, "I sympathize with campaigns that recruit two or three years before you need people to do any work. They become disillusioned and are burned out by the time the action starts."

Mahe, communications director Read, press secretary Jim Brady and George Christian, a longtime Connally aide who served as Lyndon Johnson's White House press secretary, are his closest advisers. Also high in the campaign brain trust are campaign chairman Winton (Red) Blount, who served in the Nixon Cabinet as Postmaster General; direct-mail wizzard Richard A. Viguerie, who left the Crane campaign to work for Connally; Mike Myers, another former Connally aide and Texas businessman; Charls E. Walker, a Washington lobbyist who was Deputy Treasury Secretary; and foreign-affairs specialist Sam Hoskinson, a former C.I.A. analyst and assistant to Zbig-

niew Brzezinski. The campaign staffers also say that Henry Kissinger has helped Connally on major foreign-policy speeches, but so far Kissinger has not publicly acknowledged that he is in any candidate's camp.

Connally's style on the hustings could be a problem. Campaigning among the masses does not seem to suit him. It looks like work when he presses the flesh. He is also a clumsy speech reader. That hurts him only minimally in campaigning, because he usually speaks off-the-cuff, but it did weaken even his much-rehearsed, brief appearance in his first television commercials. (Future commercials will limit his appearances to spontaneous question-and-answer sessions, a format in which his knowledge, self-assurance and contact with the audience project so well that he may be the most captivating performer on the American political stage.)

Mahe predicts that Connally will come in second or third in New Hampshire, that the race will soon narrow to Connally versus Reagan, and that Connally will "beat Reagan in the Florida primary [on March 11] and in a few other places, and that'll be the end of him. Then Baker or Bush, one or the other, will rise to the surface, and we'll get rid of them and get ready for November."

One way they'll get ready for November is by reaching out for voters they aren't worrying about now. Connally is remarkably candid about this, explaining to fund-raising audiences that the major reason he now lags in most public-opinion polls — and the reason he hasn't campaigned among blacks or urban union workers, for example — is that at this stage he has to concentrate on that small group who vote in Republican primaries.

While he is now being pigeonholed as the businessman's candidate, Connally has

demonstrated that were he the nominee he would move dramatically to broaden his and the party's base. He concedes votes to no one. In 1962, he was the archconservative candidate in the Democratic Texas gubernatorial primary runoff. Yet by gathering a few well-publicized endorsements and by organizing a group of minority politicians who'd been shut out of power in the opposing camp, and by running a clever neighborhood campaign that used leaflets picturing him with John Kennedy, Connally managed to beat liberal Don Yarborough in key Hispanic precincts and come close to matching him in some major black districts. Such is Connally's willingness to lean into his opposition's strength and to do the dramatic that, among all candidates running this time out, his selection of a black Vice Presidential running mate would be the least surprising.

Yet there is one group whom many feel Connally has lost forever: the Jews. In his Oct. 11 speech to the Washington Press Club, Connally, whose first major television commercial has him declaring that he is "the candidate of the forgotten American who goes to church on Sunday," called on Israel to withdraw from all territories gained in the 1967 war. He also urged some kind of autonomous Palestinian state. In return for that, he said, the Arab world should, and would, guarantee an uninterrupted flow of oil to the West at relatively stable prices. To guarantee Israel's survival, he also proposed a hefty new United States military presence in the region: an Air Force base in the Sinai region and a "Fifth Fleet" in the Indian Ocean.

Two days after the speech, Connally's top Jewish supporter, New York lawyer Rita Hauser, resigned from the campaign, explaining that she had not been consulted on the speech (she was given a completed copy the night before) and that she couldn't support it. Most other Jewish supporters have since left the Connally camp. In a Republican nominating contest, Jewish support isn't necessary. Nor is Jewish financial support crucial to a campaign funded in the first three quarters of the year 47 percent by Texans, and in which $1,000 is the maximum contribution. The question is whether Connally can retrieve some Jewish support by November.

"Connally simply said what a lot of people believe, but that only he has the guts to come out and say," says campaign manager Mahe. But Mahe does not believe this will be fatal in the long run to his candidate's attempt to get Jewish votes. "By November," he says, "we'll have gone back to

Connally would have us believe that blunt language coming from a self-proclaimed wheeler-dealer is the language of pragmatism. Yet his Israel-and-oil speech is widely viewed by Middle East experts as fatally unrealistic.

every Jewish group and shown them that with his realism and his guarantee of a military presence there, Connally is *the* candidate who's best for Israel."

"There is no way he'll convince us of that unless he changes his position," Rita Hauser counters, adding, "John is a brilliant guy and has great leadership qualities, but I've always been afraid that he shoots from the hip too much, and I think this is an example."

Either way, the Israel-oil speech was smart politics for the short run. It has paid off in more ink for Connally so far than anything his Republican opponents have done or said

has brought them. He has won editorial praise for being specific on a controversial issue. And it's probably a safe bet that people pulling Republican levers in primaries in New Hampshire, South Carolina or Florida won't be offended the way Rita Hauser is. In fact, Connally's aides, in background briefings, seem to welcome her opposition, as if it would actually help them with the voters they need now.

Indeed, as with the former wage-and-price-control enforcer's current evangelizing for free enterprise, Connally's Israel speech raises the question of whether it is a product of a commitment to style or to a firm belief. In the past, Connally has repeatedly assured

Jewish supporters that oil should not be tied to a Middle East settlement. Has he really changed his mind or is this simply another successful part of the effort to appeal to Republican primary voters by being uniquely "candid" and "tough" with a position most will agree with?

More important, as with his similarly tough talk about Japanese trade, or government waste, Connally seems to be assuming that voters will automatically equate his "I'll-snap-my-fingers-and-do-it" language about the Middle East with what is genuinely practical. Connally would have us believe that blunt *language* coming from a self-proclaimed wheeler-dealer is the language of pragmatism. Yet the Israel-oil speech is widely viewed by Middle East experts as fatally unrealistic, and Connally's notion that OPEC nations would ease up on the oil squeeze because the United States and Israel have made concessions to the Palestinians is seen as downright fanciful.

"John Connally is the darling of the business communi-

ty, but I have real qualms about him, which were reconfirmed by his speech on Israel," says investment banker Felix G. Rohatyn, who has previously argued that Israel should be more flexible in the peace talks. "To think you can trade Israel for oil is totally impractical, in addition to being immoral."

Like former Texas Attorney General Carr, Rohatyn doesn't see Connally as the L.B.J.-type wheeler-dealer he makes himself out to be. "The issue of leadership has become *the* issue," he says. "But leadership in a democracy is being able to sit down and get people of different views to work things out. That's not the image Connally burnishes every day. He doesn't seem to want to conciliate. He talks tough.

"I went incognito to a bond-club dinner and watched him speak," Rohatyn continues. "He mesmerized them. He's what most of them would like to be — self-confident and tough. He makes it sound easy. 'I'll do it by fiat, A, B, C and D.' It sounds terrific, but it's

just not practical. It's dangerous.

"Theodore Roosevelt," Rohatyn concludes, "was O.K. in his day because we had the muscle to back it up. At this point it's very dangerous to play that role because I don't think we can back it up."

Connally, of course, says we do have the muscle, and millions of frustrated Americans are going to be eager to believe him.

A campaign against Carter, the Connally people maintain, will be a fight against a well-intentioned, weak President who can't cut it in a tough world. A contest against Kennedy, they say, will be a battle against the tired old programs and slogans of the 1960's. Yet Connally's candidacy may, in the end, represent something more dated than Carter's good intentions or Kennedy's liberalism — the notion that Teddy Roosevelt's tough talk can cut it in the 1980's. ■

November 18, 1979

CHAPTER 9

Popular Culture

Jim Henson's Muppets are now known worldwide, having become entertainment stars in their own right.
United Press International

At Last, All Systems Are 'Go' for 'Star Trek'

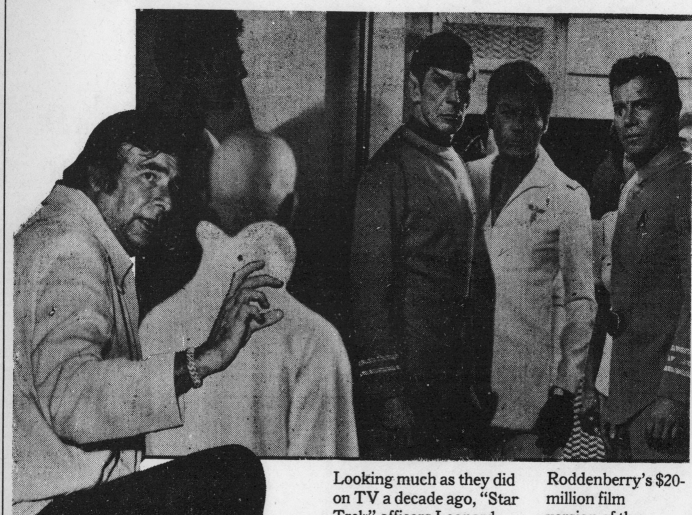

Looking much as they did on TV a decade ago, "Star Trek" officers Leonard Nimoy, De Forest Kelley and William Shatner (right) repeat their roles in producer Gene Roddenberry's $20-million film version of the television series.

By M.L. STEIN

HOLLYWOOD

Gene Roddenberry, a gangling, soft-spoken man of 57, strode through the four Paramount sound stages where "Star Trek: The Motion Picture," which he is producing from the science fiction television series he created, is being filmed

M.L. Stein is a California writer with an interest in films.

for release late this year. Mr. Roddenberry euphorically showed a visitor the bridge of a 23d-century "starship," then the craft's white-walled sickbay, then some of the crew quarters. Finally, he paused before a viewing screen that, it turned out, is part of a computerized on-board "science station."

Mr. Roddenberry smiled. "Wandering through here is like being a kid again. We used to shoot the television show for $186,000 an episode. . ." He glanced again at the sets whose lavishness befits a production with a budget in the $20-million range. "I always thought of what I could do if I had an

extra five grand a week. This is everything I've ever dreamed of."

That dream was born a long time ago — in fact, long before the release of the two science fiction films against which "Star Trek" will be measured. Mr. Roddenberry, who well knows the perils of the inevitable comparisons, countered, "My picture will not be a rip-off of 'Star Wars' or 'Close Encounters.' 'Star Trek' has a life of its own."

Indeed it has. "Star Trek" is the longest running science-fiction series in network TV history. For 79 weeks between 1966 and 1969, viewers entered

"Space, the final frontier" to participate in the Starship Enterprise's "five-year mission to explore strange new worlds, to seek out new life and new civilizations, to boldly go where no man has gone before."

(In the film, the crew of the Enterprise will speed across the galaxy in "warp drive" to confront "an alien intruder with the most incredible, most destructive power ever encountered.")

However, even though "Star Trek" introduced into the popular culture such icons as the omniscient, pointy-eared space hybrid Mr. Spock (portrayed by Leonard Nimoy) and such futuristic hardware as "phaser guns" and "transporter beams," Mr. Roddenberry acknowledged that "the series got generally bad reviews and low audience ratings.

"But after NBC canceled it," he continued, "the reruns caught on all over the world. By 1973, man had walked on the moon, and 'Star Trek' was generating a tremendous amount of interest in science fiction. Fan clubs sprang up. Colleges began offering sci-fi courses. The time seemed right for a movie."

But the enthusiasm of the "Trekkies" — as "Star Trek" fans were called — had not yet spread to Paramount, which had produced the TV series, nor were other studios interested. "They wanted things like 'The Monster Who Gobbled Up Glendale' or 'The Beach Boys Meet the Martians,'" the producer recalled ruefully. "Even '2001' was regarded as an oddity."

Then, in 1976, Paramount's new president, Michael D. Eisner, an alumnus of ABC-TV, reactivated "Star Trek." "The project should have been done in 1975, but the enthusiasm wasn't there," he said recently. "Frankly, I think my predecessors made a mistake. There was a lack of understanding of how important TV is. 'Star Trek' already had its audience."

But perhaps more important than Mr. Eisner's interest was another event that occurred in 1976.

"I was on vacation in San Diego when I saw 'Star Wars,'" Mr. Roddenberry said. "I was delighted with the picture, but I felt a gigantic frustration that 'Star Trek' was not up there on the screen." The producer did not have to contain his frustration for long, and today he credits George Lucas's smash with speeding up the decision-making process at Paramount: "When 'Star Wars' came out, they were on the edge of going ahead with my picture. That movie pushed them over."

Indeed, by then box office receipts for "Star Wars" were making science-fiction believers out of the most skeptical critics; even Charles Bluhdorn, chairman of Gulf & Western, Paramount's parent company, reportedly inquired why there was so much foot-dragging on "Star Trek."

Having become the second major studio to jump aboard the science fic-

tion bandwagon (Columbia was then preparing Steven Spielberg's "Close Encounters of the Third Kind" for release), Paramount decided not to skimp. "Star Trek," said Mr. Eisner flatly, "will be the biggest picture we have ever made."

Faced with a budget almost seven times the original, Mr. Roddenberry now had to cope with three crucial problems: direction, casting and creating special effects commensurate with the film's newly-enlarged scale.

The first director was Philip Kaufman. When Mr. Kaufman left the project before shooting began (he went on to the current critical and commercial success "Invasion of the Body Snatchers"), the job was offered to Robert Wise, who won Oscars for "West Side Story" and "The Sound of Music" and directed the science fiction films, "The Day the Earth Stood Still" and "The Andromeda Strain."

Mr. Wise, 64, admitted he had not been a "Star Trek" fan: "When they asked me about doing the picture, I looked at the TV episodes. I thought some were brilliant but others pretty awful." And in his opinion, the sets and costumes already commissioned for the film "appeared to be leftovers from the television series. They looked a little cheap." Mr. Wise said the sets and costumes have been "upgraded."

Casting was also crucial, for it was questionable whether "Star Trek" aficionados would accept new actors in such pivotal roles as Mr. Spock, Commander Kirk (William Shatner), Dr. McCoy (De Forest Kelley), Scotty (James Doohan), Sulu (George Takei) and Uhura (Nichelle Nichols). Mr. Roddenberry said it was not hard reuniting the entire TV cast, though Mr. Nimoy "hesitated a bit. But then he asked himself if he would want to see anyone but himself playing the role of Spock and decided that he would not."

The original cast members, most of whom still look as though they have just stepped out of a 1969 episode, have been joined by two newcomers, Stephen Collins (as the Enterprise's second-in-command) and the Indian actress Persis Khambatta (as a Kojak-bald denizen of the planet Delta).

•

Finally, there are the special effects. "'Star Wars' started off with a science fiction audience and then attracted a general audience," said Mr. Eisner. "That's what 'Star Trek' will have to do — if the sci-fi fans reject it, so will the mass moviegoers. So we're doing technical things that have never been done before. The appetite of our technical people is insatiable. They'd go on for 10 more years if we let them."

Mr. Roddenberry, though, becomes uncomfortable if "Star Trek" is discussed solely in terms of technology and space shoot-outs: "We try to analyze how a spaceship will be operated in the 23d Century, so the Enterprise is

bigger and more sophisticated. But while there will be plenty of special effects, they're related to the characters — they're part of the dramatic integrity, not an end in themselves. They won't take over the picture."

Mr. Wise concurred, adding he had tinkered with the script "to develop characters more strongly and establish chemistry between them. I thought it needed more emotion and feeling to make the story more believable. 'Close Encounters' had an interesting beginning, but fell apart in the middle."

"Star Trek," in various incarnations, has consumed almost half of Mr. Roddenberry's adult life. A former Pan Am pilot and Los Angeles policeman who turned to TV writing in the late 1950's, he decided "to do something that, in my opinion, no one ever had done right — turn out a science fiction story about real people. I also wanted to make a statement about the human condition.

"I really don't consider myself a science fiction writer," he continued, "but I'm interested in what's happening on this planet and what may happen. In our society, we're treating man less and less like an individual and more like a social organism."

At the time he conceived his show, Mr. Roddenberry said, "television was so tightly censored that science fiction was the only way to escape the taboos in politics, religion, or anything else that was controversial. I thought of 'Star Trek' as a 'Gulliver's Travels.' Yet, when I first went to NBC with it in 1963, they said it was too cerebral."

NBC eventually took that chance, a decision that added little to the producer's bank account (he claims he has not profited from the show's syndication) but earned him the thanks of science fiction fans everywhere. Yet though often invited to Trekkie conventions, the producer has attended only four. "They're nice folk," he explained, "but I'm not cut out for adulation. All those people pressing around me made me nervous. I spent most of the time in my hotel room, sipping Scotch and waiting for the convention to end."

His long fight to get "Star Trek: The Motion Picture" made has not dampened the producer's spirits, though he confessed that "Even after the cast party, I couldn't believe we were going into production. It was only when the weekly budget sheets began coming in that I was convinced it was all real." And he remains optimistic about the film's chances, for he views it as a basically upbeat movie that is "really about changes and values. People want to imagine man in 300 years as a sort of super 20th-century hero, fighting for the benefit of humanity."

After the film's release, Mr. Roddenberry says he plans to "try a comic novel. Just once, I must write something in which I don't have to worry about casting, actors and budgets." ∎

January 21, 1979

MERCHANDIZING DISCO FOR THE MASSES

From a network of scattered dance halls, disco has become an industry estimated to generate $4 billion annually, making it as big as network television. Soon, franchised discos will be fixtures in suburban shopping malls coast to coast.

THE INNOVATORS: *Producer Jacques Morali, far left, inventor of the Village People, accepts an award, as "disco queen" Donna Summer is kissed by her discoverer, Casablanca Records president Neil Bogart.*

SHRINE: *Disc jockey's booth at 2001 Odyssey in Brooklyn, where Travolta danced to glory in "Saturday Night Fever."*

By Jesse Kornbluth

Steve Rubell's friends warned him he was asking for trouble if he chose the old Fortune Opera House on West 54th Street as the site of his discothèque: "They said, 'Broadway? Eighth Avenue? All you'll get is the spillover from the theaters, and the hookers!' " Rubell recalled recently. A slight, hyperenergetic man not known for his attentiveness, on this one occasion Rubell took his friends' warnings seriously: When he and Ian Schrager opened Studio 54 in the spring of 1977, he devised an admission policy so restrictive that thousands of people began thronging 34th Street

Jesse Kornbluth is a journalist and screenwriter who lives in Manhattan.

At Guys and Dolls Teen Disco, near Hempstead.

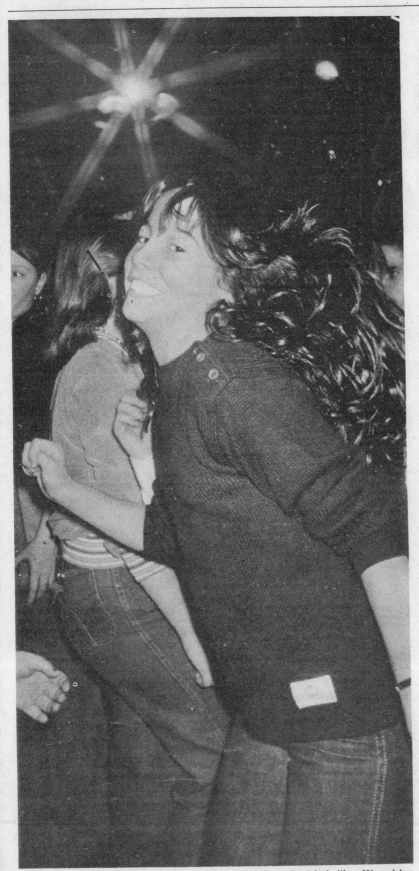

high-school students succumb to Friday and Saturday night fever. But it looks like a 50's sock hop.

PARAPHERNALIA: *Jewelry that blinks to the music, clothes as gaudy as disco lights. mood enhancers — all promote a frenetic atmosphere*

MIXED METAPHORS: *"Grease" meets "Saturday Night Fever" at the pinball machines on the sidelines of the Guys and Dolls teen disco*

SENSORIUM: *At the Limelight in Mallandale, Fla., the computerized lighting is so advanced that dancers never see the same effect twice.*

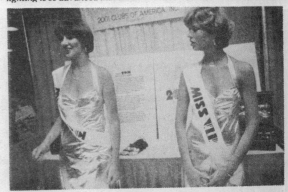

HARD SELL: *Hostesses at Disco Forum, a merchandising show in New York, attract potential franchisers for the 2001 Clubs of America.*

GIORGIO MORODER: Disco's first composer, he scored the "Midnight Express" soundtrack.

STEVE RUBELL: As co-owner of the celebrated Studio 54, Rubell is the "first pasha of disco."

RAY CAVIANO: Warner Bros. records made him the president of its brand-new disco division.

RITCHIE KAZOR: With a salary of $50,000, "54's" disk jockey is probably the best-paid in U.S.

MERCHANDISING MAVEN: At Tramp's in Washington, D.C., owner Michael O'Harro sells Tramp's T-shirts, mirrors, perfume — almost anything that isn't nailed down.

194

on weekend evenings just to see if they could get in. As a result, "Studio," as Rubell likes to call it, became disco's biggest success story, grossing an estimated $7 million in its first year. On the strength of that success, Rubell has decided to open a club in Europe, preside over a Studio 54 record label and franchise a line of Studio 54 designer jeans. For a man whose last enterprise — a discotheque in Queens — ended with his eviction, Steve Rubell has come very far very fast.

As has disco itself. Because of the publicity generated by Studio 54, "Saturday Night Fever," and the Bee Gees, it sometimes seems hard to remember that there was a world without disco, but a short five years ago disco was a phenomenon that flourished only in the homosexual, black and Hispanic neighborhoods of a few large cities. In those clubs, with relatively inexpensive equipment and a limited selection of music, disc jockeys began creating an entertainment experience that was certifiably creative. A disc jockey could move the music as if it were a tornado, sweeping the sound across the room while he intensified it with computer-coordinated lighting effects. On a second turntable, he might play another record, its beat matching the first thud for thud while faceless singers who would never perform together again shouted out duets of riveting carnality. And because the dancers were, in the main, already high on cocaine, marijuana and a variety of "poppers," they became so overwhelmed by this experience that, for hours at a time, they could not imagine a world outside the disco.

This was an entertainment formula too hot — and too profitable — to remain the sole property of an urban subculture. It would take until 1978, when "Saturday Night Fever" topped even the easy-listening charts, for disco to be acknowledged as this decade's dominant musical event. But even before Steve Rubell turned Studio 54 into a media favorite, disco entrepreneurs were thinking about exploiting the phenomenon. They did not care that rock critics suggested disco was roller-rink music, or called it, as Lester Bangs wrote in "Rolling Stone," "as identifiable a commodity as *Smile* buttons and just as vital." What they saw was a way to take the disparate elements of 1970's society — progressive sexuality, computer technology, mind-stopping drugs, the fitness craze and a nostalgia for Jazz Age decadence — and amalgamate them into one cash business.

These entrepreneurs were not particularly interested in disco's possibilities as an art form, nor did they have much affection for the urban discophile. Their market was in the suburbs, and their mission, as they saw it, was to make disco attractive to the family man, the working woman, and the student looking for an inexpensive date. If disco's homosexual origins had to be diluted in order to do that, so much the better — a fad which could be made safe as milk might hang on long enough to become a fixture in the shopping mall.

The entrepreneurs are succeeding. From a network of scattered dance

ROLLER DISCO: *Vinzerrelli performs at Brooklyn's Empire Rollerdome.*

halls, disco has become an industry estimated to generate $4 billion annually, making it as big as network television. There are now teen discos, like Long Island's Guys and Dolls, where, at 9 on a Friday evening, one might find four cars in the parking lot and a thousand 13- to 17-year-olds inside. There is all-disco radio like New York's WKTU-FM, which became the most listened-to station in America less than six months after it abandoned "the mellow sound" for non stop dance music. Old-age homes use disco to reawaken their residents' interest in life. In the Midwest, disco-hungry promoters convert supermarkets into dance halls. Even in Jonestown, children wrote the names of recent disco hits in their notebooks. Disco here, disco there, discodollars everywhere: from the boom in clothes made of Lycra spandex to the ephemera of a Louis Vuitton disco bag at $100, there is nowhere one can look without seeing some manifestation of a bread-and-disco culture.

Next week, 150 exhibitors and 2,000 disco entrepreneurs will meet in New York for the Billboard-sponsored "Disco Forum V," where they will paint a predictably upbeat picture of disco's future. But there is a darker side of disco starting to make the headlines. In Boston last summer, the owner and night manager of a disco were found shotgunned to death, allegedly because of a dispute over a cocaine deal. When Ian Schrager walked into Studio 54 during a raid by Internal Revenue Service agents last December, he was, authorities said at the time, carrying a small bag of cocaine. It is widely known that at one of New York's most fashionable discos, the disc jockey sometimes gets a gram of the drug along with his $200-a-night. At yet another discotheque, when underworld types come to collect their "skim," they make a point of not offending the homosexual clientele: They wear checked shirts and blue jeans and take a turn on the dance floor before slipping in and out of the office. Organized crime's interest in disco is hardly surprising; for a restaurateur or night club owner to prosper and never make a payoff is almost unheard-of. But even the underworld profiteers are stunned by the size of their disco haul, which can, in a successful club, reach $50,000 a week.

As an urban amusement, disco may have already crested, but for the disco entrepreneurs — the franchisers, merchandisers, record executives, and "consultants" — the suburban market, with its vast second-generation audience, is a billion-dollar one just starting to be tapped. The days of an individual club owner fighting for a share of the weekend crowd are numbered; disco has become such a big business, the chain owner is already starting to crowd him out. Those observers who believe that disco has only six months left will be as disconcerted to learn what these men have already accomplished as they will be to hear what they plan for the future.

THE FRANCHISER

The Studio 54 bouncer would not blink at his name, and he may be a celebrity only in Pittsburgh, but 35-year-old Thomas Jayson of Bridgeville, Pa., is surely the most successful businessman working the disco market. As the founder and president of seven disco-related companies that have grown at an annual rate of 500 percent for each of the last three years, he is clearly someone who knows what the suburban discophile wants: convenience, a neighborhood atmosphere, a dash of sexual possibility, and above all, a schedule that moves a neophyte dancer so smoothly through an evening that he never realizes he is being processed as methodically as a hamburger on a McDonald's grill.

When disco entrepreneurs start talking about franchising, it is usually because they have achieved some slight success and hope to capitalize on it. Not Jayson: From the beginning, he planned to franchise, and so he designed his first 2001 Club — no relation to the disco in "Saturday Night Fever" — as a prototype, with every intention of cloning it in shopping centers across America. He was not, therefore, surprised when 1974's single 2001 Club spawned 10 franchises by 1978, nor is he amazed by the 200 applications on file for the six new franchises he will grant this year. As Jayson sees it, 2001 clubs are advancing more or less on schedule. Assuming disco continues to boom, he believes 2001 should achieve his target — 150 franchises — in the early 1980's, making his Pennsylvania-based creation the entertainment equivalent of a fast-food restaurant.

This analogy is not casual: The 2001 concept has no place for the individual who wants to pay the $35,000 franchising fee (plus 6 percent of gross income) and go his own way. As Parris Westbrook, 2001's director of operations, tells it, "Our complexes are geared like I.B.M. Once you're involved with us, we'll provide you with specifications right down to the macramé wall hangings."

The first company a successful bidder for a 2001 franchise must deal with is Thomas Jayson Associates, which has designed every club in the chain. Next, the 2001 Sales Group offers furnishings and equipment — but the franchise winner, far from passively accepting 2001's bid, is required to solicit others. Similarly, he does not have to choose the 2001 Construction Company to build his facility, though if another firm gets the contract, 2001 insists on having its supervisor on site. Finally, "because there was no one around who could do it our way," the parent company founded 2001 Industries to manufacture computers and lights, 2001 Productions to supply the clubs with slide shows and film presentations, and 2001 Management to handle the day-to-day operations. The "mom and pop" types who win franchises are happy to accede.

The 2001 customer experiences a more subtle regimentation. If he arrives between 8 and 9 in the evening, he will hear the hits of the day as performed by the Boston Pops. Between 9 and 10, the computerized sound system serves up Top 40 selections. At 10, the programming breaks down into half-hour cycles, each starting with three slow songs, then some faster numbers, and finally some genuine disco music. At 11 o'clock, 2001 clubs across America present a seven-minute slide show, usually featuring scenes from a current movie followed by a visual tour of a major American city. At midnight, the disc jockey plays hits from the past five years. In the next half-hour, he punches up the evening's sole novelty song, and at closing time — generally 2 o'clock — a slide bids the dancers good night and cautions them to drive carefully.

2001 is candid about it: What the clubs offer is hardly disco. "We have a disco party on Wednesday and Thursday nights," Westbrook explains "in a nice way, that tells the real dancer, 'Don't come here on Friday or Saturday.' On these nights, we get people who only go out on weekends. And we don't want them intimidated by more professional dancers."

The admission charge at a 2001 is never more than $3, and drinks cost $1.50 at most. "We're not looking for folks who can drop $50 a night," says vice president James Kowalczyk. "We're happy with people who will spend $5 to $7 — but if we can pull 2,000 of them, we'll do all right." A measure of the company's all rightness is that earnings are expected to reach $3 million in 1980, at which time Jayson and Kowalczyk plan to go public.

'Our concept is boy-meets-girl,'' Kowalczyk says. "We know that's not a tired idea."

THE RECORD EXECUTIVE

Once the major record companies understood that disco songs could become even more popular than suggestive songs from the rock era, they started looking hard for executives who understood disco. There are not many disco purists at the executive level, but late last year, Warner Bros. Records found one, 28-year-old Ray Caviano. Warner's gave him a particularly nice Christmas present: $6 million to start a disco division at Warner Records, a label of his own, a staff of 12, and the first regional promotion unit in disco. For a disco-shy giant, this was commitment on the grand scale.

When Ray Caviano first started working in disco promotion, none of the big labels had any interest in disco. The reasons were more sociological than esthetic: Disco was a largely homosexual phenomenon. A record played in a dozen clubs might, without any radio exposure, sell 50,000 copies, but promoters from companies like Warner Bros. were still more comfortable with radio as the vehicle for "breaking" songs. "One could be analytical and say the big record companies resisted rock, too," Caviano recalls. "But in an area where one-to-one contact with the disc jockey is the most important thing, the straight record executives were offended by the disc jockey's life style. I started by breaking records right out of Fire Island. I didn't see CBS there."

When Caviano worked as national disco promotions director of a small record company operating out of a Florida warehouse, he was free to work in his own style. He test-marketed his releases at selected New York clubs. He made sure the "record pools" which service disc jockeys were supplied with his records. And he advertised his records in homosexual magazines. "The cheapest advertising in the world," he says; "$10,000 in six or seven magazines can start a trend."

While Caviano was solidifying his relationships within what he calls "the disco community," larger companies refused to give free promotional records to disc jockeys of homosexual clubs. Unwilling or unable to penetrate the disco culture, they tried to latch onto another trend: punk rock. Meanwhile Caviano's small company was grossing $10 million a year on disco alone. Clearly, once punk failed, Caviano would not remain in Florida long.

Caviano began his Warner's tenure with a program that, for the fast-buck disco market, is outright radical. He has persuaded Warner's to support promotional tours that would, for the first time, make the disco artist as visible as the producer. He intends to spend as much time in New York — "Disco Heaven" — as possible, and give one day a week to meetings with disc jockeys and club owners. And he plans to persuade nondisco Warner's musicians to cross over into disco: "That could bring in an extra 750,000 to 1,000,000 units [albums]. Which didn't hurt the Rolling Stones. No one said, 'Jagger's sold out.' "

Caviano talks like a man who believes he is riding more than a fad. "We're going to disco-ize the Warner's staff," he promises. "We're going to be aggressive. If there are mindless and repetitive records, they're not going to be ours. And people will see, as we go along, that disco gives you the opportunity to drop inhibitions, share harmonious feelings with other people, and — let's call it what it is — escape. And don't forget it takes $500,000 to break a rock act properly today. In disco, the liability's not nearly so great — if you can talk to the D.J., if you know how to advertise and promote.

"You can't come into a market as open as this and not feel bullish."

THE COMPULSIVE MARKETER

Michael O'Harro shakes the most recent envelope from his clipping service and scatters a thick pile of newspaper articles onto the desk. "Look at them!" he exclaims. "I know everything that's been written in English about disco!" Not surprisingly, that gesture is followed by another pile of clippings, this time on the subject of one Michael O'Harro.

As a Navy ensign who had trouble meeting women, O'Harro founded the "Junior Officers' and Professional Association" in the 1960's, thereby — he claims — inventing the singles bar. He turned that association into a chain of bars with 300,000 members. When singles bars fell into disrepute, he fled to Europe, where he discovered that in European discos "meeting, not the end result, was the important thing." He returned to America in 1973 to spread that gospel. In 1975, he started an easy-on-the-ears, ask - for - any - song - you - want disco called Tramp's ("the disco for people who don't like discos") in the banquet room of a Washington restaurant, and parlayed $100,000 in start-up costs into a $900,000 annual gross, 45 percent of that pure profit. In the process, he boasts, "I've tried to take the whole disco life style upon myself, like dating movie stars."

That life style does not seem to weigh heavily on the 39-year-old O'Harro; a tireless promoter, he does not shrink from beginning an anecdote, "The story I most enjoy telling . . . ." He has published a how-to guide for those about to open a disco (price: $75) and has been named Billboard's "disco consultant of the year" three times (fee: $500 a day, two days minimum.)

This, too, has nothing to do with disco and everything to do with marketing. In 1977, he created Tramp's cigarettes and sold 2,000 cartons. Last year, he experimented with a Tramp's boutique in a Washington department store; this year he hopes to sell Tramp's perfume. He has moved 1,000 Tramp's necklaces at $8, and 20,000 Tramp's T-shirts at $1.50 profit per shirt. Someone suggested mirrors: O'Harro made up a bunch and sold a hundred. Once, when people commented on his Christmas card — a picture of O'Harro, in riding togs, with his foot on the bumper of his '52 Bentley as his chauffeur pours champagne — O'Harro had the picture blown up to poster size and sold over 1,000. When a Georgetown shoe store stocked Tramp's pumps, 1,500 pairs (at $28) walked out of the shop in less than a month.

What, one wonders, is left for Michael O'Harro? Not franchising: He says he no longer wants to make America safe for Tramp's. Not more products: In O'Harro's market, there is not much he has not already sold. Not more of the disco life style: O'Harro jokes he has done everything but learn to dance.

"The Olympics," O'Harro answers. "I want to build a disco in the Soviet Union for the 1980 Olympics. Wouldn't that be a smash?"

THE MECHANICAL DISC JOCKEY

On Jan. 15, 1977, Mike Wilkinson, then a 31-year-old graduate of the Harvard Business School, stood on top of Aspen Mountain. He was, he says, "in a state of shock." He had been to three Aspen discotheques in as many nights, and "none of them knew that a certain new album even existed." As he considered this tragedy, he began to see in it an idea for a new company: a service that would provide geographically isolated club owners with 26 two-record samplers of disco music a year. Not the greatest hits of the distant past, but the very best about-to-be-released material selected by top disc jockeys and premixed into 20-minute sets.

Wilkinson returned to New York and mailed a test sampler off to Aspen; days later, the Aspen disco called to ask for another. Buoyed by that response, Wilkinson incorporated his concept as Disconet and invited club owners to subscribe for $500 a year. He did not intend that his service would replace a single disc jockey. He simply wanted Disconet to be the bible of this decade's secular churches: "The D.J. — the preacher — gives the sermon, but he quotes the bible."

But it is not hard to understand why 20 Chicago disc jockeys, upon hearing of Disconet, shipped Wilkinson a load of dead fish: Eliminating the disc jockey has long been an ambition of budget-minded club owners. Holiday Inns, for example, once spent four months exploring computer-coordinated light-and-sound devices for its 2,000 cocktail lounges, only to conclude, with Paul Gregory of Litelab Corporation, that machines wouldn't work because "machines don't feel." Wilkinson himself had briefly considered offering club owners six-hour tape cassettes to replace disc jockeys completely, but rejected the idea when he realized that a spinner who knows how to "cycle the floor" can get dancers moving toward the bar every 30 minutes and boost liquor sales by as much as 30 percent.

To provide a touch a whimsy in an increasingly profit-obsessed businesss — and to short-circuit criticism that Disconet is designed for disco entrepreneurs who prefer computers to disc jockeys — Wilkinson hired astrologer Jesse Portis Helm to provide "astrological analysis" for the newsletter that accompanies each Disconet record. Using the date a song's vocal tracks were recorded as its "birthday," Helm predicts the song's success without ever hearing it. Disconet subscribers thus knew that "Boogie Oogie Ooogie" — a Scorpio, with Venus in Scorpio, Moon in Pisces, and Jupiter in Cancer — would be a major hit, particularly for Aries listeners, long before other record industry seers had said it would make the charts.

THE DISC JOCKEY AS CLONE

On a Saturday night when as many as 5,000 people will pay $12 each to enter Studio 54, disc jockey Ritchie Kazor will make perhaps $250. This disparity does not bother Kazor, but last fall, it began to trouble some stagehands and they polled their co-workers on the touchy subject of unionization. On Nov. 5, they claimed a majority of potential members had signed Theatrical Protective Union pledge cards. The next day, they say, Steve Rubell began interrogating them, giving them the impression they were under surveillance. According to two fired employees, Ian Schrager threatened the life of a third. Studio 54 denies these charges — "Nobody ever, ever, *ever* has been physically threatened," Rubell says — but the five union-minded employees, three of whom were fired shortly after they started organizing, have filed affidavits with the National Labor Relations Board. The N.L.R.B., in turn, has charged Studio 54 and Xenon, another trendy New York disco, with unfair labor practices.

At Disco Forum V, panelists will be discussing the music in purely business terms: profitability, market share, foreign expansion. For some particapants, however, the most important panel will focus on the disc jockey, who is, increasingly, the only creative element in an otherwise choreographed environment. "Is unionization of D.J.'s imminent?" — the question before that panel — may be hotly discussed, but few insiders think an affirmative answer will mean much. Michael O'Harro doubts there will be any unionization of discos. Ray Caviano believes it is more likely that club owners will provide health plans for their employees. And looking back from the vantage point of a career that spans the disco years, Tom Savarese, one of the best of the original disc jockeys, insists that any talk of employee benefits, health plans or profit sharing is just clever public relations.

"The last time I worked a downtown club," Savarese says, "they bowed to the booth at 4 A.M. You couldn't go any further than I had taken them. At 6, I got a second ovation, and the owners started talking about having me back. As far as I'm concerned, anyone who can do what I did deserves $300. They wanted it for nothing. They're mad because they launched the phenomenon and they never get any recognition, so why should they recognize the D.J. They don't know we're just as dependent on them for the temple."

Savarese, a perfectionist who tapes his performances and keeps a list of every song he's ever played, is unpopular among club owners because he has hired an agent and insists on a contract. These days, he mostly works installing sound systems for wealthy homeowners and remixing songs for record companies.

"The disc jockey of the future? He's invisible," Savarese predicts. "Someone they'll call a D.J. will come in every 40 minutes and change a tape slug. Face it, it's Coney Island. And you can't teach art to thugs."

THE FAMILY FRANCHISE

In the cities where the phenomenon was born, disco may have become, in Tom Savarese's words, "a sleaze operation," but 90 miles from Pittsburgh, it is bringing Norman LaBıuzzo and his family nothing but all-American happiness. LaBruzzo, a 47-year-old father of five, and his 46-year-old partner, Tony Zuccaro, were once known in Meadville, Pa. (population 18,000) as the proprietors of The Pizza Parlor. No more: As the franchisees of a 2001 Club, they are Meadville's kings of disco.

LaBruzzo's transformation began four years ago, when his 18-year-old son returned from a visit to his grandparents in Buffalo. He had been to a disco there, had been fascinated by it, and thought it might be a good business for his father. LaBruzzo drove 120 miles to see his first disco. When he returned, he and Tony Zuccaro started looking for a solid franchiser.

"I didn't know Tom Jayson from a bushel of apples," LaBruzzo says. "But his 2001 in Pittsburgh blew my mind. The lights, the dance floor, the chrome — it was like a movie!" He restrained his enthusiasm for Jayson's disco when he visited his banker. "That was 1976, so we didn't even mention disco. We talked about 'entertainment complexes.' We knew the bankers thought disco was dirty and had something to do with drugs."

Six months after they opened their first club in Meadville, LaBruzzo and Zuccaro went back to the bank — Meadville was so successful, they wanted to open a second club, 60 miles away, in Jamestown, N. Y. And when their second club flourished in Jamestown, they went to their backers again, this time for financing of a third club in Rochester. This winter, they are negotiating a lease in Buffalo, and are looking at locations in Detroit, Allentown and Philadelphia.

Success has only made LaBruzzo more optimistic about disco: instead of gutting old W.T Grant stores and A. & P. supermarkets for his clubs, as he has in the past, he is now planning to construct free-standing buildings. The son who suggested that he see what the dance hall craze was all about is now the 22-year-old manager of LaBruzzo's and Zuccaro's Rochester club. LaBruzzo's 10-year-old often tugs at his father's sleeve on "Family Night" and the family drives off to the shopping center for an hour's exercise. And when LaBruzzo's daughter was married not long ago, the reception was held at his club, where, instead of the usual light show, the guests were treated to slides of the bride and groom from infancy to the present. "It was the most beautiful wedding you've ever seen," LaBruzzo recalls proudly. "The older people were particularly fascinated. I think they came more to see the disco than the wedding."

LaBruzzo sounds so cheerful about disco's future that it seemed almost unkind to wonder aloud what he would do with his 10 year leases if the phenomenon suddenly flopped. Norman LaBruzzo laughed at that idea. "The way these franchisers do it," he says, "disco can last indefinitely." ∎

February 18, 1979

THE MATURING OF WOODY ALLEN

'Until we find a resolution for our terrors, we're going to have an expedient culture.' Is this Woody Allen speaking? Yes — the *new* Woody Allen. His latest film reveals an uncommonly serious artist at work.

By Natalie Gittelson

The only time Woody Allen rises from his chair is to change the background music in his duplex penthouse living room: He switches records from Mozart to Beethoven, from Beethoven to Schumann. On the coffee table beside him, crowned with fresh spring flowers in perfect esthetic array, bowls of nuts and fruits — arranged with Cézanne-like care — look almost too good to eat. He doesn't eat. He doesn't peel a banana or nibble an almond. He doesn't smoke and he doesn't drink — not even a glass of water. He hardly even gestures. Not to put too fine a point on it, he hardly even moves.

He sits somewhat uneasily on the edge of his favorite easy chair and talks about his craft. In the eight movies he has directed, Allen says, he has been steadily "trying to advance in the direction of films that are more human and less cartoon." "'Take the Money and Run' and 'Bananas' were my first two movies," he points out. "I was

Natalie Gittelson, an editor of this Magazine, is the author of "Dominus: A Woman Looks at Men's Lives."

learning, floundering, trying to get by on my sense of humor. When it came to the moment of truth, I felt I could count on the laughs. I depended on the laughs to bail me out."

Behind the familiar, thick-lensed glasses, his eyes grow reflective as he continues to trace his evolution: "With 'Everything You Always Wanted to Know About Sex,' I struggled a lot to pay a bit more attention to technique." Of the vignettes that shaped that movie, Allen says, "It was experimental for me to do all those short pieces — but it helped me to improve a little technically.

"'Sleeper' and 'Love and Death' were cartoon-style films. I was still struggling to develop a sense of cinema, a better feeling for technique. But even though those films tried for some satirical content, they were still cartoons. I had intended to be very serious in 'Love and Death.' But the serious intent underlying the humor was not very apparent to most audiences. Laughter submerges everything else. That's why I felt that, with 'Annie Hall,' I would have to reduce some of the laughter. I didn't want to destroy the credibility for the sake of the laugh."

Woody Allen also excised many funny

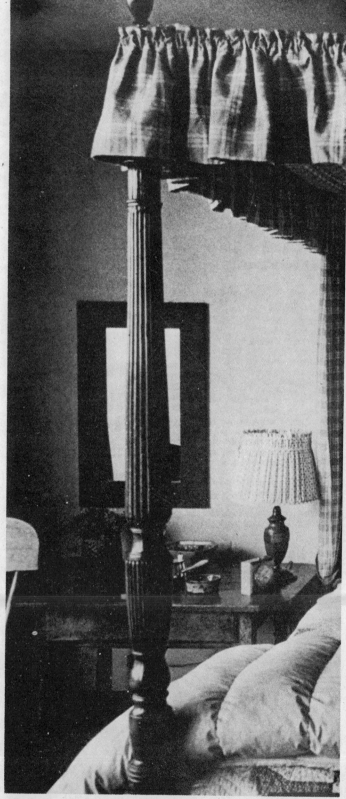

Woody Allen believes in allowing himself room to dream the largest of dreams.

Mary Ellen Mark

As a consequence, he sleeps in an enormous Early American four-poster, canopy bed. In other ways, however, his personal life style is modest — he says "dull."

scenes from his latest movie, "Manhattan," which opens Wednesday. "They were superfluous. They stopped the flow. And sometimes they were too funny," he says.

Too funny? Can this be Woody Allen, the man we once knew — or thought we knew — as a lovable, self-mocking nebbish? Well, yes and no. Both creatively and personally, this is where he's at today — more mature, less comic, than anyone might expect and, at 43, on the verge of becoming one of America's major *serious* film makers.

"Manhattan" blends the drama and high concerns of "Interiors" with the comedy and charm of "Annie Hall." The result is an incandescent comedy-drama, bristling with grown-up insight — in short, it's a showcase for the new Woody Allen. What he sees as a bad time for our culture has provided him with an excellent opportunity to view the erratic couplings and uncouplings of men and women with a wide-angled lens more intensely scrutinizing than he has put to comedy before.

"Manhattan" testifies to Allen's ever-deepening personal vision and his growth as an artist. He has shed his adolescent insecurity. He has largely given up the slapstick and sight gags which, although often hilarious, were also a defense against the confrontation with self.

Woody Allen emerges — in the character of a television joke writer who, symbolically, quits his job — as a human being with a stronger mind and also with a stronger heart. If he ever was self-deprecating, he is not anymore. And with his new self-acceptance has come a new willingness to communicate honestly his male emotions. As "Annie Hall" was Diane Keaton's movie and a story about the condition of contemporary woman, "Manhattan" is Woody Allen's movie and a story about the predicament of contemporary man.

Possibly not since Chaplin — one of Allen's idols — has there been an American film that captures with such tenderness, hilarity and virile muscle what it means to be a vulnerable, selfish, innocent, corrupt, hollow, harried, loving, frightened and terribly mixed-up human being.

□

In the Allen apartment, 20 floors above Central Park, the walls are windows and all of Manhattan seems wrapped around his living room. "It really lights up out there," he says, *almost* smiling, of that cityscape which — like the questions he raises — also seems to jab recklessly at eternity.

"There's such widespread religious disappointment, a general realization about the emptiness of everything that's very hard for the society to bear," he says. "Either we've got to accept that life is *not* meaningless, for reasons as yet unknown, or we've got to create some sort of social structure that offers us the opportunity for real fulfillment.

"It's not a good time for society," he continues quietly. "It's a society with so many shortcomings — desensitized by television, drugs, fast-food chains,

Allen's love interests in "Manhattan": (top) getting rained on with Diane Keaton in Central Park; (above) visiting a soda fountain with young Mariel Hemingway.

loud music and feelingless, mechanical sex. Until we find a resolution for our terrors, we're going to have an *expedient* culture, that's all — directing all its energies toward coping with the nightmares and fears of existence, seeking nothing but peace, respite and surcease from anxiety.

"We've got to give up the immediate, self-gratifying view," he goes on. "We've got to find the transition to a life style and a culture in which we make tough, honest, moral and ethical choices simply because — on the most basic, pragmatic grounds — they are seen to be the highest good."

This advocate of the *summum bonum* may be controlling a lot of repressed fury. The tone of his voice and the set of his mouth suggest it. It is said that on the sets of his films he politely but firmly distances himself from everyone but Diane Keaton, actor Michael Murphy and other trusted friends. He delivers most of his orders to the technical crew through intermediaries. Otherwise, his anger might explode at mortal men and blow the production sky-high.

Allen takes out some of his feelings on show business generally, which he sees as a microcosm of society: "They're a lot of worldly, sophisticated, talented people who seem to be conducting their lives reasonably well — except that they buy out the really tough choices by doping themselves up."

Woody Allen does not believe in buying himself out of tough choices. He believes, more than anything, in his work.

□

In "Manhattan," Allen plays Isaac Davis, a 42-year-old fellow whose former wife (played by Meryl Streep) "left me for another woman." Partly to rationalize her own feelings, she writes a book that exposes, graphically, why their marriage went haywire. Perhaps to console himself, Isaac takes up with a 17-year-old girl (played by Mariel Hemingway). On these fragile relationships, Allen has hung a strong, hilarious, haunting modern morality tale.

After "Interiors," his first foray into serious drama, Allen explains, "I wanted to make an amusing film, but a film with feelings that went deeper than 'Annie Hall.' It would take place in New York, again it was Diane and me [Diane Keaton does not play the wife but the *other* other woman], and again it would be a comedy. But this time, I wanted to try and go farther on the serious side than I had done before. That continues along a path I would like to follow."

What he did, he says, was to "incorporate into 'Manhattan' areas and ideas I had learned to handle a little bit in making 'Interiors.' There I was trying to deal with heavy emotions, heavy confrontations. I was also learning things about myself and my craft: how far one can go, how to go even farther. Each time, I'm attempting to express more and more feelings."

Feelings — and translating them into potent images — are Allen's primary concerns. Of the new film, he says "It's my own (Continued)

WOODY ALLEN

Continued

He goes to great lengths, and strange hats, to avoid recognition.

feelings — my subjective, romantic view — of contemporary life in Manhattan. I like to think that, a hundred years from now, if people see the picture, they will learn something about what life in the city was like in the 1970's. They'll get some sense of what it looked like and an accurate feeling about how some people lived, what they cared about."

The film is, of course, no somber documentary on our frayed morals and mores. Yet in "Manhattan," the comic mask slips one more inch — without by any means forsaking comedy. Allen still grins, if ironically, at his large philosophic questions. He still mocks what he takes seriously and takes seriously what he mocks. This is Woody Allen — dualism intact — in the full flower of his maturity as a writer, director and actor; and dealing more directly than ever with his own ambiguity, ambivalence, emptiness — and ours.

Still, the comic despair of "Manhattan," no matter how brilliantly conveyed, is not enough for today's Allen. He dreams of confronting the despair cold. "Interiors" was his first farewell to the determinedly comic vision. Now he plans to come out more often from under the cover of even dark (as against light) comedy and what he deems the comfort of "playing it safe."

"Tragedy," Allen says flatly, "is a form to which I

would ultimately like to aspire. I tend to prefer it to comedy. Comedy is easier for me. There's not the same level of pain in its creation, or the confrontation with issues or with oneself, or the working through of ideas."

☐

At Michael's Pub, a Manhattan watering spot where Allen plays the clarinet in a New Orleans-style jazz band on Monday nights, he sits down between sets to refine his self-portrait. He does not, he says, want to be thought of as a "cheerless, grim, comic personality."

Furthermore: "I'm not holed up in my apartment every night poring over Russian literature and certain Danish philosophers. I'm really hardly a recluse. When a half-dozen *paparazzi* follow me down the street, naturally I don't like that very much. But I do go out all the time — to movies, to shop, to walk around in the street, to those parties I think I'll enjoy, to Elaine's." Why to Elaine's, that most obvious of New York celebrity enclaves? Allen says he likes the food. "And besides, I'm comfortable there."

☐

Woody Allen is trying earnestly to introduce the real Woody Allen. He is, he says, an ordinary man. Bland. Unexciting. Without the "fresh, explosive, wonderful insights of,

say, Norman Mailer. I have no flamboyant side," he claims. "I'm a middle-class film maker. I'm not a druggie or a drinker. I relax with my friends in a very uninteresting way. After I play the clarinet here, I go home, have a bite to eat, and go to bed."

As proof, Allen describes the weekend just past: "On Saturday night, I went with Michael Murphy and Marshall Brickman and his wife to John's Pizzeria on Bleecker Street. [Brickman wrote "Manhattan" with Allen; Murphy plays the second lead.] But we failed to go into John's because the place was jammed to the rafters. So we came back to my apartment and I got a lot of food out of the refrigerator." Like his entire residence, Woody's eat-in kitchen has a warm, country-house look: exposed red brick wall hung with copper pots, Early American dining table.

"We watched the Bob and Ray special on TV, sat around for a while, then went to a late party in honor of Jean Doumanian, who produced the show." Allen numbers Jean Doumanian among his best friends, along with Diane Keaton, Louise Lasser, his second wife, Dick Cavett, Brickman, Murphy, Mickey Rose, a childhood chum with whom he once planned to open an optometrist's shop, actor Tony Roberts and Joel Schumacher, the screenwriter and costume designer.

Sunday morning: "Trying hard not to work, I watched 'Richard II' on TV, because I like the play. In the afternoon, there were a few little budget things on 'Manhattan' — I had to be present at a meeting with Charles Joffe and Bobby Greenhut [producer and associate producer]. At noon, before I left home, I had lunch standing over the cookie jar: chocolate-chip cookies and two glasses of milk."

☐

Contrary to legend, Woody Allen says, *not* everybody loves either Allen or his movies. "My pictures," he notes, "are *reasonably* successful at the box office, not overwhelmingly successful. I told United Artists that there was every possibility I would have to fail with 'Interiors,' that it might be a catastrophe. Yet even it turned a little profit." Particularly because of that film's heavy themes, "I worked hard to make it entertaining," he explains. "I wish I could have done better."

"I'm able to entertain audiences," Allen says with that faint but unmistakable air of deprecation he reserves for *mere* entertainment. But on the other side of the coin is pride: "If one of my pictures is playing, it's a worthwhile

gamble, as often as not, that there's going to be at least some entertainment. And audiences know I won't insult them. I may strike out — and I often do — but it won't be demeaning. They're not going to have to sit through a lot of stupid, infantile jokes."

Actually, Woody Allen almost never *just* makes jokes anymore — that trait which once caused the critic John Simon to write (about "Love and Death," Allen's favorite Woody Allen film), "We could have gotten roughly the same effect from laughing gas, sneezing powder or a mutual tickling session with a friendly prankster." Today, Allen's humor probes the deeper vein of character — basically, his own. No other American moviemaker working today uses *himself* as his source of material more consistently or to better effect.

☐

Although comedy comes naturally to Woody Allen, he makes no bones about the fact that, even so, "work is very difficult. You miss the mark so often." He admits that making movies brings him fulfillment, but he will not admit that it brings him joy. "It's hard for me to find out where the pleasure is," he says. "I'm too concerned about all the aspects of the thing coming together."

As it does often, Allen's perfectionist mind returns to "Interiors": "If you try for something and it fails honorably, it's better failing that way than to succeed with something exploitative."

Sex, per se, seems to him exploitative. Nevertheless, in "Manhattan," Allen allows the character he plays more authentic sexuality — without any taint of exploitation — than he has ever done before. He has abandoned his customary comic diffidence for a sexual posture much more forthright and unembarrassed.

"In films," he says, "what you do — in bed or elsewhere — is dictated by the tone of the piece. In 'Manhattan,' we played much less broadly, much less for laughs, than in pictures like 'Bananas' or 'Sleeper.' This could be perceived as a change in my own sexual style. But it's simply that the character I play has, hopefully, more breadth. This demands a certain sort of comedic restraint."

Allen denies having much sex appeal of his own. If women do like him — and he reluctantly concedes that some of them may — it has little to do with sexual attraction, he says. "Maybe one or two people have said that. But if you took a poll, if you did a wide survey," he suggests, "you'd find it isn't true."

☐

By the age of 21, a college flunkout, Allen had already wed Harlene Rosen, his "first girl," and entered psychoanalysis. "I was very young, writing, high-strung and married. I had a tendency to get a little depressed." At about the same time, he was also making the crucial conversion from gag writer to stand-up comic.

Of his 22 years of psychoanalysis (today he has a female analyst), Allen says: "You don't learn anything in a dramatic rush. It's very slow and unnoticeable. But an hour a day, talking about your emotions, hopes, angers, disappointments, with someone who's trained to evaluate this material — over a period of years, you're bound to get more in touch with feelings than someone who makes no effort. I think that has a tendency to liberate your natural gifts. Possibly you work more productively, too, because you don't obsess over self-destructive things."

In 1964, when he was 28, Woody Allen was performing at the Blue Angel, a New York nightclub. Film producer Charles K. Feldman entered his life on the arm of Shirley MacLaine. The MacLaine laughter helped persuade Feldman to offer Allen his first movie job as writer and actor. The eventual result was "What's New, Pussycat?" a huge commercial success that did not, and does not, please Allen. But the film grossed $17 million, more than any other comedy before it.

Palomar Pictures then put up $1.6 million for "Take the Money and Run." Allen wrote the screenplay with Mickey Rose, this time without interference; he also directed and starred. It and "Bananas," "Everything You Always Wanted to Know About Sex but Were Afraid to Ask," "Sleeper," "Love and Death," and "Play It Again, Sam," were all small (in the Hollywood lexicon) but profit-making pictures as well as critical and popular successes. Then "Annie Hall" saw Woody Allen triumphantly emergent, taking Diane Keaton with him.

Of course, he himself would never put it that way. He denies Keaton's alleged professional dependence on him. "She would accept any good, intelligent, significant role that comes along, and that she feels she can do, whether it's in my picture or not," Allen says.

The once celebrated Keaton-Allen love affair has long since lapsed, and she is now seeing Warren Beatty. But she stars with Allen in "Manhattan," and her director continues his abundant praise. "Diane's fortunate. She got hit by the talent stick in a very big way. I'm overwhelmed, I'm awestruck, by her gifts. She always makes me look good. I think the biggest contribution I've ever made is bringing Keaton as an actress to the attention of the public."

Conversations about craft absorb Allen intellectually, but Diane Keaton enlists his emotions. They lived together for a year in Allen's penthouse, which Keaton helped to decorate. "We were real friends when we split up," he says. "There's never been any strain between us; no fights, no falling out, almost never a bad word."

Allen adds that he and Keaton have never discussed Warren Beatty, whom Allen says he does not know. Of Beatty, Allen remarks, "He's probably very sweet and nice. In fact, I'm sure he's wonderful. He certainly *seems* to be intelligent."

□

The two most important women in Allen's life have both become part of the contemporary American legend — Diane Keaton, with "Annie Hall," and Louise Lasser, television's "Mary Hartman." How alike are they in other ways?

"They're two huge talents — unique stylists in the entertainment world," Allen says. "They both had more to give me than I had to give them. I couldn't have accomplished half of what I accomplished without Louise . . . and Diane." Like Keaton, "Louise is a great woman, a great companion. Her comprehension of life is very great," he adds. "Both of them are hilariously funny. Both of them are highly, *highly* perceptive. And both of them have very sweet faces. I've had other nice relationships in my life, but these two women I have lived with spoiled me to a very great degree. I'm crazy about both of them."

In the end, however, Woody Allen returns to Diane Keaton, her artistic intuition and her taste. "If she tells me something's no good, I may not understand why at the time, but eventually I come to realize it's no good. That's some strange form of genius."

□

"Manhattan" testifies with eloquence and candor that Allen also may have a soft spot in his heart for young, young women. But there is little of Humbert ("Lolita") Humbert here. Although sex is by no means devalued, the real attraction lies between kindred spirits. The older Allen grows, the more he seems to value innocence in women — not sex-

ual innocence, but that shiningness of soul that age so often tarnishes. Nevertheless, he says decisively, "There's no significant woman in my life at this time," either young or older. "But I do take women out pretty often," he explains. "If either of us doesn't have a nice time, we don't do it again."

He has various platonic "letter-writing relationships" with young women, "when they read out as serious and substantial." His youngest correspondent is an 11-year-old girl, whose letter was precocious in the extreme, Allen says. "I wrote her back, 'If you're really the age you say you are, it's phenomenal. But if you're not, don't write to me again and waste my time.' Finally, I met her whole family. They all came to see me, including her mother. She's a nice, intelligent girl. She's 11."

In "The Whore of Mensa," one of Allen's most wickedly funny and endearing New Yorker "casuals," the private eye, Kaiser Lupowitz, says to an 18-year-old girl who, "for a price, will come over and discuss any subject — Proust, Yeats, anthropology. Exchange of ideas . . . 'Suppose I wanted to — have a party?'

'Like what kind of party?'

'Suppose I wanted Noam Chomsky explained to me by two girls?'

'Oh, wow!'" the young woman sighs. "Mensa" may express the dichotomy that Allen apparently cannot resolve regarding the female body and the female brain.

"I really do get hooked into bright, highly educated girls," he admits. "It must be some carry-over from my teen years."

□

Woody Allen likes success as much as any other man, but perhaps does not court it with quite the same fervor. The luxury of failure is always on his mind. "It's very important that there's a certain amount of stuff one fails with. That's *very* important," he emphasizes. Failure, to Allen, is a sure sign that "you're not playing it safe, that you're still experimenting, still taking creative risks. It's frightful to get into the habit of trying to make hits. Compromises and concessions begin to turn up in the work.

"Bergman once asked me," he goes on, " 'Have you ever had a picture that simply nobody liked, a total disaster?' " Allen actually looks sheepish. "I hadn't. And it causes you to doubt yourself. If you're succeeding too much, you're doing something wrong."

This need to reach, to stretch, to adventure *out*, he says, "has to do with putting a

value on developing and trying to grow artistically. Some artists don't have that value. They find something they can do superbly, and they love doing it, and they keep on doing it, and the audiences grow to depend on it. I think that's a mistake."

Allen points again to Chaplin: "He was preoccupied with developing. He was willing to take many chances and he often failed. Some of his movies were quite terrible. But he was always trying to grow. And whatever he did, you felt in contact with an interesting artist."

Chaplin was not only willing to fail; he was also, Allen says, "willing to disappoint his audience." This, he thinks, may be the highest act of courage for a star. "Even though the public loves you the way you are, and counts on you, and showers you with love and affection, you've got to risk having people come into the theater expecting everything and walk out disappointed because, this time, you've overreached yourself. If you don't take that risk, you find yourself modifying your potential in order to ingratiate yourself with your fans."

Much of Allen's artistic independence grows out of his iconoclastic attitude toward money. He seems challenged by a modest budget. "Money is an important creative tool," he says. "For instance, if I make the decision for 100 extras, I shoot for one less day. If I go for 50 extras, I buy one more day of shooting. Money becomes a part of the art."

United Artists has financed all of his films except "Take the Money and Run." "I never ask U.A. for a lot of money," Allen says. "My own salary is exactly the same today as when I made 'Bananas.' It's only been in the last couple of movies that I've made some money on the percentage end. A picture's really got to soar for me to make anything at all that way." So far, the only Allen picture that has actually soared is "Annie Hall." And, he says, he also made "a tiny bit of money" on "Sleeper."

To write, direct and star in a film takes more than a year, Allen notes. "I get less, all told, than what Marshall Brickman would get just to write a script. I could probably make more playing four weeks in Vegas than I make on an entire picture." The writer-director-star has never made even close to a half-million dollars on any film. Put against the million or more many actors make for eight weeks' work, this is indeed a modest sum.

But money is of no intrinsic interest to him, he maintains. "Of course, in terms of the

standards I was brought up with, I do make an awful lot. But many an average movie actor makes much more money. In fact, I make nothing but a pittance compared to most actors working today.

"Jack Rollins and Charles Joffe [his managers and also his producers] are smart enough to know that money is a fortunate byproduct, that's all. What interests me is solely the work itself. I could probably put myself to hire and direct other people's scripts and make a lot more than I do, but that's not my intention. I didn't enter the film business and then ask myself, 'What can I buy? What can I option?' I have ideas that I want to express."

Allen explains how the idea for "Manhattan" grew: "Marshall and I walk around the city. We review the experiences of my life. We talk about the ideas and feelings that are meaningful to us, that we would like to express. It happens in an amorphous way. 'What if my ex-wife were writing a book about me?' we ask each other. Characters begin to appear: people I know, people we both know. Then others appear that are totally fabricated, sheer flights of fantasy."

Once they develop the basic screen play, Brickman disappears and Allen puts on his director's hat. "On the set, many things change," Allen says. "One actor may turn out to have a slightly different quality than I expected: I rewrite. I change tons of dialogue. Keaton doesn't feel comfortable with something I've written, but she feels very comfortable with something else. That goes in. I never thought of Meryl Streep as a possibility for the role of my wife in 'Manhattan,' but as soon as it was suggested, it was clear to me that she would be great. More rewriting. I met Dustin Hoffman's wife, Anne Byrne, at a basketball game; I saw her as correct for Emily. [Emily is the wife of Isaac Davis's best friend.] I changed several things to suit her. The movie was always growing, being rewritten and rewritten on the set." Allen's productions are notorious within the industry for "wrapping" promptly at 5 P.M. He goes home every night — to continue to rewrite.

The score for "Manhattan" is by George Gershwin. Allen has photographed the film in black and white. "The Gershwin music fits in with Isaac's comprehension of the city," Allen explains. "He *sees* Manhattan in black and white and moving to the tunes of George Gershwin. The score helps to vibrate several themes in the picture: the passing of time;

the poignancy of where the city's gone; the fact that, in a certain sense, Isaac's living in the past, when things seemed better to him."

Allen returns to his own creative process: "As I edit the film, it continues to metamorphose. Sometimes concepts get lost, or I abandon them completely, and stronger concepts emerge. Finally, I finish the rough cut, the first draft, of the film. It's always 45 minutes to an hour longer than it ought to be. It's full of things I hate."

Allen pauses to lament: "There's such a disparity between what you set out to make and what, at last, you do make that you just want to . . . die. You want never to see it again."

But even after the film is "frozen," Allen "sees it five or six more times — to check the prints, to even it out, to fix the sound." He sighs.

"You get an idea for a funny New Yorker piece on Monday; by Thursday, you've written most of it. But a film is a gigantic undertaking from the point of view of sheer quantity. There are tons of decisions to make in areas where you can only hope your instincts will guide you properly." As an *auteur* with total control ("You're only aware of total control when you don't have it," he says), Allen must make decisions about everything from the color of the couch to the clothes the stars wear and whether the dress looks right with the wallpaper. "You want different throw pillows," he says, "different ashtrays. Then it comes down to the music behind the scene: Do you want them to hear just the music or do you want them to hear the faucet dripping and the wind rustling in the trees, or *just* the faucet dripping or *just* the trees?"

Finally, the picture winds up in a screening room for the first time. "Something happens there that's really nonverbal and preconscious," Allen says of the reactions of the few intimates who preview his films. "It's never what they *tell* you. What they verbalize is very deceptive. You feel the way they're responding.

"The more I can keep people separate from the project up until this point, the fresher they come to it. Marshall sees it quite freshly now. There are some surprises in it for him, some disappointments. I get a fresh feeling toward it through him and a few friends. [Allen's younger sister, Letty, is always one of the early viewers.] Then I go out and shoot some more. Little by little, I round it out."

This postproduction work can be both determining and

agonizing. "In the theater, when you have a scene that's wrong, you can rework it in a hotel room. But film is footage in a tin can. It waits around for four months while you're shooting the thing and it's all done out of sequence. When you start the editing, your heart sinks. That joke doesn't work, this is no good. You've guessed wrong on a lot of stuff. The relationships between the people aren't clear, therefore, they're not believable. You've got to go back and redo those scenes." Allen always makes sure that his budget allows for between two to four weeks of reshooting. "I've never not needed it," he says. "I've never gone from start to finish with nothing to redo."

When asked, Allen adds that the fact of being Jewish never consciously enters his work. "It's not on my mind; it's no part of my artistic consciousness. There are certain cultural differences between Jews and non-Jews, I guess, but I think they're largely superficial. Of course, any character I play would be Jewish, just because I'm Jewish. I'm also metropolitan oriented. I wouldn't play a farmer or an Irish seaman. So I write about metropolitan characters who happen to be Jewish." But these, to Allen, are surfaces; and, as he has let us know, it is interiors that interest him.

Although he is now writing another screenplay for a movie scheduled to begin production around Labor Day, theater — specifically, Lincoln Center's Vivian Beaumont — is much on his mind. As one of the Beaumont's new five-member directorate, "I hope to write some plays for it: a funny play to help get the Beaumont off the ground, and then some more serious works." He also may try to stage a play or two. While Allen hopes to "have some voice in what goes on at the Beaumont," he sees his role there only as "formalizing a commitment to devote some of my time to working at Lincoln Center." Theater intrigues him, he says, because, "like literature, it deals primarily with words. Film is *about* photography. It's a very different thing."

He prefers writing, he says, to performing or directing. "I go upstairs to my bedroom and close the door. I don't have to go anyplace. I can knock off and play the clarinet for an hour. I can go out and hit a couple of tennis balls. It's a bum's life," he declares. "If you don't like what you're doing, the work never has to meet the test of reality. When I'm making a film, I can't tear it up and throw it away if I don't like it. I've already spent United Artists' $3 million."

But the sheer complexity of

moviemaking keeps Allen continually tuned in to that medium. "In 'Sleeper,' I became more aware of visuals," he says. "Since then, I've gotten deeper and deeper into visually arresting films. It's not just decorative, of course, but hopefully part of the storytelling. In 'Manhattan,' the black-and-white photography, the Panavision, the romanticized view of New York — all that's part of the story. In 'Annie Hall,' Gordon Willis, the cinematographer, used the gray, overcast skies of this city against the unremitting, harsh blue sunlight of Los Angeles. With each film, one struggles to utilize photography, light, colors, sound track, performances more effectively and integrate them better into the character of the film."

The moviemaker Allen ranks highest is Ingmar Bergman. Bergman's gloom is, spiritually, Allen's gloom: "I have a personal taste for the mood he sets. That's my kind of good evening. He makes innovative, cinematic, magnificent, strong, high dramas. In this film or that film, he may have missed out. But 12 or 15 masterpieces, in varying degrees, out of 40 pictures? That's astounding!"

Of course, Allen's creative thinking is not stimulated only by tragedy. "As great as I find the Marx Brothers," he says, "in the final analysis, I still get more emotionally involved with Chaplin. Many of my friends prefer Buster Keaton to Chaplin. Was he more sophisticated, more inventive, more superb? Perhaps so. But after all is said and done, I still prefer Chaplin because he's a funnier human being. As soon as Chaplin comes down the street, I start to laugh — that really primitive, unmotivated hostility, pulling other men's beards, blowing into their bowls of soup! Sometimes he slobbers into sentimentality, sometimes he makes it. Many friends of mine prefer Keaton's clean, artistic, American interpretation better. I do appreciate that. But Chaplin I must say I love.

"He presented the human condition, so that the laughter resonates on another, deeper level than it does when you watch the Marx Brothers, who always appeal to your intellectual, cerebral appreciation. You laugh, but you don't feel the pain. In Chaplin, pain has a lot to do with it: to win the girl, to feed the kid. His humor is more dimensional."

Woody Allen's ever-growing and all too attentive public sometimes causes his private life to take on the aura of a Woody Allen comedy. To him, it's a mixed blessing. One afternoon recently, he returns, perturbed, from his own front

hall, throwing his arms into the air. The elevator man has just delivered a message. "I've got the worst problem," Allen moans. "There's a live white rabbit waiting in my lobby. Somebody left me a rabbit." He considers making room in his life for a white rabbit. "No, it would really be cruel to bring it up here. It would starve to death. I don't know how to take care of a rabbit." Accustomed to making life-and-death decisions, he opts for the A.S.P.C.A.

He says fans leave shirts, layer cakes, candy, records, photographs, books, but this is his first rabbit. Other fans, spying him in the street, follow him all the way home. Recently, one young woman, driving a car, paced him slowly down Fifth Avenue.

While a companion waited in the automobile, Allen says, "she got to me as I was coming into the building. She was dressed strangely and she was a little noisy, but the guys in the lobby restrained her."

"I never know who's following me, I'm doomed to be recognized and I handle it badly," he admits. "I depend a lot on hats for disguises." One is a battered gray panhandler's fedora, another a taxi-driver's cap which he wears pulled down over his eyes. "If I remember to wear one of these hats, I am not recognized 75 percent of the time," he confides.

Woody Allen's existential anxieties, as he has called them, follow him home, too. On a recent afternoon, during a photographer's sitting, two young women set up equipment outside on his terrace. Watching them through a living-room window, Allen's pale face grows paler. "I'm convinced that one of them is going to fall off. I don't want any guests to fall off my terrace," he says, going outside to police the operation. Later, the photographer puts one foot on a very low ledge, no higher than a street curb, trying for an interesting angle. "Don't do that!" he demands, refusing to continue to pose until she returns to safer territory. "Do you know what kind of red tape would be involved for me if you fell off?" he asks, appealing to her human sympathy. There is a tall fence around the Allen terrace. No one could fall off, not even a giraffe.

"Early in life," Allen says, "I was visited by the bluebird of anxiety." That anxiety has informed his work and his persona to extraordinary, droll effect from the early slapstick to the vintage wit and rue of "Manhattan." It may be a symptom of his new maturity that rue is the herb of grace. ∎

April 22, 1979

THE MUPPETS IN MOVIELAND

Having created America's top syndicated television show, Jim Henson is ready to expand his Muppet empire. This month, he takes his magical Muppets from the small screen to the large one with 'The Muppet Movie.'

Center: Jim Henson surrounded by his Muppet characters. Below, counterclockwise: Guest star Steve Martin with Miss Piggy and Kermit; Bob Hope and Fozzie; Miss Piggy in Hollywood; Miss Piggy prepares for a boating scene; Crazy Harry works the movie lights; movie guest stars Edgar Bergen and Charlie McCarthy; Fozzie and Kermit going to Hollywood; mad scientist Mel Brooks about to fry Kermit; Kermit's swamp serenade; guest star Charles Durning kidnaps Miss Piggy.

By John Culhane

Last autumn, when Edgar Bergen died, and his voice and the voices of Charlie McCarthy and Mortimer Snerd were stilled forever, the great ventriloquist's widow, Frances, and his daughter, Candice, asked Jim Henson and Kermit the Frog to say a few words at the funeral.

"There seems to be something strange about having a puppet in this situation," began tall, lean, 42-year-old Henson, the brown-bearded, gentle-voiced, ever-calm creator of The Muppets.

"I've never appeared at a funeral before," said the hand puppet frog on Henson's hand.

"But the family asked me if I would bring Kermit and . . ."

"Charlie would have liked it," Kermit interrupted Henson — and only then did it sink in for some that Charlie and Mortimer were gone, too.

"I think of all these guys as part of puppetry," Kermit told the mourners. "The frog here — and Charlie and Mortimer — Punch and Judy — Kukla and Ollie. It's interesting to note that there have been puppets as long as we have had records of mankind. [See page 72 for some notes about the modern technology of puppetry.] Some of the early puppets were used by witch doctors — or for religious purposes. In any case, puppets have often been connected with magic.

"Certainly, Edgar Bergen's work with Charlie and Mortimer was magic," Henson said. "Magic in the real sense. Something happened when Edgar spoke through Charlie — things were said that couldn't be said by ordinary people . . . We of the Muppets, as well as many others, are continuing in his footsteps. We're part of the cycle. We take up where he left off — and we thank him for leaving this delightful legacy of love and humor and whimsy."

In appreciation, Frances and Candice Bergen gave Henson a photograph of Edgar and Charlie that they set in a silver frame. The frame is engraved: "Dear Jim — Keep the Magic Alive."

This spring, Henson set that photograph on the mantelpiece facing him in his new office. The office is at the top of a four-story oval staircase in the Manhattan mansion that Henson had just bought to use as the United States headquarters of his rapidly expanding Muppet empire. For the magic is alive and thriving. Puppetry — the art of manipulating inanimate objects — has not been so popular in the United States (nor so profitable) since the heyday of Edgar Bergen and Charlie McCarthy.

Indeed, no puppets in the history of the world have achieved the global popularity of Henson's Muppets. Puppet shows are found in almost every civilization and in almost all historical periods. The ancient Greek historian Xenophon mentions one in the 5th century B.C. and a 17th-century Chinese scroll shows children playing with

John Culhane has been a regular Muppet-watcher since he wrote "Report Card on Sesame Street" for this Magazine in 1970.

marionettes. Pulcinella, a human character in the Italian commedia dell'arte, began to appear as a puppet on stages in the 17th century. Around the time of the French Revolution, national puppet heroes displaced the descendants of Pulcinella: in France, it was Guignol; in Germany, Kasperle. But in places where Punch and Judy, Kukla and Ollie, Kasperle — and even Charlie McCarthy — and even Charlie McCarthy — are unknown, Kermit the Frog and Miss Piggy are easily recognized.

"The Muppet Show," which begins its fourth season in September, is the top syndicated television program in the United States, according to a Nielsen analysis. It has some 235 million television viewers in 102 countries. Significantly, more than half of that audience is estimated to be adult (75 percent in Britain).

Millions of these fans are buying Muppet merchandise — making Henson and his wife and partner, Jane, millionaires through the international franchising by Henson Associates (letterhead: ha!) of puppets, toys, dolls, books, apparel, jewelry, household items, art objects, record albums, a $17.50 Harry N. Abrams art book called "The Muppet Show Book," a Miss Piggy Doll that will debut this summer, and, for 1980, a calendar in which Miss Piggy will re-create in color photographs a dozen of the roles made famous by other international sex symbols.

This summer, "The Muppet Movie," their first feature film, will play on theater screens in the United States and abroad. " 'The Muppet Movie' is the reverse of 'The Muppet Show,'" said Henson, who created his first puppets in 1955 as a student at the University of Maryland. "On the television show we invite one guest into the world of the Muppets. In the movie, we are taking the Muppets out into the real world."

Henson may see all this as continuing in the footsteps of Edgar Bergen, but in many ways he is much more reminiscent of Walt Disney. At least, the externals of his business are beginning to take on a Disneylike aspect.

"I'm slightly uncomfortable with all the people who want to say things like that about me," Henson responded to the comparison, " 'cause I like Disney but I don't ever particularly want to do

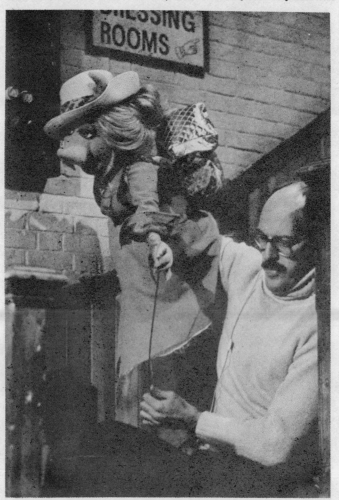

The wizardry of Oz: Frank Oz studies the nuances of a gesture as he manipulates Miss Piggy's arm. He is also the source of her seductive — and tough — voice.

what he did. He built this great, huge empire. I'm not particularly inclined to do that. You get that large a thing going and I'm not sure that the quality of the work can be maintained." He smiled sheepishly. "It seems that I'm bigger now than I thought I would be."

In fact, Henson is bigger now than he thinks he is. In discussing his situation recently, he said he thought he had "between 40 and 50" employees. In fact, since the success of "The Muppet Show" became obvious in 1977, Henson Associates has expanded from 28 full-time staff members to 71. There are now 11, rather than 6, puppeteers. There are now 30 designers, builders and other artisans who work in the Muppet workshop (there were 12 in 1977). And, most significant, 30 people are now involved in administration (there were 10 in 1977).

Walt Disney Productions in the 30's was only a little company in which Walt Disney ran the creative end; his brother, Roy, ran the business end, and an attorney named Gunther R. Lessing did their legal work. Henson Associates, too, is run by a triumvirate. Henson is free to control the creative end, he said, "because of working with lawyer Al Gottesman and producer David Lazer, a former I.B.M. executive. "The two of them run the business side. It's wonderful for me."

"I like creating different worlds of puppet characters," said Henson, when the Muppets were already a unique world of puppets with very definite personalities. Today, Kermit and Miss Piggy are as much stars in their own right as Mickey Mouse and Donald Duck. Indeed, when Oscar-winner Rita Moreno appeared on "The Muppet Show" in 1977, she won an Emmy for "outstanding *supporting* actress in a variety or music program."

Whether there will ever be a Hensonland or Jim Henson World, his achievement as creator of entertainment of global appeal and as a highly ingenious merchandiser of his wares already reminds even artists who worked with Disney on the Mickey Mouse films of their legendary boss.

□

"Jim loves what's going on here this week," said Jerry Juhl, in Hollywood. He is head writer of "The Muppet Show" and co-writer with Don Hinkley of "The Muppet Movie" and an hourlong CBS television special, "The Muppets Go Hollywood," which celebrates the completion of their first movie.

We were standing out by the pool of the Ambassador Hotel, watching Miss Piggy being filmed for the special as she reclined on a chaise longe — a tad overweight, but with light blue eyes to die for. Henson wandered over to see if the towels draped over the chaise completely hid puppeteer Frank Oz from the camera. Oz was lying flat on his back on the pool deck underneath Miss Piggy, operating his *cochon couchant* through a hole in the *chaise.*

"This is one of the craziest weeks I've ever lived through," said the 40-year-old Juhl. "The Muppets have just finished taping

'Sesame Street' back in New York for the start of its 11th season next fall. Then they flew out here. On Monday, Kermit substituted for Johnny Carson as guest host of the 'Tonight' show, and they all entertained there. All week, we've been rewriting and rehearsing — and redecorating the Cocoanut Grove — for the party Jim's giving on Friday. That's tomorrow night! A huge guest list — some of the biggest stars in Hollywood — will be showing up. And in the midst of that party, while the Muppets are entertaining all these celebrities, we're going to shoot this hour-long television special.'' (CBS broadcast it in prime time on May 16.)

"The PA's [production assistants] are running around screaming, 'How are we ever going to do all this?' And Jim is wandering around in the middle of it all, perfectly calm, perfectly content," Juhl went on. "You go to him and ask, 'How's it going?' And he says, 'Oh, fine. There were hardly any airplanes overhead when we filmed Miss Piggy by the pool.' He's just like Kermit — if 'The Muppet Show' had a basketball team, the score would always be Frog 99, Chaos 98."

But Frog wins — perhaps because Henson takes for granted his offbeat but amazingly effective organizational abilities in evaluating the Muppets' popularity. His eye is on the essence.

"It's been a topic of conversation among our people as I try to zero in on what's important," Henson told me. "And I think it's a sense of innocence, of naïveté — you know, of a simple-minded young person meeting life. Even the most worldly of our characters is innocent. Our villains are innocent, really. And it's that innocence that I think is the connection to the audience."

Innocence is certainly the essence that unifies "The Muppet Movie." The story is amiable nonsense — a Muppet sendup of the American Dream as we humans dream it. Miss Piggy is first seen winning the Bogen County Beauty Pageant, a feat made plausible by the fact that the only judges are the late Edgar Bergen and Charlie McCarthy, in their last screen appearance.

When Miss Piggy fixes her famous light blue eyes on "Kermie — my frog" (she is the only major Muppet with irises — and they have highlights), it is one-sided love at first sight. She bursts into a Paul Williams love song, "Never Before, Never Again," that fits her piggy falsetto as snugly as "Evergreen" fit Streisand's voice.

The conflict is a sly little satire on America's fast-food fixation: Doc Hopper (Charles Durning) caricatures Colonel Sanders as effectively as he terrifies Kermit with his growing chain of frog's legs joints. There is also a Mad Scientist, played by Mel Brooks. But the Muppets prevail, because they are fueled by the kind of star-struck devotion to the American Dream that is epitomized by the rise from the swamps of Georgia to the swamps of Hollywood.

Mississippi-born Henson understands the dreams of a young frog from the provinces because he's been there himself. "The most sophisticated people I

know — inside they're all children," said Jim Henson, father of five. "We never really lose a certain sense we had when we were kids."

The child inside Jim Henson was born in the rural South during the Great Depression — Sept. 24, 1936. He always had a vivid imagination, taking delight in the names of things and of people — such as his grammar-school classmate in Leland, Miss., a kid he lost track of called Kermit Scott. (Dr. T. Kermit Scott, Henson will learn if he reads this article, is now a philosophy professor at Purdue University in Lafayette, Ind.) And in those days Henson liked to imagine the way that Edgar Bergen and Charlie McCarthy and Mortimer Snerd must look as they cracked jokes over the radio with big Hollywood stars.

"Bergen goes back longest in my memory," said Henson of his fellow Scandinavian-American. "But I wasn't thinking of any of those people as puppets. They were human to me."

Time magazine, in putting Bergen and McCarthy on its cover when Jim Henson was 8, said that Charlie was "probably more human to a larger number of people than any inanimate object in history. It takes only the mildest indulgence in the world of fantasy to be persuaded that Charlie . . . is actually alive."

It is Jim Henson's historical distinction to be the first to adapt the ancient art of puppetry to the 20th-century medium of television. (Burr Tillstrom and the Bairds had originally developed their own art for puppet theater, not for the screen.) And because of television's commonly recognized power to reduce the world to a global village in a way that radio — owing to the language barrier — cannot, it is now the Muppets who are more human to a larger number of people than any inanimate object in history.

Appropriately, the advent of television was the biggest event of Henson's adolescence. "My mother told me I drove 'em all crazy until they bought a television set," he said. "That was in seventh or eighth grade, when I was 13 or 14 — about 1949. Burr Tillstrom's 'Kukla, Fran and Ollie' were on when we got our set. They were on half an hour every night about dinner time. They made an impression on me. So did Bil and Cora Baird's 'Life with Snarky Parker.'

"Burr Tillstrom and the Bairds had more to do with the beginning of puppets on television than we did," Henson quickly pointed out. "But they had developed their art and style to a certain extent before hitting television. Baird had done marionette shows long before he came to television. Burr Tillstrom's puppets were basically the standard hand puppet characters that went back

Zoot, the hipster saxaphone player, takes a break between sets.

to Punch and Judy. But from the beginning, we worked watching a television monitor, which is very different from working in a puppet theater."

□

The beginning was the summer of 1954, when 17-year-old Jim Henson, inspired by the puppets he had seen on television, built his own puppets to audition for a local television show. This was in Washington, D.C., where Henson's father was involved in agricultural research. In 1955, toward the end of his freshman year in theater arts at the University of Maryland, Henson was offered a late-night five-minute television show of his own called "Sam and Friends." Meantime, Henson had also been noticed by a fellow freshman from New York named Jane Nebel. "It was admiration at first sight," she later recalled.

The admiration was mutual. Jim asked Jane to join him in operating the puppets on his television show. In 1956, Henson built a frog hand puppet and called him Kermit. "I suppose that he's an alter ego," said Henson, "but he's a little snarkier than I am — slightly wise. Kermit says things I hold myself back from saying."

Kermit, in short, was a less-controlled version of the super-controlled Henson. Jane found Jim's personality to be "calm and unbelievably patient. He's not always lovable, though. In fact, sometimes he's so patient you want to kick him."

But Henson had to be patient. Kermit would be 20 years old before Henson's

original conception of the Muppets as "entertainment for everybody" would be appreciated enough for them to get their own television series. All along the way, however, there were those who saw the Muppets' potential. Among the earliest was Steve Allen, who provided them with their network television debut when Allen was host of the pre-Carson "Tonight" show.

That was in 1957, when the Muppets were also beginning to be noticed for their ability to make television audiences remember commercial messages through delightful nuttiness. Jim and Jane decided it was time to form a legal partnership. "Janie and I own the company," said Henson. "We set it up that way in 1957, before we were married. We were married in 1959."

Those dates are correct, although Henson has been known to get confused about such dates as what year he created Kermit. "Jim hardly ever gets the past straight," said Jerry Juhl. "That's because he's completely future-oriented."

Nothing is better proof of this than Henson's gathering of his team. In 1961, Henson saw Juhl and Frank Oz working together at the National Puppetry Convention in Carmel, Calif., and immediately realized their potential. Jane was pregnant with their first child, so Henson recruited the 23-year-old Juhl to take her place with the Muppets. "I was competent," says Juhl today, "but I never felt that I was a really first-class puppeteer. I thought of myself more as a writer."

'I suppose that Kermit's an alter ego,' said Henson, 'but he's a little snarkier than I am.'

Frank Oz was the first-class puppeteer, but he was then barely 17. So Henson waited patiently for Oz to grow up, finally recruiting him in 1964. Juhl was freed to write full time for the Muppets. (Henson's instincts were right there, too. A script by Juhl, Henson, Chris Langham and Don Hinkley for "The Muppet Show" in which Miss Piggy — in a wedding gown designed by Muppet costumer Calista Hendrickson — almost tricks Kermit into marriage, won this year's Writers Guild award for the best-written television variety show.) And when Oz went into the army, Henson had a replacement ready: Jerry Nelson, an actor with a huge repertoire of voices, who had become a puppeteer by touring with Bil and Cora Baird.

In the 60's, the Muppets appeared regularly on "The Ed Sullivan Show," but they didn't become household words until the huge success of "Sesame Street" during the 1969-70 season — and even then their audience was primarily preschoolers. The Muppets' ability to make children remember letters, numbers and concepts has been clinically affirmed. But, ironically, the Emmys won by "The Muppets of Sesame Street" in 1974 and 1976 for "Outstanding Achievement in the Field of Children's Programming," and the gold record for selling a million dollars' worth of a Muppets children's record, were setbacks to Henson's original conception of the Muppets as entertainment for everybody. "They transcend all age groups," he insisted to United States network executives. "Their satiric comment on society seems to delight all ages."

But none of the three major television networks in the United States wanted a series starring the Muppets. "They always said the Muppets were child-oriented and adults wouldn't watch," Henson recalled.

The entrepreneur who took the risk was England's Lord Grade who invited Henson Associates to be his partners in producing "The Muppet Show" at his London studios for the five CBS-owned stations in the United States and for United States syndication and world distribution by his ITC Entertainment Inc.

"The Muppet Show" was the Muppet breakthrough: the American Guild of Variety Artists gave them its 1976 award as "Special Attraction of the Year" — not "special children's attraction." And Edgar Bergen himself agreed to appear on "The Muppet Show" with Charlie and Mortimer during the second season.

"The only reason some people still think of Jim Henson as a children's entertainer," Bergen told me in 1977, just before leaving for England to tape the show, "is because 'Sesame Street' is so popular. You know, I was offered juvenile hours on television and I turned them down because I didn't want to be identified as a children's entertainer. Audiences of all ages believe in both Charlie and Kermit. My act and the Muppets are both sophisticated and adult, but children love them, too, because we give children a chance to use their imaginations. They complete the illusion that our characters start."

☐

"Basically," said Henson, of his own puppets, "it all begins with those little sketches of characters that I or one of my associates make on people we know. They're based on a personality type or an attitude more than anything else. I look at the sketches until one seems to have the whole quality of the personality. Then we begin building."

When a Muppet face begins to express a personality, a puppeteer starts working with that new Muppet, to see how best he can communicate his or her personality through a voice and/or movements. On "The Muppet Show," there are five first-string puppeteers: Henson himself, Frank Oz, Jerry Nelson, and two who got all their training at Henson Associates: bearded Dave Goelz, 32, and Richard Hunt, at 27 the youngest of the group.

"Frank Oz is responsible for much of what's funny about the Muppets," says Henson, but in fact, the best comedy seems to come from the peculiar chemistry between the exuberant Oz and the more subdued Henson. "Frank tends to overdo a part and Jim tends to underdo a part," says Jane Henson; yet, together, they strike a funny balance.

Oz is also responsible for the first puppet Superstar since Charlie McCarthy — the complicated Miss Piggy.

"I remember that Miss Piggy used to be, in the very first year of the show, just a nondescript pig puppet that we had originally used prior to 'The Muppet Show,' in a half-hour pilot called 'Sex and Violence,'" said Oz. "There was a bit in that pilot with a whole bunch of pigs and she was just one of the pigs. But in one rehearsal, I was working as Miss Piggy with Jim, who was doing Kermit, and the script called for her to slap him. Instead of a slap, I gave him a funny karate hit. Somehow, that hit crystallized her character for me — the coyness hiding the aggression; the conflict of that love with her desire for a career; her hunger for a glamour image; her tremendous out-and-out ego — all those things are great fun to explore in a character."

Since the Muppets are a team effort, others also explore Miss Piggy's surprisingly deep personality. Back in New York, in the Muppet Work Shop, Calista Hendrickson dressed Miss Piggy from a

George the Janitor and Mildred enjoy a tete a tete over Oz's head.

MOVING MUPPETS

Jim Henson himself coined the term "Muppet" to describe his own individual combination of marionette and hand puppet. The heads of most of the Muppets, such as Kermit the Frog, are basically hand puppets, but the hands and other parts of many of their torsos may be operated by strings and rods and other controls, like marionettes.

Most Muppets are hand-sculpted out of foam rubber, then covered with wool, flannel or some other fleecy material from drawers in the workshop labeled "NEW FLEECE FLESH." "Fleece is the fabric we use for the skins of a lot of characters," explained head Muppet designer Caroly Wilcox. "They're mostly fake fabrics. We don't use real things almost at all except for Big Bird's feathers — which are turkey feathers that are bleached and then dyed two-tone yellow."

There are more than 400 distinct Muppets now, but Henson's puppet builders work with about 15 basic types, creating new characters by recostuming and changing varicolored noses, ears, eyebrows and other appendages, which are applied with adhesive fabric. Unlike most old-fashioned puppets, Muppets do not have a hard, wooden quality, and their flexibility — especially in their unusually wide, overbiting mouths — makes them strikingly lifelike.

As the puppet becomes a personality, it may be changed and refined. Miss Piggy herself, for example, has changed somewhat since its inception. "The design has changed a little bit — little tiny things in the corners of Miss Piggy's mouth," said Henson. "They are small things. But the result, if you look at photographs of the old and new Miss Piggy side by side, is a self-satisfied smirk more in keeping with Miss Piggy's oversize ego.

"We operate the Muppet's mouth with one hand," said Henson. "For Muppets like Kermit, who cannot grasp things with their hands, the right arms are moved with thin rods attached to their hands. We paint the rods to match the background so that they are hard to see. It's surprising how many people never notice them.

"A Muppet who can hold things takes two people to operate. One puppeteer will do his mouth and one hand with *his* two hands, and a second puppeteer will do his other hand." Other characters have electronic remote control eyes and ears, and the larger ones — like Big Bird — even have television monitor sets built in.

In short, Muppetry is puppetry made for a screen. "Often, on 'The Muppet Show,' we'll have all five puppeteers on a single number," said Henson. "We're surrounded by television monitors so we can see what the audience is seeing. We do bump into each other occasionally. But the audience shouldn't be aware of your problems or they'll lose the thread of the story."

In "The Muppet Movie," more than ever before, mechanical effects were used to make the puppetry more lifelike. "Still, every time we use mechanics," says Henson, "we try to keep them very, uh, unmechanical. As soon as the audience starts thinking about the cleverness of it all, then they're not thinking about the performances. When the Muppets are on the screen, I want the audience to believe in the moment. The audience can see that most of the characters end at the waist most of the time, and they can know who talks for them — none of that seems to kill the moment. But when they're watching us perform, believing the moment is everything." —J.C.

drawer labeled "POSSIBLE PIGGY." It is filled with fabrics from secondhand shops plus clothing that the costume designer's children have outgrown.

"Jim feels strongly that the clothes not be so strong that they take away from the character," said Calista Hendrickson. "But I think clothes can add a dimension to the personality. For instance, Miss Piggy's not aware of the fact that she's overweight — she dresses as if she's 30 pounds lighter. So she has a lot of fantasy."

Frank Oz explained: "The tremendous advantage of doing a series of 24 shows a year is that you have all that time to develop a character. And although Miss Piggy's essentially humorous, to me she's had a sad, difficult, painful life. This is not for the audience to know, but the puppeteer should know the background of any good character in order to be able to improvise."

Oz wrote a four-page Stanislavskian analysis of Miss Piggy's life and hard times and shared it with head writer Juhl. The essence of it, says Oz, is that "she grew up in a small town; her father died when she was young; her mother wasn't that nice to her. She had to enter beauty contests to survive. She has a lot of aggressiveness, but she needs a lot to survive — as many single women do. She

Ralph, the piano-playing dog, turns cameraman in Hollywood .

has a lot of vulnerability, which she has to hide, because of her need to be a superstar."

"Oh, Danny, *Danielle*," Miss Piggy recently cooed, after singing "Cheek to Cheek" with Danny Kaye, "why did *vous* want to do this particular song *avec moi?*"

"I heard you sing it once before," Kaye answered, "years and years ago, when you were thin."

The hurt conveyed by Miss Piggy's posture as she received his answer was an awesome achievement in the art of puppetry. Somehow, without her speaking a word, we could almost see her think and feel — and take his insult like a karate chop to the heart.

"Remember," says Henson, "an actor has an enormous range of expressions on his face, but most of the Muppets can only open their mouths. So the angle at which the head is held, how it's moved in relation to the body, or where the puppet is looking creates the expression. It's in the way you hold the puppet. Five degrees of tilt can convey a different emotion."

Miss Piggy's emotional range may be the broadest of any puppet in history. In fact, at her best, she may be the first round character in puppet history — in that she is capable of surprising audiences in a convincing way.

Indeed, the day may be coming when adult audiences will weep over the fates of puppets as Disney caused them to weep over the fate of cartoon characters in "Snow White" and "Dumbo" and "Bambi."

If it's possible with puppets, Henson will probably try it. "Jim's passion," said Jerry Juhl, "is to push the art form as far as he can," and Henson's next project will be his most daring use of puppetry so far. He refers to it only as "the fantasy film" and will say of it only that there will be no humans in it and that "it will derive from the vision of Brian Froud," illustrator of "The Land of Froud" and co-illustrator of the Abrams's best seller "Faeries."

Meantime, the money rolls in from Muppet merchandise — though no one will say how much — nor need they, because Henson Associates is a private company. When CBS's "60 Minutes" devoted a segment to "The Muppet Show" last March, Henson said that he would rather not have the coverage if it meant talking money, because "that's not what the Muppets are about. Indeed, we do earn money and we like to be successful and profitable, but the Muppets are there for creative pur-

poses and to entertain people." Nevertheless, he did concede that they "probably" made millions. "It's not what matters to me, it's not where I put my attention," he added.

This may be true enough, but it does not mean that commerce is neglected at Henson Associates. Seventeen countries now have active Muppet licensing programs, including 18 top manufacturers (Arista Records, Hallmark Card, Fisher-Price Toys, etc.) in the United States.

Of course, these revenues do not go to Henson Associates alone. The company splits with Lord Grade on a 50-50 basis. Grade's company (ITC) participates in the licensing of the characters on "The Muppet Show." And Henson Associates furnishes characters to "Sesame Street" for a fee. All of these arrangements make an exception of Kermit the Frog, who "pre-existed" "The Muppet Show" and "Sesame Street" and belongs solely to Henson Associates.

Beneath his controlled and guarded surface, Jim Henson seems to be trying to decide whether he will continue in the footsteps of Edgar Bergen and the young Walt Disney as master entertainers, or take the road that Disney took in his 50's — and become a modern media emperor. It may be significant that Muppet merchandise now refers to the characters as "Jim Henson's Muppets," and that Al Gottesman says they are thinking of calling the Muppet mansion "Henson House."

It is certainly significant that Henson is voicing his concern over whether "the quality of work can be maintained" when an organization becomes an empire.

□

So here was Jim Henson at the Cocoanut Grove, which he had redecorated for "The Muppets Go Hollywood" into an Art Deco dreamland. Tonight, the artificial coconut palms sheltered giant screens on which the guests would see an 18-minute condensation of "The Muppet Movie." Jenson had come a long way to this room from the family parlor in Leland, Miss., its radio tuned to Bergen and Charlie in Hollywood.

Tonight, Miss Piggy would be introduced as "the newest of Hollywood sex symbols," and she would be borne onto the dance floor by four muscle men, à la Mae West, with Frank Oz prone beneath her in the coffinlike couch, his right arm sticking up through the lid to keep her magic alive.

The show would begin late, and one number would have to be started over three times, apparently because technical people kept missing cues. But the number would finally be well received by many of the funniest people in Hollywood: Mel Brooks, Carl Reiner, Red Buttons, Kaye Ballard, Dom De Luise, Vincent Price, Dudley Moore. At last, human stars and Muppet stars would join a Conga line led by Rita Moreno which would be videotaped to form the climax of "The Muppets Go Hollywood." At the end of the evening, the score would be: Frog 99; Chaos 98.

But before all this hoopla commenced, there was a quiet moment. Jane Henson had flown across the country that day with their eldest daughter, Lisa, who had left Harvard for the occasion. Back home in the big house in Bedford, N.Y., were Heather, 8½; John Paul, 14; Brian, 15; and Cheryl, 17. Their mother hadn't even had a chance to change into evening clothes yet, but she stopped into the Cocoanut Grove on

her way upstairs in the hotel to make sure that their father had everything ready.

Holding hands, Jim and Jane Henson walked slowly across the dance floor where Valentino danced the tango and Joan Crawford danced the Charleston and Scott and Zelda danced the Black Bottom and Chaplin did clog dancing and William Randolph Hearst and Marion Davies did the Waltz. For the length of that walk the Hensons looked like two children allowed to go downtown alone only if they promised to hold hands crossing busy streets.

"Inside, we're all children," Jim Henson had told me two weeks before, riding in a New York taxi to his new mansion on East 69th Street. "You know, everybody identifies with that feeling of looking around at this big world and not knowing who you are and what you're supposed to be doing here." These days, of course, Jim Henson seems to know exactly who he is and what he's supposed to be doing here.

Yet that day in the taxi, beneath the controlled calmness of Henson's conversation, one could detect an undercurrent of anxiety lest the unity that he and Jane have so carefully maintained over the 23 years of Kermit's magical life be undermined, not by failure, but by a sudden flood of big successes.

"This coming year my family and I will move to England for a year," he said. "I just bought a house there. We'll do the next two seasons in London. I've been away so much. It'll pull us together a little bit more as a family." Back in New York, he would leave much of his organization quartered in the mansion.

"I pretty much bought this place for the stairway," Henson told me. "Our last location had offices in one building and the shop in another. When I saw this building with a great beautiful oval staircase right in the middle, I said, 'That's just what I need, because I want to unify everybody.' And that's the way it's working. Just running up the stairway to my office, I see six or seven people I normally wouldn't see. . . ."

At the Cocoanut Grove, the quiet moment had passed. Jane let go of her husband's hand and went upstairs to put on a bright red, floor-length formal gown. Jim went backstage to put on a green frog in a tiny tux. Before they rolled the mini-movie, Jim and Kermit had an announcement to make. "We're dedicating 'The Muppet Movie' to Edgar Bergen, who worked with us on it. . . ."

Frances and Candy Bergen would be pleased by that — and so would the child inside Jim Henson. ∎

June 10, 1979

The group: (left to right) Clem Burke,
Debbie Harry, Jimmy Destri, Chris
Stein, Frank Infante, Nigel Harrison.

Dennis McGuire

A COOL BLONDE AND A HOT BAND

By Ann Bardach and Susan Lydon

July 13 is a hot and airless Friday in Philadelphia. Inside the sold-out Tower Theater, tempers and temperaments are running high among 3,100 young fans eager to see the hottest new presence on the pop-music scene — the six-member rock band Blondie on its summer-long national tour.

A foot-stomping, standing ovation greets the band and, in particular, lead vocalist Deborah Harry, the blonde of Blondie. (Friends and fans call her Debbie; it incenses both her and the band when she's called Blondie. She speaks of the band as "Me and Blondie.") Debbie's drop-dead brand of glamour coolly evokes — and satirizes — hallowed images of Hollywood sex queens and Blondie's sophisticated "new-wave" music does the same for early 60's "girl-group" rock-and-roll, characterized by elaborately harmonious love songs sometimes exaggerated to a point of hilarity.

A year ago, most of these kids probably never heard of Blondie, but by now the album "Parallel Lines," Blondie's third, latest and best, has sold more than five million copies worldwide; and sales of the album's hit single, "Heart of Glass," exceed two million (at a time when the record industry reports slumping sales for the first time in 25 years). "Heart of Glass" is the most significant example of a song by a new-wave band crossing over to capture both the disco and MOR ("middle-of-the-road") audiences. Record-industry seers expect equal success for Blondie's new album, "Eat to the Beat," scheduled for release in mid-September.

Now, in the Tower Theater, when Debbie sings, the audience responds with familiarity to every song. Debbie stalks across the stage like a tigress staking out her territory. Dressed in a one-piece strapless white cotton jump suit, she smiles and blows kisses, camps and vamps her way through Blondie's hit repertory. Her movements, in the words of the Times critic John Rockwell, reflect "the awkwardly endearing klutzishness so charmingly at odds with the glossy sexuality of her photos," and her voice, which critics agree has steadily strengthened and improved, belts out the ending to "11:59," now on almost every jukebox in the country. "I'm still alive, still

alive," she exults, the band punctuating the phrases with staccato bass riffs, "still alive, still alive, bomp bomp, bomp bomp."

It's even hotter onstage than in the audience and Debbie's disregard for underwear soon becomes evident as sweat soaks through her jump suit, turning it semitransparent. The band swings through the synthesized, subdued, not-quite-disco strains of "Heart of Glass," their platinum hit ("platinum" being a flashy industry term signifying more than a million units sold), priming the audience for the show's climax. On the next number, the band rocks out as Debbie, snarling and sneering, eyes growing big and demented, her gestures predatory and provocative, bites off the words that half-threaten, half-promise her hungry fans: "One way or another, I'm gonna find ya, I'm gonna getcha getcha getcha getcha."

She is, too. This 5-foot-3 powerhouse who gives her age as 34 (insiders place her on the other side of 35) shows no signs of an energy crisis as she blasts off into mainstream celebrity. But backstage she collapses on a sofa after the show, her head wrapped in a towel. Here she looks tired and somehow vulnerable, not quite up to greeting the well-wishers who crowd around her. Chris Stein, Blondie's 29-year-old, prematurely graying lead guitarist and Debbie's live-in friend and collaborator, still dressed in the Doctor X T-shirt, black leather pants and dark glasses he wore onstage, wanders about the scene, videotaping everything. Recovered, Debbie clowns for the camera, slipping a carrot and celery stick in her nostrils and guzzling Perrier from a giant bottle until it drools out of her mouth. "As you can see," she announces solemnly, "I'm a health-food freak."

Described variously by trade and national publications as a "punk Garbo," "the Marilyn Monroe of punk rock" and "the most photographed woman in rock-and-roll," Debbie is a strikingly good-looking woman. Her broad face,

Debbie Harry singing on a television show in London.

Ann Bardach and Susan Lydon are freelance writers.

Sonia Moskowitz

Debbie at a disco with friends Andy Warhol (to her right), Truman Capote, Jerry Hall (far left), Paloma Picasso.

BLONDIE

Continued

with its high cheekbones, boasts translucent skin, big blue eyes and the most perfect lips since Clara Bow's. Chameleonlike, she can break into a wide smile that shows off her dimples and lights up her face (an aspect of her almost never seen in photographs) or she can pucker those sensuous lips into a sudden, petulant sulk; one moment she can stare at you in almost total innocence and in the next look as if she had just stepped out of the cast of "Marat/Sade." Quite simply, she has old-fashioned star appeal. Some entertainment moguls believe that Debbie has what it takes to go the route of a Streisand, a Diana Ross, or a Cher.

☐

But Blondie is more than Deborah Harry. The band's music, melodically diverse, is characterized by a compulsively throbbing beat, ululating guitars and celestial keyboards that sound like futuristic carrousels. Unlike other bands in which every member flaunts his ego, Blondie flaunts Debbie, the obvious visual focus of each performance. Their repertory includes updated versions of "golden oldies," sci-fi and horror-movie songs and original satire like "Rip Her to Shreds" ("Red eyeshadow, green mascara/Yeccch! She's too much!"). Each member of the group is a talented songwriter, their supply of catchy melodies and commercial "hooks" — musical devices that catch and hold attention — seemingly inexhaustible.

Blondie began to receive attention in this country early this year when news spread that "Parallel Lines" had gone platinum. By spring, "Heart of Glass" topped the charts. A cynical love song ("Once I had a love/and it was divine/ soon found out I was losing my mind"), featuring Debbie's dreamy, disembodied vocals over a hypnotic, roller-rink organ background, "Heart of Glass" appalled Blondie's fans and fellow musicians on the underground new-wave scene. But it got the band the AM radio airplay which translates into more fans and big bucks.

The new-wave (formerly known as punk) bands had long complained that they couldn't get their music heard on the radio; although they voiced scorn for Blondie's pragmatism in "selling out to disco," they subliminally hoped that its breakthrough would pave the way for their own entrance into mainstream acceptance.

Blondie grew up with punk,

which began in the mid-70's as a loose movement of bands wanting to recycle the raw, high-powered energy of 50's and early 60's rock-and-roll. They were reacting against the overproduced, too-slick sound of most pop music — the mindless, repetitive rhythms of disco and the bland creaminess of studio-created pop, both of which monopolized the airwaves.

Bands like the Ramones toured England, where their punk attitude — they behaved onstage like truculent street toughs — was adopted by English pub bands as a vehicle for the political outrage expressed by the theater's "Angry Young Men" two decades earlier. Punk music was minimalist, some said chaotic; its lyrics emphasized alternating currents of nihilism and sentimentality. When more musicianly and cerebral groups, such as Talking Heads, came onto the scene, English music writers invented the umbrella term "new wave" to encompass all late-70's groups, however disparate, which were aiming to restore gut feeling to rock-and-roll.

American music critics also touted new wave as an intelligent alternative to disco and the music flourished in the suburbs and on the Lower East Side's avant-garde art scene. Blondie was one of the first and flashiest New York new-wave bands, although Debbie never liked the "punk" label. "Punk is a time signature," she says. "Punk to us is a time in New York, a time in the world. Acid is before; glitter is before; R 'n' B [Rhythm and Blues] is before, and now it's punk. But it's all rock-and-roll, straight down the line."

☐

Blondie successfully toured Europe, where the new wave has wider acceptance, before "Heart of Glass" broke open the American market last spring. Debbie's face has appeared on the cover of almost every European magazine. Careful packaging and promotion helped the success of "Parallel Lines" in the United States. So did the unusual pairing of the band with a "bubblegum" hit maker, the producer Mike Chapman, a wizard at capturing the teen-age audience — a move engineered by Blondie's former manager, Peter Leeds. Leeds also paid $500,000 to buy back the group's record contract, shifting Blondie to the Chrysalis label, which he felt would serve the group better.

The emergence of Deborah Harry in her own right is due to her special personal style.

Playing clubs like CBGB's and Max's Kansas City in New York in the early days of punk, Debbie created a unique blend of what she called "trash, flash and freak chic." Against the prevailing mode of black leather, safety pins and chains, Debbie stuck with her miniskirts, spike heels and teased beehive hairstyle, exploiting every successful feminine stereotype of the 50's and 60's. Making it as a female rocker in a man's world, she presented herself almost as a cartoon character with a blatant, ironic sexiness that proved surprisingly unthreatening, even winning, to the women in her audiences. "I wish I had invented sex," she told an English interviewer. "Sex is everything. It's Number 1."

Sex sells and the new-wave bands — despite their rebellious stance — did want commercial success. And so the punk attitude manifested itself in New York as a psychodramatic fashion show, emulating the Andy Warhol-dominated world of pop art and art rock. A friend introduced Debbie to Warhol years ago. Last June, after she had appeared at Max's, Warhol gave a party in her honor at Studio 54. Although Debbie's face was then gracing the cover of Warhol's Interview magazine, and although the disco's huge sound system played her records, few there over 25 knew who she was. Near the rear door, Deborah Harry smiled faintly to appease the paparazzi. The reason she was at the rear door was that she was trying to find a way out of her own party.

"She is shy," Debbie's mother told a Rolling Stone reporter. "When she's not performing, she's quiet, with a very pixie sense of humor. She's not real outgoing or loud. She's sort of retiring." Richard and Catherine Harry adopted Debbie in New Jersey when she was 3 months old. They raised and schooled her with their other daughter in the quiet, middle-class suburb of Hawthorne, N.J. The Harrys are modest Episcopalian people who have since moved to Cooperstown, N.Y., where they own a gift shop. Debbie sang in church and acted in school plays. "I used to always mimic everything I saw," she says. "Whatever was there, I was copying it."

Debbie's adoption has perplexed and disturbed her. In the summer of 1978, she was asked by the photographer and film maker Sam Shaw to provide biographical information for a one-hour documentary film on Blondie. Interviewed by the screenwriter/novelist Ted Allan, Debbie mentioned liking a play, "Fame," loosely based on the life of Marilyn Monroe. The conversation continued:

ALLAN: Do you have an

affinity for Marilyn Monroe?

HARRY: Tremendous. I always thought she was my mother.

ALLAN: Did you ever seriously think that you'd go and meet her and say, "You might be my mother."

HARRY: No! God! Well, you know, there's that kind of admiration, I guess. They say that most adopted children now, in their adult life, look for their real parents . . . I sort of have my wild imaginings. Like she [Monroe] had wild imaginings about Clark Gable being her father . . . See, my mother did keep me for three months, and I have memories, a visual memory of when I was 3 months old when I was adopted.

Such fantasies percolated through Debbie's adolescence. She says that she felt "different," worried about being crazy. As late as last year, Debbie reminded a reporter that "Marilyn was also an adopted child."

At any rate, the suburban rituals of Hawthorne and its public-school system left much to be desired for the adolescent Debbie. Early on in high school, she says, she embarked on weekend sojourns to New York's Lower East Side, where she eventually moved after two years at a finishing school — "a reform school for debs" which she "wasn't rebellious enough to leave."

On her own finally, Debbie tried writing, painting, acting, singing, waitressing and, not least of all, drugs. Asked if she took a lot of acid in the 60's, Debbie replies, grinning, "I can't remember." In 1968 she was a backup singer in a folk group called Wind in the Willows. In photos from this period, she is almost unrecognizable, dressed in loose Indian-bedspread clothes with her mousy brown hair hanging straight from a center part. When the band broke up, she went through a difficult period, ending up with a heroin habit. Drugs weren't all that new to her; she recalled taking "speed and ups all through high school," but this was worse.

"For a while I was so pent-up inside I couldn't sing," she recalls. "Like, everytime I would sing I would cry. It was really horrible." Debbie credits her abandonment of drugs to a "semireligious experience," which she does not elaborate upon except to say that at the time she was "studying Indian music, working as a Playboy bunny, and doing yoga." When her doctor observed that she was in exceptionally good health for an addict, she attributed it to the yoga. "I'm glad I did it," she says now of her drug days, re-

calling that she ended up "coming back to life, wanting to live, developing faith. I mean not faith in God, but just faith, generally."

Debbie knocked around the art scene for a while, appearing in underground movies and plays, eventually teaming up with an all-girl singing group called the Stilettos. At their second show, in 1973, a friend brought Chris Stein to see the group, and Debbie recalls sensing "this guy staring at me real intensely. I couldn't see him, but I could feel his eyes looking at me." Chris is the son of intellectual parents who were radical activists in the trade-union movement in the 30's. He attended New York's School of Visual Arts. He and Debbie became friends; he joined the Stilettos' backup band, and after their relationship became romantic, they formed Blondie.

☐

Despite success, all is not well in the Blondie camp. Many fans who were loyal to them in their lean years are deserting because they resent Blondie's "commercial sellout." In addition, the group has spent the last year in acrimonious and costly negotiations (Blondie gave up a share in its future gross, plus an undisclosed sum) to break from its first manager, Peter Leeds. Shep Gordon, Alice Cooper's manager, was chosen to replace Leeds, and he took over on Aug. 1. In the interim, Debbie and Chris were managing the band themselves and on occasion — because press attention seems naturally to focus on Debbie — that proved to be a can of worms. When asked for interviews, Debbie and Chris tactfully insisted that the other members of the band be interviewed first.

The marked internal discord in the band expresses itself in the well-distributed button declaring "Blondie is a group," an irate reminder that Blondie has six members. However, many people in the industry agree with Rolling Stone's judgment that Debbie Harry is "the only one in the hot new rock package who can't be replaced." A possible exception is Stein, lead guitarist and Blondie's principal composer, who has been described as Debbie's éminence grise. The boys in the band — Clem Burke, drummer; Frank Infante, guitarist; Nigel Harrison, bass; and Jimmy Destri, keyboardist — all want their place in the sun, too. For the time being, Debbie and Chris are trying to appease them.

While the band members are telling reporters how each joined the group, Debbie, who has changed into a black dress slit up the sides and high-heeled backless mules, stares uninterestedly at the television set or at the wall. Her hands fidget nervously with her room key and her face is suffused with despair and boredom. Stein, who speaks with a slight lisp, mutters about the "misquoting and distortions" of the press. When the reporters suggest shifting the focus to Debbie, the normally reasonable Burke, declaring the interview "irrelevant and boring," dumps the contents of a wineglass and an ashtray on a reporter's head. ("I'm not surprised," a record company executive at their label, Chrysalis, commented later. "It's happened before. It could have been worse.")

☐

Because Blondie's rise was a long time in coming, Debbie — a survivor — is not about to jeopardize it now. This seemingly spontaneous, casual performer takes nothing lightly. Alone in her room later in the evening, she relaxes, begins to talk: "I was singing at a time when Janis Joplin was really out there, and then she died. I kept thinking about Janis and Billie Holiday and blues. It just kept going through my head — all that sadness and tragedy. I love the blues, but I didn't wanna sing them. I wanted to be up and happy and entertain people and have a good time. Those blues singers were forced to live out the reality of blues. I had to make sure that I was strong enough and felt good enough about myself to avoid that."

Debbie speaks with a street accent, redolent of Brooklynese and punctuated with flat, nasal "yeah's" and "y'know's." As a performer, she has remarkably broad appeal; her campy sexuality attracts young and old, homosexual and heterosexual, men and women. Asked if it is hard to keep herself together on the road, she replies, "It's hard. Life is hard. It's hard to get up at 8 in the morning and type 60 letters in a day. It's hard to work at Woolworth's. Everyone's life is hard." She feels that many people can relate to Blondie's music because "all of our songs aren't about love and hate relationships and broken hearts

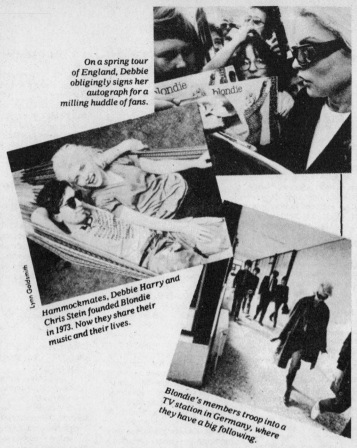

On a spring tour of England, Debbie obligingly signs her autograph for a milling huddle of fans.

Lynn Goldsmith

Hammockmates, Debbie Harry and Chris Stein founded Blondie in 1973. Now they share their music and their lives.

Blondie's members troop into a TV station in Germany, where they have a big following.

and stuff like that. We have songs that are about cars and minimovie themes and different subjects — the gas shortage, everything."

Deborah Harry, the singer, delivers her songs in a flat, almost uninflected manner, but the actress in her infuses the songs (which she characterizes as "high-energy music") with theatrical

drama. "Part of a singer's job is to make a song believable," she says, "to make it come alive and to transmit a story. If I'm singing about surf's up, I'm in the sun and I'm waiting for my wave, and while I'm really in New York, it's believable. It's so much easier with music than with acting. Different tonal qualities express dif-

ferent emotions and get different feelings going, and the beat and the excitement and all the electricity just make it that much hotter."

☐

For the past six years Debbie has maintained a stable relationship with Chris Stein. They live together in a penthouse apartment above Sixth Avenue and are described by themselves and others as inseparable. Their life together, except when touring or recording, is quiet, and Debbie has even described herself to the press as "a housewife." Stein has his own brand of radical anarchist politics, strongly based on the belief that "we are entering an age of apocalypse to be followed by an age of enlightenment."

Debbie, like Stein, has been interested in making films since her East Village days. Though she may now speak of having only a casual ambition to be a film star, few believe it to be so. Only a year ago she spoke of "maybe breaking out on my own, definitely toward film."

In the meantime, Debbie is staying ready for the future, whatever it may be. She tries to take care of herself, for one thing. Now she neither drinks nor smokes, often leaving rooms of smokers to protect her throat and sinuses from inhaled irritation. She has no strong antidrug stance, but says she doesn't much use them anymore. "I like to get high, but I don't like the hangover. That's what it amounts to." Offstage and on, she exudes a rare wholesomeness, all the more startling in its juxtaposition with the brassy bleached hair (she has a beautician's license, but she deliberately lets the dark roots show) and 50's downbeat clothes. She has intelligence, wit and resolve. One thing she has resolved *not* to be is "some kind of a victim." Deborah Harry has better things in mind for the 1980's. ■

August 26, 1979

Unmasked Lone Ranger Pledges Life to His Role

By ROBERT LINDSEY
Special to The New York Times

LOS ANGELES, Sept. 6 — "I've been the Lone Ranger for 30 years," Clayton Moore said in a drawl that seemed to have originated somewhere West of the Pecos and South of Yesteryear, "and I intend to be the Lone Ranger for the rest of my life; I've decided I'll stay the Lone Ranger until I'm called."

His blue eyes are partly hidden by wide, dark sunglasses that vaguely resemble the Lone Ranger's mask, of which Mr. Moore was officially stripped recently by a Superior Court judge. Through the glasses, the eyes seem to flicker with outrage whenever he contemplates the prospect that any one else would ever think of playing the masked rider of the plains.

"They tried to shoot me down, but I stood up to them," Mr. Moore said in an interview. "I've lived

by the Lone Ranger creed for 30 years and I'm fighting now for fair play, justice, law and order."

From TV's Limelight to County Fairs

As a television actor, Mr. Moore portrayed the fictional hero in almost 200 half-hour dramas from 1949 to 1956. Since then he has been earning his living much of the time by making personal appearances at county fairs, shopping center openings and other events, wearing a leather mask, a white hat and six-shooters.

Last month he was ordered into court to answer charges by Lone Ranger Television Inc. and the Wrather Corporation, which owns the copyright to the character and is producing a new movie based on the Lone Ranger stories. The companies charged that, at the age of 64, Mr. Moore was "no longer an appropriate physical representative of the trim 19th-century Western hero."

At a hearing on the dispute, a lawyer for the Wrather Corporation argued that Mr. Moore had "no right to the Long Ranger mask."

"It's our mask," the lawyer said. "By wearing the mask, Moore is appearing as the Lone Ranger. But in spite of what Mr. Moore feels in his heart, he is not the Lone Ranger. We own the Lone Ranger."

A Difficult Decision for an Actor

Mr. Moore now bristles when he thinks of this and other remarks made at the hearing, which ended with a court order that he remove his mask. As he discusses his fight to wear the mask, his conversation drifts, almost seamlessly, from talking about himself in the first person to talking about himself in the third person and then, in a continuous fabric, he seems to begin talking about himself as the Lone Ranger.

In 1956, he said, after the television series ended, "I had a decision to make about what to do, whether to go back into motion pictures as an actor, which is what I did before, or spend the rest of my life appearing as Clayton Moore, who portrayed the Lone Ranger."

His decision, he said, was to appear as the Lone Ranger, and now he had no intention of giving up that role. Mr. Moore said he would continue to be the Lone Ranger, "without the mask, if I have to — this is my life."

"I've always believed in law and order and I'm abiding now by the court" by not wearing his mask, he said. But he added that his fight was not over: The ruling to unmask him would be appealed. "I want a jury trial," Mr. Moore said. "We're going to fight it in an honorable way."

"I have no animosity toward the Wrather people, but I just hope and pray that they live up to the moral code of the Clayton Moore for the last 30 years."

As a child growing up in Chicago, he said he decided early that he would become a policeman or a cowboy. Thus, in many ways, he said, it was not surprising that he became television's Lone Ranger. "There are a lot of good guys who wear white hats," he said, "and I'm one of the good guys; the police are white hats; John Wayne was a good guy, so was Tom Mix, my hero. This is Americana."

Living Amid Memorabilia of TV Days

Mr. Moore and his wife, Sally, to whom he has been married for 36 years, live in what acquaintances describe as a large but not lavish home in the rural outskirts of the San Fernando Valley here. The home is decorated with memorabilia of Mr. Moore's television career, including, of course, a silver bullet.

"I try to lead a good, clean life; I don't smoke; I don't drink; it's not that I'm trying to be a goody-good, but I just don't want to," he said.

He said that news about his court fight has generated "a worldwide response — I've had 95 radio interviews since this started; it's been 18 days

United Press International

Clayton Moore, who portrayed the Lone Ranger in more than 200 half-hour TV dramas, displaying new sunglasses.

straight, clear from Australia and London, all over."

Mr. Moore, who left Los Angeles today for a weekend convention of nostalgia buffs in New York City before beginning a series of personal appearances in Texas, said: "This has made me feel 10 feet tall. I'm just elated by my fans from all over responding like this."

Guardedly, he acknowledged that the recent publicity had helped his income from making personal appearances, but he declined to answer questions about his income. "Actors have a great deal of expense," he said. "We must put on a good front and buy costumes; the public expects a lot from actors and we can't disappoint them."

Mr. Moore said he was moderately pleased with the sunglasses because of their vague suggestion of a mask: "I searched this town for the proper glasses." But he said the glasses didn't make up for the loss of his real mask.

"I miss that mask, I really miss it," he said, "I want that mask and I'm going to fight to get it back."

The time for the interview was up.

Mr. Moore paused a moment and said:

"Adios, kemo sabe."

September 7, 1979

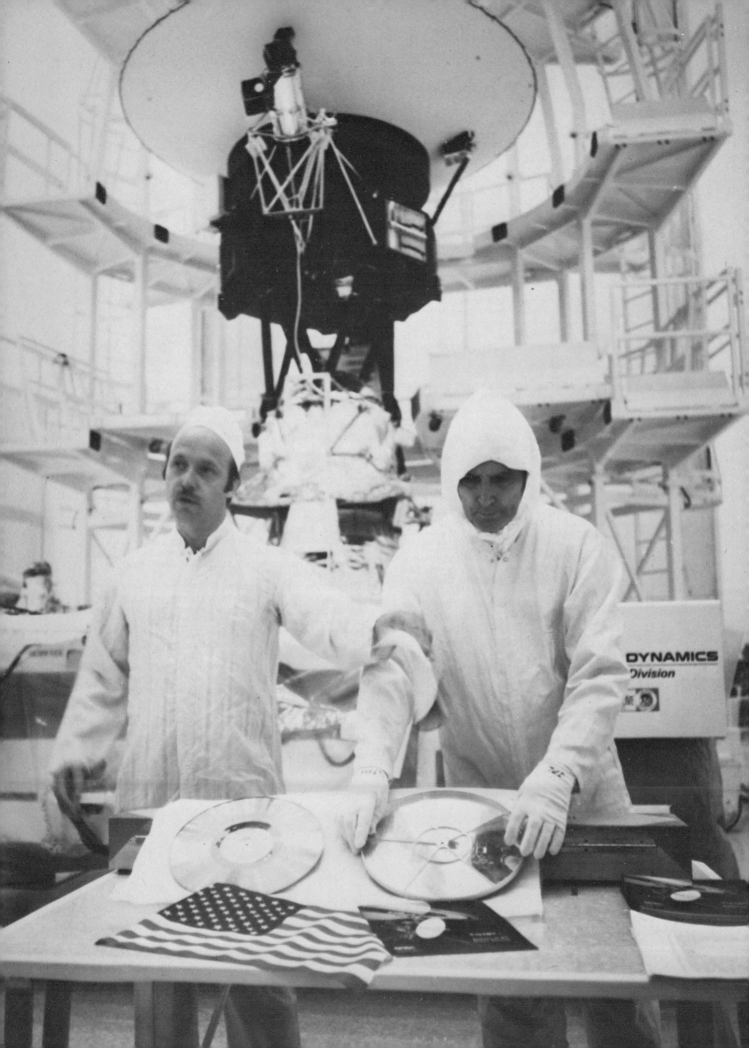

CHAPTER 10

Science in the Twentieth Century

The "Sounds of Earth" record was mounted on the Voyager 2 spacecraft prior to Voyager's rendezvous with Jupiter in July, 1979.

NASA

CRUCIBLES OF THE COSMOS

Astronomers who are studying galaxies are discovering unexpected activity that may yield answers to the most important questions in cosmology: the size, the age and the rate of expansion of the universe.

By Timothy Ferris

Dr. Allan Sandage sat up late one night recently at the Kitt Peak National Observatory in Arizona exposing photographic plates on the giant four-meter telescope and talking about galaxies. It was a dark, moonless night. It always is when Sandage is to be found observing, for his interests lie with starlight that has been journeying in space for millions of years and arrives in so feeble a condition that it can be analyzed and recorded properly only by the dark of the moon. Sandage is one of the blue-water sailors among astronomers.

The sky over Kitt Peak was full of stars. Sandage could see none of them. He was sealed in an observer's cage suspended within the telescope tube high above the observatory floor, his vista of the cosmos limited to its close black walls and to a single blob of light in the eyepiece that he scrutinized to assure himself that the telescope was tracking correctly while the plate was exposed. The plate is what matters in viewing galaxies. The human eye is insufficiently sensitive to view galaxies in detail and deep-space astronomers rarely *look* at them through telescopes. Nor do they spend much time admiring the resulting photographs, though galaxies are indisputably beautiful and I have seen an ordinarily sober astronomer jump up and down with excitement at a photograph of a particularly striking one. Instead, they analyze the plates, counting and measuring thousands of star images on them, seeking to discern the anatomy of these cities of stars. As to the stars in the Arizona sky, Sandage appreciates their spectacle as do the rest of us, but he finds them rather ... local. Astronomers like Sandage enjoy gazing at them casually but find them of such peripheral concern that many never bother to learn the constellations. Some would have trouble locating the Big Dipper. These stars are our neighbors. They crowd the foreground like leaves in a celestial tree that makes up the Milky Way galaxy. The deep-space astronomers are interested in peering out beyond, to the other galaxies. *(Continued on Page 39)*

Timothy Ferris is the author of "The Red Limit" and co-author of "Murmurs of the Earth: The Story of the Voyager Interstellar Record."

Astronomer Allan Sandage in the observer's cage of the four-meter telescope at Kitt Peak. Top, overlooking its Arizona site.

Steve Northup

"Trees aren't important when you're interested in the forest," Sandage said over the intercom from the observer's cage. By "forest," he meant the universe.

Now astronomers are finding that galaxies are stranger than anyone had suspected. Nesting in them may be creatures as provocative as black holes feeding on stars, and quasars, pouring out power with a vigor that astonishes physicists. They are enormous physical laboratories; by better understanding what goes on inside them, astonomers expect to gain fresh insights into the age, size and dimensions of the universe.

A team of researchers including Wallace Sargent of the Hale Observatories in Pasadena and C. R. Lynds of Kitt Peak recently found what they believe to be a huge black hole in the nucleus of the massive galaxy M-87, making the galaxy resemble a sort of cosmic sink with an infinitely deep drain in its center. Earlier this year, astronomer Donald Hall of Kitt Peak, studying how stars form in galaxies, discovered a star so young that it has been shining, in Hall's estimate, only since Homeric times, a mere instant in astronomical terms. A host of investigators in galactic dynamics, among them Beatrice Tinsley and Richard Larson of Yale University, James Gunn of Cal Tech, and the veteran observer Halton Arp of the Hale Observatories, are finding that galaxies undergo startling transformations. Big galaxies swallow little ones. Galaxies collide. Some, it appears, explode. Some dance grand passacaglias, swapping stars by the millions. These transformations are less violent than they sound — we who are riding aboard a galaxy need worry little more about them than Jonah worried about the perambulations of the whale — but they add up to a picture of a universe in evolution, in whose story our own evolution plays a part.

It is beginning to look as if we are deeply connected with our cosmic surroundings, like birds perched in a galactic tree, our lives involved with the fortunes of the starry forest. Some of these connections may dictate matters as mundane as the weather. Several British astronomers, among them Sir Fred Hoyle, R.A. Lyttleton and William McCrea, suggest that the Ice Ages may have been triggered by the solar system's having passed through a cloud of dust and gas associated with one of our galaxy's spiral arms. A Colorado

A "small" galaxy passes near a large galaxy, tugging stars by the millions into the void of intergalactic space between them.

physicist, Lars Wahlin, has proposed that lightning bolts, those symbols of the caprice of fate, may be produced by cosmic rays — particles accellerated through interstellar space by our galaxy's magnetic field. If Wahlin is right, then the old commonplace is true, and lightning bolts really are, in a sense, dispatched from heaven.

Still deeper connections between ourselves and the galaxies are being discerned. Astrophysicists studying the chemical composition of stars, and biologists investigating the chemical composition of our bodies, have found that we are made up of much the same allotment of elements as is our galaxy: The metals found in trace elements in our bodies appear to have been formed in the explosions of stars that died before the sun was born, seeding space with the metal-rich dust and gas from which our solar system and, eventually, ourselves were formed.

The story of galaxies is beginning to reveal itself to a new generation of astronomers, as well as to their elders, through the eyes and ears of new astronomical tools. X-ray satellites have made it possible to study the tenuous intergalactic gas

through which galaxies pass, permitting analysis of how entire clusters of galaxies behave. Ultraviolet sensors, working in realms of the spectrum beyond the range of the human eye, have yielded clues to the puzzling quasars, which may represent galaxies in their early stages of development, shimmering with light that started on its voyage to our telescopes billions of years ago. Researchers aided by computer simulation techniques are taking steps toward cracking the highly complicated problem of how galaxies herded their billions of stars together in the first place. Says Gunn: "We've only just started looking at galaxies in analytical ways, but I have every faith that some real understanding will be reached and that it won't take terribly long. Tremendous effort is being dedicated to it." And Sandage, as we talked, said, "We see the story of the evolution of galaxies coming to a head in a remarkable way. Problems that had seemed impossible to understand are yielding, and a general picture of sorts is emerging. The subject is going to be very lively in the next few years."

The astronomers who know galaxies best tend to be care-

ful in choosing their words about them. They are mindful of the contrast between the grand reality of the galaxies and the elementary nature of our understanding of them. Sandage, who in the estimation of many colleagues possesses the most accomplished mental picture of galaxies of any man alive, can be circumspect in the extreme. Sometimes he begins a conversation by denying he knows anything at all, a characteristically scientific sort of defense, the social equivalent of calibrating all meters to zero. That is what happened when I appeared at Kitt Peak and announced myself through the intercom.

"I don't work on galaxies anymore," said Sandage from his perch up in the dark. "I gave it up."

The night assistant sitting beside me grinned and pointed to the coordinates glowing in ruby numbers on the control console. They indicated that Sandage was photographing M-81, a spiral galaxy six and a half million light-years away.

"That looks like M-81 you're on now, Allan," I said.

"Oh, well, it's not a big galaxy. Only 90 billion stars."

"The residents of M-81

might not be pleased to hear that. It's a beautiful galaxy."

Yes. You know, beauty in galaxies appreciates according to their mass. The more stars a galaxy has, the prettier its spiral arms."

We talked through the night. Like a guide spotting fishing holes, Sandage can show you where in a galaxy to find nurseries churning out young stars, or retirement homes of old stars nearly spent, or slowly winking variable stars that will help you chart the galaxy's distance. He is familiar with giant elliptical galaxies larger than our Milky Way, with paltry irregular galaxies that look like little more than thimblefuls of sand, with spirals of elegant beauty and galaxies whose eventful past has left them tattered. He can show you two spiral galaxies cocked to each other like an open pocket watch, and note that in such pairs, the two spiral in opposite directions. (Our own galaxy and the great spiral in Andromeda form such a pair.) "Galaxies are what they are," Sandage said, quoting a Zenlike remark of his mentor Edwin Hubble, one of the discoverers of the expansion of the universe. "You try to learn from them."

Sandage searches for similarities in galaxies, seeking to improve our estimates of the dimensions of the universe and the rate at which it is expanding; these numbers in turn ought to yield a prediction of the fate of the universe — whether expansion will go on forever. The approach stands in contrast to that of some younger researchers whose curiosity leads them to be concerned more with individual galaxies — how they formed and why they look the way they do. Many are physicists rather than astronomers. "We who came into the field from physics are not so much grand builders," says Wallace Sargent, a physicist at Cal Tech. ("Grand builder" in astronomical circles is more or less a code name for Sandage.) "The big picture is far too uncertain as yet. We tend more toward being explorers, toward looking for an interesting particular problem and going after it. We look for galaxies whose stars are bluer than average, or redder, or fainter than average, and try to see what the hell is going on there." A colleague put it more baldly and less seriously: "We physicists are opportunists."

Recently several of the physicists have been looking into the question of why some galaxies seem to be producing no new stars. In a "healthy" galaxy like ours, new stars form with regularity, but some spirals appear to have closed up their star-making shops long ago; they contain only old stars. Our sun is itself a relatively young star. Had our galaxy stopped forming stars six billion years or more ago, we wouldn't be here to ask why. Two young Kitt Peak astronomers, Karen and Stephen Strom, suggest that the "unhealthy" galaxies may have suffered collisions with other galaxies or with clouds of intergalactic gas, collisions that swept them clean of the dust and gas they needed to make more stars. The Stroms and others have found evidence that "unhealthy" galaxies tend to be found in the crowded inner regions of galaxy clusters, where collisions would be most likely.

The question would be interesting to residents of an "unhealthy" galaxy if they were interested in developing technology, because metals apparently are available only to inhabitants of planets of younger-generation stars. According to this theory, widely accepted in astrophysics, metals are forged in the explosions of dying stars, which seed the interstellar medium with heavy elements that in turn condense to form new stars and planets. In galaxies where star formation ceased before the metals seeded space, all the planets would be metal-poor. Astronomers there, studying the skies with telescopes formed of

bamboo or lignum vitae or whatever other light materials had been left them by their cosmic fortunes, presumably would ask themselves the same sorts of questions the Stroms are asking here: Why do galaxies turn out differently from one another? Says Stephen Strom, "We're trying to unravel what might be called the genetics of galaxies, their evolutionary processes."

One obstacle has been that galaxies conduct their affairs over such long periods of time that on a human time scale they appear frozen. If you want to know whether hawks nest on a particular cliff you can sit in a blind for a month or two and see if any hawks show up. But if you want to see two galaxies interact as they pass each other, you'll have to wait a few hundred million years, and that is more than even the most patient astronomers can manage. Recently scientists at several institutions have employed computer simulations to look into the past and future of galaxies. The computers are programmed with data on the mass and size and relative locations of galaxies. Then the programs are run to re-create events that took eons to unfold.

A pioneer in this endeavor, Alar Toomre, showed me a computer-generated film one afternoon in a darkened office at M.I.T. The subjects were M-51, a lovely spiral galaxy, and a small nearby irregular designated NGC 5195. The symmetry of M-51 is broken by one distended arm that reaches out toward the smaller galaxy; astronomers had wondered just what was going on there. Toomre's film appears to have solved the riddle.

Toomre started the film at a point millions of years ago. As the millenniums sped by (a counter in the corner off the screen, calibrated in millions of years, flickered faster than the eye could follow), the two galaxies churned aross space toward each other. At first they looked as unperturbed as strangers approaching on a city street. Then their gravity began to make itself felt as the space between them narrowed. Each galaxy contorted like the face of a man who has been punched in the stomach. Their arms, each home to billions of stars, waved like tentacles, reached out almost to touch — and the film froze. We were at the present.

The computer rotated the model in one dimension, and we were treated to the sight, unprecedented on earth, of how a couple of galaxies look from another perspective. If Toomre's film is correct, the two galaxies have experienced their nearest encounter and NGC 5195 is continuing off into space, followed by one beseeching arm of M-51. A few million stars have been left homeless by the episode. They

now ride alone in intergalactic space. If astronomers evolve there one day, they might, by recourse to the sort of analysis Toomre has applied, learn how they came to find themselves dwelling under starless skies, drifting between two galaxies and claimed by neither.

Computer analysis by Toomre, Roger Lynds of Kitt Peak and others has enjoyed at least preliminary success in explaining the curious ring galaxies. Dozens of these puzzling objects have been found, each resembling a smoke ring. The computer studies suggest that ring galaxies are formed when a smaller galaxy passes through the center of a large one, punching a hole in it. And indeed, near every known ring galaxy is found a smaller galaxy, slinking away from the scene of the accident.

Galaxies are mostly space, and collisions between them may take place with few if any stars running into each other. The exceptions to this rule are the nuclei of galaxies, their centers, the Grand Central Terminals of stars. What goes on there is not yet well understood. Some galaxies emit considerable light and natural radio noise from their nuclei, and our own galaxy's core rumbles a bit in radio wavelengths. (Intervening dust clouds prevent our seeing the center of the Milky Way galaxy in visual light, but radio telescopes can "see" it.) A number of astronomers and physicists are interested in galactic nuclei, not least because what goes on there may tell us something about two of the most provocative subjects of modern astronomy, black holes and quasars.

Black holes were predicted indirectly by that cornucopia of gravity physics, Einstein's theory of general relativity. The theory implies that if mass is concentrated to a high enough density — as in the case of a giant collapsed star — its gravitation will be so intense that nothing, not even its own light, can escape from it. In Einsteinian terms, the black hole occupies a well in space-time whose sides are so steep that radiation cannot "climb out." Most theorists today take seriously the possibility that black holes exist, but actually finding one poses difficulties, as black holes are invisible. A black hole swallowing up gas from an interstellar cloud, or tearing it off the surface of a nearby star, might betray its presence by a sort of scream released by the doomed gas. Calculations show that the scream ought to be detectable in the zone of the electromagnetic spectrum known as X-rays, and observations of X-ray satellites have revealed several such sources that may signal the presence of genuine black holes.

Quasars, discovered by Sandage in 1960, appear to be very remote objects glowing with furious brightness. They are so bright that astrophysicists have trouble imagining how they turn

out so much energy. One popular speculation is that they represent the nuclei of young galaxies, where the collapse of gas clouds formed massive black holes; their intense energy would result from tormented dust and gas heated as it spins into the black hole. More than one hypothesis urges that stars, too, are consumed by black holes at the center of these galaxies. Some quasars flicker; this might signal that they are gorging themselves on stars. "It would be quite a sight," says one physicist, "stars spiraling down into the hole like bowling balls, first dismembered and then disappearing in sheets of light."

Virtually all quasars lie at distances of a billion light-years or more, meaning that it has taken their light a billion years or more to reach us, so we are seeing a feature of the universe as it was that long ago. No quasars are known to exist in the modern universe. But some relatively nearby galaxies exhibit signs of lingering violence at their centers, like the thunder in a subsiding storm. If these nuclei once were quasars, and if the black hole concept of quasars is correct, we should find black holes at the nuclei of many galaxies today. With this in mind, astronomers in the United States and England recently examined two large galaxies, whose great mass makes them likely candidates, and found evidence of giant black holes sitting in the nucleus of each. Observations by an orbiting space telescope, scheduled for launch in 1982, should help decide the matter.

An alternative to looking for relics of ancient cosmic violence is to search out galaxies still in their birth throes. The more conspicuous galaxies, ours among them, were born billions of years ago and seem to have settled into calm middle age, but young galaxies may yet be condensing out of raw dust and gas, their violent adolescence still in the future. Looking for them will take time — the number of available telescopes is small, galaxies are strewn across the sky like beach sand, and very young ones are likely to be dim — but astronomer R. B. Larson of Yale calls it "the most intriguing and perhaps most promising possibility" in the field.

At Kitt Peak, Sandage and I talked on past midnight. His voice grew weary.

"I don't know what I'm accomplishing here, letting starlight put marks on a silver nitrate photograhic plate," he said with a sigh. "But it sure looks pretty. I don't know why people have to try to do what they couldn't do before. I don't know what it's all about. All I know is that I feel awful bad if I don't keep working."

At 1 A.M. Sandage had exposed all the plates in the observer's cage. He ordered the telescope tilted down to the floor so he could reload with fresh film to keep him supplied till dawn. In the dark interior of the dome I watched the telescope heel over and down until the observer's cage reached a white steel scaffold. A door in the cage opened, spilling red night-vision light. Sandage's booted feet emerged. He climbed out, a case of plates under his arm. We shook hands. "This will take about 10 minutes," he told the night assistant. He hurried toward the darkroom to reload the plates. He did not quite run. ■

January 14, 1979

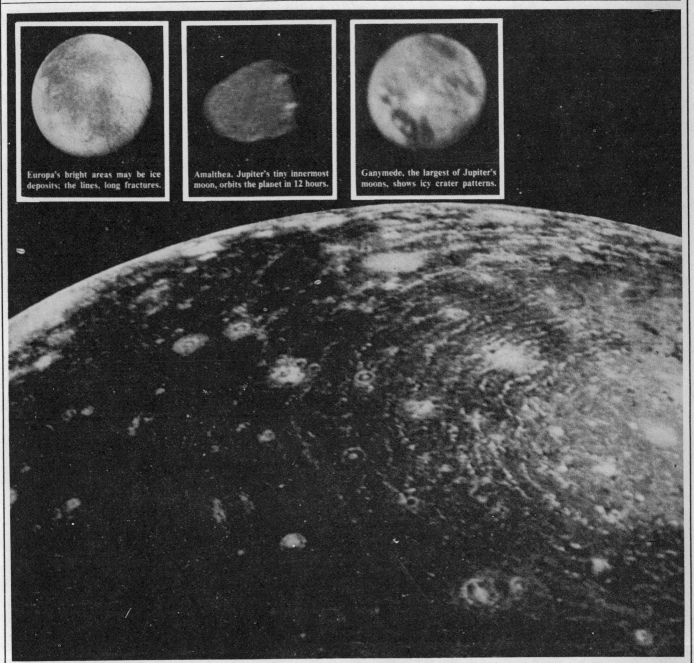

Europa's bright areas may be ice deposits; the lines, long fractures.

Amalthea, Jupiter's tiny innermost moon, orbits the planet in 12 hours.

Ganymede, the largest of Jupiter's moons, shows icy crater patterns.

Scientists believe Callisto, above, resembles an oversized drop of water capped by a thin crust of rock and ice. Insets: Jovian moons Europa, Amalthea and Ganymede.

NATURE'S TURBULENT TEST TUBE

Jupiter's atmosphere is a massive laboratory for chemical evolution.

By Robert Jastrow

Three mysteries make up the riddle of Genesis: The origin of the universe, the origin of the solar system, and the origin of life. If we can resolve these cosmic mysteries, we will have science's answer to the age-old question of religion and philosophy: How is man to understand his place in the cosmos?

Astronomers are grappling with the first of the three great mysteries; the exploration of the Moon has shed some light on the second. Jupiter seems destined to illuminate the third. At this moment, nature is conducting experiments in the atmosphere of Jupiter of the kind that were conducted on earth more than four billion years ago, when our planet was young and life had not yet arisen. We cannot reconstruct those terrestrial experiments because all traces of the first molecules on the threshold of life are lost; we only know the experiment succeeded. But there is every reason to believe that Jupiter is nature's laboratory for chemical evolution. It is the crucible of life.

The Voyager spacecraft reached Jupiter at the beginning of March after covering 625 million miles in 18 months. It flew over the planet's surface at a distance of 170,000 miles, photographed the mysterious Red Spot, and reconnoitered the four earthlike moons of Jupiter before continuing on to a rendezvous with Saturn in 1980. The spacecraft is continuing to send information back to us as it leaves Jupiter. The mission has already produced some of the most astonishing results in the short history of planetary exploration.

To many, the most remarkable of those results are the images transmitted back to earth by the Voyager cameras. [See facing page for an article about the scientific team that received the Voyager messages.] The images are beautiful; but what do they tell us about the cosmic mysteries?

219

Above, Jupiter's full face from 20 milliom miles away shows banded atmospheric structure and the Great Red Spot, lower left.

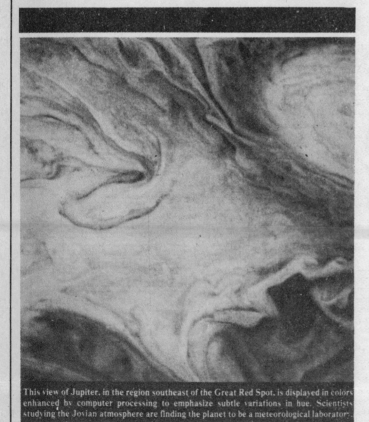

This view of Jupiter, in the region southeast of the Great Red Spot, is displayed in colors enhanced by computer processing to emphasize subtle variations in hue. Scientists studying the Jovian atmosphere are finding the planet to be a meteorological laboratory.

Jupiter was known to be a very strange object even before Voyager reached it. The planet is a huge, rapidly spinning ball of liquid and gas 90,000 miles in diameter — more than 10 times the size of the Earth — and 318 times as massive. It is larger than all the other planets combined. Because of Jupiter's enormous mass, gravity is much stronger than on earth.

An astronaut landing on the giant planet would be plastered to the spot by its gravitational force, which would give him a weight of 600 pounds. But the landing can never take place because Jupiter has no surface to stand on. Beneath its swirling clouds there is only a vast sea of molten hydrogen — hot, corrosive, destructive to life.

Jupiter is made mainly of hydrogen and helium. It also contains some iron and rock, enough to make up several earths. The rocky materials are probably concentrated at the center of Jupiter, forming a core several thousand miles in diameter. If the core were exposed, it would look like the kind of planet with which we are familiar. However, it is buried beneath a mantle of molten hydrogen 40,000 miles thick, which makes up the bulk of Jupiter.

The molten hydrogen merges imperceptibly into Jupiter's dense atmosphere. The atmosphere contains, in addition to hydrogen and helium, appreciable amounts of noxious gases such as ammonia, methane and acetylene. Water also occurs, in the form of steam in the lower atmosphere where the temperature is high, and there are droplets of liquid water at higher altitudes where the air is cooler.

The water droplets form a thick layer of clouds covering the entire planet. A second layer of clouds, composed of tiny crystals of frozen ammonia, floats above the water clouds. The clouds of ammonia form the visible face of Jupiter; it is these clouds that appear in the striking photographs of the planet taken by the Voyager cameras.

The most extraordinary feature in the Voyager photographs of Jupiter is the Red Spot, an orange-red oval 15,000 miles long and 8,000 miles wide, roughly the size of the Pacific Ocean. The photographs reveal swirling motions in the vicinity of the Red Spot, which suggest that the Spot is a mammoth hurricane. However, the Red Spot has persisted for at least 300 years, whereas hurricanes and storms on our planet play themselves out and disappear after a few days at most, because their energy is dissipated into turbulence in the surrounding air. What force has fed energy into the Red Spot to keep it intact for centuries? No one knows.

There is another puzzle in the Spot: Hurricanes and storms tend to wander over the surface, whereas the Red Spot is anchored in one place. A huge object, lying beneath the clouds on Jupiter, must be the Red Spot's anchor. What is this object? The *(Continued)*

Robert Jastrow is director of NASA's Goddard Institute for Space Studies, and the author, most recently, of "Until the Sun Dies" and "God and the Astronomers."

question is also unanswered. Although the Voyager photographs show the Red Spot in more detail than it has ever been seen before, it remains a mystery.

Throughout the period in which Voyager was photographing Jupiter, the planet's gravity pulled at it steadily, causing it to pick up speed. When the spacecraft whipped around Jupiter at its closest approach on March 5, it was traveling at 80,000 miles per hour, more than double its speed a month earlier. The boost in speed acquired from Jupiter's gravity was an essential part of the scientific plan; it would reduce Voyager's travel time to Saturn — the next stop, 500 million miles farther out in the solar system — from three years to 20 months.

The spacecraft was now 422 million miles from the Earth, and its radio signals, traveling through space at the speed of light, required 38 minutes to reach the Jet Propulsion Laboratory. At this time, the scientific team sent instructions to the craft to swivel its cameras around for a look at Jupiter's moons — and made their first big discovery: Jupiter, like Saturn, is surrounded by a huge ring of rock fragments, 160,000 miles in diameter.

The existence of the ring had been unsuspected because, unlike Saturn's rings, it is too faint to be visible through earth telescopes. The Voyager cameras were so close that they couldn't miss it.

Although the discovery of Jupiter's ring was a surprise, its occurrence is not difficult to explain with hindsight. Suppose you were in orbit around Jupiter with your feet pointed toward the planet. Since a planet's gravitational pull is stronger on objects that are closer to it, your feet would feel a stronger force toward Jupiter than your head. The result would be a tendency to separate your head from your feet, i.e., to pull you apart.

In the same way, a moon circling Jupiter must feel a stronger gravitational attraction on its near side, closest to the planet, than on its far side, tending to pull it apart. However, this tendency is resisted by the gravitational attraction of the moon itself, which tends to draw all parts of it together.

If the moon is at a considerable distance from Jupiter, its internal force of gravity, holding it together, will be stronger than the disruptive effect of Jupiter's gravity, and no damage will occur. But if the moon approaches too close to Jupiter, the planet's gravity will break it up into many small fragments. The fragments gradually spread out into a ring of debris circling the planet. This is the explanation for Saturn's rings, and presumably also the explanation for the ring around Jupiter.

How close could one of Jupiter's moons approach the planet without being broken up? A formula based on a comparison between the moon's gravity and Jupiter's gravity tells us that if the moon is more than 100,000 miles from the center of Jupiter, it is safe; if it is within that distance, it will be broken up. Since Jupiter's ring is about 80,000 miles from the center of the planet — well within the critical distance — its existence is accounted for nicely by the theory. The rock fragments in the ring are simply pieces of moon that came too close. (They could also be fragments of rock that were close to Jupiter initially, and never had a chance to become a full-sized moon because of Jupiter's disruptive gravity.)

Just outside the critical distance for breakup into a ring, a small, red satellite called Amalthea orbits Jupiter at a distance of 110,000 miles. Voyager captured a somewhat blurred image of Amalthea as the little moon raced around its parent planet at a speed of 70,000 miles per hour. The Voyager photograph reveals Amalthea to be a potato-shaped chunk of rock about 100 miles long and 80 miles wide. This little satellite may have been captured from the asteroid belt, or it may have condensed out of the cloud of gaseous dust surrounding Jupiter at the same time that the planet was forming.

Beyond Amalthea lies the satellite Io, named after one of the mythological lovers of Jupiter. Io was also photographed by Voyager, and yielded the second big surprise of the mission. Io is alive! It has actively erupting volcanoes; in fact, one volcano erupted while the photographs were being taken.

Our Moon is dead; it died three billion years ago. Mars is dead; its last volcanoes died out a billion years ago. Io and the Earth are the only volcanically active planetary bodies known in the solar system. Why is this important? Because vapors released by erupting volcanoes are the source of all the water in the Earth's oceans. A planet with actively erupting volcanoes

has water and warmth — two elements essential for life.

When our Moon was young and volcanically active, it may also have had the potential for life; but when its volcanoes became extinct, the opportunity vanished, and our Moon evolved into the bone-dry, lifeless body we know today. Volcanism lasted longer on Mars, which is why the prospects for finding life, or the remains of life, are much greater on Io than on the Moon.

Volcanoes and life — the two are inextricably linked. A student of the evolution of planets and life feels a warmth in his heart as he contemplates the picture of an active volcano on Io. Io has no atmosphere in the ordinary sense, and no warm, shallow seas, conducive to nature's experiments with evolution; yet it is a friendly planet, more like the Earth in some ways than any other we have visited.

The evidence for active volcanoes on Io is convincing, but it is very hard to understand why this moon is volcanic. Why is the interior of Io so hot? Io is a small body, only slightly larger than earth's Moon, and small moons and planets, like small animals, lose their internal heat quickly. That is why our Moon's fires went out three billion years ago. Io's fires should also have died out around that time. Why do they still burn?

The answer appeared in a research paper published in the journal Science on March 2, 1979, three days before anyone knew about Io's volcanoes. Drs. Stanton J. Peale, Patrick M. Cassen and Ray T. Reynolds showed that Jupiter's gravity, tugging at the near side of Io harder than the far side, would cause the interior of the moon to yield somewhat, creating friction and heating. According to their calculations, the heat accumulated from this friction, produced by the repetition of orbit after orbit over millions of years, would be sufficient to melt the interior of Io.

The effect proposed by the three scientists is similar to the one that causes moons to break up into fragments and form rings. The difference is that Io, being a little farther away from Jupiter than the critical distance for forming a ring, was not broken up, but only twisted and pulled about somewhat. The scientists suggested that Io "might be the most intensely heated terrestrial-type body in the solar system." Their conclusion: "Widespread and recurrent surface volcanism [would occur]. . . . Voyager images of

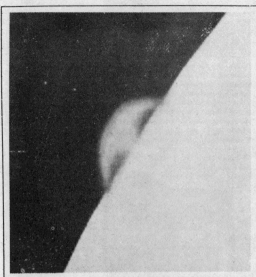

As it passed Io, Voyager took this picture of an active volcano.

THE IO INFERNO

The Voyager photographs of Io show a brightly colored surface with large fields of yellow, orange-red and white. The yellow color may be produced by sulfur; the red is probably iron oxide, or rust, but could also be a sulfur compound or some unknown exotic material. The white suggests evaporites — salts produced when water loaded with minerals wells upward and accumulates on the surface of a planet as a byproduct of volcanic activity. On airless Io, the water would evaporate immediately, leaving the deposits of white salt on the surface.

The suggestion of a hot, molten interior for Io seems to be confirmed by the fact

that no craters were visible in the Voyager photographs of this moon. Great numbers of these craters, caused by the impact of meteorites with the surface, are visible on our moon and were also seen in the Voyager photographs of Ganymede and of Callisto. Their absence from Io indicates that the surface of this particular moon is new and fresh, presumably because streams of hot lava, loaded with water and salts, flowed over it recently.

These arguments for volcanoes on Io are based on indirect evidence. They were confirmed dramatically when Voyager photographed an erupting volcano on the satellite. — R.J.

Io may reveal evidence for a planetary structure dramatically different from any previously observed." It was a tour de force of scientific prediction.

Beyond Io lie three large moons and eight small ones. The small moons, probably captured from the asteroid belt, were not investigated on this Voyager mission. However, the three large moons were studied carefully. Their names, in order of increasing distance from the planet, are Europa, Ganymede and Callisto — all lovers of Jupiter in classical myth. They were examined, along with Io, by Galileo in 1610 when he first turned his telescope on Jupiter. Their relatively rapid motions around their parent planet suggested to Galileo that Jupiter was a solar system in miniature. He concluded that the earth could not be the cen-

ter of all motion in the universe. This was the most powerful argument yet discovered for the theories of Copernicus. Jupiter's moons were responsible to a large degree for the trouble Galileo found himself in a few years later.

Ganymede, Callisto and Europa are known as the icy satellites because they appear to have substantial amounts of ice and water. Ganymede and Callisto are made largely of water; each can be described as a huge drop of liquid water with a core of rock or mud, and a thin crust of ice mixed with rock. Europa has less water or ice; it consists mainly of rock, with ice making up perhaps 5 to 10 percent of its bulk. Much of Europa's ice is lying on its surface; the Voyager photographs of Europa reveal large, bright areas, which are enormous ice sheets covering a substantial part of the planet.

The photographs also show marks on the surface which may be fractures or faults, which is why the prospects for finding life, or the remains of life, are much greater on Io than on the Moon.

Volcanoes and life — the two are inextricably linked. A student of the evolution of planets and life feels a warmth in his heart as he contemplates the picture of an active volcano on Io. Io has no atmosphere in the ordinary sense, and no warm, shallow seas, conducive to nature's experiments with evolution; yet it is a friendly planet, more like the Earth in some ways than any other we have visited.

The evidence for active volcanoes on Io is convincing, but it is very hard to understand why this moon is volcanic. Why is the interior of Io so hot? Io is a small body, only slightly larger than earth's Moon, and small moons and planets, like small animals, lose their internal heat quickly. That is why our Moon's fires went out three billion years ago. Io's fires should also have died out around that time. Why do they still burn?

The answer appeared in a research paper published in the journal Science on March 2, 1979, three days before anyone knew about Io's volcanoes. Drs. Stanton J. Peale, Patrick M. Cassen and Ray T. Reynolds showed that Jupiter's gravity, tugging at the near side of Io harder than the far side, would cause the interior of the moon to yield somewhat, creating friction and heating. According to their calculations, the heat accumulated from this friction, produced by the repetition of orbit after orbit over millions of years, would be sufficient to melt the interior of Io.

The effect proposed by the three scientists is similar to the one that causes moons to break up into fragments and form rings. The difference is that Io, being a little farther away from Jupiter than the critical distance for forming a ring, was not broken up, but only twisted and pulled about somewhat. The scientists suggested that Io "might be the most intensely heated terrestrial-type body in the solar system." Their conclusion: "Widespread and recurrent surface volcanism [would occur]. . . . Voyager images of Io may reveal evidence for a planetary structure dramatically different from any previously observed." It was a tour de force of scientific prediction.

Beyond Io lie three large moons and eight small ones. The small moons, probably

Ganymede's icy crust is deformed into a grooved landscape.

luminate another mystery: How did life arise on the Earth? Although Jupiter is a strange, unearthly body, and hostile to most terrestrial organisms, it has a special property that may lead to the solution of this problem.

As I mentioned earlier, the atmosphere of Jupiter contains abundant amounts of the gases ammonia and methane, in addition to hydrogen and water. Laboratory experiments have shown that these four gases — ammonia, methane, hydrogen and water — when mixed together and energized by an electric spark, produce copious amounts of amino acids — the molecular building blocks of living matter. This fact has given rise to a theory of the origin of life which is held in high regard by most scientists. According to the theory, the Earth's atmosphere was composed of substantial amounts of methane and ammonia 4.5 billion years ago, when our planet was newly formed. Strokes of lightning — represented by the electric sparks in the laboratory experiment — energized the gases of the earth's atmosphere and produced a rich yield of amino acids. The amino acids, accumulating in the oceans, built up a nutritional broth. Random collisions in the broth, occuring again and again over millions of years, linked small molecules into long ones, and finally produced a molecule on the threshold of life.

Once the threshold of life was crossed, evolution commenced and the laws of natural selection came into play to produce the variety of plants and animals that now exist on our planet.

The life-giving gases — ammonia, methane and hydrogen — have long since vanished from the earth, except for trace amounts released from barnyards and rotting vegetation. (Pockets of methane — known as natural gas or coal gas — are still present underground.) The reasons for their disappearance is that they contain atoms of hydrogen. Ultraviolet rays from the Sun break up the molecules of methane and ammonia and release the hydrogen atoms, which then escape from the Earth because the pull of its gravity is too weak to hold them. Jupiter's gravity, however, is so strong that not even hydrogen atoms can escape. This is the explanation for the presence of hydrogen compounds, such as methane and ammonia, on Jupiter.

Science has never been able to check its theory of the origin of life by searching the fossil record, because the earliest forms of life, lying on the threshold between the living and the nonliving worlds, were very fragile, and all traces of them have disappeared. And, of course, no molecules on the threshold of life are being formed today, because the essential gases — methane and ammonia — also disappeared a long time ago.

Perhaps the theory can be checked in the laboratory, by putting together amino acids and other molecular building blocks and trying to create life out of nonliving matter. This prospect fires the imagination of the scientist. He looks into his flask and sees the lightning of a storm in the young Earth's atmosphere; he smells the pungent aroma of ammonia; amino acids are forming; soon the first living organisms will emerge. . . .

But they never do. The elements of living matter accumulate in the flask, but no life climbs out. Why does the experiment fail? Because it lacks a key ingredient. The missing ingredient is time — millions on millions of years of ceaseless experimentation.

But on Jupiter, which still has its ammonia and methane, experiments on the origin of life have been going on for four billion years, and are still continuing. Even the electric sparks are present; the Voyager photographs showed a brightening in some of the stormiest regions of the Jupiter atmosphere that suggested lightning discharges on a scale dwarfing terrestrial experience. With all the conditions of the laboratory experiments duplicated in the atmosphere of Jupiter, we can almost be certain that the building blocks of life exist there in great numbers. Many interesting products of their collisions may also be present. Jupiter affords a unique opportunity to find out how life arose on the Earth, and how it may arise elsewhere in the cosmos. It is indeed the crucible of life.

Could life on Jupiter have progressed past the threshold and evolved to higher levels? Any such life would be small, squat and toadlike, adapted to withstand the crushing force of Jupiter's gravity — yet, perhaps, with a highly developed brain. Unfortunately, this prospect seems less likely. On the Earth, evolution proceeded rapidly in its early stages because the amino acids and other building blocks of life became concentrated in the shallow seas of our planet, and created a fairly thick soup in which collisions occurred frequently. These collisions, which linked small molecules together to form large ones, were the first step in the evolution of life. But Jupiter has no seas to catch the amino acids: They remain dispersed throughout the atmosphere, colliding infrequently, and the pace of evolution is correspondingly reduced. We would expect evolution to reach the threshold of life on the giant planet, and possibly to cross it, but not to progress far beyond that point. Still, Jupiter has given us many surprises. We may be in for another. ■

VOYAGER'S PROGENY

Two Voyagers were launched in 1977. The second will reach Jupiter this July and focus its attention on the moons of the planet before following its twin to Saturn. One of the two Voyagers may be targeted for an encounter with Uranus later in the 1980's.

An even more extraordinary spacecraft called Galileo is scheduled for launch to Jupiter in 1982. Galileo will reach the planet in 1985, and drop into an orbit around it, from which it will then eject a probe designed to fall toward Jupiter, radioing back information on the conditions that it encounters as it descends through the atmosphere. The probe will carry a mass spectrometer (an instrument that measures the masses of atoms and molecules accurately) which may provide the first direct information on the presence of the large molecules that could be building blocks of living matter. This instrument should be able to get down at least to the level of the water clouds on Jupiter, where the density of the atmosphere is high, collisions are more frequent, the temperature is comfortable, and conditions for the onset of chemical evolution are fairly favorable.

Galileo will bring us to the threshold of a serious search for life in the outer reaches of the solar system. A plan on the drawing boards calls for the subsequent exploration of Titan, largest moon of Saturn, which is important to that search because, unlike any other moon, it possesses its own atmosphere. Titan, like Jupiter, is also rich in methane and other gases helpful to the evolution of life. It may be even more favorable than Jupiter for the onset of evolution because it is likely to have a solid surface on which pools of liquid can accumulate for more rapid progress in nature's experiments.

N.A.S.A. and the scientific community are very interested in the mission to Titan, but no funds exist for it at the present time. The exploration of the outer planets ends with Galileo. — R.J.

April 1, 1979

NAVIGATORS WHO PROBE THE MYSTERIES OF DEEP SPACE

Despite the success of Voyager thus far, the Jet Propulsion lab is still struggling to obtain funding from Congress for future interplanetary missions.

By Timothy Ferris

The reconnaissance of Jupiter by Voyager 1 has yielded a treasure of scientific discovery and sheer spectacle. As it sailed past Jupiter the spacecraft took a close look at what is virtually a mini solar system — the lordly planet itself, which harbors 90 percent of the mass of all the planets, its moons, some of them almost as hefty as planets, and (as Voyager discovered) a ring of particles that calls to mind the rings of Saturn and Uranus, and the vast belt of asteroids that circle the sun.

Data from instruments designed to assess the composition of Jupiter and its surroundings will take time to be fully analyzed. The impact of the photographs is more immediate. Scientists cried out in delight as the Voyager cameras presented them with sights no eye had ever seen. The gaseous outer layers of Jupiter were resolved into a roiling stew of salmon pink, brick red and ice blue whose intricate filigree offered fresh evidence that the more closely one looks at nature, the more beautiful it becomes. The moons, previously known as little more than dots of light swimming in a telescope's field of view, turned out to be as individualistic as flowers. Ganymede is an agate wreathed in veins of white, Europa a cracked ocher ball that bears a passing resemblance to Mars. Callisto, a lonely outpost, is the color of a corroded Roman coin, while Io, a garish orange and cream, looks as whimsically decorative as a hot-air balloon. Armed with fresh questions engendered by the flyby study of Jupiter and its surroundings, project personnel are busy determining how best to employ the camera and sensors aboard Voyager 2 when it arrives at the giant planet in July. Meanwhile, Voyager 1 is hurrying off to keep a date with Saturn in 1980, after which it may be dispatched still farther toward the outer reaches of the solar system for an encounter with Uranus.

The success of the Jupiter encounter produced elation at the Jet Propulsion Laboratory, the organization that managed the mission and guided the twin spacecraft on their long journey for the National Aeronautics and Space Administration. J.P.L. people have a unique specialty: They take over responsiblity for interplanetary spcecraft once they have been launched into space and guide them out into the realm of the planets, thus realizing a dream that goes back to the time of Kepler. They are very good at what they do. They have tracked the probes which were dropped into the maelstroms of Venus, dispatched others to take close-up photos of Mercury and to brave scorching near encounters with the sun, have assayed the solar system out almost to the orbit of Saturn and have helped soft-land the mechanical emissaries on Mars, 300 million miles away, to within 6 miles of their target. (Pioneer 11, also guided by J.P.L., will reach Saturn this September.) "Life seems very rich, just now," said the forceful, youthful-looking director of

Timothy Ferris is professor of English at Brooklyn College and author of "The Red Limit: The Search for the Edge of the Universe."

J.P.L., Dr. Bruce Murray, 47, as praise for Voyagers's success poured in from around the world.

Dr. Murray felt he could use a little richness. When he took over J.P.L. just three years ago, the lab had fallen on hard times. So few interplanetary space missions were planned for the 1980's that the survival of the lab as a space-faring center stood in jeopardy. Now, however, he hopes that J.P.L.'s fortunes are improving.

The Voyager mission itself had begun in a climate of uncertainty. The launch of the first of its twin spacecraft was attended by a series of near disasters, the most disturbing of them a crisis that came to be known ruefully as the "software affair." "It was no idle matter," Dr. Murray recalled. "It was a $400 million mission and we were in deep trouble."

"Software" is computer jargon for the myriad instructions programmed into a computer. For a "smart" spacecraft like Voyager, software constitutes what might be called the computer's world view — its sense of where it is, what is happening, its alertness to problems and its acumen in solving them. Properly programmed, Voyager can carry out complex assignments — such as locating and photographing Jupiter's moons — and can manage its affairs with little human help. Improperly programmed, the spacecraft computer can become, in effect, psychotic. To a psychotic computer, perfectly normal events can take on a sinister cast, like that of a human psychotic who, say, interprets the dripping of a leaky faucet as the ticking of a bomb.

That is what happened on the occasion of the first Voyager launch. Because of a programming oversight, the spacecraft interpreted its own launch as a disaster, and duly reported to the ground that it had suffered a massive electrical breakdown. The mission seemed to be over almost as soon as it had begun.

The thunderous launch of Voyager atop a Titan Centaur rocket from Cape Canaveral, Fla., overwhelmed the delicate sensibilities of the spacecraft's internal guidance system, a package of gyroscopes designed to cope with the far less violent accelerations Voyager would experience once it had left Earth behind. This raised no concern on the part of humans involved in the mission, who knew that the gyroscopes would sort themselves out as soon as the launch phase ended and Voyager was cruis-

Dr. Bruce Murray, the J.P.L.'s high-powered director, sits in the lab's briefing room in front of a full-scale model of Voyager 1.

ing in space. But it raised grave concern on the part of the computer. The software people had neglected to tell the computer that there would be such a thing as a launch. Voyager didn't know that it was being swamped by the acceleration of its powerful booster rocket. All it knew was that it couldn't tell where it was.

In mounting alarm, the main Voyager computer tested each set of gyros in turn and found that all were behaving crazily. Since the simultaneous failure of all the gyros was extremely unlikely, the computer checked to see whether the fault might lie in the electronic circuits that reported on the condition of the gyros. The circuits seemed fine. If the problem wasn't the gyros and wasn't the circuits, the computer was obliged to reach the galling conclusion that it must itself be to blame. Voyager carries two identical master computers, one of which lies dormant and waits to take over should the other fail. With the mission only an hour and 13 minutes old, Voyager's A computer diagnosed itself as faulty and abdicated control of the spacecraft to the B computer. The B computer in turn performed the same sequence of tests, arrived at the same conclusion, and reported to the ground that all was calamity.

It took time for Voyager personnel to untangle this story and satisfy themselves that the spacecraft's problem had been emotional rather than physical. Meanwhile, more tangible problems appeared as the spacecraft sped out beyond the orbit of the moon. An on-board radio receiver failed. A sensor designed to signal the deployment of a boom carrying scientific instruments failed to report: Had the boom itself failed to deploy or, which turned out to be the case, had only the sensor failed?

Problems like these are diagnosed at J.P.L. by reference to a vast set of records that detail the history of each of the space vehicle's 61,000 parts. Ultimately, a technician who suspects an obscure spacecraft subassembly of misbehaving millions of miles out in space can look up the names of the people who built that particular part, call them, and ask them how the unit "seemed" back when they held it in their hands. So thorough a system works well, as witness the sterling performance of Voyager today after its shaky start — but it takes time and money to operate. As they untangled Voyager's problems, the project people got farther and farther behind

schedule. NASA, footing the bill, expressed concern. With the help of a J.P.L. review board made up in part of veterans of the spectacularly successful Viking mission to Mars, the Voyager problems were ironed out and the mission was running smoothly by the time Voyager 1 approached Jupiter.

But the memory of the software affair lingers, a symbol of the problems faced by the interplanetary navigators in the days of the shrinking dollar. Voyager's cry of alarm came at a time when the fortunes of J.P.L. had ebbed. The laboratory had but a single iron in the fiscal fire, a proposal for a mission known as the Jupiter Orbiter Probe (now renamed Galileo) and that proposal had been voted down by the House Appropriations Committee. The Viking landing on Mars the summer of 1976 was a brilliant success, and the data being returned from Mars were adding volumes to human understanding of the solar system. But Viking had cost nearly a billion dollars, and the word from Washington was to bring off subsequent missions on the cheap. "Voyager was an understaffed, underplanned mission," said Dr. Peter Lyman, one of the Viking veterans drafted to help straighten out the Voyager mission. "They had been pressured to do it as cheaply as possible. As a result, in the long run, it cost more to set things right than it would have to have done them right in the first place."

Having achieved the opportunity to realize the ancient dream of understanding the solar system, the nation seemed uncertain whether it wanted to go through with it. The White House was at best lukewarm on the subject of unmanned space exploration. Congress was wary of appropriating money for space probes at a time when many Americans were finding it difficult to put beef on the table. And NASA, which funds J.P.L., had amassed a notoriously poor record of selling the public on unmanned space exploration, an effort it undertook by arguing the unlikely hypothesis that taxpayers should support the investigation of other worlds because the resulting technology might produce wrist radios and efficient coatings for frying pans.

"It's remarkable when you think about it," said Gentry Lee of J.P.L. We sat talking in a sunlit central plaza at the laboratory, a cluster of steel and glass buildings nestled in the foothills in Pasadena that looks like a junior-college

campus but for the nests of microwave dish antennas. Dr. Lee looked upward, rolling his eyes in a gesture of incomprehension. "Imagine that you were living in the Pleiades star cluster, writing a history of our world in the second half of the 20th century," he said, "and you found that here an intelligent species developed the capacity to extend its senses out to touch all the planets of its solar system. And yet for some strange reason, that species decided not to employ that ability. Instead it turned its back on this adventure."

As we sat up one night in a building that hummed with the musings of computers alert to the status of spacecraft from the sun to Saturn, another top J.P.L. staff member admitted, "I'm not optimistic about the future of planetary exploration. I think the public appreciates its value — understands its grandeur and excitement — and perhaps Congress understands it. But NASA doesn't seem able to understand it, and President Carter seems to have all but written it off. I get discouraged sometimes. But then I come into the lab and a new photo of Jupiter is on the screen, and we're seeing things that no human being ever saw before, and I get such a boost."

Dr. Murray admits to similar bouts of pessimism. A tall, square-jawed man more comfortable giving orders than listening to advice, he brought to the lab an aggressive — some would say abrasive — style of leadership under which its fortunes have sharply improved. Dr. Murray calls the process "climbing out of the bathtub," referring to a chart of laboratory funding that descended to a low plateau by 1977 and now is slowly ascending. The ascent began when Congress, in a House floor vote, rare in such matters, finally approved Galileo, a mission to drop an instrumented probe into the seething atmosphere of Jupiter in 1985. There are two further missions in the wings for J.P.L. One is a comet-chasing probe powered by an ion engine driven by sunlight. The other, Solar Polar, funded jointly by NASA and the European Space Agency, calls for looping a pair of spacecraft high above and below the plane in which the planets orbit to survey these previously unexplored regions of the solar system. With the bottom of the bathtub behind him, Dr. Murray might be expected to rejoice. Instead, he is worried about the future of unmanned space flight, frustrated by the limitations of bureaucracy, and impatient with

the rate at which humans are piecing together an understanding of their cosmic surroundings.

Dr. Murray recalls his early years at J.P.L. with the air of someone remembering a bad dream. When he took over the lab in 1976, his previous administrative experience was limited to having run a geology project at the California Institute of Technology that employed six persons and spent $200,000 a year. J.P.L. employs 4,000 on a budget of roughly $250 million a year. The lab is a hybrid, technically part of Cal Tech but funded primarily by NASA, a situation that some at J.P.L. view as resembling the grip of contending talons. Yet Dr. Murray found that he lacked for advisers on either side. The president of Cal Tech, Harold Brown, had gone off to become President Carter's Secretary of Defense. The head of NASA, James Fletcher, a Nixon appointee, had resigned and was yet to be replaced, J.P.L.'s only source of substantial income aside from space exploration was a contract to help the Energy Resource Development Administration research cheap solar energy, but President Carter, soon after entering office, eliminated E.R.D.A. (It later became the Department of Energy.)

Feeling rather lonely in his lofty new post, Dr. Murray was subjected to a prompt education in the hazards of administration. Much of it came in the form of what he called "torpedoes in the night," bureaucratic warheads that reached his desk in the morning mail. The Internal Revenue Service audited Cal Tech and challenged the nonprofit status of J.P.L. The Federal affirmative-action program charged the laboratory with hiring too few members of minorities and threatened to cut off its Federal funds. (This is a perennial problem in the aerospace industry, where educational requirements are stratospheric and starting salaries usually unimpressive.) There followed the earthquake "torpedo." A group of consulting engineers warned that J.P.L. was vulnerable to earthquakes. The director of the Viking project, Jim Martin, a crew-cut perfectionist with the bearing of a career military officer, got wind of the report and ordered top mission personnel to keep their passports and vaccination certificates up to date so they could fly to deep-space tracking stations in Australia or Spain and carry on with the Mars mission should J.P.L. crumble into rubble. This

struck Dr. Murray as amounting to something less than a vote of confidence. "After all, I'm a geologist — God, this is so agonizing to talk about — and I know about these things," he said recently as we talked in his home in Pasadena. "J.P.L. did not appear to me to be in any particular danger. At any rate, we drilled holes all over the place. And we earthquake-proofed the buildings. Now it's better understood geologically than almost any place in the region."

Dr. Murray was discovering that space exploration had its unglamorous side. As the Viking spacecraft approached Mars, a bomb scare forced evacuation of the entire laboratory. Bubonic plague was found being carried by squirrels in the foothills near the lab. J.P.L. began losing personnel to sects, notably Scientology. "The problem isn't when they quit," Dr. Murray grumbled. "The problem is the ones that don't quit — if they stay on, but stop thinking."

In Washington, Dr. Murray lobbied from a final redoubt: Unless further unmanned space missions were funded, he said, the laboratory would be unable to continue to employ the men and women who had learned how to navigate to the planets. Some would be laid off; others, assigned to nonspace projects. As Gentry Lee put it, "The engineering feats involved in interplanetary space flight can be compared to the building of the Roman aqueducts or the medieval cathedrals. If you dismantle the teams that landed Viking on Mars and steered Voyager to Jupiter, then try to pick up where they left off again in a few years, you'll find that it's no mean trick. You may well not be able to do it."

Is interplanetary space flight as practical as the Roman aqueducts, or as fanciful as the cathedrals? Both, Dr. Murray answers: "We do it for international prestige. What has reflected better on the United States than Viking and Voyager? This has a practical result — other nations here on Earth must believe that we have a good and positive side to us, and they have to believe in our technical prowess. We do it for the sake of our domestic self-esteem; unmanned space exploration is a popular thing with Americans, especially when you remember that the cost of doing it amounts to about a dollar or so per person per year. We do it because by setting such challenging tasks for ourselves we drive our tech-

'If astronomers at the beginning of the 20th century had listened to arguments like Senator Proxmire's,' argues Dr. Bruce Murray, 'the most important discoveries — puslars, quasars, and radio astronomy — would never have happened.'

nology to otherwise unattainable accomplishments, and this in turn strengthens our entire technological and economic system. And finally, exploration has scientific and philosophical significance. The expansion of man's consciousness through space exploration will have no less significant effects on mankind's view of itself and of the reality in which we exist than have the explorations from the 14th to the 19th centuries when we explored the Earth and opened our minds to our environment."

This theme runs through many lines of conversation at J.P.L., that space ought to be explored because it is ultimately our environment. "Are we doing space research in order to understand the earth?" asks Moustafa Chahine, head of the earth and space sciences division of J.P.L. "No. We're trying to better understand our solar system, seeking knowledge for its own sake, for its own merit. Are we benefiting here on earth from this research? Are we coming to understand our planet better? Yes."

But given the economic climate of the times, the people at J.P.L. are not bashful about arguing in terms of cost-effectiveness, even when discussing missions whose public-relations appeal might seem to be chiefly romantic. Ask J.P.L.'s resident comet expert, Ray Neuburg, why we should spend money to send probes chasing after comets and he will note that they are made of material left over from the time when the sun and planets formed, that they are in a sense scraps from the cosmic workbench, and that understanding them could help us solve one of the great riddles of modern astronomy, that of how the solar system came into existence. Having set forth the scientific arguments in favor of the mission, he then promptly adds a remark in a cost-conscious vein: Since comets are buffeted by the solar wind, like sailboats blown off course by a terrestrial gale, the orbits of comets could tell us about the solar wind if we understood in detail just how they are affected by

it. Says Dr. Neuburg, "If we can observe just one comet up close and understand how the solar wind acts upon it, then each subsequent comet will become a free interplanetary probe."

Arguments like these helped Dr. Murray and his colleagues persuade Congress to sustain the nation's commitment to a continuing reconnaissance of our solar system. But Dr. Murray's one disappointment in Washington involved the least expensive of his pet projects. This was the Search for Extra-terrestrial Intelligence, or SETI, a plan developed jointly by J.P.L. and the NASA Ames Research Center near Palo Alto to use radio telescopes part-time to search the heavens for messages possibly dispatched by inhabitants of other worlds. Given its relatively modest cost of $20 million (as compared with more than $400 million for Galileo) and its enormous potential (if successful in receiving an alien signal, SETI could be expected to alter the course of human thought), Dr. Murray hoped that SETI would get through Congress. Instead, it attracted the unwelcome unattention of Senator William Proxmire of Wisconsin, who awarded his "Golden Fleece" award as a "ridiculous example of wasteful spending." Senator Proxmire noted that "there is not now a scintilla of evidence that life beyond our solar system exists." His remarks infuriated many advocates of SETI, since searching for such a "scintilla of evidence" was the whole idea of SETI. "The detection of evidence of extraterrestrial intelligence is the most significant fact that could be discovered," says Dr. Murray. "We are not likely to find something unless we look for it. If astronomers at the beginning of the 20th century had listened to arguments like Senator Proxmire's, many of the most important discoveries of this century — the discovery of pulsars, of quasars, the development of radio astronomy — never would have happened."

Whatever becomes of SETI and of more ambitious plans being hatched at J.P.L. —

plans that include snatching up pieces of comets and asteroids and returning them to Earth, Dr. Murray has managed for the present to keep J.P.L. in business as the world's only full-time center of interplanetary exploration. Symbolic of this commitment is the Deep Space Network, the division of J.P.L. that stays in touch with interplanetary sapcecraft as they wander, year after year, through the void.

The network stands as a monument to audacious precision. Its radio receiving stations in Australia, Spain and the Mojave Desert can pick up signals as weak as the energy given off by a safety match struck on the surface of Mars. Its computers sustain an awareness of space and time so precise that they can report on the location of a probe hurtling through space 100 million miles away with an accuracy of a few yards. J.P.L. navigators have found the very stars insufficiently constant to steer by — stars slowly swarm as our galaxy turns — and they now navigate by sighting quasars, brilliant beacons at the outposts of the cosmos whose light in many cases has been traveling through space since before the earth was formed. With their sights set on quasars and their clocks ticking with a precision in comparison to which the spinning of the earth is a mere drunken wallow, the D.S.N. navigators deserve much of the credit for space spectaculars like the Voyager encounter with Jupiter. Curious about what sort of people had built this monument to precision, I paid a visit to the D.S.N. headquarters at J.P.L.

To visit William Bayley who manages the D.S.N., and Mahlon Easterling, its architect, one negotiates a building filled with the trappings of futurism — the low hum of computers, the march of data like colored confetti across television screens, the perambulations of preoccupied technicians. But what one finds in the middle of all this is an office devoid of electronic gear, its lights dimmed, occupied by two men who would not be out of place playing checkers by the stove in a country store. Dr. Easterling is red-faced, with flowing white hair and a voice that ripples with the sound of a mountain stream just above a good distillery. Dr. Bayley carries himself with the quiet calm of a man whose most serious technical problem is deciding whether to bother his sleeping hounds long enough to get up and fetch his fly rig. Both were

wearing sweater vests and plaid shirts, a theme Dr. Easterling had extended to include a pair of clashing plaid pants.

Dr. Bayley began his explanation of how the Deep Space Network works by jotting a few figures on a blackboard and frowning at them. "Now these figures are wrong, you understand," he said, "but they help me keep it in my head where everything is. Take Pioneer 11, for example. Pioneer 11 has a round-trip travel time of five hours . . . ."

"I think you mean Pioneer 10, Bill," said Dr. Easterling.

"Yes, right, Pioneer 10. Anyhow, it takes two and a half hours for radio signals traveling at the speed of light to get to Pioneer 10, and another two and a half hours for it to reply, so if you want to have a conversation with Pioneer 10 it takes five hours to send a command and find out if the spaceship has heard you."

Dr. Bayley grinned and shook his head. "Its hard for me to think of something a billion and a half miles away," he said. "I find it even harder to think of all those data streaming back from the spacecraft, that long thin trickle of data stretching across all that space."

Dr. Easterling chuckled. "There aren't too many of those bits of data, but they're sure going like the devil," he said.

"Yes, and that old Pioneer 10 is just going on out now," Dr. Bayley replied. "You know, it's an old spacecraft, and it has only a 10-watt transmitter. That's about like a refrigerator bulb. If you want to see something operating at 10 watts, look at the light in your refrigerator. So we get the data back pretty slowly from Pioneer, only about one thousand bits per second. We're rather proud of the way we've been able to improve the rate that we get data back from our spacecraft. When Mariner 4 flew past Mars in 1964, its photographs and scientific data came back at the rate of only 8.3 bits per second. With Voyager, we're getting 117,000 bits per second, and that from way out at Jupiter. Pretty soon we hope to be able to go as high as eight million bits per second."

"We're always improving the accuracy of our clocks, too," said Dr. Easterling. The two spoke in the tone of old friends swapping stories they had exchanged many times before. "Right now we operate at an accuracy of one second in three million years. That sounds pretty accurate, but believe it or not we have people who complain bitterly that

it's not good enough."

"Soon we will have improved that to better than one second in three billion years," Dr. Bayley said. He hesitated. "Or is it one second in 300 million years? I forget." He went to the telephone and dialed his secretary. "Barbara? Will the next generation of D.S.N. clocks get us to one second in three *billion* or one second in three *hundred million* years?" He returned with a scrap of paper. "It's three billion years," he said happily.

I asked Dr. Easterling how he had come to be involved in building the Deep Space Network.

"To understand that, you should know that I have every issue of Analog science-fiction magazine since 1948," he said. "In 1952 I read about interplanetary navigation for the first time. It was an article in Analog by John Pierce titled 'Don't Write: Telegraph.' There had been some science-fiction stories about using rockets to deliver the mail, and these had got John angered because they were so inefficient. So he worked out the design of an interplanetary communication system, using microwaves. It employed exactly the techniques we use today.

"I went to Columbia University as an electronics engineer, and then I worked on military radar. When Sputnik was launched I said, 'Whoops, now's the time.' I wanted to work on deep space. It took me a year to find the right place, which was J.P.L. I've been here ever since." Dr. Easterling looked out the window at the San Gabriel Mountains. "So you see, when it came to designing an interplanetary communication system, it was no trouble, because I'd read how to do it."

I left Drs. Bayley and Easterling, each of whom has logged more than 20 years at J.P.L., joking comfortably with each other. Fresh streams of data were pouring in from Voyager 1 and marching across the screens. While we had been talking, Voyager had traveled another 25,000 miles. In a few months it will reach Saturn, then it will leave our solar system. In 40,000 years it will have passed abeam of another star, a white dwarf designated AC+79 3888. Unless someone or something snares Voyager at some future time and place, it will still be journeying among the stars billions of years from now, when the sun expands and consumes the earth. ∎

April 1, 1979

THE RUNAWAY USE OF ANTIBIOTICS

The abuse of antibiotics over the last 30 years is resulting in an alarming and ever-increasing number of bacteria that are resistant to these drugs.

By Marietta Whittlesey

Over the past 30 years, antibiotics have permitted a control of infectious disease that would have been considered miraculous by earlier generations. And until recent years, the medical world made the comfortable assumption that antibiotics were a sure-fire weapon against bacterial infections. Now, however, doctors are beginning to have problems treating some of the diseases they thought they had defeated with antibiotics. There are, for instance, two new strains of gonorrhea that are resistant to even megadoses of penicillin.

What has been happening is that the indiscriminate prescribing and taking of antibiotics for anything from minor colds to venereal diseases have led to an alarming increase in drug-resistant bacteria. Highly controversial at the moment is the question of antibiotics as feed additives for livestock; antibiotics are fed to these animals in part to promote rapid growth. Critics of this practice contend that these additives only generate more resistant organisms.

The most dramatic demonstration that resistance to antibiotics could be life-threatening to large populations was a typhoid epidemic in 1972-73 in Mexico. Of the 100,000 who were afflicted in what became the largest typhoid outbreak in recorded history, says Dr. Ephraim S. Anderson, 20,000 died because the typhoid bacteria had already picked up resistance to chloramphenicol, the drug normally used to combat it. Dr. Anderson, now retired, who was at the time of the outbreak director of the Enteric Reference Laboratory in London — which identifies

Marietta Whittlesey is a freelance science writer and author of "Killer Salt."

and records bacterial samples from all over the world — ran tests to determine which drugs were still effective against this resistant strain, and found that ampicillin was.

Epidemics in Guatemala and Brazil soon followed. In all three countries, the use of antibiotics is poorly understood and inadequately controlled. Though recently there has been an attempt to regulate all drugs in South America, outside the large cities it is still possible to obtain virtually any antibiotic over the counter; many pharmacists often dispense antibiotics for a variety of ailments, such as headaches and colds, for which the drugs are totally ineffectual.

Self-medication is not just a matter of economics (taking a pill is cheaper than seeing a doctor); it is a product of ignorance. And the ignorance isn't confined to developing countries. Doctors often dismiss the problem of resistance because, they insist, new drugs discovered daily can be relied upon to overcome resistance to antibiotics.

Resistance is not a new problem. Sir Alexander Fleming encountered it with strains of *Staphylococcus aureus* as far back as 1942. He issued a surprised warning at the time that some strains of that species of bacteria seemed "almost insensitive" to penicillin.

Originally it was thought that bacteria acquired resistance only through mutation. In most populations of bacteria there are only a few naturally resistant variants; in the presence of an antibiotic they flourish because the antibiotic kills or inhibits the drug-sensitive bacteria. Furthermore, in species of bacteria where resistance has not been known, researchers found something quite unexpected.

In 1959 it was discovered by a medical research team in Japan that a strain of *Shigella dysenteriae* that was resistant

to four antibiotics could transfer its resistance to other bacteria. This type of resistance was not the result of a new mutation but the result of an actual transfer of genetic material from one bacterium to another (even of a different family) through physical contact. Resistance genes are transferred on a structure called a plasmid. Thus, a benign but resistant organism could transfer its drug resistance to virulent pathogens such as salmonellae, or to resident bacteria such as *E. coli*.

As antibiotic usage continued to rise, it became apparent that the plasmids could carry more than one resistance gene. While in the early 1960's plasmids carrying more than a couple were uncommon, it is now commonplace to see bacteria carrying genes for resistance to as many as 10 different drugs. Moreover, one research team has isolated an *E. coli* plasmid that carries genes for drug resistance *and* the production of enterotoxin (the substance that causes traveler's diarrhea).

This form of "natural" genetic recombination is far more real a threat to the general population today than the research in recombinant DNA. According to Dr. Anderson, "we have accomplished in half a century evolutionary changes that couldn't have happened in a million years

under natural circumstances. Recombinant DNA research is tiddlywinks compared with what has been done ecologically to bacteria with the emergence of antibacterials."

The mechanics of drug resistance vary with the bacteria and the drug involved. One way bacteria become resistant is through changes in the cell membrane so that the antibiotic cannot enter the cell and take effect. Researchers in the microbiology department at Cornell University Medical College are working on a method of overcoming this type of resistance by using surfactants — surface-active agents similar to detergents — to break down the cell membrane. Dr. William M. O'Leary, professor of microbiology at Cornell, feels that this method is applicable to a large number of resistant bacteria.

Another way bacteria become drug resistant is through the production of enzymes that destroy the drug molecule. One such enzyme is beta-lactamase, which destroys penicillin by breaking up one bond of the penicillin molecule while it is in the body. An example of beta-lactamase-producing bacteria is the so-called "incurable" form of gonorrhea. Until recently, two new strains of gonorrhea that are resistant to penicillin and several other drugs could be controlled by spectinomycin, which is 10

times as expensive as penicillin. However, in 1973, spectinomycin-resistant strains were reported from Denmark.

While thousands of cases of penicillin-resistant *N. gonorrhoeae* have been reported in 18 countries around the world since 1976, when the bacteria acquired the penicillinase plasmid, there have been only 368 cases in the United States and Guam. For the most part, the American carriers were servicemen who had been stationed in the Philippines. It was discovered that Filipino prostitutes had been routinely taking low levels of penicillin as a prophylactic against the very disease they were incubating in their bodies.

□

What is responsible for the rise in resistant bacteria? There are those who believe that antibiotic feed additives play a major part. Antibiotics are fed to livestock for two reasons: (1) to promote rapid growth; (2) as prophylaxis against diseases, such as shipping fever, which can ravage herds in no time. When antibiotics are used in this way, they are said to cut down on the amount of space and feed needed per animal.

The "growth promotion effect" was discovered in 1949 by Dr. Thomas H. Jukes, now professor in residence at the medical physics division of the University of California at Berkeley. He found that the drug residue from the fermentation of antibiotics increased weights and size by about 10 to 20 percent. To this day it is not known how this effect works. The antibiotics may be suppressing the detrimental effects of certain resident bacteria, or they may be stimulating beneficial resident intestinal microflora, such as coliform bacteria and yeast cells, which synthesize nutrients. It is known that levels of these resident flora increase when an animal is fed by this method. These animals also show a thinner gut wall, which may be more capable of absorbing nutrients than that of untreated animals.

The growth effect does not yet appear to be jeopardized, as the bacteria involved in it have been slower to acquire resistance. Yet public-health experts warn that the plasmids have turned out to be very adaptable. "They were able to get into all sorts of bugs which never had them before," says Dr. Richard P. Novick, chief of the department of plasmid biology of the Public Health Research Institute of the City of New York Inc. "It's taken a very long time," he continues. "What we've seen is a very rare event that is statistically improbable, but if you put the pressure on high

The past decade has seen a clear slowing down in the discovery of usable new antibiotics. Above, an antibiotic production laboratory.

enough you get very rare events, and once you have the first one, the rest can take off from there. So when people say that it's been 25 years without any trouble and we can be confident of safety, I would argue that no one has the right to make that assumption. In fact, it's just now that the trouble's starting."

Part of the trouble is the number of resistant organisms in these animals' bodies that could be transferred to humans. In a study of farm families who raised chickens that had been fed tetracycline-supplemented feed, Dr. Stuart Levy of Tufts University Medical School in Boston demonstrated that resistance plasmids and *E. coli* in the chickens were picked up by some farm handlers. In addition, the tetracycline-supplemented feed caused an increase in plasmids that imparted resistance not only to tetracycline but to several other antibiotics as well.

Many supporters of a ban by the Food and Drug Administration on antibiotic additives base their arguments on what happened in Britain in 1964-68 when an outbreak of salmonellosis swept through the cattle herds. Thousands of animals as well as seven humans died as a result of this epidemic.

Dr. Ephraim S. Anderson, who brought the situation to the Government's attention, traced the epidemic to one dealer, who denied having any infection among his livestock, but who was later found to have sent more than 800 carcasses to the knacker's yard—where old or sick animals are sent to be slaughtered and sterilized for nonhuman consumption. His business was, in effect, the salmonellosis distribution center, as his stock was shipped all over England. The strain of *S. typhimurium* acquired resistance to each drug the veterinarians used to treat the disease. By the time the epidemic died out (mainly as a result of the dealer's death in an automobile accident in 1967), the strain was resistant to eight different drugs.

The Swann Committee was convened in London to discuss the additives problem, and out of it came a law that banned antibiotics as feed additives if they were of primary importance in treating humans. Other antibiotics require a veterinarian's prescription and are used as medicines only.

Influenced by the Swann Committee, the F.D.A. in 1977 and 1978 proposed rules that would make antibiotic additives available only on a veterinarian's prescription and made only from drugs with no use in human therapy. The F.D.A. also proposed that feeds with additives could be purchased only from F.D.A.-licensed feed mills, and that the diagnosis be stated on the prescription form. Right now, a veterinary medicine can be sold over the counter if the directions for use can be explained adequately to a layman in writing.

The proposed rules are not without their problems. In the vast meat-producing areas of the country, there are not always veterinarians within a few miles. Most smaller feedlot operators cannot afford their own veterinarians, and complying with F.D.A. rules, they say, would mean a lot of added costs, which might put them out of business, and which would certainly mean a rise in meat prices.

No action has yet been taken on the proposed rules, which are generally opposed by the drug industry. Dr. John Farnham, manager of quality assurance at American Cyanamid's agricultural division, points out that meat from animals not given antibiotics could be contaminated with pathogens, such as salmonella, which he feels would rise dramatically were prophylactic use of antibiotics to be curtailed. Members of the drug industry are also quick to point out that the British model has not worked well. There is a thriving black market in additives, and many farmers are able to evade the rules that the drugs not be given for growth-promotion purposes by giving the drugs as prophylactics.

The F.D.A., supported by large numbers of medical researchers, feels that the more resistant organisms are let into the environment, the more resistance will spread.

□

Once viewed as wonder drugs and miracles, antibiotics are now called overworked miracles. Perhaps more than any other group, the drug industry has been responsible for the overuse of these valuable drugs. All too often, pushy sales representatives offer free samples and glossy brochures promoting unneeded items to their customers. Not surprisingly, antibiotic use increases year by year. Americans use up enough penicillin to provide a weekly dose for every man, woman and child.

Patients are also responsible for antibiotic abuse. Many patients are resentful if their visit to a doctor does not result in a prescription for medication. And many doctors, to avoid losing these patients, have learned not to argue; they will dispense drugs for their placebo effect. However, in the case of antibiotics, this is not a harmless practice. To prescribe antibiotics for flu and viral infections, against which the drugs are useless, is to create a public-health hazard by raising the number of resistant resident bacteria.

Another dangerous practice is self-medication by patients. Often, drugs left over from previous illnesses are put to use two years later. And even though there should be no antibiotics left over from a prescription, too many patients have the habit of taking an antibiotic for a few days until they feel better and then halting medication. This is the perfect way to cause a relapse. The antibiotic is present just long enough to slow down or kill most of the sensitive ones, leaving intact the minority of resistant organisms. If the antibiotic is withdrawn at this point, the bacteria will rebound, more resistant for having to adapt to the antibiotic.

Self-medication is also unwise because not all antibiotics are effective against all bacteria. Some are known as "broad-spectrum antibiotics," and they are active against a wide range of bacteria. Broad-spectrum drugs kill off everything in their path, including normal intestinal flora. When this happens, one is open to invasion by other bacteria.

Furthermore, self-medication can be hazardous because of drug-food interactions. Until recently, little was known about this subject, and what information there was often wasn't passed on to patients. For instance, the combination of tetracyclines with milk or antacids can reduce drug absorption up to 100 percent; streptomycin and diuretics can cause temporary or permanent deafness.

Many physicians themselves often show a fundamental lack of understanding of recent findings in bacteriology and internal medicine. This was clearly shown by Dr. Harold C. Neu of the division of infectious diseases at Columbia Presbyterian Medical Center. Dr. Neu devised a self-assessment test on the diagnosis and treatment of infectious diseases, and he found that out of 4,513 physicians the median score was only 68 percent; residents and those in practice for fewer than five years scored best.

According to Dr. Georges Causse of the World Health Organization in Geneva, part of the reason for the spread of the beta-lactamase-producing strain of gonorrhea is that physicians have not kept up with medical research, and they continue to prescribe doses of penicillin that don't reach a high enough concentration in the body to eliminate even the sensitive bacteria. These patients become sources of highly resistant organisms and are threats to those around them.

In 1976, Dr. William Schaffner and his colleagues at the Vanderbilt University Medical School in Nashville did a study of office prescriptions of chloramphenicol by the state's doctors. Even though this antibiotic has been found to be so highly toxic to human tissues that it is indicated only rarely in office practice, 6 percent of the doctors prescribed it frequently, and most commonly for colds. For rural doctors, the figure was 21 percent. Not only were the drugs dispensed unwisely, but in cases where they were indicated, they were often given in too low a dosage to treat an infection adequately.

□

One of the more serious manifestations of the resistance problem is that of nosocomial, or hospital-acquired, infections. Bacteria are present everywhere, and hospitals are no exception. While sanitation can do much toward keeping these bacterial populations under control, bacteria can never be totally eliminated, and optimal sanitation is far from universal. During the 50's and 60's, the most common hospital bacterium was a strain of *Staphylococcus aureus*. Then, without anyone's knowing why, *Serratia marcescens, Klebsiella, Proteus* and *Pseudomonas* bacteria began to prevail. *Serratia*, which used to be considered so harmless it was given to medical students to learn from, is now epidemic in hospitals. It is often resistant to all drugs.

Dr. Richard B. Dixon, chief of the hospital infections branch at the Center for Disease Control in Atlanta, estimates there are two million nosocomial infections a year. In some instances, patients infected with these bugs cannot be treated, and in their weakened condition their bodies cannot fight off the bacterial invasion and they die. According to Dr. Schaffner, these infections prolong the average hospital stay by a week. At a minimum of $150 a day, another $1,000 is added to the patient's hospital bill. Nationally, the figure comes to $1.5 billion.

Dr. Edward Lowbury at the University of Aston in Birmingham, England, is emphatic in stating that in order to cope systematically with hospital infections, a strict antibiotic control program must be set up by each hospital.

He recommends that the hospital's antibiotic policy be run by an infection-control team that surveys the rate and type of infection present in the hospital, as well as the shifting patterns of resistance. Antibiotics are to be classified in categories of availability:

(1) *Relatively unrestricted drugs* which have shown little resistance, and drugs which are inexpensive or which have a low toxicity.

(2) *Relatively restricted drugs* which are available on general prescription but which have shown increasing resistance, or drugs which are toxic.

(3) *Reserved drugs* which are used only for severe infections where other antibiotics are inadequate, and are available only on a consultant's signature. Drugs are placed in this category to preserve their high value or to restore their activity when there has been widespread resistance to them. When a drug is removed from use, the resistant bacteria gradually fall off.

What is the future of antibiotic drugs? There is a great divergence of opinion. Some feel it is too late to undo any of the damage, although we will probably not be wiped out by bacterial epidemics simply because we have improved our sanitation greatly since the early part of the century. Others feel that with the proper controls, the drugs can be made to last.

The outlook is generally bleak. Over the past decade there has been a slowing down in the rate of discovery of usable new antibiotics. Although "new" drugs are continually being found, these drugs are often chemical variants of known drugs and do not represent the major advances that penicillin, the tetracyclines and the synthetic penicillins did. According to Dr. Novick, "No major new class of antibiotics has been discovered in at least the last 10 years." Structurally and mechanically, the new drugs are more and more similar. Many of them are also more toxic to the patient's tissues.

Many see education and regulation as the most important and necessary weapons with which to fight resistance. Dr. Novick feels that antibiotics should be as strictly controlled as narcotics. He proposes that physicians prescribing the drugs should be required to indicate the diagnosis on the prescription form itself. Under such a system, the information could be stored in a computer for periodic review by authorities, such as a committee of infectious-disease physicians.

Although the American Medical Association and most of its members would probably respond unfavorably to such a proposal, Dr. Novick argues that "the howls of anguish that will be raised about interference with the 'right' to prescribe can be answered with the comment that the medical profession has totally failed to develop and enforce even minimum standards during the 30 years of antibiotic usage." ■

May 6, 1979

227

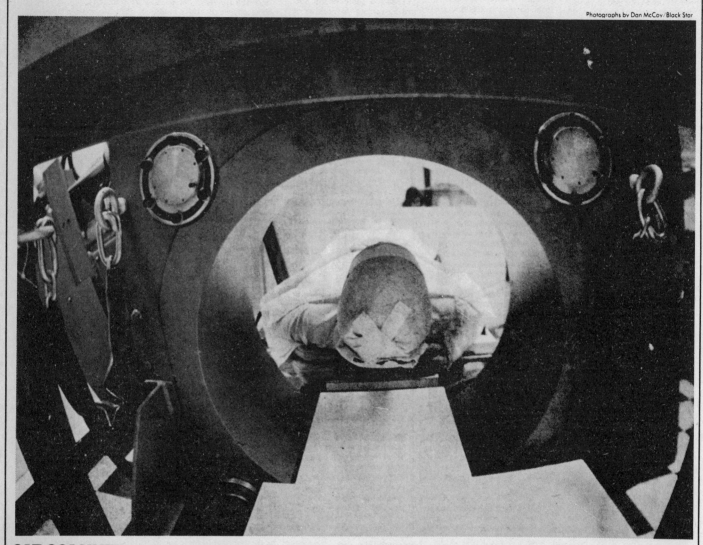

Photographs by Dan McCoy/Black Star

CAT SCANNER *A patient lies inside a Computerized Axial Tomography scanner. Possibly the most significant diagnostic tool of the last 70 years, it provides information about organs formerly available only through surgery. It is also one of the most expensive medical machines ever made.*

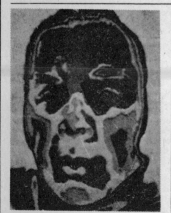

THERMOGRAPHY

Sensitive detectors convert infrared rays from the body's heat to electronic signals displayed on a screen. This "heat map" pinpoints malignant areas.

MEDICAL TECHNOLOGY: THE NEW REVOLUTION

By Laurence Cherry

If you have to go to the hospital 10 years from now, your visit may run something like this:

Past the inevitable receptionist, your street clothes exchanged for a shapeless hospital gown, you enter a small, antiseptic room. You take a seat and offer your arm to a rectangular machine. Painlessly the machine draws a blood sample and within seconds has analyzed it down to its smallest mean-ingful platelet. The information is flashed to a central computer deep within the hospital, where it is compared with previous readings to detect an infection anywhere in your body.

You move to an adjoining room and sit under a massive apparatus. Silently, your body is probed by X-rays or microwaves, while sensors inspect the surface of your skin. A thermogram will show areas of raised temperature, which can reveal incipient disease. Although you feel nothing, see nothing, the inner workings of your body are being deeply scrutinized. A developing gallstone, still barely larger than a grain of sand, is noted and appraised; like a tiny white clot lodged within a coronary artery or a polyp hidden in your nasal cavity, it may be harmless, but nevertheless deserves watching.

Within moments, a minutely detailed workup has been prepared by computer and sent to your doctor. It contains both

Laurence Cherry writes frequently about medicine.

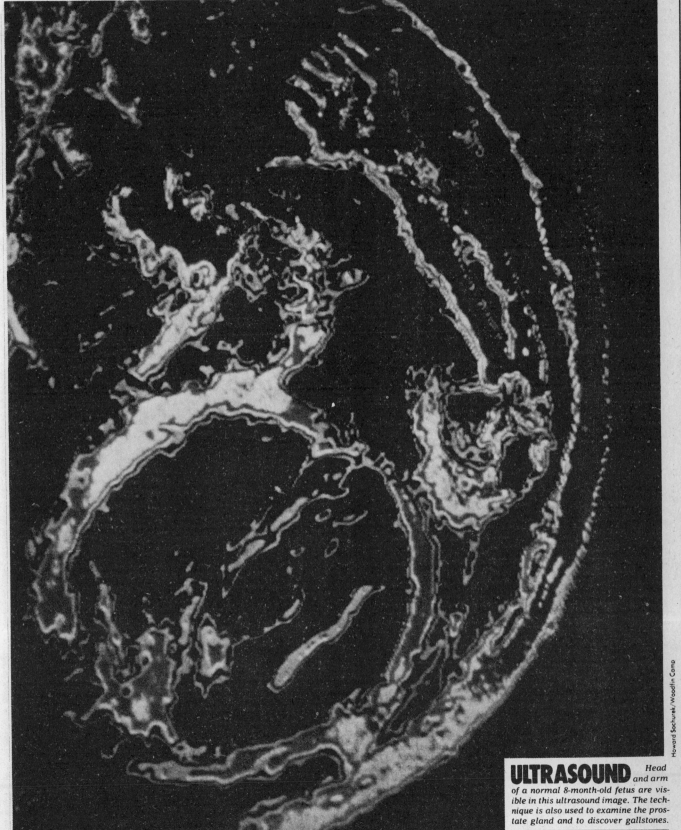

Howard Sochurek/Woodfin Camp

ULTRASOUND *Head and arm of a normal 8-month-old fetus are visible in this ultrasound image. The technique is also used to examine the prostate gland and to discover gallstones.*

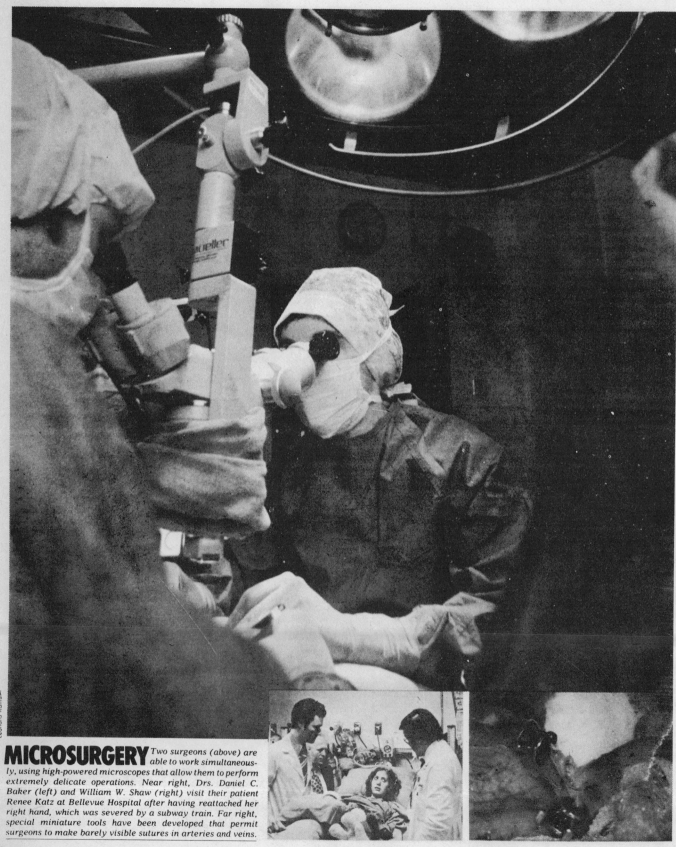

Leonard Kamsler

MICROSURGERY
Two surgeons (above) are able to work simultaneously, using high-powered microscopes that allow them to perform extremely delicate operations. Near right, Drs. Daniel C. Baker (left) and William W. Shaw (right) visit their patient Renee Katz at Bellevue Hospital after having reattached her right hand, which was severed by a subway train. Far right, special miniature tools have been developed that permit surgeons to make barely visible sutures in arteries and veins.

diagnoses and suggested methods of treatment.

□

Medicine today is in the midst of a technological revolution, a transformation that is already changing the art of healing. New instruments and machines can already glimpse a 3-month-old fetus curled within its mother's womb, painlessly diagnose brain damage, help surgeons fuse damaged nerves and blood vessels.

And still other seemingly miraculous inventions are almost at hand. At Rockefeller University and New York Hospital, an experimental "artificial pancreas" can instantaneously measure glucose levels and adjust the amount of insulin fed into the bloodstreams of selected diabetic patients, preventing wild daily fluctuations in glucose levels. Experiments with sophisticated electronics at the University of Utah and New York's Columbia Presbyterian Medical Center have meant sight for the blind and hearing for the deaf. At the National Institutes of Health in Bethesda, various medical teams are refining a complicated technique used in chemistry called nuclear magnetic resonance to track the chemistry of living cells. In the not very distant future they hope to be able to detect changes in heart muscle cells before a heart attack occurs and then quickly arrest them; they should also be able to detect early rejection of transplanted organs and take preventive action.

But along with this dramatic new technology and its apparent potential for solving many of our most perplexing medical problems, controversy has developed about its benefits and costs. Critics contend that the increasing use of machines is dehumanizing the practice of medicine and creating intolerable medical expenses as well as hidden medical dangers. This past November, at the behest of top Congressional leaders, a National Center for Health-Care Technology was established in Washington to take a closer look at the new technology. "Medical instrumentation has reached the point and expense where it must be closely scrutinized," says Dr. Seymour Perry, the acting director of the new center. "We know there have been truly important advances. Now we have to decide which of them are appropriate and which are not."

The Korean and Vietnam Wars both played a part in creating the new technology. Experimental physicists and mathematicians first employed to develop deadly new electronic gadgets for warfare (such as improved radar to track enemy missiles or infrared sensors to spot troop movements through nighttime jungles) began to look around for other areas in which to practice their skills. Advanced computers arrived on the scene in the early 1960's. By the early 1970's, the amalgamation of medicine and physics began to be known as bioengineering and departments of bioengineering cropped up in medical schools around the country.

In a small pastel room at New York Hospital, a 36-year-old woman, six months pregnant with her first child, lies on an examining table. While music plays in the background, a young technician swabs her belly with a thick gel, then holds a beige wandlike probe, not much longer than a toothbrush, directly over it.

Nearby, an image of the fetus in the mother's womb suddenly swims into view on the screen of what looks like a small portable television set. Though murky, the image is clear enough to show the baby curled within the womb, and, at the center, a smudge of pulsing gray, its rapidly beating heart.

"Is that my baby?" the woman excitedly asks, craning her neck for a better look at the screen.

"That's your baby's chest, and there's a foot," answers a radiologist standing next to the screen, snapping a lever that will record the view in still photos to be examined later. "He's lying with his head down."

"Is everything normal?"

"Everything looks fine."

□

Ultrasound is one of the most promising developments in the new field of bioengineering. Conventional X-rays reveal little about the organs deep within our anatomy; for example, until recently the only way to obtain truly accurate information about the heart was to insert a snakelike catheter through a vein in the arm and gradually work it into the chambers of the heart, a risky and sometimes even fatal procedure. But now with ultrasound a picture can be obtained of the organs of the body at work.

Ultrasound, using the same principle developed years ago to locate enemy submarines, employs ultrasonic waves. Sonar signals too high for humans to hear are bounced off the organs of the body and received by a tiny instrument that translates them into pictures. In patients with gallbladder disease — a leading cause of surgery in adults — ultrasound has now been refined to the point where it can detect gallstones. It is also beginning to be employed to examine the prostate gland, malfunctions of which affect nearly half of all men over 50. "Nowadays, instead of going through time-consuming and uncomfortable tests, the patient goes straight to ultrasound," says Dr. Joseph P. Whalen, radiologist in chief at New York Hospital and professor of radiology at Cornell University Medical College. "It's a whole new dimension in diagnosis."

But probably the most widespread use of ultrasound has been in obstetrics, where it is now employed to monitor almost half of all pregnancies and deliveries. "In the first three months we can use ultrasound to determine if indeed the woman is pregnant," says Dr. Heidi Weissmann, a radiologist and ultrasound specialist at Montefiore Hospital and Medical Center. "Calculations and chemical tests are by no means always reliable." The probe can detect if the fetus is developing outside the womb, for example — a serious hazard to the mother's health.

Ultrasound has also made amniocentesis much safer. This procedure involves the insertion of a hollow needle into the amniotic cavity and the removal of some amniotic fluid. This contains cells sloughed off by the growing baby, cells that can be used to determine if it has certain genetic abnormalities. (In the case of a deformed or Mongoloid fetus, the mother might elect to have an abortion.) "A few years ago, before ultrasound, we used to have to stick the needle in without knowing exactly where," says Weissmann. "There was always the chance of jabbing it into some vital area. Now you can see where the head, heart, umbilical cord are."

□

The woman in her early 60's is rushed to Yale-New Haven Hospital, showing all the symptoms of a massive stroke. Her eyes are glazed, her speech slurred; she is unable to lift her right hand. On an impulse, the emergency-room doctor calls the hospital's radiology department. The woman is quickly wheeled to a well-lighted room dominated by a large, arching apparatus. She is placed on a special couch or "gantry" and slowly moved by conveyor belt into what looks like the hatch of a space capsule. A thin beam of X-rays makes a lateral sweep of her head, collecting more than 10,000 readings; it then rotates 10 degrees for another sweep. As the angle changes, a completely new set of data is collected and stored.

Across the way, in a darkened control room, a radiologist peers at the screen of a computer console, which provides a detailed outline of the woman's brain. The results of the scan indicate a dark gray circle on one of her pulsing cerebral lobes; the woman has not had a stroke, she has a brain tumor.

In a matter of hours the woman has surgery for what turns out to be a benign tumor. In a few weeks she is released from the hospital.

□

The Computerized Axial Tomography (CAT) scanner is an imaging device closely allied to ultrasound. In fact, at some places the two have been combined in a single treatment unit. New York Hospital, for example, created the Department of Body Imaging for the machines. But unlike ultrasound, the CAT scanner uses a battery of X-rays to take a cross-sectional picture — a tomogram — of the patient's body. Then computers put the millions of bits of information together in a clear picture on a computer screen. "I think most physicians would agree that it's one of the most revolutionary advances in almost 70 years," says Dr. Whalen.

In many cases, the CAT scanner has already replaced other forms of diagnosis. Instead of having to inject air into the brain — a sometimes excruciating procedure called pneumoencephalography — doctors can now obtain an even clearer view of a suspected brain tumor with a CAT scanner. "Another example is cancer of the pancreas," says Dr. Whalen. "This is extremely difficult to diagnose without surgery. In

the olden days — that is, five years ago — you might have had to put a catheter into the arteries of the pancreas and inject it with constrasting solution. Now the patient can come in to one of our three scanners, spend possibly half an hour, and the only discomfort involved is having to hold his breath for a few seconds. Sometimes we've found inoperable cancer, but even then we've saved the patient needless additional suffering and expense."

□

But diagnosis is only one area in which the new medical technology has had a striking impact. Important gains have also been made in treatment.

For 32-year-old fireman James Spink, a pleasant afternoon has become a nightmare. While doing some carpentry work at home, his electric saw slipped from his grasp, slashing his left hand and severing his thumb. He is rushed to Montefiore Hospital in the Bronx, where his bleeding is stopped, but his thumb cannot be saved. Since Spink is left-handed, the loss of his thumb could mean the end of his 11-year career as a firefighter.

But plastic surgeons at Montefiore have a different idea. With Spink's eager permission, they use microsurgery — surgery conducted with the aid of high-powered microscopes — to remove the fireman's big toe, along with its attached blood vessels, nerves and tendons, and transfer it to his hand to replace the missing thumb. With sutures almost invisible to the naked eye, blood vessels must be sewn together, nerves reattached so that the toe-be-come-thumb will have enough sensation to function.

Fourteen months later, James Spink returns to his job as hook-and-ladder man on Engine 46 with Ladder Company 27 in the Bronx, having passed the Fire Department's rigorous physical examinations. The remaining stump of his big toe gives him enough stability to walk, and his new thumb, while it looks a little shorter and stubbier than the original, works almost as well. "I didn't think I'd ever be able to work as a fireman again," says Spink. "Going back to work was the happiest moment of my life."

□

"Microsurgery is probably the most important thing to come along in surgery in decades," says Dr. Laurence LeWinn, assistant professor of surgery at New York Hospital-Cornell Medical Center. "It has opened up new vistas on just about everything."

The basic tools for this advance are microscopes capable of magnifying nerves and blood vessels up to 40 times. Usually operated by zoom pedals to make clumsy manipulation unnecessary, they are often connected to television screens so that other members of the operating team can see what the surgeon is doing. Almost as important are the instruments — from scissors with tiny blades to miniature forceps. Surgical thread is so thin that it is practically invisible to the naked eye. Used together, these have allowed specially trained surgeons to accomplish feats impossible a few years ago.

Within the past year, microsurgical cases have regularly made headlines. One that attracted particular interest was that of Renee Katz, a 17-year-old student and flutist at New York's High School of Music and Art. In May, she was pushed by a stranger into the path of a subway train. She was taken to Bellevue Hospital with her severed hand, and surgeons there managed to reattach it in a long and grueling operation. When Renee Katz left the hospital last month, Bellevue doctors were hopeful that she might even be able to play the flute again.

The high success rate for reattachment is due in part to better understanding of the complicated structure of nerves and how one can be grafted onto another that is partially damaged.

Microsurgery is also being used for such things as replacing a cancerous esophagus with a stretch of the patient's intestine, removal of tiny pituitary tumors that can turn children into ungainly giants, and restoring mobility to a partially paralyzed face by transferring nerves. But perhaps the most exciting use of microsurgery is in the prevention of strokes. When the brain is cut off from the flow of blood by an obstruction in an artery, the result may be irreversible damage. "Stroke is by far the commonest serious disease of the nervous system and one of the most important health problems in the United States," says Dr. Jack M. Fein, associate professor of neurosurgery at the Albert Einstein College of Medicine. There are half a million new stroke victims every year; a third die within a month, while half of the rest become permanent invalids. "Contrary to what many people think, this is by no means entirely a geriatric problem," says Dr. Fein. "Several of our patients have been in their 30's, and probably the greatest number are in their 50's and early 60's. This terrible affliction cuts down people in the prime of life."

But before a major crippling stroke, a victim usually experiences brief "warning strokes" that can last anywhere from several minutes to a whole day. "The symptoms can range from numbness in one side of the body to difficulty in speaking or writing or even temporary blindness, depending on which part of the brain is affected," says Dr. Russel H. Patterson Jr., professor of surgery-neurosurgery at New York Hospital-Cornell Medical Center.

Not long ago, a 40-year-old executive was talking to a client when suddenly he began to use the wrong words (since blood was unable to reach the speech center of his brain). Within two minutes, his right side was paralyzed. Within an hour, however, the symptoms had passed. The next day, his doctor referred him to the Albert Einstein College of Medicine, where he was found to be an excellent candidate for an exciting new treatment made possible by microsurgery: a stroke-prevention operation.

First, a CAT scanner may determine the extent of permanent tissue damage; no operation can help those whose brains have already been massively,

With electrodes implanted on the visual areas of his brain, this totally blind man is enabled to "see" crude images. The television camera feeds information to the computer, which then sends it to the man's brain through a disconnectable cable.

permanently affected. If the obstruction is in the brain, the surgeon drills a hole about the size of a silver dollar through the skull and, under the high magnification of the microscope, hooks one nearby artery to another, thus bypassing the obstruction.

Although only a few years old, the operation has already proved its effectiveness. The executive, who had two more warning strokes in quick succession while awaiting stroke-prevention surgery, has had no recurrence in the two years since. Out of one group of 400 high-risk patients of all ages studied for almost three years, only four of those who had the operation suffered strokes, in marked contrast to a control group, where almost half did. "I think this is going to have tremendous impact," says Dr. Fein. "Up until now, we had really no way of preventing this truly terrible affliction. Almost every family in this country has at least in some way been affected by the day-to-day agony of a relative who has experienced a stroke — what one neurologist in Boston calls 'super death.' Now, at least with a large number of people, we can try to do something about it. In the future, this kind of surgery should greatly reduce the personal, social and economic toll of this dreaded disease."

Another experimental but promising area of the new medical technology is the use of complicated electronics to restore vision to the blind and hearing to the deaf.

Five years ago, a team of scientists at the University of Utah, working with other scientists at the University of Western Ontario, electronically stimulated the brains of blind volunteers in the cerebral area known as the visual cortex so that the volunteers "saw" letters of the alphabet.

An even more dramatic experiment took place four years ago. A 33-year-old man had been blinded 10 years earlier in a gunshot accident. When a television camera was connected to elec-

trodes implanted on his brain, he was able to detect white horizontal and vertical lines on a dark background. "The experimental system functioned as a complete, though crude, artificial eye," says Dr. William H. Dobelle, director of the Utah project and currently both a member of the faculty at the University of Utah and head of the Division of Artificial Organs, Department of Surgery, at Columbia Presbyterian Medical Center in New York. In a later experiment, six electrodes on the man's brain were stimulated to form an image of letters of the Braille alphabet. With this stimulation, he was able to "read" phrases and short sentences, such as, "When the crow went into . . ." and "He had a cat and ball." "Without practice, the man — who was an extremely poor Braille reader — was able to read Braille at 30 characters per minute, or five times faster than he could using his finger tips," says Dr. Dobelle. This past December in New York, two new volunteers received implants, and so far the electrodes are functioning and they too seem to be able to visualize Braille letters.

But Dr. Dobelle and his collaborators are hoping that their system will do more than help the blind to read. Essentially, our eyes are like incredibly intricate cameras that catch light, transform it into electrical impulses, and then relay it to the visual cortex in the brain, where it is translated into the complex phenomenon called vision. "As far as we know now, there seems to be no reason why this can't be artificially duplicated," says Dr. Dobelle. He and his colleagues have already made plans for a special miniature television camera to be implanted in the eye of a volunteer and connected with wires to several hundred tiny electrodes on the brain. A CAT scanner will first be used to create an image of the volunteer's visual cortex, since the shape of each person's brain differs markedly; then the individually designed array of electrodes will be implanted, and the camera in the eye will

transmit information about light patterns to the brain — the artificial equivalent of vision. "From computer simulation studies we've done, we know that with anywhere from 250 to 500 electrodes you can start getting useful results," says Dr. Dobelle. Although he does not expect to restore normal vision completely, he nevertheless hopes that the blind will be able to read and, most important of all, to recognize people and things around them. "I hope we'll be able to give them the same sort of murky, black-and-white vision that you can see in the scratchy television pictures that early spacecraft sent back to earth," he says.

A related area in which Dr. Dobelle and his colleagues at Columbia and Utah are working intensively is artificial hearing. A quarter of a million Americans are totally deaf; one way to restore their hearing is to stimulate auditory nerve fibers electrically. Dr. Dobelle and his group have already implanted electrodes in the inner ears of a few volunteers — with encouraging results. One 47-year-old social worker, totally deaf for many years, was able to recognize such melodies as "Mary Had a Little Lamb," "Yankee Doodle" and "Twinkle, Twinkle, Little Star." "I believe society can be assured of the ultimate success of this research for the deaf, although of course many unforeseen obstacles doubtless remain," says Dr. Dobelle. "Naturally what they will hear is going to be pretty tinny. Walter Cronkite may sound like Donald Duck to these people, but the important thing is that they'll be able to recognize sounds and, above all, speech."

Dr. Dobelle is the first to admit that problems remain to be faced before that goal is realized. "The last thing we want is to have hordes of people lined up outside our hospital, hoping for help that we can't yet give them," he says. Vision and hearing are enormously complicated areas; nevertheless, on the basis of work already done, Dr. Dobelle is optimistic about the future. "I do really think that this is going to happen, that this technology is going to work," he says.

Other advances in medical technology also seem to hold promise for the future.

The "artificial pancreas," developed recently by Miles Laboratories, duplicates one of the main functions of the pancreas: it regulates changes in levels of blood glucose by injecting tiny amounts of insulin into the bloodstream. Still highly experimental, it is currently being tested at a few medical centers and has not yet received final F.D.A. approval.

Looking rather like a large stove on wheels, the machine is able both to monitor the glucose level in a diabetic patient and then, if it senses too much of a deviation, dole out tiny amounts of insulin for up to four days. The device has already aided several carefully selected diabetic women, who were helped through labor while attached to the machine. Doctors look forward to the day when smaller and more portable versions of the machine may help the country's hundreds of thousands of diabetics lead more normal lives.

A technique called nuclear magnetic resonance (N.M.R.) is also promising. Basically, the procedure involves putting groups of living cells in a strong magnetic field and then bombarding them with radio waves. A computer analyzes the amount of energy absorbed by different molecules within the cell to provide information about various events occurring within it. "Nuclear magnetic resonance allows us to see minute changes taking place within cells almost as they happen," says one researcher at the National Institutes of Health. "We can see all sorts of things that X-rays won't show — tracking several chemical reactions at the same time, for example." The technique is still too new and experimental for scientists to be willing to predict many of its potential uses. But several are hopeful about the new information it will give about the differences between normal and diseased cells — how a healthy cell turns into a dangerous, malignant one, for instance, as well as what kinds of treatment work best in either halting or reversing the process. "There's a lot of excitement," admits one scientist. "The living cell is where the action is, and N.M.R. offers one of the best ways of finding out exactly what's going on."

And yet, despite its promise, the new medical technology has come under increasingly sharp criticism. "Medical technologies are neither perfect nor risk-free," says Dr. Ivan L. Bennett Jr., provost/dean of the School of Medicine at New York University, who blames "the Marcus Welby syndrome" for the assumption that doctors — and the machines they use — can cure every case "in 30 or 60 minutes, less time out for commercials." The history of modern medicine is crowded with techniques that were loudly praised and then quietly discarded, from using X-ray machines in the 1940's to irradiate the thymus gland as a disease-prevention measure (but which, in fact, has caused increased risk of thyroid cancer) to the vogue for gastric freezing as a cure for ulcers in the 1960's, long since dismissed as either ineffective or downright harmful.

One much-discussed example of potential abuse is ultrasound. Although few deny the great benefits it can bring, several in the medical community consider it almost alarmingly over-used, particularly for fetal monitoring. All too often, they charge, technicians have misread images and imagined complications where none in fact existed; the result has been needless, and potentially harmful, Caesarian sections. "Although everyone agrees it's a great device, I think there's generally a greater sense of caution now," admits one New York radiologist. "We do recognize that this technological advance can be misused, and that if you use a lot of ultrasound, you can cause abnormalities of fetal development."

More subtle risks may be involved in the over-use of medical technology, too. The patient who spends more than a few days in the hospital may soon have the discouraging feeling that he is being treated by an array of machines. This is particularly true of the patient in an intensive-care unit, where he is attached, day and night, to beeping, buzzing newly developed electronic monitors. "One of the things that happens to many of these people is that they develop what psychiatrists call an 'I.C.U. psychosis,'" says Dr. Andrew A. Sorensen, associate professor of preventive medicine and community health at the University of Rochester Medical School. "They just stop reacting and retreat into their own world. In these cases, there are unquestionably harmful effects associated with medical technology."

Other critics predict these effects will become even more common. Despite their grudging respect for the healing potential of the new technology, their vision of the future is not a splendid one, where the ailing can be promptly and painlessly cured by a host of beneficent machines. Instead, they foresee a time when doctors will be largely transformed into technicians, operating apparatuses the layman cannot hope to fathom. The gulf of suspicion between patients and their healers, already distressingly large, will grow even wider. The general practitioner, made newly important within the past 10 years by medical schools anxious to emphasize medicine's "human face," will be unable to compete with the sophisticated diagnostic machines that only large hospitals can afford, and may well disappear. In this gloomy view of the future, the patient arriving at a hospital will not have a single doctor at all, but will simply be led, like a reluctant child, from one machine to the next, each manned by a detached and indifferent staff. "I'm not even sure that — for many patients — this will mean better medicine," says one critic. "We're finding out more and more how important the psychological element is in healing. If the patient feels anxious, resentful, caught in an alien and hostile environment, for many the end result of that stress may be to negate whatever superior medical advantages the machines offer."

Moreover, some see the possibility of greater geographic imbalance in the availability of treatment than exist now. "There are already massive inequities in our national health-care system," says Dr. Andrew Sorensen. "Places like New York City, Chicago, San Francisco are very well served in some areas, but there are still many sections of the country, particularly rural ones, where the level of medical care is shockingly low." Since only the largest urban medical centers will be able to afford the expensive new technology, in the future the disparity between the privileged metropolis and medically neglected small town is likely to grow rather than to diminish.

The overall price tag of the new technology has already stirred angry debate. "Health technologies contribute enormously to the inflationary spiral," Senator Edward M. Kennedy, chairman of the Subcommittee on Health and Scientific Research, has recently said. "As much as 40 percent of the annual increase in the cost of a hospital day can be attributed to their use." Total medical costs have soared to $182 billion (from less than a quarter of that 10 years ago), while the cost to the average person of a one-day stay in the hospital has risen from $35 in 1963 to $195 today. The medical-technology industry, responsible for manufacturing the new devices, is now valued at over $7 billion in this country.

One of the favorite targets of critics has been the CAT scanner, which — at $750,000 per unit — is one of the most expensive medical machines ever manufactured. Some see hospitals' eagerness for the device — what is called CAT fever — as a perfect illustration of what Dr. Ivan Bennett has dubbed "the technological imperative": the compulsive urge among hospitals to acquire the latest gadgetry in order to impress patients and the medical community. "If one hospital has it, another wants it," says one expert. "But is it the patient who really benefits?"

President Carter himself recently criticized hospitals' attitudes and confessed that he had once shared them. Mr. Carter said he now realizes that when he was a member of the Sumter County (Ga.) Hospital Authority (along with his uncle, mother and brother), he had "ripped people off." "We were naturally inclined to buy a new machine whenever it became available," he admitted, "and then to require that every patient who came into the hospital submit a blood sample or some other aspect of their body to the machine for analysis, whether they needed it or not, in order to rapidly defray the cost of purchase of the machine."

The response of many doctors to the President's increasingly strident criticism of expensive medical machines has been swift — and angry. "The new devices have saved lives and relieved suffering in countless cases," says one. "How does the Government intend to reckon that up in dollars and cents?" Even before the CAT scanner became generally available five years ago, some point out, hospital costs were steadily rising. "Technology is being made the whipping boy by politicians looking for votes," insists one physician at a large New York medical center. "But waste has always been a problem in hospitals. How about all the millions of dollars spent on unnecessary linen and paper slippers for patients, all the piddling expenses that keep adding up?"

But in a time of soaring inflation and general retrenchment, the cost of the new technology seems to affront more than just politicians. The CAT scanner, admittedly, is super-expensive, but few of the other devices are cheap. An ultrasound unit, for example, costs up to $100,000. The usual defense offered by hospital officials, that older procedures — since they required more personnel — often cost more, is usually ignored or derided: The mood in Washington and elsewhere today is stern and unforgiving. The more than 200 health-systems agencies scattered through the country, responsible for approving large-scale hospital expenditures, have already begun to clamp the lid down. Only about 200 hospitals a year now have their requests for CAT scanners approved — a small fraction of those who apply. "One result of this in the corporate world is that some companies have begun to edge away from expensive technical research in the health area," says one executive with a large medical-equipment firm. "Other companies have squelched plans to research improved versions of the CAT scanner. "Why should we invest $5 million in researching it if the Government won't let us sell it to customers?" says one official.

But few believe that a Government policy of strict across-the-board cost containment can really work without affecting health care. "The current system is ridiculous, but most plans around to change it aren't much better," says Dr. Murray Eden, chief of the Biomedical Engineering and Instrumentation Branch of the National Institutes of Health. "There's a great need for new direction and a settling of priorities. From a purely technological point of view, the prospects for the future are glittering, but the drive to contain health-care costs will obviously have an effect. As consumers, most of us want, demand, better health care, but we are offended when we get the bill. Somewhere there has to be some clearer thinking about where we're going to spend our money."

In just such an attempt to introduce clear thinking into the technology debate, Congress last year created the National Center for Health-Care Technology, a part of the Department of Health, Education and Welfare. "We intend to assess the value and cost of various aspects of the new medical technology and tell the public, practicing physicians, medical centers and big third-party payers like Blue Cross-Blue Shield what we find," says Dr. Seymour Perry, the center's acting head. "We aim to act as a catalyst, primarily through consensus conferences held every few months or so about different devices and procedures. Our findings won't have the force of law, but they will have the prestige of prominent members of the medical community behind them. That will count for a lot more than if some bureaucrat in Washington has the final say."

Whether the center can truly help to settle the debate remains to be seen, but its establishment has been generally — if hardly lavishly — welcomed in the medical community. "Most of all we hope to influence attitudes," says Dr. Perry. "Americans have long had a love affair with technology. As a people, we are finally going to have to take a good hard look at our limited resources and see which of the new devices around should be allowed to survive and spread. Our aim should be to make sure that medical technology is our servant and never becomes our master." ∎

American and Briton Get Nobel Prize for X-Ray Advance

By LAWRENCE K. ALTMAN

An American and a Briton won the 1979 Nobel Prize for physiology and medicine yesterday for developing a revolutionary X-ray technique that gives doctors an astonishingly clear look inside the living human body.

In the six years since its introduction, the technique, known as computed axial tomography, or the CAT scan, has been used in the diagnostic evaluation of the ailments of millions of patients.

The $190,000 award, which the two men will share, was made by the Nobel Assembly of the Karolinska Institute in Stockholm to Allan McLeod Cormack, 55 years old, a physicist at Tufts University in Medford, Mass., and Godfrey Newbold Hounsfield, 60, an electronics engineer at the British company EMI.

Mr. Cormack, who was born in South Africa, came to this country in 1956 and is now an American citizen.

The CAT scanning technique, in which a fully conscious patient lies on his back while an X-ray tube rotates around his head or other parts of the body, allows doctors to take pictures that reveal specific slices of the anatomy in more detail than possible through any other nonsurgical technique.

Mr. Hounsfield and Mr. Cormack will receive the prize along with other laureates yet to be named on Dec. 10, the anniversary of the death of Alfred Nobel, the inventor of dynamite, who left his fortune for these prizes. The other awards will be in chemistry, physics, economics and literature.

Yesterday's award was one of the most unusual in the 78-year history of the prizes. Among the reasons for that were the following:

¶Neither laureate has a doctoral degree in medicine or any field of science.

¶After a long debate that delayed the announcement for an hour, the 54 voting members of the Nobel Assembly vetoed the choice of their own selection committee. The identity of the original nominee or nominees was not made known. Swedish national television reported last night that the delay was presumably due to a split between two factions within the Nobel Assembly, with one favoring discoveries in basic science and the other discoveries with more direct application to everyday medicine.

¶The two researchers have never met. They did their research independently of each other.

First Physics Prize Recalled

The prize recalls the first Nobel Prize in physics, in 1901, which went to William K. Roentgen for his discovery, in 1895, of the X-ray, which also revolutionized the practice of medicine.

In the interval, medical physics has become a specialized field of medicine, and the CAT scanner was awarded a prize in the medical category because it reflected

Godfrey Newbold Hounsfield

Allan MacLeod Cormack

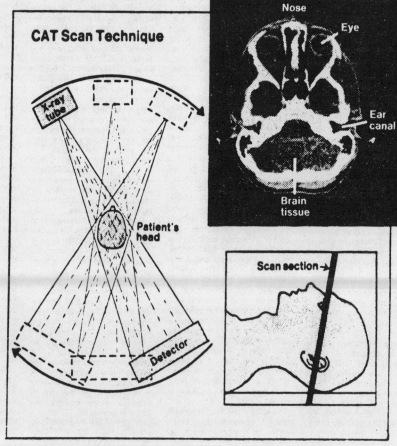

CAT Scan Technique

X-ray tube

Patient's head

Detector

Nose

Eye

Ear canal

Brain tissue

Scan section →

The New York Times / Oct. 12, 1979

In computed axial tomography, developed by Mr. Cormack and Mr. Hounsfield, an X-ray tube is rotated around a specific area of the body, such as the head, to yield a highly detailed photographic "slice" of the anatomy.

developments in mathematical physics that had their greatest application and significance in medicine.

The CAT scan has been called the greatest advance in radiology since the discovery of X-rays. But the award is sure to set off a controversy among health care experts and government officials over the use of the new device. Critics say that the equipment is too costly for many hospitals and contend that it is sometimes used when simpler and cheaper methods might suffice.

Proponents say that it has revolutionized the practice of neurology and other fields of medicine since it was introduced in 1973.

The Nobel Committee said in its citation: "It is no exaggeration to state that no other method within X-ray diagnostics within such a short period of time has led to such remarkable advances in research and in a multitude of applications" as CAT scans.

Two thousand CAT scanners, which can cost $500,000 or more each, have been sold in more than 50 countries, and 1,110 are in use in this country. A spokesman for EMI said that that company had sold more than 1,100 units, including more than 600 in the United States. In addition to EMI, there are four other chief manufacturers of CAT scanners: General Electric, Technicare, which is a division of Johnson & Johnson; Pfizer, and Siemens.

Dr. Ronald G. Evens of the Mallinckrodt Institute in St. Louis said that "modern neurology and neurosurgery cannot be practiced without a CAT scanner" because it has so changed the care of patients.

It has drastically reduced the number of painful, and sometimes dangerous, tests that were needed to diagnose brain tumors, birth defects and other brain conditions. CAT scans can diagnose strokes, but the diagnosis can also be made through other tests and therefore their cost-effectiveness has been questioned.

It is the application of the CAT scan to detection of problems elsewhere in the body that is the subject of considerable controversy. Because the body scanner was introduced in 1975, and because advances in ultrasound and nuclear medicine have been made in recent years, sufficient data have not been collected to adequately compare the techniques and to answer key medical questions about their relative values.

Dr. Evens said that there was little, if any, controversy about the usefulness of the CAT scanner in diagnosing a number of disorders, among them those affecting the kidney and certain lymph nodes near it; abnormalities of the spine, such as those resulting from injuries resulting in bone fragments that might tear the spinal cord and cause paralysis, and for cancers and infections and other problems of the face and the pelvis.

He said that the CAT scan shows "high promise" for determining whether surgery could be done for lung cancers and disorders of the liver. However, he said, questions remain whether CAT scans will replace other studies for these purposes.

As studies have shown when and for what conditions CAT scans are medically useful, the key question in the controversy has come down to: How many machines should there be in each geographic area?

Asked how he viewed the debate, Mr. Hounsfield said in a telephone interview from London that the answers to the economic questions must await further studies. "We are still learning," he said, adding that he was studying even newer uses for the device, which he began developing in 1967.

October 12, 1979

Nobel Prizes Are Awarded to 3 Physicists and 2 Chemists

Americans at Harvard and Purdue Win — German and Pakistani Cited

By MALCOLM W. BROWNE

Chemical research as down-to-earth as arthritis medicine and a theory of physics so profound as to affect man's perception of existence were both honored by Nobel Prizes yesterday.

The chemistry and physics prizes, each worth approximately $190,000 this year, in addition to the valuable gold medals themselves, were shared among three Americans, a Pakistani and a West German.

The chemistry prize was won by Herbert C. Brown of Purdue University and Georg Wittig of Heidelberg University, and the physics prize was awarded jointly to Steven Weinberg and Sheldon L. Glashow of Harvard University, and Abdus Salam, who works at universities in London and Trieste, Italy.

Aid in Drug Production

The two chemistry prize winners were honored for developing a group of substances capable of facilitating otherwise very difficult chemical reactions. This discovery has made possible the mass production of hundreds of important pharmaceuticals and industrial chemicals that would have been prohibitively expensive otherwise, including the arthritis medicine hydrocortisone.

While such practical applications have no part in the work of the physics prize recipients, many scientists regard their work as fundamental to understanding nature.

Although Drs. Weinberg, Glashow and Salam always worked separately, they were honored for their complementary research on a theory known to colleagues for the past decade as the Weinberg-Salam Theory of Weak Interactions. The theory is regarded by most scientists as a major step toward the goal of finding a unifying thread holding together the four fundamental forces of nature.

Two of these forces, gravitation and electromagnetism, had been known for many centuries. The two others were discovered only after science began investigating the atom, since they operate only within an environment as small as an atomic nucleus. One is known as the strong force, or, to use a phrase preferred by physicists, the strong interaction, which holds atomic nuclei together. The other force, the weak interaction, causes radioactive decay in certain kinds of atomic nuclei.

Many scientists, including Albert Einstein, sought a system of mathematical equations that could explain all four interactions as separate manifestations of a single underlying principle of nature. While these attempts have defied complete solution, yesterday's Nobel award recognized that there was now general acceptance of a principle, developed by the three physics prize winners, unifying the weak and electromagnetic interactions.

Working separately, Dr. Weinberg, 46 years old, and Dr. Salam, 53, developed a system of equations known as a "gauge theory." Gauge theories are based on the mathematical equivalents of telescopes and microscopes, serving to change the scale of one frame of reference so as to compare it with a completely different frame of reference. In this case, the two different frames of reference were electromagnetism, which operates between large,

235

easily observed objects, and the weak interaction, which is a nuclear force.

Among other things, the Weinberg-Salam theory states that weak interactions involve a flow of "neutral current" in an atomic nucleus. The neutral current is somewhat analagous to an electromagnetic current but does not carry any electrical charge. It relates to beta decay, a kind of radiation known since 1896, in which negative or positive electrons are ejected from the nuclei of radioactive atoms as they break apart.

Having developed their gauge theory, Drs. Weinberg and Salam were dependent on the work of a number of other investigators to fill in the necessary mathematics needed to equate theory with reality. Furthermore, predictions made by their theory had to be confirmed by experiments.

One of the difficulties of the Weinberg-Salam theory was resolved by the practical work of the 46-year-old Dr. Glashow, the third winner of yesterday's physics prize.

In the early 1960's, Murray Gell-Mann of the California Institute of Technology introduced the quark theory, which stated that all the large fundamental particles of nature, including protons, neutrons and mesons, are made up of combinations of still more fundamental particles named quarks. Three basic types of quarks, known as up quarks, down quarks and strange quarks were known at the time Drs. Weinberg and Salam announced their theory. But one aspect of their theory could not be explained by these three quarks alone.

Role of 'Charmed Quark'

The research conducted by Dr. Glashow on a new type of quark known as the "charmed quark" overcame this seeming difficulty in the Weinberg-Salam theory.

Experiments in Europe and the United States in 1972 and 1973 effectively confirmed the existence of neutral current predicted by the theory, and last year, scientists working with Stanford University's linear particle accelerator capped the confirmation of the theory with an experiment hailed by high-energy physicists throughout the world.

In the Stanford experiment, electrons were hurled at atomic nuclei and examined as they rebounded. Electrons may spin either in a left-handed or right-handed direction, and it could be expected that there would be equal numbers of left-handed and right-handed electrons rebounding from nuclear collisions.

But the Weinberg-Salam theory predicted that because of the weak nuclear force, there would be a slight but significant difference in the number of right-handed and left-handed electrons emerging from the collisions. This effect, known as "parity violation," was actually found in the Stanford experiment.

Although the Weinberg-Salam theory is not regarded as absolutely proved,

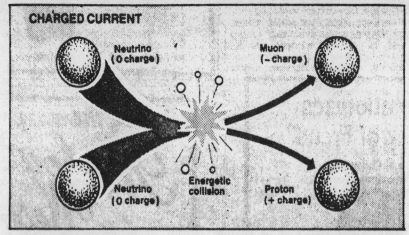

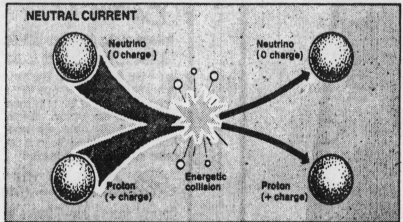

New York Times/Oct. 16, 1979

Chart shows theory about the weak force in subatomic behavior. It was known that the collision of two separate neutrinos, each with no electrical charge, could produce a negatively-charged muon and a positive proton, the result of a "charged current" flowing through the reaction. To unify theoretically the force of electromagnetism with these weak forces, Dr. Weinberg and Dr. Salam proposed a type of collision producing no change in charge, in which this "neutral current" is the carrier of the weak force.

the Stanford experiment lent it such weight that few if any scientists challenge the main outline of the theory at this point.

Yesterday's chemistry prize grew out of efforts that spanned much of the careers of the recipients.

In Germany Dr. Wittig, who is 82, based the main body of his work on compounds of phosphorus, while his American cowinner, Dr. Brown, 67, worked primarily with compounds of boron. But both had a common aim.

Many complicated molecules are extremely difficult to join, but the chemical linking of two or more large molecules is often vital to industrial processes, including the manufacture of pharmaceuticals.

Large Molecules Joined

A breakthrough in chemical technique permitting the comparatively easy joining of certain large molecules was the result of the work of Drs. Wittig and Brown. The compounds they have developed serve as temporary chemical links holding these large molecules together until permanent chemical

joints are inserted to replace them.

The "links" used by Dr. Brown are known as boranes — compounds of boron and hydrogen that readily join with many of the large molecules on which life is based. Dr. Wittig uses chains of carbon, hydrogen and phosphorus atoms in somewhat the same way.

These links have facilitated large-scale manufacture of many pharmaceuticals and a compound of uranium developed during World War II by Dr. Brown as a possible way of refining fuel for atomic bombs.

"There are many interesting applications," he said in an interview yesterday. "I feel that we have uncovered a new continent, just beginning to explore its mountain ranges and valleys. But it will take another generation of chemists to fully explore and apply this new chemistry of boron hydrides and organoboranes."

Several of yesterday's winners said they had not been very surprised to learn of their awards, while one, Dr. Brown, who was on a business trip to

New Jersey, said: "If I'd known I'd have stayed home to be with my wife, who's also a chemist." All the new laureates, who were selected by the Nobel Committee of the Swedish Academy, were deluged with congratulatory telephone calls and visits from friends and colleagues.

The Nobel Prizes, which have been awarded since 1901, were established by the will of Alfred Nobel, the Swedish chemist and industrialist who invented dynamite. Nobel, who died in 1896, stipulated that there should be only three categories of science in which prizes would be awarded: physics, chemistry and medicine. The two other prizes are for literature and peace. The Nobel Memorial Prize in Economic Science was established by the Central Bank of Sweden in 1968 as a memorial to Nobel.

Nobel has been criticized for his failure to recognize other branches of science, notably mathematics and astronomy. To remedy this, leading mathematicians created the Field Medal, which is regarded by mathematicians as roughly equal in prestige to the Nobel Prize.

Apart from its cash value, the Nobel Prize is especially valuable to an American scientist this year. Although the financing of scientific research has increased, the cost of research is growing so rapidly that many scientists find themselves constantly forced to justify their work to obtain renewed research grants. A Nobel Prize winner is among the few normally exempt from this requirement.

October 16, 1979

High-Level Balloon Detects Antimatter Flow From Space

By MALCOLM W. BROWNE

Using a high-altitude balloon carrying a 5,000-pound package of ultrasensitive instruments, a team of scientists at New Mexico State University has detected a stream of antimatter from interstellar space — the first such matter ever found outside a terrestrial laboratory.

The discovery of interstellar antiprotons, made by Dr. Robert L. Golden and his colleagues under a $5 million National Aeronautics and Space Administration program, was reported Monday in the scientific journal Physical Review Letters. It is expected by scientists to have an important impact on theories about the origin of the universe.

Scientists who study the particles that make up atoms and their nuclei concluded decades ago that for every type of normal particle, a corresponding particle of antimatter can exist, carrying an opposite charge. Thus, the negatively charged electron has the positively charged positron as its antiparticle, and an antiproton is the negatively charged antimatter equivalent of a normal proton.

When matter and antimatter are made to collide in laboratory experiments, the result is that both particles are annihilated and a burst of energy and new particles emerge.

Finding Undercuts Theory

One of the puzzles of modern astronomy and cosmology has centered on the apparent absence of antimatter in the universe. According to one view of the universe, equal amounts of ordinary matter and antimatter should have been created in the genesis of the universe. One theory attempting to explain the apparent absence of antimatter was that antimatter is intrinsically unstable and had somehow decayed almost to the vanishing point since the universe was born.

But Dr. Golden's discovery appears to have demolished that theory by showing that the antiproton, a key form of antimatter, is stable and capable of living for many millions of years.

Dr. Golden and an assistant, Brad Mauger, described their research in a telephone interview yesterday.

The experiments that detected interstellar antiprotons were carried out in a flight made last June 21 and 22 by a balloon inflated with 28 million cubic feet of helium, which reached an altitude of 112,000 feet, above the atmospheric shield that blocks antimatter particles.

Dr. Golden said that observations would be clearer aboard a space satellite, but until an antimatter detection array can be sent into space aboard the planned space shuttle, the New Mexico team must use balloons. The space shuttle is not expected to be launched before the middle of 1980.

Prediction Proves Correct

The New Mexico group had theorized that cosmic rays originating in the cores of stars and galaxies, which include high-energy protons, must collide with clouds of interstellar hydrogen from time to time. Some of the collisions between protons and the nuclei of hydrogen atoms in these clouds would produce antimatter in the form of antiprotons. Calculations showed that for every 10,000 ordinary protons, 4.4 antiprotons should be produced by such collisions.

The actual proportion of protons and antiprotons detected by Dr. Golden's balloon array almost perfectly matched this theoretical prediction.

"This also told us that antiprotons in space are stable," Dr. Golden said. "The age of a cosmic ray can be determined by measuring the radioactive decay of some of its atoms, especially beryllium-10. We therefore also know the age of the antiprotons in the beam. They are many millions of years old."

At the core of the antimatter detection system was a superconducting magnet chilled by liquid helium to only a few degrees above the absolute zero.

"That magnet is so powerful," Dr. Golden said, "that it turns the entire 5,000-pound instrument package to point north — surely the biggest compass ever built."

The purpose of the magnet is to deflect the course of heavy particles reaching the detector from outer space, so that the charge and momentum of each particle — its characteristic signature — can be measured. Most of the instruments in the package were designed to measure and identify all the extraneous background particles which must be deducted from the overall results to arrive at a measurement of true interstellar matter.

"These findings throw the problem of explaining the origin of the universe back to the particle physicists," Dr. Golden said. He said his former teacher, Steven Weinberg of Harvard University (who shared the Nobel Prize in Physics Monday) had studied the problem, proposing that in the first instants after creation, most matter and antimatter would have annihilated each other, leaving just enough excess normal matter to make up everything we now see in the universe.

October 17, 1979

237

CHAPTER **11**
The Middle East

UNTOLD STORY OF THE MIDEAST TALKS

Few diplomatic actions have received as much attention as the extraordinary bid for peace between Israel and Egypt, but much of the story remains undisclosed. What follows is an inside account that sheds light on how the Camp David agreement was reached — and why it fell apart. *First of a two-part series.*

By Sidney Zion and Uri Dan

On Friday, Sept. 16, 1977, Rachel Dayan, the blond, elegant wife of Moshe Dayan, boarded an airliner at Brussels for New York. By itself, this would hardly be worthy of notice: Mrs. Dayan was accompanying her husband to the United States. As Israel's new Foreign Minister, he was to meet President Carter and prepare for the opening of the United Nations General Assembly. What attracted attention at the airport was that she was alone.

At first, reporters assumed that Moshe Dayan had boarded earlier, for security reasons. After takeoff, they discovered he had not. By the time the jet landed at Kennedy, his whereabouts had become an international mystery.

In Israel, the rumor market had him in Paris — or in Geneva or Bucharest or Vienna or Teheran — conferring with King Hussein or Andrei Gromyko or King Khalid or the Shah. From the State Department, an urgent cable went to the American Embassy in Tel Aviv: "Where Moshe Dayan?" The embassy buttonholed, importuned and pressed its usually reliable sources, but to no avail. That was puzzling in itself.

This article was reported by Sidney Zion, a columnist for New York magazine, and Uri Dan, a correspondent of the Israeli paper Ma'ariv, and was written by Mr. Zion. The article is adapted from a book to be published by Times Books.

Israel seldom kept anything from the Americans; its top secrets were passed on to Washington almost as a matter of course. Was the new hard-line Government of Prime Minister Menachem Begin off on some hazardous tangent?

The next night, Dayan landed in Tel Aviv. Turning aside all questions with his enigmatic smile, Dayan slipped into a limousine and told newsmen: "I'm going home to sleep." He went, instead, to Begin's house in Jerusalem, where the Prime Minister and the chief of Israeli intelligence were waiting. No one in Israel had been in contact with Dayan since he left Brussels the day of his scheduled flight to New York. Only a handful of top officials knew where he had been.

He had been to Morocco — flown there in a private jet supplied by King Hassan II, his identity disguised by oversize dark glasses and a fedora. There he had met an Egyptian, a calm, quiet man with an odd title — Deputy Prime Minister at the Presidency. The Egyptian's name was Hassan el-Tohamy and his title was a convenient camouflage. Tohamy is President Anwar el-Sadat's closest confidant, his *éminence grise*. The word in Israeli intelligence is that "if Mubarak [Vice President Hosni Mubarak] is Egypt's Number 2, Tohamy is Number 1½." Dayan was engaged in a dialogue with a man who could speak for the Egyptian President. And the man, it turned out, had plenty to say. Their conversation took a philosophical, then a religious turn; it was now sharp, now humorous; finally, and essentially, it was pragmatic.

In Jerusalem, Begin had no way of knowing what was said at the meeting in Rabat until Dayan returned: The mission was too delicate for telephones or telex. When Dayan arrived at Begin's house, the Prime Minister skipped the amenities — no small thing for the "Polish gentleman" whose Continental manners stand out in contrast to the brusque ways of the native-born sabras. He asked Dayan to get right to the punch line.

Dayan, generally imperturbable, was flushed with excitement. His news:

Sadat was willing to meet secretly with Begin. He was willing to negotiate a nonbelligerency agreement, and perhaps much more.

Sadat wanted no part of a Geneva conference, an approach resuscitated by the Carter Administration — and strenuously opposed by Israel, since it would bring in the Palestinians and the Russians and would be dominated by the thorny issue of a "Palestinian homeland."

Sadat wanted no part of the Russians, either, and was not interested in a Palestinian state on the West Bank. Furthermore, he did not rule out a full-fledged peace treaty between Egypt and Israel. The price: return of the Sinai to Egyptian sovereignty and some "Palestinian arrangement" linking the West Bank and Gaza with Jordan, perhaps in some form of confederation.

Was the price too high? Probably. But after four wars and 30 years of tension and distrust, it was enough that, for the first time, there would be negotiations with an Arab country, the most important Arab country of all, and on

the highest level. It was astonishing. And, considering the background of the prospective peacemakers, it was irony at its height.

Menachem Begin, commander of the terrorist Irgun Zvai Leumi during the British Mandate; perceived by the Arabs as the "butcher" of the Arab village of Deir Yassin; a superhawk with visions of a return to the biblical lands of Judea and Samaria — Begin was the man the Arabs feared and hated most.

And Anwar el-Sadat, the one-time political hothead (the British jailed him during the Second World War for connections with German agents); long regarded in Israel as strongly anti-Jewish; the architect of the sneak attack on Israel on Yom Kippur, 1973 — how different was he from his intransigent mentor, the late Gamal Abdel Nasser?

Could these two leaders really bring peace to the Middle East?

The risks were monumental. If this initiative failed, the worst that could happen would be all too likely to arrive.

What follows is the untold story of the search for peace. It is a tale filled with intrigue, romance, jealousy, pettiness, gallantry — and, always, danger. Over a period of 16 months, the authors talked to more than 80 informants in Egypt, Israel, the United States, France and Morocco. Among them were political leaders, diplomats, military men, intelligence experts and those anonymous people who work in the shadow of the great. The inside view of what passed among them sheds light on why the historic accord reached at Camp David came apart after the negotiators of Israel, Egypt and the United States went home, and why the attempt to put the agreement together again has proved so difficult. If the story sometimes defies logic, it is well to remember that we are in the Middle East.

The story begins in July 1977, when Menachem Begin received a visit in the Prime Minister's office from the men who run Israeli intelligence. They came with a thick file containing details of a startling plot: The Libyan dictator, Col. Muammar el-Qaddafi, had organized an operation to assassinate his antagonist and rival for Arab leadership, Anwar el-Sadat.

Israeli agents had uncovered the conspiracy by accident. This was not unusual. The Palestine Liberation Or-

Stars of diplomatic drama: Carter (center) and, clockwise from top left, Dayan; Sadat and Begin; Ariel Sharon; Hassan el-Tohamy; Generals Ali, Gamasy and Weizman.

ganization had connections with terrorist groups around the world, and wherever the P.L.O. went, there went Israeli intelligence. And Colonel Qaddafi's assassination team was composed mainly of Palestinian extremists.

Begin, less than a month in office after 30 years in the political wilderness, was uncertain how to react. "What did my predecessors do in such situations?"

The intelligence men filled him in. When the information concerned non-Arab governments, it was passed on to them. When it concerned the Arab neighbors — as with the plots of recent years to assassinate King Hussein of Jordan, the late King Faisal of Saudi Arabia, and Faisal's successor Khalid — the files were turned over to the United States Central Intelligence Agency and the C.I.A. transmitted the material under its own byline to the government involved.

Begin, formal as always in his dark business suit and starched collar, thumbed through the file and put it down on his desk. He wiped his horn-rimmed glasses and went through the file a second time. Without looking up, he said, "Why don't you give this directly to the Egyptians?"

The intelligence chiefs were hard put to hide their incredulity. Was the Prime Minister telling them to bypass the United States? And on a case as big as this? They were so shocked they were still in their seats when he stood up. He smiled faintly, as though to say, "Well, what are you waiting for?"

Morocco was a good channel; King Hassan was delighted to be of help. For years the Moroccan monarch had tried to bring Israel and Egypt together. Several secret meetings had been held in Rabat between emissaries of the two countries when Yitzhak Rabin was Prime Minister, and Rabin himself had flown incognito to Morocco in 1976 to meet with the king, but these contacts had come to nothing. This Israeli request for another session smacked of something more promising. The Israeli intelligence chief flew to Casablanca. There he met with his Egyptian counterpart, Lieut. Gen. Kamal Hassan Ali (today the Minister of Defense).

General Ali was back in Cairo in short order, his suspicions aroused. True, Qaddafi was a wild man; *magnoun,*

Sadat called him — crazy. Qaddafi, in his own eyes, was the savior of the Arab Nation, the legitimate heir of Gamal Nasser, while Sadat was the pretender, a traitor to pan-Arabism. But a plot to *murder* Sadat?

And why wouldn't have Ali's people known about it? The Egyptian intelligence service was first-rate; even the Israelis said so. The Jews could be up to something, some trick to cause further havoc in the Arab world.

And yet there was this file — the names and addresses of the hit men, their arms drops, all in the heart of Cairo. The secret police moved in and caught the conspirators red-handed. Everything was there — the arms, the plans, Libyan documents, Libyan and Palestinian gunmen. If Ali still had his doubts, they were allayed when the prisoners were interrogated. It was the real thing, all right; the Israelis weren't playing games.

Sadat kept the whole thing secret. Five days later, he launched a lightning border war against Libya. His bombers blasted Qaddafi's Soviet-constructed radar installations, while Egyptian commandos engaged the Libyans in a series of bloody clashes. To governments unapprized of what had happened behind the scenes — and this included Washington — his action was a bolt out of the blue. Begin took the Knesset podium to declare that Israel would do nothing to disturb the Egyptians in the Sinai while they were busy in Libya. Israelis shrugged, having no idea why Begin would bother to make such a statement.

For Sadat, the whole incident was a stunning revelation. He had been profoundly depressed by Begin's upset victory at the polls. If he had been unable to make peace with Golda Meir, if he had managed to wring only modest territorial concessions out of Rabin, what chance would he have with Begin, whose signature was "not an inch"?

Begin's choice of ministers had deepened his gloom. General Dayan, the man who had commanded Israel's forces in both the 1967 and 1973 wars — Foreign Minister. Maj. Gen. Ezer Weizman, the creator of the modern Israeli Air Force — Minister of Defense. Gen. Ariel Sharon, the Israeli Patton, who had crossed the Suez Canal in the October war — Minister of Agriculture.

Sadat was not fooled by the pastoral title: Sharon would be Minister of Bulldozers, Creator of Settlements, Chief of Occupation. There would be no peace as long as these men ran Israel. And if there was no peace, there would be war. Sadat could not, would not, sit by forever while a foreign nation held on to Egyptian land.

Now all of this had to be reassessed. Begin had acted to save an Egyptian President's life. Sadat, master of the *beau geste,* appreciated this more than might have been true of another man in his place. And he respected Begin all the more for doing it on his own, without consulting the Americans. It did not necessarily mean that Begin was Israel's Charles de Gaulle, the one man with hard-line credentials impeccable enough to make peace. But it had to mean something, and it might even mean that.

So this time it was Sadat who approached King Hassan — and, after several preliminary middle-level meetings in Rabat, the Dayan-Tohamy tête-à-tête was the result.

I f Sadat was depressed by Begin's election, the C.I.A. was dumbfounded. The agency had been so sure the leader of the Likud opposition had no chance to unseat the Labor Party that it hadn't bothered to work out an analysis of the situation to face the United States should Begin win. The omission was one that the C.I.A. director, Stansfield Turner, had to try to explain personally to the President. The trouble — as in other foreign countries before and since — was that American officials in Israel talked mainly to the "ins," who assured them, as they assured one another, that what they didn't want couldn't occur.

When it did, the Carter Administration was dismayed. A Government headed by Menachem Begin — could there be a bigger obstacle to peace?

At this point, there was a flurry of activity in Washington and Tel Aviv. Top members of the former Labor regime shuttled between the two capitals, huddling with American officials in both places — a curious development, considering how little attention the Americans had paid to the Israeli opposition

when it was headed by Begin. Some of the meetings were with Zbigniew Brzezinski, the President's national-security adviser.

"Brzezinski saw more ex's than a gigolo in those days," an Israeli insider recalls. "Only you couldn't tell who was doing the romancing."

At the same time, American officials engaged in "background" talks with American and Israeli journalists and publishers, promoting an image of Begin as a bellicose fanatic, a man who would never make peace, and who would endanger the "special relationship" between Israel and the United States. The same message was put across at kaffeeklatsch sessions with influential Zionists.

What was going on?

One analysis was conveyed to Begin by Israeli intelligence experts. They told him that the Americans were trying to undermine him by turning public opinion in both countries against him; and that the Carter Administration was being aided in this endeavor by certain top Labor officials in Israel, as well as by some government bureaucrats still in office. All concerned, Begin was told, would be only too happy if he were forced to step down.

Begin, interestingly enough, did not ask for a full report. He did not want to know the full details. He was hurt by the American attitude, his intimates say, but not overly surprised.

He had never been popular in the United States. During his underground period in the 1940's, he was condemned by the American-Jewish establishment. When, after the creation of Israel, he visited New York in 1948, he was attacked in some respectable circles as a Fascist. In the years to follow, he was cold-shouldered by American Jews (except for a small band of followers) and ignored by the United States Government.

This history of hostility and scorn, topped by the unsubtle campaign against him in his moment of victory, very likely entered into Begin's thinking when he decided to bypass Washington in his dealings with Sadat. The fundamental consideration, however, appears to have lain elsewhere.

During his years as opposition leader, Begin had hotly criticized what he viewed as Israel's evolution into a satellite of the United States. He recognized Israel's need for American arms and money, but he argued that it was a two-way street: The United States needed Israel as an anti-Soviet bulwark in the Middle East. Now, as Prime Minister, Begin faced an American Administration that had revived an approach to a regional peace settlement that was beset, from the Israeli standpoint, with potential dangers. Reconvening the Geneva conference, as proposed by Washington, would subordinate everything to the Arab demand for the creation of an independent Palestinian state, a scheme viewed by the vast majority of Israelis as threatening to create a new and potentially aggressive power center on the other side of an indefensible borderline.

The offer by Sadat — a nonbelligerency pact with *(Continued)*

At a closed meeting with Dayan, Carter delivered a virtual ultimatum: Israel would have to accept a Palestinian homeland on the West Bank. If Israel refused, the President would address himself to 'friends of Israel in the United States,' warning that Iraeli intransigence was endangering vital American interests.

Egypt and perhaps even a separate peace — could not be more timely. Taken at face value, it was a heaven-sent alternative to the diplomatic campaign launched by the United States. In the circumstances, the offer had best be explored without consultation with third parties — a preference Sadat seemingly shared.

There was something in Begin's makeup the C.I.A. specialists had missed when they whipped up a profile on him after his election victory. Begin, it seems, aspired after the mantle of peacemaker. "I must bring peace to my people," he told his intimates when he assumed power. His predecessors, who he felt had maligned him down the years, had failed to make peace. Now he would do it — somehow.

The Carter Administration took office with a plan for the Middle East. Both in the White House and the State Department, the prevailing view was that Henry Kissinger's step-by-step approach had gone as far as it could; now, the experts felt, the time was ripe for a comprehensive settlement. The policy they adopted was modeled largely on a 1976 report by the Brookings Institution, a Washington think tank, whose signatories included Zbigniew Brzezinski, now the head of the National Security Council, and William B. Quandt, now the N.S.C.'s top Middle East expert. One of the report's conclusions was that a general settlement could not be had without agreement on the establishment of a Palestinian political entity in the West Bank.

Two months after his inauguration, Carter said at a town meeting in Clinton, Mass., that "there has to be a homeland provided for the Palestinian refugees." The Israelis were shaken — no ranking American official had come out publicly for a Palestinian homeland up to then, and Carter had taken a strongly pro-Israeli position in his Presidential campaign. But Carter repeated the statement several times in the ensuing weeks, so it was obvious that a new policy was in the works.

On Oct. 1, 1977, the United States joined the Soviet Union in a joint declaration supporting the Palestinians' "legitimate rights." By now there was an uproar in Israel and in the American-Jewish community. The phrase "legitimate rights" had come to be viewed by many as a code word for a Palestinian state. And bringing in the Russians, after Kissinger had so deftly excluded them from the Mid-

dle East negotiating process, struck many as a cardinal blunder.

It is highly unlikely that Carter would have proceeded as he did had he known of the contact between Begin and Sadat; he undoubtedly would have waited to see what came of that dialogue. But the Dayan-Tohamy meeting of two weeks earlier was still unknown to American intelligence.

The political storm aroused by the Washington-Moscow statement was accepted by the White House as inevitable, something to ride out. Carter's inner circle was sure that in the country at large an American President was in a better position to rally public opinion than an Israeli Prime Minister, particularly this one. Carter did appear surprised by the intensity of the outcry among Jewish leaders in the United States, but he regarded it as an emotional reaction that would pass. Begin, the White House felt, was in no position to rally the American Jews.

On Oct. 5, at a closed meeting with Dayan at the United Nations Plaza Hotel in New York, Carter delivered a virtual ultimatum. The Palestinians, he said, would be represented at Geneva, and Israel would eventually have to accept a Palestinian "entity" or "homeland" on the West Bank and Gaza. If Israel refused to go along with this, the President would address himself to the "friends of Israel in the United States" (read "American Jews"), warning that Israeli intransigence was holding up peace and endangering vital American interests. And although the United States would honor its commitment to Israel's security, not everything promised to Israel was "in the box."

Carter was unprepared for Dayan's reply.

Israel, Dayan said, would never accept an independent Palestinian state under any name. Israel would not sit with any P.L.O. representative in Geneva. Israel considered the American-Soviet declaration "totally unacceptable."

"Tell me," the President said, "what makes you dissatisfied with the declaration?"

"Everything."

"Let's study it clause by clause," Carter suggested.

"There's no need. The mere fact that you introduced the Russians into the arrangement is enough."

When Carter played his ace — "our promises to you are not in the box" — Dayan responded with a little monologue.

"In the life of a nation," he

said, "there are situations when leaders must decide whether they prefer promises to their deepest beliefs. This is particularly so when there is reason to suspect that the promises will not be fulfilled."

Coolly, with the half-smile that is his hallmark, he reviewed Israel's experience with American Presidential pledges.

"During the Holocaust, President Roosevelt saw fit not to open the gates to Jewish refugees.

"In 1948, while our nation fought for survival against all the Arab states, President Truman — though he had recognized Israel — refused to give us even one rifle, not one rifle!

"In 1956, after the Sinai campaign, President Eisenhower promised us free navigation in international waters in return for our unilateral evacuation of Sinai. But in 1967, when Egypt blockaded the Straits of Tiran, we had to break the blockade ourselves.

"When we came to President Johnson in May of 1967 and asked him to fulfill Ike's promise, he said to us, 'The U.S. is not the policeman of the world.' "

So, Dayan concluded, Israel did not expect anyone to save its people from another holocaust. "But it will not happen, because we will prevent it."

The meeting, which lasted six hours, ended with Carter easing his stance. He instructed Secretary of State Cyrus Vance to draw up a mildly worded working paper for the Geneva conference.

According to a participant at the meeting, Carter asked Dayan where he'd be going next; when Dayan told him he'd be speaking to Jewish leaders in Chicago, the President said with a smile, "Do me a favor. Don't attack me."

Unbeknown to Washington, which continued to regard him as an exponent of the pan-Arab viewpoint, Sadat was just as disturbed by the American-Soviet declaration as Begin.

He had kicked the Russians out of Egypt in 1972 and did not want them back in the picture — certainly not in diplomatic tandem with the United States.

He knew that Israel would never accept a Palestinian state — and, whatever their rhetoric, that neither would Jordan nor Saudi Arabia.

He was far from enthusiastic about another Geneva conference, fearing that Egyptian interests would be submerged. He had seen much more merit in Kissinger's shuttle diploma-

cy. It was the Carter Administration's abandonment of that technique that persuaded him to make his own secret contact with Begin.

For Anwar el-Sadat wanted peace, and wanted it badly. He had to get the Sinai back. It was a matter of national honor, and no leader in Cairo could long survive if he came to be seen as reconciled to so huge a loss of Egyptian territory. And Sadat knew there was no way of getting Israel to unhand the Sinai except through a peace settlement. The Egyptian Army had surprised the world, and itself, by pushing the Israelis away from the Suez Canal, but it could hardly expect to improve on that achievement.

Then there was the economic situation at home. The Egyptian people were used to privation, but things had gotten so bad that in January 1977 large-scale riots took place in many parts of the country. Sadat took pains to deny that he had to have peace in order to free resources for a more effective attack on his domestic problems, but the relationship between the two was inescapable.

On Nov. 9, 1977, with P.L.O. chieftain Yasir Arafat among the honored guests, Sadat made a major address opening a new session of the Egyptian Parliament. He spoke emotionally of the need for peace for the whole region, accused the Israelis of putting up stumbling blocks, and added: "I am ready to go to the Israeli Parliament itself and discuss it with them."

Later, Tohamy was to tell an Israeli official how Sadat made up his mind to undertake the journey. "It took one minute between his voicing the idea and his decision to do it."

The Egyptian Government was stunned. Foreign Minister Ismail Fahmy resigned; in his ministry, career diplomats trained in the pan-Arab tradition of the Nasser era muttered and sulked. In Cairo's political salons, heads shook. All, of course, to no avail. Sadat does what Sadat does, with the help of Allah and a strong army.

To Begin, the Egyptian President's words were doubly astonishing because of a military buildup on the western side of the Suez Canal early that month — a massing of Egyptian troops that threw the Israeli military into a frenzy of reconnaissance flights and had the Prime Minister wondering anew if the Moroccan talks had been a ruse. Now

these fears were allayed. Sadat could hardly have made a more convincing gesture. An untried Prime Minister whose hopes for Rabat had never gone beyond the slim possibility of a secret meeting with Sadat, Begin now had to find an adequate response to one of the most dramatic peace offers of our time.

His initial move was to invite Sadat to address the Knesset. Then he called the Cabinet into emergency session. Among his ministers, only Dayan, Sharon, Weizman and one or two others were aware of the Rabat contacts, and Begin preferred to keep it that way. The immediate concern was with Israel's world posture. Sadat had enhanced his international standing; wasn't there a way for Israel to pick up some points?

General Sharon, the Minister of Agriculture and the Government's fiercest hawk, suggested, surprisingly, that Begin top Sadat's speech before the Knesset with an offer to return El Arish, a strategic point on the Sinai's Mediterranean coast, to Egyptian civilian administration. The motion was supported by only one man: Menachem Begin. (Months later, Sadat was to say to Defense Minister Ezer Weizman: "They say the Jews are a very clever people. How come, then, you didn't do a gesture when I came to Jerusalem? Let's say if you gave me El Arish . . . you would have taken all the wind out of my sails.")

On Nov. 19, a jet emblazoned with the Egyptian flag landed at Ben-Gurion International Airport and, to a fanfare of trumpets, Sadat stood on the tarmac facing Begin. As they shook hands, Begin said, "Mr. President, welcome. All of the Israeli Government is here to welcome you."

"Also General Sharon?" Sadat asked.

Going down the reception line, he stopped when he reached Sharon. "I wanted to capture you in the canal."

His meeting with Ezer Weizman was one of the high points of his visit. Weizman was prevented from attending the airport ceremonies by injuries sustained in an automobile accident the previous week. That evening, however, he barged into a select gathering at the home of the then President of Israel, Ephraim Katzir. Seated on a sofa were Katzir, Begin and Sadat. Weizman arrived in a wheelchair. He stood up, faced Sadat, held his cane up in a present-arms position, and said: "Mr. President, I salute you."

With a kind of roar, Sadat jumped up and embraced Weizman. Thus was struck an affinity between the two that continues to this day.

Weizman met with Sadat again the next morning. In the clipped English accent he acquired while serving in the Royal Air Force during World War II, he told the Egyptian that, as a "military man," he respected him for the surprise attack of 1973.

Sadat, obviously flattered, said, "Let's not talk about wars any more."

"Before we stop talking war, why don't you tell me why you lately massed your troops on the Suez?" Weizman had a pretty good idea by then that the troop movements had been maneuvers, but perhaps they had been "political maneuvers" as well?

Sadat gave him an enigmatic look. "Why did you mass yours?" The Israelis had sent their tanks poking into forward positions: Had it been just a response?

In the Middle East, ambivalence is never entirely absent, even when enemies become friends.

"Perhaps we should avoid such misunderstandings in the future," Weizman suggested.

Sadat was emphatic: "For sure!"

The Egyptian President's speech before the Knesset was enthusiastically applauded, which was odd, considering that he was demanding complete Israeli withdrawal from all Arab lands occupied in the 1967 war. But the Israelis were not listening to his words; they were enthralled by his presence;* and Begin's response — largely a review of the tragedies that had befallen the Jewish people across the centuries — bore little relevance to what Sadat said.

Anyway, it made no difference: Sadat was speaking mainly to the Arab world, pouring on the rhetoric to deflect accusations that he was suing for a separate peace. The one thing he did mean was that the Sinai — all of the Sinai — had to be returned to Egypt. And this he thought he had in hand before he came to Jerusalem; indeed, without it he would not have come. What grounds did he have for his confidence? Simply that Dayan had told Tohamy in Morocco that Israel was willing to restore to Egypt "sovereignty" over the Sinai.

Sadat was to find out later that the exact meaning of sovereignty can provide nego-

*While Sadat is thought to be first Arab leader to visit Israel, actually King Hussein of Jordan visited Tel Aviv incognito in 1974 for secret — and abortive — talks with Prime Minister Yitzhak Rabin.

Qaddafi: The Egyptian attack on him was a bolt out of the blue.

tiators with plentiful material for deadlock, but in the euphoria of the moment all that counted was the big picture. "No more war," said Sadat. "No more war," echoed Begin. The Palestinian problem would somehow be solved. Jordan, even Syria, would eventually follow Sadat's lead. With this historic breakthrough, peace would surely come. Nothing was impossible.

Now the pressure was on Begin to come up with some gesture comparable to Sadat's. In mid-December, the Prime Minister drew up an offer of "autonomy" for the Arab residents of the West Bank and Gaza. He did so to ward off some stiffer possible demand by Cairo or Washington — and to seize the initiative. "Let's *areinhopen* it," he said to his closest aides — meaning, in Yiddish, "Let's grab it." Without consulting his Cabinet, he stuck the plan in his briefcase and took it — uninvited — to the White House.

Why the decision to bring in the Americans after excluding them thus far? One reason could be found in Begin's sense of personal vindication. At long last, the despised "terrorist" was a figure on the world stage, a respected man in the American-Jewish community. He had accomplished what none of his predecessors had come close to, and he would be meeting with the American President not as a vassal but as a partner, the proud head of the Jewish Nation. More than

that, Carter was having his own problems at home, and by giving him a peacemaking role in the Middle East, Menachem Begin, as one of the Prime Minister's aides flatteringly suggested, "would make Carter into a President."

Pleasing as these thoughts were, they would not have been enough to send him off to Washington. He went at the urging of the man he relied on most in foreign affairs — Moshe Dayan.

Dayan had cut his military teeth under David Ben-Gurion, and Ben-Gurion had taught him that Israel dared not make a move without America. If "B.G." hadn't taught him that, "B.G.'s" biggest mistake in foreign affairs would have. In 1956, Ben-Gurion launched the Sinai campaign against Nasser without clearing it with Washington. The expedition, led by Dayan, turned to ashes when President Eisenhower forced Israel to give back the Sinai — for nothing.

It was one thing to fool around with the Egyptians in Morocco, but now, Dayan counseled, it was the big leagues and there was only one thing to do. The plan would never work without Washington.

Washington awaited the Prime Minister without visible enthusiasm. All its well-laid plans for Geneva lay shattered under the impact of the developments of early November. The President had added his voice to the international chorus of praise for Sadat, but his words

had barely concealed his shock over a turn of events that seemed to rob Washington of all initiative. Now this puzzling visit from Begin, who claimed he had something important to show.

That something proved to be a plan for the West Bank. Begin proudly unfolded it at a meeting with top American officials. The scheme provided for limited Arab self-rule, with retention of Israeli settlements and an Israeli military presence.

This was a far cry from the kind of Palestinian entity that the White House regarded as essential for any kind of peace agreement. Carter, Brzezinski and Vance listened politely, and the President complimented Begin's legal adviser for his fine draftsmanship, but that night Vance's lawyers went over the 26-point document and the next morning the Secretary was ready with his questions. These were meant as a critique of the plan's political viability, but Vance put the questions so diplomatically that his objections were lost on Begin, who, in any case, seemed to be in a state of euphoria. When Carter said publicly that the plan was "encouraging," Begin left town certain that the President had endorsed it.

Back in Israel after a stopover in London, where Prime Minister James Callaghan also had nice things to say about his offer, Begin was ecstatic. He told his Cabinet, only a few of whose members had seen the plan, that everybody loved it.

"Jimmy Carter said it's *beautiful*," Begin reported. "Callaghan said it's *beautiful*."

Amy Carter, who presumably didn't comment, was beautiful, and Rosalynn was beautiful, too.

Sadat was nonplused. He had gambled on his journey to Jerusalem in an attempt to bypass the United States, with its insistence on a general conference focusing on the Palestinian dilemma, and he had assumed that Begin understood this. Now Begin had linked their bilateral talks with the American approach. And he, Sadat, was being presented with what Begin was advertising as virtually an Israeli-American plan for the Palestinians — a plan whose inadequacy, from the pan-Arab viewpoint, stood out in bold relief.

Nonetheless, the Israeli-Egyptian talks went forward. On Dec. 20, 1977, Weizman, as Defense Minister, flew to Alexandria in an American plane to confer with his Egyptian opposite number, Gen. Mohammed Abdel Ghany el-

Gamasy, on the establishment of a hot line between Cairo and Jerusalem. Sadat took the opportunity to welcome his new Israeli friend on Egyptian soil. The two conversed briefly in Alexandria before Weizman proceeded to a military camp outside the city for his conference with General Gamasy. Weizman brought Sadat two presents from the Israeli Government — a pipe and a clock with the inscription: "To President Sadat, the man who put the clock hands forward in the Middle East."

On Christmas Day, Sadat met with Begin in Ismailia, beside the Suez Canal. At a private tête-á-tête, he agreed to the Begin autonomy proposal. Writing out his own summary of the plan, he read it to the assembled aides of both sides.

When he finished reading, Ahmed Esmat Abdel-Meguid, Egypt's permanent representative at the United Nations, said: "It is unacceptable."

In a chorus, the other top Egyptian diplomats backed him up.

To see Sadat contested by his own Foreign Office in front of an Israeli delegation was startling; what happened next came as a shock. Anwar el-Sadat beat a hasty retreat.

He would have to reconsider, he said. He would have to study the plan in detail.

The next item on the agenda was the Sinai. Begin began with a sweeping statement. Israel would deliver the whole of the Sinai back to Egypt. Egypt would have sovereignty over the entire peninsula, up to the recognized international borders. Then he added a qualification. Israel, of course, would retain its air bases in the Sinai, as well as its string of settlements in the Rafah salient bordering the Gaza Strip.

Sadat gave him a cold stare. He was incensed at what he saw as trickery. At the Moroccan talks, and again in Jerusalem, he had been promised sovereignty over the Sinai. What kind of sovereignty was this? Israeli settlements, Israeli air bases? Did Begin think him stupid?

Rather than invite a showdown, he kept his options open. The conference ended with the announcement that the talks would resume in January on a double track: a meeting of political committees of both sides in Jerusalem and a meeting of military committees in Cairo.

"I am a happy man," Begin told reporters as he left for home.

His experts merely shrugged. They saw more clearly than their boss that happiness was premature. ■

January 21, 1979

UNTOLD STORY OF THE MIDEAST TALKS

At one point in the Egyptian-Israeli negotiations it seemed that full accord was only days away. Then the talks collapsed. This article reveals the hidden reasons for the turnabout — and for hope that peace may yet be won. *Second of a two-part series.*

By Sidney Zion and Uri Dan

For the statesmen engaged in the momentous effort to bring peace to the Middle East, 1978 opened on an uncertain note. Behind them — as recounted in last week's installment — were President Sadat's journey to Jerusalem, the establishment of a hot line between Jerusalem and Cairo, and the direct talks between Sadat and Prime Minister Begin at the Egyptian city of Ismailia by the Suez Canal. But at Ismailia, basic disagreements emerged — and the momentum faltered.

President Anwar el-Sadat had gambled too much on his peace initiative to be discouraged by the gulf between the Egyptian and Israeli negotiating positions. The demands presented at Ismailia were, after all, only the first moves. The situation had to be explored in greater depth. He now sought to do that with an invitation to the one Israeli he felt most at home with — the ebullient Defense Minister, Maj. Gen. Ezer Weizman. He proposed that they meet in Aswan for a round of talks.

Weizman arrived at the Aswan airport on Jan. 11. Hobbling down the ramp — he was still recovering from an automobile accident of two months back — he was greeted effusively by the Egyptian President. Their conversation, according to an Israeli present, went as follows:

"So, Ezra!" — Sadat had taken to

This article was reported by Sidney Zion, a columnist for New York magazine, and Uri Dan, a correspondent of the Israeli paper Ma'ariv, and was written by Mr. Zion. The article is adapted from a book to be published by Times Books.

calling him Ezra — "I see you're still with the cane."

"Screw the cane!" exclaimed Weizman, tossing it into the air.

Delighted, Sadat hugged him. "You're looking very smart, Ezra."

"One must *be* very smart when coming to visit the smartest leader in the Arab world."

This was their first opportunity for an extended talk. Weizman came away from it greatly impressed. Sadat had virtually disregarded the Palestinian question; he needed only a "fig leaf" to cover him with the Arab world; he had ditched pan-Arabism and wanted a separate deal with Israel — there was no doubt about any of this in Weizman's mind.

Gen. Mohammed Abdel Ghany el-Gamasy, the Egyptian War Minister, also in Aswan for the talks, had struck Weizman as even firmer in his views —

contemptuous of the Palestine Liberation Organization, sharply critical of King Hussein of Jordan, expressing a wish for an eventual federation between the Palestinians and Israel.

Weizman returned to Jerusalem convinced that peace was at hand.

□

Menachem Begin wasn't so sure. One thing that *was* clear to him after the Ismailia conference was that Sadat wanted Sinai back — all of it, and unconditionally, save for demilitarized zones and the like. Begin didn't intend to give it to him that way, of course — chances were that Israel would have to give up either its air bases in Sinai or the Israeli settlements in the Rafah salient just beyond the Gaza Strip *(see map)*. But he was prepared for a compromise. What he was not prepared for was the heated dispute this issue would engender in his own ranks.

Gen. Ariel Sharon, the Minister of Agriculture, insisted that whatever was to be returned in Sinai, it could not be the settlements. The air bases could easily be moved back a few kilometers into the Negev, but the settlements were crucial as a buffer zone between Sinai and the heavy Arab population in the Gaza Strip. If anything, the settlements should be "thickened."

General Weizman, on the other hand, objected that if it came to a choice, the settlements had to go — one brigade on the Israeli side of the border could do the job and do it better. Establishing

new air bases, however, was an expensive proposition. The air bases had to stay.

Faced with a spreading quarrel, Begin appointed a committee of three — Foreign Minister Moshe Dayan, Sharon and Weizman — to come up with a recommendation. A meeting was held at Dayan's home outside Tel Aviv. Sharon and Dayan sat at a table covered with maps of Sinai. Weizman walked around the room, picking at cakes, and poured himself a Scotch.

Dayan said, "Ezer, take a look at this."

"Go ahead," Weizman said, "bulldoze the whole Sinai, do what you want. You're all crazy."

The committee — Weizman abstaining — recommended that the settlements be augmented. The proposal was put to the Cabinet — Weizman sat, saying nothing — and was adopted.

Sharon asked, "Shall I send in the bulldozers?"

"I don't think," Begin said, "the Minister of Agriculture should take the time of the Government to ask technological questions."

"It is a political question, Mr. Prime Minister," Sharon stressed.

The Cabinet approved. The tractors were on their way within hours — and kicked up an international storm. The Carter Administration condemned the action — as well as the existing Sinai settlements — as illegal. While he was at it, the President had his spokesmen make clear, first, that Israel's settlements on the West Bank were also regarded by Washington as illegal; and, second, that Begin's proposal for the West Bank — limited Arab self-rule, with retention of Israeli settlements and an Israeli military presence — had never been accepted by the United States except as a basis for negotiation.

"What has Arik [Sharon] done to me?" Begin complained privately. "Until the bulldozers, Carter said my plan was *beautiful.*"

But the tractors plowed on.

□

In mid-January, in accordance with the agreement in Ismailia, talks between Israel and Egypt were resumed on a political plane in Jerusalem and on a military level in Cairo. The political talks were a fiasco. To start with, a toast by Begin contained something the Egyptians found slighting. And the chief Egyptian delegate, Foreign Minister Mohammed Ibrahim Kamel, bombarded Sadat with cables complaining that he was being humiliated and that the Israelis were arrogant, and, by implication, not interested in peace. Yielding to pressure, Sadat recalled the delegation 48 hours after its arrival in Jerusalem.

Begin's impulse was to call off the military talks in Cairo as well. But he was dissuaded by Weizman, who headed the Israeli delegation in the Egyptian capital — and who reported excellent progress.

After several long talks with Sadat and General Gamasy, Weizman was convinced that the Egyptian leaders were finished with pan-Arabism, that they wanted out of the cycle of recur-

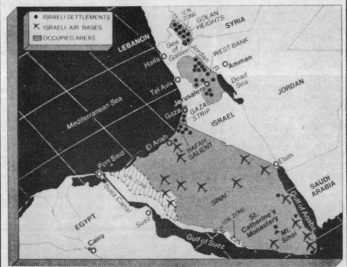

Key issues: the West Bank's status and Israel's settlements and air bases in Sinai.

rent wars, and that they required little more than rhetoric on the Palestinian question. "Give me a cover for the Arab world," Sadat had told Weizman, over and over again.

Now, Weizman reported, he had a concrete proposal from General Gamasy: In return for an Israeli agreement to evacuate the Sinai, Egypt would sign a separate peace. The treaty would contain provisions giving Israel veto power over any future solution of the Palestinian problem.

The reaction in Jerusalem was skeptical. Dayan, in particular, had his doubts. The formula proposed for the West Bank was unclear — and everything depended on what physical means was left to Israel to prevent the West Bank from evolving into an independent Palestinian state. Such a development, in the opinion of the vast majority of Israelis, would create a gaping breach in their military security, giving them an indefensible border with a potentially dangerous new power center.

Having begun as a farmer living among Arabs, Dayan considered himself an expert on Arab mentality. He was not ready to discard his lifelong belief in the influence of pan-Arab ideology on events in the Arab world. Too much of the Middle East's violent history could be explained only as the product of Arab solidarity, at key junctures, against the common enemy. Weizman's faith in Sadat's change of heart — of the Egyptian President's readiness to sweep the Palestinians under the rug — was seen by Dayan as the enthusiasm of a cosmopolite ignorant of Arab ways.

Begin, too, held to an image of the "Arab Nation," and found it difficult to conceive of an Arab leader who would ditch that ideal for the sake of narrower interests, no matter how much the Arabs might squabble among themselves. Weren't the Jews their own worst enemies, too, and yet wasn't there, in the same sense, a Jewish Nation to which they remained loyal in the end? Therefore, while he was willing to accept that Sadat would not insist on a Palestinian state, he readily agreed with Dayan that Sadat was not simply interested in a "fig leaf" — that the agreement would have to be a package deal, with a place for Jordan in any arrangement for the West Bank.

On the basis of this reasoning, the offer relayed by Weizman from Cairo was rejected. This proved ironic, since the offer was somewhat more advantageous to Israel than the accord reached at Camp David later that year.

□

Within the Israel Cabinet, the argument on the Sinai — whether to hold on to the settlements or the air bases — was settled in favor of the settlements. Taking his defeat in good grace, Weizman, in Cairo, suggested that the set-

tlements be placed under United Nations jurisdiction. Sadat had been insisting that both the air bases and the settlements had to go. General Gamasy proved more flexible. He said that if the the settlements stayed, they would have to stay under an Egyptian flag.

Weizman and Gen. Avraham Tamir, a career officer serving as his top aide in the talks, replied that Israel couldn't let its nationals live in Egyptian territory.

"How can you ask a million Palestinians to live under your sovereignty," Gamasy countered, referring to the Arabs residing in the West Bank and Gaza, "if you won't permit 2,000 Jews to live under ours?"

Weizman and General Tamir smiled wryly, as though admitting that Gamasy had a point. In their report to Jerusalem they didn't seem to think the idea was all that bad. But to the Israeli Government it was heresy, and on Dizingoff Street in Tel Aviv the wags began referring to the two negotiators as "Mustapha" Weizman and "Ibrahim" Tamir.

By late January, 90 percent of the military issues concerning the Sinai had been solved. As for the Sinai settlements, Sadat was sure the Israelis weren't serious about holding on to them.

"We will have no trouble about it," he said. "Israel should have never raised it. It was a mistake. For sure, it will not stop anything."

But the Israelis were dead serious, and Weizman was ordered to raise the question directly with Sadat.

"Don't mention these settlements to me or I'll chase you to the ends of the world!" Sadat exclaimed.

"Let's not talk about chasing each other," Weizman said. "We've had chases before, and you know the results."

Sadat laughed and embraced "Ezra," as usual.

But that didn't stop the tractors from continuing to break new ground in the Sinai. Leaving Cairo for Jerusalem in early February, Weizman said to a reporter, "Oy, what Arik's bulldozers are doing to me."

□

Now Sadat himself turned to the United States. The Cairo talks had gone as far as they could; 99 percent of the cards, Sadat said, were now in American hands. On Feb. 2, 1978, he flew to Washington, where, in his new guise as a man of peace, he received a warm welcome. He also got a commitment of jets for his air force — as did the Saudi Arabians.

Weizman followed up in early March, to tie up an earlier American commitment to sell sophisticated jets to Israel. While in New York en route to Washington, he got word that Begin had authorized Sharon to send the bulldozers to a new settlement site

at Nebi Selach on the West Bank. In the presence of Israeli reporters, he phoned Begin and said, "With one tractor in Nabi Salah you're going to plow up America." When this message made headlines, Begin called off the bulldozers.

Begin's retreat came against a backdrop of a rising "Peace Now" movement in Israel. Started by several hundred reserve army officers, the movement cut across political and social lines. Its hero was Ezer Weizman. While maintaining a correct distance, Weizman encouraged the movement. But any plans Weizman may have had to topple Begin were pushed aside by a major P.L.O. attack inside Israel. On March 11, a P.L.O. dinghy landed near the Tel Aviv beach and a group of terrorists killed 35 Israelis and one American on the Tel Aviv-Haifa highway.

Before ordering a retaliatory invasion of southern Lebanon, Begin informed Sadat through the Israeli military mission in Cairo — an unprecedented act. Sadat raised no objections. He was, just then, embroiled in his own fight with the P.L.O. In retaliation for what they considered his betrayal of their cause, Palestinian terrorists struck in Cyprus, assassinating Sadat's close friend Yousef el-Sebai, editor of Al Ahram, and hijacking an airliner with Egyptians aboard. Egyptian commandos flown to the Nicosia airport to free the Egyptian hostages in an Entebbe-style rescue had run into heavy fire in which 15 Egyptians were killed. Contrary to most reports, the fire came not only from the Cypriot forces but from P.L.O. gunmen. On the streets of Cairo, crowds screamed "Death to the Palestinians!"

□

The Israeli invasion of Lebanon was seen around the world, and by many in Israel, as overkill, and when Begin made his second visit to the White House late in March he was given the cold shoulder. Carter, it seemed to the Israeli delegation, wanted him out of office. Begin told his aides his meeting with the President was the "worst moment of my life." And on his return to Jerusalem, he faced a new challenge to his leadership.

While Begin was in flight, Weizman had called for a "National Peace Government," including the Labor Party, to break the deadlock. Confronted by Begin at a private meeting, Weizman claimed he had been quoted out of context. Yet few insiders doubted that Weizman's appetite for the Premiership was growing. This became increasingly evident as Begin, acutely depressed by the failure of his Washington trip and troubled anew by his heart ailment, seemed to retreat from active leadership. The manoach, Weizman called him, within earshot of a reporter, meaning one who has passed away.

In July, proceeding to Vienna for a conference of the Socialist International, Sadat invited Weizman to meet with him in Salzburg. A few days after their meeting, Sadat met in Vienna with the Labor Party leader, Shimon Peres. Rumor had it that the United States had swung its support behind a prospective Weizman-Peres coalition. Begin, back in firm command, was affronted by the whole performance. When Weizman returned to Jerusalem, the Cabinet was "too busy" to hear his report. He stormed out of the Cabinet meeting, ripping a peace poster off the wall of an adjoining office. "What do you need this for?" he shouted. "This Government doesn't want peace."

He was permitted to brief the Government the following week. Sadat, he said, was still pressing for a separate peace. On the Palestinian issue, he still wanted nothing more than a "cover" for the Arab world. Weizman quoted Sadat as saying: "I will give you water from the Nile, oil from Sinai — let us make peace." He asked of Israel a gesture to get things going again — the return of El Arish, on the Sinai coast, perhaps, or of St. Catherine on Mount Sinai. On the latter site he would build great temples to Moses, Jesus and Mohammed.

Dayan asked, "Do you have a protocol of this conversation?"

Weizman said, "You don't trust me?"

It was not a question of trust, Dayan said. He had just met in Leeds Castle, near London, with the American Secretary of State, Cyrus Vance, and the Egyptian Foreign Minister, Mohammed Ibrahim Kamel, and Kamel didn't talk about any "cover"— he still talked straight pan-Arabism, demanding real self-determination for the Palestinians and an Israeli retreat to the 1967 borders. There would be no separate peace, Kamel made it clear; there would be a package deal or nothing. And Vance, Dayan said, was in full agreement.

What, then, was Egypt's true policy? Once again, Sadat was speaking one way and his Foreign Minister another. Dayan, as before, believed the Foreign Minister; Weizman, as before, believed Sadat.

In either event, there would be no "gesture" on St. Catherine or El Arish. Begin had been for the idea of giving El Arish back to Egypt when Sadat was making ready to appear in Jerusalem. But that was then; now was now; and Sadat, the Prime Minister said, would get "nothing for nothing."

He said so publicly. Whereupon Sadat evicted the Israeli military mission from Cairo. It looked like the end of the road.

□

Early in August, Cyrus Vance went to the Middle East with handwritten invitations from President Carter. Would

Prime Minister Begin and President Sadat join him at Camp David for an effort to break the stalemate? Both men accepted immediately. How could they refuse the President of the United States? Sharon proposed that Israel prepare a detailed working paper for Camp David. "It's time we stopped pulling little notes out of our pidjaks," he said, lapsing into the Russian word for jackets. "It's time we did it right."

At his suggestion, General Tamir was appointed to head a committee to draw up the document. The 90-page working paper (called the Blue File, because Tamir happened to put it in a blue folder) was largely a reflection of views the general shared with Weizman, in the wake of their Cairo experience. It recommended the following negotiating position:

The target for Israel was a separate peace. To achieve it, Israel had to be prepared to give back the entire Sinai — air bases, settlements and all. As to the West Bank and Gaza, the key was verbal flexibility — both to fulfill American expectations and to satisfy Sadat's need for a politically viable position vis-á-vis the Arab world. Whatever the final arrangements, they would have to include three fundamental provisions. Israel would maintain its military forces and settlements in the West Bank. Israel would not be required to cede its claim to sovereignty over the area. There would be no independent Palestinian state.

Dayan read the Blue File for the first time on the flight to Camp David.

"Who authorized this?" he asked. Tamir pointed to Begin.

"Well, it's nonsense," Dayan said, "pure nonsense."

What he meant was that the Blue File portrayed an unreal world — that the Egyptians would never accept an accord that left the Palestinians with a lick and a promise. At the opening session at Camp David on Sept. 4, his view seemed to be substantiated.

Sadat submitted a paper demanding that Israel withdraw everywhere to its pre-1967 borders and recognize the right of self-determination for the Arabs on the West Bank and Gaza. It was a blueprint for a Palestinian state, confirming Dayan's contention that Weizman and Tamir had been conned all along. Begin was incensed. If this was Sadat's idea of negotiation, he told his delegation, he would walk out.

Then, quietly, things began to move. Carter went from cabin to cabin, conducting long talks with both sides and taking copious notes. He spent an entire evening with Tamir working over Sinai, virtually hill by hill. Leaving, he said, "I learned more about Sinai tonight than I have since I took office."

Sadat told Weizman and Tamir to drop in on him "any time," and they took him at his word, visiting him in his cabin at all hours of the day and night. Weizman wanted to get Sadat together with Dayan. Sadat was reluctant; Dayan was too stark a symbol to be casually invited into the Egyptian President's sanctum. But Weizman, though Dayan's rival, insisted that Dayan was the key figure on the Israeli side and that Sadat *had* to see him if there was to be any deal.

It took more than a week for Sadat to relent; when he did, it paid off.

Once the meeting was set, Carter took Dayan aside and urged him not to try to corner Sadat but to talk in broader terms. But when the two men met alone in Sadat's cabin, the Egyptian went straight to the essense of the problems. After hearing him out, Dayan came to believe for the first time that a de facto separate peace with Egypt was possible and that an understanding on the West Bank could be reached without King Hussein. Sadat's earlier demands on this issue had been no more than bargaining chips.

The meeting ended with some light bantering. "When you visit Egypt," Sadat said, "I've got a lot of archeological treasures to show you. But you must promise you won't take any of them away."

Dayan, known for his sticky fingers when it came to archeological pieces, laughed and said, "I can't give you my word on that."

But serious problems remained. The State Department and the President's national-security adviser, Zbigniew Brzezinski, were still flatly opposed to anything that smacked of a separate peace and left the Palestinian problem unresolved. Begin was particularly suspicious of Brzezinski, describing him to his aides as "*ocher Israel*" — roughly, "a hater of Israel," the harshest pejorative in the Hebrew language.

That left the United States as the primary obstacle to an agreement, and *that*, for Carter, was politically untenable. Just as President Truman overruled the State Department and recognized the State of Israel in 1948, so Carter overruled his advisers. The Sinai and the West Bank issues would be separated, with only some ambiguous language as a link.

But now another crisis blew up.

Israel had agreed to a staged withdrawal from Sinai, including evacuation of the air bases, but the settlements still had to be discussed, and on this issue Begin dug in his heels. He refused to remove the settlements — even though General Tamir's Blue File had envisaged the necessity of such a concession. Like Dayan, he had not really believed that the issue would be put to the test — that the Egyptians would accept the Blue File's nebulous West Bank formula.

Advised of this, Sadat ordered his bags packed.

But Carter was not to be denied. He told Begin that if Israel remained adamant, the conference would collapse and there would almost certainly be war. The Egyptians, he said, would move their armies into the demilitarized zone and Israel would be forced to fight or back down. The unstated implication was that, in such an event, Israel could not depend on full-scale American aid.

Begin was persuaded. But he feared the political repercussions at home. If General Sharon, with his immense authority on security matters, said that giving up the Sinai settlements would endanger Israel's safety, there could be grave domestic difficulties. The peace plan itself might be rejected in the Knesset.

Begin phoned Sharon in Israel. It had come down to the Sinai settlements, he said. Would Sharon agree to their removal?

"If this is the price for peace, do it," Sharon said.

The Egyptian Foreign Minister was less obliging; Kamel said he would not be a party to a sellout of the Palestinians. Sadat, who has less trouble dealing with domestic discord, accepted Kamel's resignation.

☐

The breakthrough at Camp David made peace seem imminent. The negotiators had the makings of two separate accords. The first was agreement on a peace treaty establishing normal relations between Israel and Egypt and providing for a phased Israeli withdrawal from Sinai. The second was an understanding that all parties concerned would work toward some arrangement for giving the Palestinians in the West Bank and Gaza limited self-rule — while safeguarding Israel's security interests. On Oct. 12, an Egyptian team headed by Kamal Hassan Ali, the new Defense Minister, and an Israeli team headed by Dayan and Weizman, met at the Madison Hotel in Washington to work out the final details.

At Camp David, the negotiators had set Dec. 17 — three months off — as the target date for the signing of the peace treaty. At the Madison Hotel (dubbed Kibbutz Madison by the Israelis), it took only four days to produce a virtually completed draft document.

"We'll have it wrapped up in a week," Weizman told a friend. Two days later he was even more optimistic: "We'll have a treaty within hours."

Dayan was more cautious. "There are difficulties," he told reporters. Most observers thought this was just meant to put Weizman in his place. The major issues seemed to have been resolved; the only remaining question was how to create the best public-relations atmosphere for the signing. Would it be done on Nov. 19, the first anniversary of Sadat's visit to Jerusalem? Would it be done on Christmas Day, with Jimmy Carter receiving the

"tablets" from Begin and Sadat on Mt. Sinai?

But "hours" passed, and "days" passed, and the treaty still wasn't "wrapped up." Dayan had been right: There were difficulties. But what were they?

Jerusalem was upset over a few words in the preamble to the draft treaty that seemed to make the treaty dependent upon negotiations on the future of the West Bank and Gaza. This apparent link accentuated a grave concern that only now was making itself strongly felt in Israel. The concern was with the very principle of autonomy, the genie that Begin had let out of the bottle when he offered the Palestinians some undefined share of self-government in a bid to counter Sadat's Jerusalem journey with a gesture of his own.

At first imperceptibly, then with mounting force, Begin had come under domestic attack from both left and right for taking this initiative. His critics charged that just as the Balfour Declaration of 1917 had laid the basis of a Jewish state, so Begin, with his autonomy plan, would become the Balfour of a Palestinian state.

That, of course, was the very opposite of Begin's intentions. His 26-point proposal had envisaged a kind of town- or village-level autonomy for the Palestinians, with the Israeli military government still in place. At Camp David, it was true, he accepted a "framework" that contained the possibility of broader autonomy, with a Palestinian governing authority for the whole region. But this theoretical possibility was dependent on so many other things — agreement among the faction-ridden Palestinians, for instance, or Jordanian willingness to sign a peace treaty with Israel — that Begin did not view the change in wording as presenting much of a risk. Fundamentally, the Begin Government still held to its claim to sovereignty over the West Bank — a claim based not only on biblical history but on its interpretation of the Palestine Mandate and of international law governing territory taken in the course of a defensive war. And in any case, under the Camp David accord, the final status of the West Bank and Gaza would not be determined until five long years after autonomy began.

By mid-October, however, the domestic critics of the West Bank accord had picked enough holes in it to make Begin nervous. The result was a month's haggling over wording. In Washington, this had the effect of confirming long-held American suspicions that Begin wasn't really serious about granting the West Bank any kind of autonomy worthy of the name. Washington tried to force his hand.

In late October, Assistant Secretary of State Harold Saunders went to the Middle East. He talked with King Hussein of Jordan, in an effort to bring him into the negotiations. And he talked with Palestinian leaders in the West Bank in an effort to convince them that autonomy, though far short of their goals, was a viable interim program. The implication was that such a program could eventually lead to a Palestinian state. Saunders also said that Israel would dismantle its settlements in the West Bank — something that Israel had never agreed to. While in East Jerusalem, he plucked at a particularly sensitive Israeli nerve by calling that sector "occupied

territory."

Saunders's statements had a critical impact. To the Israelis it appeared that the United States was once again working toward a comprehensive settlement for the Middle East — with an independent Palestinian state as its centerpiece. Jerusalem's reaction was in character. The Government announced plans to "thicken" the Israeli settlements on the West Bank.

After that, a hardening of Sadat's attitude was inevitable. The Egyptian President could not afford to seem less solicitous of Palestinian aspirations than the United States. Satisfied hitherto with only the loosest connection between the timing of the peace treaty and the first moves toward autonomy on the West Bank, he now demanded that the two be tightly linked.

It all seemed like an instant replay of the pre-Camp David period. The United States, having meddled, came to the rescue, and, having rescued, meddled again. And it was ironic that, just as hopes for an early accord began to wane, Sadat and Begin were awarded the Nobel Peace Prize for 1978.

☐

Ezer Weizman was still the incorrigible optimist. Informed during a negotiating session at the Madison Hotel that he had become a grandfather, he turned to the Egyptian Defense Minister and said, "You know, General, maybe there is a symbol here. The boy was delivered by Caesarian *shnit* [cut]. And so will this peace."

Shortly after Thanksgiving, Washington produced a compromise draft. It was less advantageous to Israel than the one that Egypt had been ready to accept — but Israel had balked at — a month earlier. Now Israel snapped it up and it was Sadat who demurred. In the wake of the Saunders episode, he felt it necessary to hold out for firmer "linkage" to progress toward Palestinian autonomy.

Dec. 17 came and went, and the signing was no closer. Washington thereupon encouraged Egypt to put forward a series of "side letters" to the accords. Israel rejected them as distinct departures from what had been put down on paper. Carter, irked, placed the onus of the new stalemate on Israel. Begin, in turn, charged that America had tilted toward Egypt. By the end of 1978, chances for an agreement looked bleak.

☐

Yet too much had been achieved for any of the parties to the negotiations to accept final collapse. With the New Year, all three capitals were in a mood to get the talks started again. The question was how and when.

January 28, 1979

ARMED IRANIANS RUSH U.S. EMBASSY; KHOMEINI'S FORCES FREE STAFF OF 100

American Envoy Shot Dead in Afghanistan

SHARP TEHERAN CLASH

Americans Trapped More Than 2 Hours—Attack Is Laid to Leftists

By NICHOLAS GAGE
Special to The New York Times

TEHERAN, Iran, Feb. 14 — Armed urban guerrillas attacked the American Embassy today and trapped Ambassador William H. Sullivan and about 100 members of his staff inside for more than two hours.

Ambassador Sullivan called Iran's revolutionary authorities for help and the Americans were freed by forces of Ayatollah Ruhollah Khomeini, led by a Deputy Prime Minister of the provisional Government.

The assault, in which two Iranians were reported killed and two United States marines were wounded, underscored the problems that Ayatollah Khomeini's forces face in trying to bring the country under their control.

Several Pitched Battles Fought

Rival groups battled at several points in Teheran tonight, including Iran's national television center.

"They attacked the compound from three sides," Ambassador Sullivan said later of the assault on the embassy. "They shot up my home, my office and the chancery. We telephoned the Khomeini people, and they arrived in the nick of time — an interesting Valentine's Day."

[Reacting to the attack, officials in Washington said the Carter Administration planned to begin evacuating most of the remaining 7,000 Americans in Iran as soon as the Teheran airport was reopened. That is expected to occur on Saturday. Page A16.]

Attackers' Identity Unclear

It was not clear who started the attack on the embassy. Some supporters of Ayatollah Khomeini accused the Communists' Tudeh Party of having been responsible for the assault, while others blamed

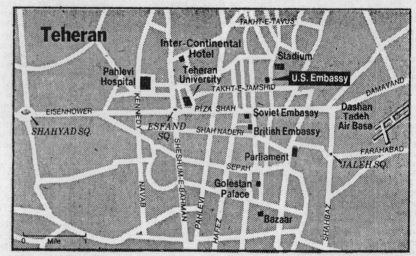

Teheran — TAKHT-E-TAVUS — Inter-Continental Hotel — Pahlevi Hospital — Teheran University — TAKHT-E-JAMSHID — Stadium — U.S. Embassy — DAMAVAND — EISENHOWER — KENNEDY — PIZA SHAH — Soviet Embassy — Dashan Tadeh Air Base — SHAHYAD SQ. — ESFAND SQ. — SHESHUM-E-BAHMAN — SHAH NADERI — British Embassy — FARAHABAD — NAVAB — Parliament — SEPAH — JALEH SQ. — Golestan Palace — PAHLEVI — HAFEZ — Bazaar — SHAHBAZ — 0 Mile 1

The New York Times/Feb. 15, 1979

the People's Fedayeen, a Marxist guerrilla group. Later the People's Fedayeen denied responsibility

Embassy employees credited the leadership of Ambassador Sullivan with having prevented heavy American casualties despite the severe fire with which the guerrillas attacked. The Ambassador, his aides said, ordered the 19 marines in the compound not to engage the attackers and collected his staff in protected rooms of the chancery.

Arrest of Bakhtiar Denied

A friend of former Prime Minister Shahpur Bakhtiar, meanwhile, denied the reports that circulated yesterday that the ousted leader had been arrested. The friend said that he had talked with Dr. Bakhtiar today and that the former Prime Minister was still hiding.

"He will probably appear on his own when the situation improves," the friend declared.

Many of Ayatollah Khomeini's followers did not heed his order of yesterday to turn in weapons seized when the country's armed forces capitulated last weekend.

In an effort to end the chaos in the country, the religious leader issued a proclamation today asking all Iranians to return to their jobs on Saturday, the beginning of the Moslem week.

"Now that you have demolished the pillars of the Pahlevi dynasty and the revolutionary temporary Government has been established, all workers, merchants, office staffs, students and teachers are hereby requested to resume work on Saturday," he said.

The Ayatollah said that because of the strikes that were necessary to the revolution, "the country is now in such a state" that the economy must be immediately revived.

"Those who disobey the revolutionary Government will be regarded as opponents of the revolution," Ayatollah Khomeini said.

The response to the back-to-work order should show whether Iranians are as obedient to the Ayatollah now that their revolution has triumphed as they were during the struggle.

A Radical Government Demanded

Leftist groups have already challenged the Khomeini forces. The People's Fedayeen, which has established itself at Teheran University, held a news conference today and called for a radical government and a people's army.

In several cities outside the capital, armed opposition to the Khomeini forces has been reported. In Tabriz, the capital of Azerbaijan Province in the northeast, street battles have been fought for two days. Reports said that both supporters of Shah Mohammed Riza Pahlevi and left-wing Azerbaijan separatists were fighting revolutionary forces. Witnesses in Tabriz said there were many casualties in today's fighting.

The attack on the American Embassy in Teheran began at about 10:15 A.M. when guerrillas began scaling the high, brick walls and iron gates of the embassy compound in downtown Teheran, firing at the chancery, the main embassy building, as they dropped to the ground inside.

According to two reporters who were in the chancery at the time, the 19 United States marines guarding the embassy fired tear-gas shells at the attacking guerrillas and pulled down the heavy iron shutters of the chancery windows.

Shortly after the initial attack, truckloads of pro-Khomeini forces carrying a variety of automatic weapons arrived to surround the compound, and the firing grew more intense.

A 32-year-old leader of the commandos, who said he had returned to Iran from the United States six months ago to take part in the revolution, insisted to three reporters pinned down in a guardhouse in the compound that the embassy had been abandoned and had been taken over by agents of Savak, the disbanded Iranian secret police.

"We got a call at our command post that the Shah's people are inside and shooting," he said. "We are not attacking the Americans — we have nothing against them."

The embassy had operated with only a skeleton crew during the recent bloody battles in the capital, but because the shooting abated somewhat yesterday more than half of the staff of 185 had reported for work.

When the shooting at the chancery mounted, Ambassador Sullivan told the marines not to fire at the overwhelming force of guerrillas outside but instead to empty their weapons, put them down, fill the ground floor with tear gas and lead other people to the communications room upstairs. This is the safest chamber in the embassy because of its windowless and reinforced walls. Meanwhile, Mr. Sullivan called the revolutionary committee for help.

The force that arrived was headed by Dr. Ibrahim Yazdi, a Deputy Prime Minister in the new Government. Dr. Yazdi halted the shooting and then made contact with the Ambassador, who walked out of the chancery's side entrance followed by the rest of his staff.

People in House Told to Come Out

They were led by Dr. Yazdi in a column to the entrance of the Ambassador's house in the rear of the compound. Using a bullhorn, Dr. Yazdi called on everyone in the house, including any secret police agents who might be inside to come out.

Only bewildered servants in uniform apppeared from the residence. Dr. Yazdi then told all the Americans to go inside the house and asked his commandos to escort all reporters out of the compound.

When the various accounts of the incident are pieced together, the most likely explanation is that extremists launched the initial attack after calling commando posts of the pro-Khomeini forces and saying that secret police agents had occupied the embassy and were firing from it. It would have been assumed that the commandos would rush to the scene and open fire themselves.

There was speculation that the extremists had expected the incident to result in the killing of a number of Americans, further widening the breach between Washington and the Khomeini Government.

Guerrillas Providing Protection

TEHERAN, Feb. 14 (AP) — The security chief of the American Embassy told reporters after the assault today that the Americans there had been placed under the protection of guerrillas loyal to the Khomeini Government.

The officer, Lieut. Col. Leland Holland, said that staff members could move about freely and that most had returned to their quarters, remaining inside the embassy compound on the guerrillas' advice.

The embassy staff had been taken hostage inside the chancery during the assault when Ambassador Sullivan ordered the marine guards to cease resistance and to surrender with the other embassy aides.

Pro-Khomeini guerrillas rushed to the scene. The young irregulars, many wearing Khomeini badges, fired hundreds of shots from roofs surrounding the compound and then rushed into the complex to drive out the attackers.

Spent machine-gun cartridges littered the walkways inside the compound after the assault. The main building was pockmarked with bullet holes and the iron gratings over its windows had been bent or ripped aside.

An embassy official said that top-secret radio equipment worth $500,000 had been blown up by staff members as the attackers closed in on the embassy's communications center. The official said that employees also had set fire to secret files but that many classified documents had survived and been left unprotected in areas penetrated by the attackers.

February 15, 1979

EGYPT AND ISRAEL SIGN FORMAL TREATY, ENDING A STATE OF WAR AFTER 30 YEARS; SADAT AND BEGIN PRAISE CARTER'S ROLE

CEREMONY IS FESTIVE

Accord on Sinai Oil Opens Way to the First Peace in Mideast Dispute

By BERNARD GWERTZMAN
Special to The New York Times

WASHINGTON, March 26 — After confronting each other for nearly 31 years as hostile neighbors, Egypt and Israel signed a formal treaty at the White House today to establish peace and "normal and friendly relations."

On this chilly early spring day, about 1,500 invited guests and millions more watching television saw President Anwar el-Sadat of Egypt and Prime Minister Menachem Begin of Israel put their signatures on the Arabic, Hebrew and English versions of the first peace treaty between Israel and an Arab country.

President Carter, who was credited by both leaders for having made the agreement possible, signed, as a witness, for the United States. In a somber speech he said, "Peace has come."

'The First Step of Peace'

"We have won, at last, the first step of peace — a first step on a long and difficult road," he added.

Later, at a state dinner, Mr. Begin suggested that Mr. Carter be given the Nobel Peace Peace, and Mr. Sadat agreed.

At the signing ceremony, all three leaders offered prayers that the treaty would bring true peace to the Middle East and end the enmity that has erupted into war four times since Israel declared its independence on May 14, 1948.

By coincidence, they all referred to the words of the Prophet Isaiah.

"Let us work together until the day comes when they beat their swords into plowshares and their spears into pruning hooks," Mr. Sadat said in his paraphrase of the biblical text.

'No More War,' Begin Says

Mr. Begin, who gave the longest and most emotional of the addresses, exclaimed: "No more war, no more bloodshed, no more bereavement, peace unto you, shalom, saalam, forever."

"Shalom" and "salaam" are the Hebrew and Arabic words for "peace."

The Israeli leader, noted for oratorical skill, provided a dash of humor when in the course of his speech he seconded Mr. Sadat's remark that Mr. Carter was "the unknown soldier of the peacemaking effort." Mr. Begin said, pausing, "I agree, but as usual with an amendment" — that Mr. Carter was not completely unknown and that his peace effort would "be remembered and recorded by generations to come."

Since Mr. Begin was known through the negotiations as a stickler for details, much to the American side's annoyance, Mr. Carter seemed to explode with laughter at Mr. Begin's reference to "an amendment."

Minutes later, Mr. Begin was deeply somber as he put on the Jewish skull cap and quoted in Hebrew from Psalm 126.

The signing was followed by an outdoor dinner on the South Lawn at the White House for 1,300 guests.

The treaty was the result of months of grueling, often frustrating negotiations that finally were concluded early this morning when a final compromise was reached on the last remaining issue — a timetable for Israel to give up Sinai oilfields.

Under the treaty, Israel will withdraw its military forces and civilians from the Sinai Peninsula in stages over three years. Two-thirds of the area will be returned within nine months, after formal ratification documents are exchanged. The ratification process is expected to begin in about two weeks.

In return for Israel's withdrawal, Egypt has agreed to end the state of war and to establish peace. After the initial nine-month withdrawal is completed, Egypt and Israel will establish "normal and friendly relations" in many fields, including diplomatic, cultural and economic relations.

Breakthough at Camp David

The outline for the peace treaty was achieved in September when Mr. Carter, Mr. Sadat and Mr. Begin met at Camp David, Md., for 13 days. In addition to the treaty, they also agreed on the framework for an accord to provide self-rule to the more than one million Palestinians living in the Israeli-occupied areas of the West Bank of the Jordan and the Gaza Strip.

The Camp David accords were opposed by most countries in the Arab world for two reasons. The Arabs regarded the decision by Mr. Sadat to sign a peace treaty with Israel as a betrayal of the Arab cause, since it suggested that Egypt would no longer be willing to go to war against Israel to help Syria, Jordan, and the Palestinians regain territory.

Arabs also viewed the self-rule agreement for Palestinians as insufficient because it did not guarantee the creation of a Palestinian state.

As a result of that opposition, today's signing was greeted by criticism throughout the Arab world. Echoes of that were heard in Washington, where about a thousand Arabs demonstrated in Lafayette Park, several hundred yards from the signing ceremony. Their anti-Sadat chants could be heard at the White House.

"We must not minimize the obstacles that still lie ahead," Mr. Carter said. "Differences still separate the signatories to this treaty from each other and also from some of their neighbors who fear what they have just done.

"To overcome these differences, to dispel those fears, we must rededicate ourselves to the goal of a broader peace with justice for all who have lived in a state of conflict in the Middle East.

"We have no illusions — we have hopes, dreams, prayers, yes — but no illusions."

Mr. Carter read out a long passage that turned on a metaphor of peace being waged like war. It was later disclosed by the White House that the section was quoted from an essay written by the Rev. Walker L. Knight in the House Missions Magazine of the Southern Baptist Convention.

At the end of the ceremony Mr. Carter, Mr. Sadat and Mr. Begin grasped each other in a three-way handclasp. Despite the show of cordiality, there were signs that differences between Egypt and Israel were far from over.

In his speech, Mr. Sadat never referred to Mr. Begin, whom he reportedly does not like personally. By contrast, Mr. Sadat praised Mr. Carter as "the man who performed the miracle."

"Without any exaggeration, what he did constitutes one of the greatest achievements of our time," President Sadat said.

In the printed text of his speech, Mr. Sadat made a strong appeal to Mr. Carter to lend "support and backing" to the Palestinians and reassure them that they would be able "to take the first step on the road to self-determination and statehood."

Sadat Cites 'a Grave Injustice'

The following was in the text of Mr. Sadat's address, but he did not read it publicly:

"No one is more entitled to your support and backing than the Palestinian people. A grave injustice was inflicted upon them in the past. They need a reassurance that they will be able to take the first step on the road to self-determination and statehood.

"A dialogue between the United States and the representatives of the Palestinian people will be a very helpful development. On the other hand, we must be certain that the provisions of the Camp David framework on the establishments of a self-governing authority with full autonomy are carried out. There must be a genuine transfer of authority to the Palestinians in their land. Without that, the problem will remain unsolved."

The remarks about the Palestinians would have been provocative to Mr. Begin, who has declared he will never permit a Palestinian state to be established. He has called the Palestine Liberation Organization the most "barbaric" group since the Nazis.

Later, Mohammed Hakki, the Egyptian Embassy's spokesman, said that the section on the Palestinians, which was on page seven of the printed text, had been "inadvertently" omitted because Mr. Sadat had turned two pages, instead of one, and accidently skipped that portion.

Mr. Begin's speech seemed highly charged with personal emotions, especially in two separate allusions to Jerusalem. These amounted to a reassertion of the Israeli stand on Jerusalem, in a context that was likely to prove embarrassing to Mr. Sadat.

The Israeli Prime Minister said that it was "the third greatest day in my life." The first, he said, was the day of Israel's independence, May 14, 1948, and the second "was when Jerusalem became one city and our brave, perhaps most hardened soldiers, the parachutists, embraced with tears and kissed the ancient stones of the remnants of the wall destined to protect the chosen place of God's glory."

This was a reference to Israel's capture of East Jerusalem from Jordan in the 1967 war and Israel's subsequent annexation of that part of the city to become part of Israeli Jerusalem.

A major point of difference between Israel and the Arabs is the future of Jerusalem, with the Arabs, including Egypt, insisting that Israel must relinquish control over the eastern sector, and Israel's declarations that it will never yield it.

Last night, Mr. Sadat underscored the continuing problem when, in the course of a 90-minute meeting with Mr. Begin, he invited the Israeli Prime Minister to make a one-day trip to Cairo next Monday but declined an invitation to visit Jerusalem.

Mr. Sadat visited Jersualem in November 1977, and it was that dramatic trip that started the process leading to today's treaty signing.

Egyptians officials said that Mr. Sadat wanted to put off another trip to Israel until progress was achieved on the Palestinian negotiations, which are to start in about six weeks.

The peace treaty negotiations went through a series of ups and downs and surprises.

They began in October in Washington with expectations of an early conclusion. Although the basic treaty text was approved by both Egypt and Israel by early December, three months more were needed to obtain agreement on differing interpretations of the treaty — the subject of a separate document of "agreed minutes" — and over issues such as when ambassadors would be exchanged and target dates for beginning and concluding the Palestinian self-rule negotiations.

Mr. Carter finally resolved most of the questions during a weeklong trip to the Middle East earlier this month.

Even though both Governments approved the treaty, it was not completed until late last night when Mr. Begin and Mr. Sadat agreed that the Sinai oilfield would be returned to Egypt seven months after the treaty was ratified, instead of the nine months Israel had preferred and the six months Egypt had earlier asked.

In addition, Mr. Begin agreed to turn over the El Arish area within two months instead of the three months originally proposed by Israel.

An arrangement was also made to insure Israel a right to buy oil from the fields without interruption.

Even this morning, in the final drafting, differences arose over whether to call a body of water the Gulf of Aqaba or the Gulf of Eilat. The Arabic and English texts refer to it as "Aqaba," the name of the Jordanian port by that name. The Hebrew version calls it Eilat, after the Israeli port adjacent to Aqaba.

The White House made public the texts of all the documents included in the peace treaty package. These include the actual preamble, nine articles, three annexes and one appendix that comprise the actual treaty text. In addition, there is a document of "agreed minutes" covering differing interpretations of the treaty.

A letter signed by Mr. Begin and Mr. Sadat and covering the controversial "linkage" question of when negotiations on the Palestinian self-rule questions would begin — one month after ratification of the treaty — and when the negotiations would conclude — about a year afterwards — was also released, as were certain clarification letters from Mr. Carter and maps.

March 27, 1979

MAN ON THE SPOT
SADAT'S PEACE BECOMES HUSSEIN'S TRIAL

By Christopher Wren

The weathered face, framed by the soft red and white checked cotton of an Arab kaffiyeh, belongs to a survivor. The once cocky young smile circulates these days only on stamps and bank notes in the Hashemite Kingdom of Jordan. The dark eyes look sadder, more pensive. The clipped military mustache has given way to a full beard that seems excessively gray for its owner's 43 years.

There is little in the way of wars, attempted coups and assassinations, intrigues and personal tragedies that Hussein ibn Talal has not experienced in his 26 years on the throne. With the fall of the Shah of Iran, King Hussein endures as the longest reigning monarch of a major Middle East country.

The peace treaty concluded between Egypt and Israel last month has placed Hussein on the spot. His small country has become the next logical pressure point for both the United States and the radical Arabs opposing the treaty.

Hussein's refusal to join in negotiating the future status of the West Bank of the Jordan River has undercut his old image as America's friend in the Arab world. Egypt's Anwar el-Sadat has replaced him, although the King, with a British education and an American wife, probably fathoms the Western mentality better than any other Arab leader.

Hussein originally had called Mr. Sadat's trip to Israel "courageous" but his enthusiasm chilled after the Egyptian leader signed the Camp David accords last September without consulting him. The King explained that just before the summit "we received a letter from President Sadat emphasizing his determination to work for a compre-

Christopher Wren is The New York Times Cairo bureau chief.

A case of pragmatic coexistence: Hussein (left) greets P.L.O. leader Yasser Arafat.

hensive settlement and outlining the position that he made clear in his visit to Israel, which we had no objection to. But then what emerged from Camp David was something totally different. When we examined it and sought answers [from the United States], we found there was so much vagueness that we could not possibly involve ourselves."

☐

Jordan has been maneuvered into the ranks of rejectionist Arabs as much by the pressing of the United States and

Egypt as by the hard-line tugging of Iraq or the Palestine Liberation Organization. After National Security Adviser Zbigniew Brzezinski visited Hussein last month, the King angrily complained about American "arm-twisting." Washington denied it but there was implicit pressure in the unresolved questions of future deliveries of American weapons, on which the small Jordanian Army depends.

Hussein has parried the entreaties of a half-dozen American Presidents but he still worries that President Carter does not realize how limited Jordan's

options are in giving the Israeli-Egyptian peace treaty a chance.

"Right from the beginning, from 1967 onward, our position was very clear," the King explained wearily on a rainy evening as we talked in his office at Basman Palace. "We could not bargain over an inch of occupied territory, could not bargain or make concessions regarding Palestinian rights. If we were assured these rights and territories would be recovered, we'd be willing to involve ourselves in anything that would achieve those ends. But we must know what the ends are."

Israeli Prime Minister Menachem Begin has said that Arab Jerusalem, which was captured from Jordan and annexed in 1967, is part of Israel's "eternal capital" and will never be returned. The Israelis have redesignated the West Bank, seized in the same war, by its biblical names of Judea and Samaria and are populating it with settlements. Israel refuses to consider anything more than carefully supervised Palestinian home rule (see following story, Page 20).

Having watched Egypt's problems getting back the arid Sinai peninsula, to which Israel lays no biblical claim, Jordanians see little point to similar negotiations over the West Bank and Arab Jerusalem. The King has not lost his nerve: He has invested too much in building up his nation to gamble his stake away. Were he to plunge into a dialogue with Israel as President Sadat did, what would be his chances of bobbing to the surface in the sea of internal and external political currents?

☐

The constraints upon Hussein are evident in the nature of the hatchet-shaped kingdom over which he reigns. Its border with Israel, longer than that of any Arab state, runs over 200 miles. To the north, east and south lie Syria, Iraq and Saudi Arabia, which have definite views about how their weaker neighbor should behave. Lacking Egypt's relative insularity, Jordan must accommodate to survive.

Unlike most other Arab desert countries, Jordan has never struck oil. The prime agricultural land on the West Bank was lost to Israel and the kingdom has begun spending $1 billion over a period of 10 years to develop what is left of the Jordan Valley. Because of the prevailing desert, 90 percent of Jordan's population lives on 10 percent of the land.

Since its independence in 1946, Jordan has relied upon foreign aid, first from the British and more recently from the oil-rich Arabs. This year, 62 percent of the Government budget will be subsidized from abroad. The country runs a chronic trade deficit.

To these geographical and economic vulnerabilities must be added a precarious demographic structure. Palestinians make up more than 50 percent of Jordan's population of 2.2 million, not including another 750,000 on the West Bank. Some have risen to positions of private wealth and public responsibility, but others still nurture the dream of their own home some day.

Jordan's Palestinian majority has prompted some Israelis to suggest that a feasible Palestinian state already exists. Hussein refuted this in 1970 when his Bedouin combat troops began clearing Palestinian guerrillas from the country in eight months of sometimes cruel fighting. Though Hussein and the P.L.O. have undertaken a reconciliation in recent months, some Palestinians have still not forgiven him for what they call Black September, when the killing swept into the refugee camps. Still, Hussein's doggedness has not only held his small nation together but given it a high literacy rate and a bustling laissez-faire economy.

Jordan's state of belligerency with Israel has dragged on for lack of other options. "The only card we hold is that we can deny Israel a straightforward peace," conceded one Jordanian official. "By accepting negotiations, we would have discarded the last card in our hand, replacing the situation with a tacit peace which would exist as soon as negotiations started."

Of all the issues in dispute, the most sensitive for Hussein is that of Arab Jerusalem, which went unmentioned in the Camp David framework peace accords. Many Moslems revere Jerusalem as Islam's third holiest city after Mecca and Medina, and the King, who claims direct lineage from the prophet Mohammed, feels its loss particularly keenly. With the other Arabs flailing Sadat for having disregarded the city in making peace with Israel, a Western diplomat in Amman noted that "Hussein is not going to be the one who gives up Jerusalem. He is not going to take the blame."

□

Jordan's West Bank role, outlined at Camp David, upset Hussein mightily. He told me that the notion of helping Israel police the West Bank during a transition to self-government was "totally unrealistic and unacceptable" because Jordan was being asked to safeguard "the security of the occupying power against the people under occupation."

Partly because he feared American pressure even then, Hussein welcomed

There is little in the way of wars, attempted coups and assassinations that Hussein has not seen in 26 years as King.

As commander in chief, King Hussein (standing, right) visits his frontier troops.

the Arab Baghdad summit last November. His credibility as a moderate was hurt abroad by Jordan's acceptance of a promise of more than $1 billion a year in subsidies to reward "steadfastness." When I observed that the Egyptian press had accused him of going to Baghdad for the money, the King replied icily, "I have always stood for peace, sir, and I always will — a just and lasting peace." The King insisted that the Baghdad summit was a victory for moderation because it averted the splintering of the Arab world and ac-

knowledged the United Nations resolutions that recognized Israel's right to live in peace in return for its withdrawal from occupied Arab lands.

President Sadat once privately disparaged Hussein by quipping to a visitor that the Jordanian monarch had jumped into the 1967 war when he should have stayed out and was staying out of Sadat's own peace initiative when he should jump in.

But Hussein described his defeat in the Six Day War in 1967 as the price of Arab solidarity, "knowing full well

what the results might be, yet having to live through them and seeing one's worst suspicions and fears come true." Though it tore his kingdom apart, he said that he joined the losing Arab side because "we had to live up to commitments as we would have expected others to live up to their commitments."

He has continued to honor what he sees as his obligations on the West Bank. Over the first decade of Israeli military occupation, Jordan paid out more than $171 million for office rents, pensions and salaries of school teachers and other municipal employees so as not to leave a vacuum for the Israelis to fill. Jordanian tourist brochures describe the West Bank and Arab Jerusalem as though they were never lost. Hussein prudently passed up the 1973 war, though he did send some tank units to bolster Syria's southern flank.

"In his younger days he was very impulsive," said Abdul Hamid Sharaf, Hussein's foreign-policy adviser. "But it was a strange combination of impulse and strategic thinking. He acts on impulse a lot, but thinks in broad strategic terms."

□

When I first visited Jordan 10 years ago, I returned to Amman one night after spending a few days in the mountains with a guerrilla band from the Popular Front for the Liberation of Palestine. At the city's northern outskirts, our Peugeot station wagon, laden with guerrillas toting Kalashnikov assault rifles, ran into a Jordanian Army roadblock. In those days, Palestinian guerrillas were forbidden to take weapons into Amman. They habitually ignored the ban, sometimes shooting it out with Hussein's loyal Bedouin troops.

A Jordanian soldier, covered by a machine gun atop an armored personnel carrier, thrust his head inside the window and stared into the barrel of a guerrilla's Browning 9-millimeter pistol. His hesitation seemed to take forever, though it must have lasted but seconds. He abruptly backed off and waved us through.

In this climate, few reporters gave Hussein much hope of facing down the well-armed, increasingly hostile guerrillas. The Palestinian cause that he had embraced so fervently with the loss of the West Bank seemed bound to consume him.

Later the following year, however, he showed that he could still strike back. In retrospect, he justified the move as "not Jordanian against Palestinian but one that could be characterized as law and order and hope of a people, Jordanians and Palestinians alike, as opposed to chaos and anarchy."

His fear, he told me, had been that, once control was lost to the guerrillas, "Israel would have probably been in a position to move and Jordan as a country — what remained of it — would have been totally destroyed."

Although Hussein insisted that no single event forced his hand, he vividly recalled having been informed one evening that one of his heavy-artillery units in the Jordan Valley had deserted and was heading for Amman without its officers. In those turbulent days, army morale had dropped so low that entire units broke away from the front with Israel and went looking for revenge against the guerrillas.

"I drove out in the night and I met them just coming over the mountains," the King related. "I tried to stop them but they avoided my car and continued. I turned around and tried to overtake these big, heavy gun tractors, and on one or two occasions, they nearly drove me off into the valley. Eventually, I did manage to pass them and stop them in the middle of the road.

"They were wild. They were firing in the air. Some N.C.O.'s on leave had gone to see their families and the unit had received word that one of them was killed and his family and house burned.

"We had a very tough few moments. I remember one N.C.O. with a grenade in his hand telling me to clear the way or he was going to blow himself up. It took quite a bit of time to calm him down. I remember others — one in particular — saying, 'Get out of our way and let us get on with it. For months, we haven't been able to visit our families with our uniforms on.'

"Well, eventually we did calm them down but I couldn't get them back to where they should have been in the valley. At least I was able to get them to camp where they were."

☐

A few days later in September 1970, Hussein unleashed his army against the Palestinian guerrillas, though it took until the following July to expel them from their bases. The showdown remains significant because the price he paid in the Arab world for salvaging Jordan's national identity has severely limited his political options today.

To regain the good graces of the rest of the Arabs, at the Rabat summit in October 1974 the King submitted to the humiliation of renouncing his claim to speak for the Palestinians and of recognizing the P.L.O. as their sole representative. In Jordan's view, any transitional role on the West Bank properly belongs now to the P.L.O.

Thus, Hussein gave up Jordan's ambitions of regaining the West Bank, concentrating instead on developing the remaining East Bank region. There is little doubt that he would prefer some federation with the West Bank to an independent Palestinian state, but when I asked about this, he insisted that the choice was up to the Palestinian inhabitants.

Hussein's response seems disingenuous in view of the problems that a radical Palestinian neighbor could pose. His strategy has been to neutralize any long-term threat by championing the Palestinian cause even while allowing no further breaches of Jordanian sovereignty.

Western diplomats in Amman tend to feel that Hussein is not overly worried because he thinks the West Bank Palestinians would choose some link with Jordan rather than go it alone. His reputed concern is not that a militant Palestinian state would turn on him but that it might invite a reprisal from Is-

Hussein
is less worried that
a Palestinian state might turn on him
than that it might invite
reprisal from Israel.

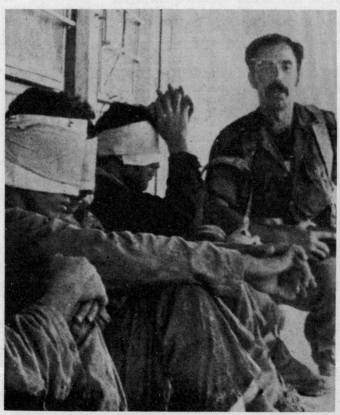

Israeli guards P.L.O. fighters fleeing Jordan's 1971 "Black September" carnage.

rael that could damage Jordan, too. But the King has sidestepped committing himself.

"I do not believe," Hussein told me, "that it is the right of any in the area — Prime Minister Begin or Sadat or myself or any other leader, or even the United States — to sit and decide what the future of the Palestinians should be. They must be involved in conditions of freedom to exercise their rights of self-determination."

Perhaps because he cannot afford the luxury, the King displays no visible

grudge against those Arabs who have worked against him. I asked if he foresaw a way of coexisting with Israel.

"Put it another way," he said. "I had the feeling, to be very honest, that it might have been possible some years ago, even in the period between 1967 and the time in which we live now. But as time has passed, I've begun to realize that there are some ingredients missing to make Arabs' and Israelis' living together a reality.

"There is a need for us to do all we can to present our case to the world in

an adequate way and to build our strength, sadly, to a point where there is a degree of parity between us and the Israelis. Because, short of that, they are still living within a fortress and with a fortress mentality. They are still bent on conquest, on establishing a country whose final shape one does not really see in terms of geographical boundaries, at the expense of the rights of others.

"It is not the lack of good will within the Arab world for the people of Israel," the King maintained.

"They are still living from day to day and not with enough vision, enough ability to look into the future. But their way has been the wrong way. Peace can come to this area when people on both sides feel that they've got something that's worth keeping, worth building upon, worth protecting. Not positions on the ground. Not territory or a denial of people's rights in their land and the land of their ancestors."

I asked if he had in mind the growing number of Jewish settlements on the West Bank. "It is against this kind of mentality that one doesn't see how peace is possible in the immediate future, sir," he replied.

☐

Hussein's success in forging a disparate population with limited resources into one of the Arab world's most literate and progressive societies prompts comparison with the failure of the wealthier, more powerful Shah of Iran to rally his people.

The difference is pointed up in an anecdote that one European diplomat tells about the two rulers at some past meeting. When Hussein took out his cigarettes, he passed around his pack to the others before lighting up. When the Shah smoked, he extracted a cigarette for himself and pocketed the pack.

When Hussein visited a Palestinian refugee camp recently, an old man tried to kiss his hand and the King, according to an observer, recoiled in embarrassment. In interviews with reporters, the King acts far less imperious than, say, Egypt's Anwar el-Sadat. Queen Noor has said of her husband that "his strength is that he doesn't have a facade. He doesn't have one face at home and another for the country."

Is this any way to run a monarchy? A Jordanian acquaintance argued that it is, at least in the final quarter of the 20th century. "You can't help but like the guy," he said. "You can't help but feel he's sincere. Some Palestinians may question his sincerity, but with foreigners there is instant rapport. This is the key to his success on a personal and world level."

The King still falls short of becoming an intellectual. He remains fascinated by aircraft and weapons — he personally tests anything new that the Jordanian armed forces acquire — and his literary taste runs heavily to military matters. Collections of Luger pistols, swords and model airplanes are evident in the royal residence at Hashemiah, outside Amman. But he has grown more introspective and, the Queen says, he often stays up late at night and reads.

HUSSEIN

Continued

Perhaps he is more cautious, too, though he retains his old appetite for physical risk. Abdul Hamid Sharaf, who is Chief of the Royal Court, reports that the King, a seasoned pilot, still does part of the flying on trips abroad. When I asked which part, Mr. Sharaf replied wryly, "The takeoffs, the landings and the rough weather."

Queen Noor says that her husband had taken up long-distance water-skiing in the Gulf of Aqaba. "He likes to go out and push himself to see what his physical endurance is," she says. When they visited the Bavarian Alps last November, the King impressed her by flinging himself down the ski slopes as a 43-year-old novice.

"No, I do not put restraints on myself, not to fly or not to do this or that," the King told me.

"I believe, sir, that if one began to think of all these dangers and what they would mean, one wouldn't be able to do very much in life. One has to be a bit of a fatalist. I've seen too much in my life to believe otherwise. When my time comes, it comes."

☐

The world attention lavished on Egypt's interplay with Israel has left Hussein in the shadows. Now that a peace treaty has been signed, Jordan in theory should be the next confrontation state to come to terms with Israel. But it would

have to be on Hussein's terms, not as an extension of the Israeli-Egyptian peace settlement, and even then he would need all the ingenuity he can muster.

The King seems too much the pragmatist not to accept the reality of an Israeli-Egyptian peace eventually, but only after Arab hostility dissipates. "If he can't forge a consensus, he is perfectly happy to sit where he is," a Jordanian observer said of Hussein's style. "He can't go out on a limb like Sadat."

For all the suspicions that have arisen between Amman and Washington, Hussein has a potentially valuable asset for the Carter Administration. Unlike Sadat, he has been careful not to cut himself off from the Arab mainstream. In time, the King could play a significant role in brokering a reconciliation between the embittered Arab majority and the

United States, but only if he has enough room for maneuver.

Whether Hussein's friends, rivals and historical circumstance will let him do so remains to be seen. "I get frustrated because the game is being played outside," Queen Noor said. "We are never sure what the stakes are."

But anyone with Hussein's gift for survival cannot be discounted. In our final conversation, I asked him what changes he had seen in himself over the years. He gave it some thought and responded, "Well, I suppose one recognizes more and more the limitations that are before one and understands matters in a deeper way. In the last few years, I've noticed an ability in myself to concentrate on some problems with greater depth rather than looking at many matters and trying to resolve them.

"But throughout my life, sir,

Hussein as a Harrow student.

I've never taken a decision in a hurry. I've always tried to hear and look at matters from every possible angle and then do my very best as it appears to me. And I don't think there are very many decisions that I would have changed as a result." ■

HUSSEIN AND HIS QUEEN

King Hussein and Queen Noor after their wedding last summer.

The harshest blow that King Hussein has suffered in his uneasy 26-year reign was probably the death of Queen Alia in a helicopter crash on Feb. 9, 1977. Two earlier marriages had ended in divorce but King Hussein remained devoted to Queen Alia. The trauma of her

death submerged him and he visibly aged. His solitariness frightened some acquaintances who recalled his father's losing bouts with depression. "He was really a man all alone in the world," one Jordanian remembered.

That changed with the courtship of Lisa Halaby, a

1974 architecture graduate of Princeton University. Her father, Najeeb Halaby, had been the head of the Federal Aviation Administration and the president of Pan American Airways and she was raised in the suburban affluence of Washington, D.C., and New York. When Mr. Halaby, as an aviation consultant, arranged the delivery of the first Boeing 747 jumbo jet to the Royal Jordanian Airline, ALIA, he brought his daughter to the ceremony and introduced her to the King.

She had been living in Beirut and she moved to Amman to work for ALIA as an interior designer. The king began dating her. For six weeks, she remembers, "we had dinner every night, speaking comfortably. It was a wonderful period when we both unwound."

When he proposed, she says, "I knew what he was saying but I was very unsure what he was saying. I was unsure I would be exactly what he needed, that I wouldn't be a hindrance, being relatively new to Jordan and because it did happen fairly quickly.

"In asking me, he was offering me an opportunity to not only love him and his children but to put what I had inside me to the best possible use."

They were married last June in a brief Moslem ceremony with a reception more like a Scarsdale garden party than a royal affair.

King Hussein proclaimed

his bride a queen and renamed her Noor, which means "light" in Arabic.

His emphasis on her Arab ancestry — a paternal grandfather had emigrated to Texas from what was then Greater Syria — struck some as a trifle far-fetched in view of her American blond, blue-eyed good looks. Still, when an American diplomat in Amman explained her nationality to a Bedouin camped in his backyard, the Bedouin scoffed: "Everybody knows that the Halabys come from Syria."

Queen Noor is a slender woman several inches taller and 16 years younger than her husband, with a quick mind, strong, capable hands and a hearty laugh. Like her husband, she seems a bit of a loner. Her admiration for him reciprocates his own devotion. On a hill behind their modern limestone residence, he has had built for his wife, who once ski-bummed in Aspen, Colo., a private ski lift with artificial snow and an artificial slope.

Though she has matured in the months since I covered their wedding, the Queen seems still somewhat dazed by the direction that her life had taken. The decisions one has to make are often difficult, or at least unfamiliar, she explains: "Sometimes I have to say, 'Please slow down and let me catch my breath.' It's a great compliment for him to assume it is second nature to me."

One of her hardest adjustments, predictably, has been

to the security constraints. "When we left for our honeymoon, we spent a few days in Scotland. From Day One, we were in the car with someone sitting in back. It was something I found very difficult in the beginning. I didn't know what was amiss until I realized there was somebody in the car all the time."

In the beginning, she says, "I worried for him because I knew he was more vulnerable than anyone I had known before. But like the car on the honeymoon, that's part of the life he's led."

When he returns home in the evening from his office at Basman Palace, they sometimes go for an evening spin on one of his B.M.W. motorcycles. "I ride on the back and we have helmets that let us talk," the Queen says. Once they appeared unannounced at a local Chinese restaurant. "I had to drag him because he's not big on Chinese food," she laughs.

The Queen, who recently suffered a miscarriage, says that she and the King discuss domestic concerns like any couple — she wants to build a cozier home. But they also talk over problems like Jordan's future with Israel.

"I trust his instincts so much, though I may not always agree. They have a pattern and a rhythm that he may not understand," she says. "He's lived on the threshold all the time. All aspects of his life are on the edge. He's constantly being tested. At the same time, he's terribly strong." —C.W. ■

April 8, 1979

In Tel Aviv, Israelis welcomed the signing of the peace treaty with candles and feelings of ambivalence, the reactions of a people to whom living in peace is an unknown.

William Karel

ISRAEL: LEARNING TO LIVE WITH PEACE

By Lesley Hazleton

"There never was a good war or a bad peace," said Benjamin Franklin. It is one of those stunningly simple statements that, if you are endowed with the grace of historical perspective, express a perfect truth. But when you are personally involved in the process of war and peace — when you are part of the flesh and blood of the process as I, an Israeli, am right now — you become painfully aware of the complexities that Franklin ignored, of the ambivalence and confusion that get lost in historical perspective but that the people involved have to suffer at the time.

The high tension in which we have lived here in the Middle East for the last 16 months has taken a strange toll. On March 26, when President Sadat, Premier Begin and President Carter fi-

Lesley Hazleton is the author of "Israeli Women: The Reality Behind the Myth." She was born in London and has lived in Israel since 1966.

nally signed the peace treaty between Egypt and Israel in Washington, the mood in Israel was one of restraint rather than elation. Yes, of course, most people were pleased that peace was finally being signed with Egypt. But there was also a marked lack of excitement, a sense of strain instead of the joyful buoyancy that it seemed should be there. It was as though we were all burned out.

Three times in recent years, President Sadat has made me cry. And I am not easily given to tears. The first time was a few months after the Yom Kippur War of 1973. During that war and the weeks that followed it, as the news trickled in of friends and acquaintances killed, wounded and "missing in action," I had no tears. There was just the automatic busy work of the reporter as I let daily action blind me to my stunned emotions. But sometime after that, a friend came to visit from America. We sat in my study on a sunny February afternoon, as the light played through the pine trees outside the window and gently flickered into the room. She asked how it had been these last few months, and I began to tell her about the killed and the wounded, the

waste of it, the inevitability of it after the 1967 Six Day War. As I began talking about all these things, to my utter surprise the tears began flowing, as though an outside hand had finally pressed a release switch.

The second time was on Nov. 19, 1977, as I watched Sadat land in Israel. Like a child watching a suspense movie, I stared hard at the television screen as the Egyptian plane landed at Lod airport. I made unspoken bets with myself as to whether it was all a bluff, whether the door of the aircraft would swing open to reveal . . . nothing, a blank, cruel joke. And then, there he was. The impossible was actually happening. He came down the steps and stood beside Begin on the red carpet, and as I held back tears in ecstatic disbelief, the Israeli Army band started playing the Egyptian national anthem and the guns began their salute. Like so many Israelis that day, witnesses to the old cliché that this is the land of miracles, I stopped holding back the tears.

Nearly a year was to pass before I was to cry again. Two days before my birthday, on Sept. 18, 1978, I was staying with a friend, and we had risen early to start on a long trip south. My

friend was out of the room when I turned on the radio to hear the 6 o'clock news. As I sat there on the bed, I heard the newscaster say: "Israel and Egypt will sign a peace treaty in three months' time. The treaty will include . . . ." I barely heard the rest of the broadcast, for the tears seemed to mist my hearing as well as my vision. The miracle that had been created and then seemed to fade into nothingness was back, this time at Camp David. That was the only item on the news, and my friend came into the room to find me sitting there, tears flowing down my face, clutching the radio to me as the morning hymn poured from it. His eyes opened wide in alarm. "What happened?" And I, incapable of coherence, laughing and crying in sheer happiness, just gestured at the radio and said: "The news, the news . . . ." ·

And then came the time when it seems that I should have cried and didn't. It was March 26, 1979, the day the peace treaty was actually signed. No one in Israel, or in Egypt, it seems, cried that day, either from sadness or from happiness. There were no tears left.

As I watched the ceremony on the White House　　　　*(Continued)*

255

ISRAEL

Continued

lawn by satellite, it was suddenly all very small to me, reduced to the size of the screen. It seemed symbolic that this signing, which would so deeply affect our lives, was taking place 5,000 miles away, as if to emphasize the unreality, the distance of it, for us in Israel. There was none of the high emotion here that came over the satellite so clearly from Washington. Even the weather seemed to indicate caution; it had suddenly turned raw and blustery, with storms and squalls throughout the country, so that few people turned out for most of the public celebrations. Even in Tel Aviv, where 100,000 people gathered in the main square, there was no dancing, no spontaneous celebration. Everyone was watching the screens, making sure it really was happening, and yet unable, somehow, to absorb it.

It seemed terribly unfair that we, the most needful of peace of all three parties signing in Washington, should be less aroused by it than the mediator, the United States. The irony was pointed up in the entertainment program broadcast later that night by Israeli television. The most joyful item was Zubin Mehta, American flag held high instead of a baton, conducting the Israel Philharmonic in an ebullient rendering of "Stars and Stripes Forever." It was wonderful, and yet sad. Where were we, the Israelis? Why were we not ecstatically happy? The impossible had happened. Peace with Egypt, a laughable proposition just two years ago, had been signed. Yet we seemed to be capable of only a sober confirmation that the signing had taken place — a good thing, most people agreed, but no reason to lose one's head. Why were we being so blasé, as though we had signed peace treaties countless times before in our short history as a state?

We were in other, dark places. We were remembering the dead, among other things, and seven hours before the treaty was signed, a red rose was placed on the grave of every soldier fallen in action. But more than that, we were remembering the future.

Two nights before the treaty was signed, I had a dream: I was living in a small shanty village on top of a volcanic mountain, sharing a house with Menachem Begin. That

afternoon, the volcano began to erupt, and many of the villagers packed and fled. I tried to persuade Begin to do likewise. "No," he said, "this is my village, this is my house. I will not leave." I felt it would be unfair to leave without him, and, in any case, by late afternoon it seemed that the eruption was ending; the rumblings had quieted down, and although there were still a few sparks in the air there was no sign of lava overflowing.

So, in the dream, I stayed. Then suddenly I heard a horrifying sound, a roaring and crackling that surrounded the house. Looking out of the window, I saw rivers of red-hot, spitting lava flowing over the mountaintop. The fierce colors of the lava — orange, scarlet, purple, gold — lighted up the night sky, radiating heat and danger as the mass rolled inexorably over houses and gardens. There were few houses left standing, and the remaining villagers had all fled. I banged on the door of Begin's room. "We have to leave now, quickly," I shouted. "The lava has practically cut us off." He came to the door in a dressing gown, standing erect and proud. "I will never leave," he said, "not even for a volcano." Exasperated, I dashed out of the house, and, seeing a narrow opening through the swirling lava, I ran down the mountain ahead of that all-consuming molten river, running for my life, until I reached the bottom of the mountain and emerged through a tunnel. It was daylight, and people were waiting for me. And as I walked toward them, the mountain behind me became a huge house . . . and collapsed.

I awoke in shock. To the best of my knowledge, I had been one of the happiest people in Israel about the impending peace treaty, my elation contrasting with the subdued mood of most. Israel and Egypt had negotiated, and come out with a treaty that was good for both of them. True, the problem of the Palestinians loomed large over the treaty, but still, I felt, we should celebrate what had been achieved. The largest step, the first one, had been taken, for we had proved that Arab and Israeli could agree. Where, then, had this dream of disaster come from?

I fought to deny it, refusing to acknowledge that subcon-

On the eve of peace, flowers were placed on the graves of Israel's soldiers who were lost in war.

sciously I might be as wary as most Israelis about the treaty. Tentatively, ashamed of my own dream, I told it to a few friends. Like me, they were longtime members of Peace Now, the grass-roots movement that sprang up two years ago to demand that the Government take positive action toward peace. The Peace Now sticker was still on the back window of their cars, as it was on mine, a happy reminder of what had been achieved.

To my surprise, my friends were far more accepting of my dream than I. They too, they said, had had visions of disaster to come, not in dreams but in waking moments throughout the day — visions that peace would be signed and war would break out the next day, visions of Israel lulled again into the false security that blinded her to the Arab buildup for the Yom Kippur War, visions all of disaster being born out of peace.

These were strange visions for us. For they seemed to reflect the attitudes of the ultraright wing, the stolid antitreaty minority of Israel's

population that is convinced Israel has surrendered her own best interests to foreign pressure for peace. They were the visions of Geula Cohen, the Knesset's unruly firebrand, once a firm supporter of Begin and now, with the peace process, his most intransigent opponent within his own party. They were the visions of Ariel Sharon, the general who led part of the Israeli strike force in the Suez Canal in the last days of the Yom Kippur War. Now Minister of Agriculture and the Minister in charge of settlement in the occupied territories, he is a fine soldier and wild civilian who treats territory in this part of the world as if it were the Wild West and occupation 10 parts of the law. Theirs were not my visions or those of my friends. Why, then, were we having them?

Though they do not explain the semiconscious fears of impending disaster brought on by the peace treaty, there are solid political reasons for Israelis to have a subdued reaction to it. The "refusal bloc" in the Arab world, which is ada-

mantly opposed to the peace treaty, is expanding; there is a fear of Saudi Arabia becoming another Iran. And toughest of all is the complex and painful problem of the Palestinians, with long and undoubtedly difficult negotiations to come on autonomy in the West Bank and Gaza, the inevitable disagreements over the meaning of autonomy, the expected increase in the number of Israeli settlements in the West Bank. All these emphasize that though we have signed a peace with Egypt, Israel's problems with its other neighbors are still acute. The day after the treaty was signed, a two-part cartoon ran in the influential morning newspaper Ha'aretz. The first part showed the hand of Israel in a stormy sea. The second part showed Israel in a paper boat labeled "Peace Treaty with Egypt," tossing helplessly in the same stormy sea.

And there are economic reasons for mistrusting the treaty that are even stronger than the political ones to most Israelis. While peace is still an intangible — no more than three loose-leaf binders as far as most Israelis are concerned — the price of peace is staggeringly heavy, and is already being felt. The week before the treaty was signed it was clear that the withdrawal of Israeli forces from the Sinai into the Negev would tie up manpower and equipment of the country's construction industry, thus exacerbating an already severe housing shortage. Just one month before the treaty was signed, fuel prices, already among the highest in the world, increased another 40 percent. And in the coming 10 years, Israel will have to repay the vast loans from the United States for the military pullback and redeployment, including the building of two major airfields. If war has not quite bankrupted this country, many Israelis fear that peace will. We are already among the most heavily taxed people in the world — and the day after the treaty was signed, the Finance Minister warned that taxes would go up yet again.

But the real reasons for the extraordinary ambivalence about peace, I think, are neither political nor economic but psychological. Two of the most common expressions in Israel right now are "We have to give peace a chance" and "Well, at least it's better than war." The grudging acknowledgment of the advantages of peace and the concentration on its risks are the reactions of a people who have never lived in peace, so that paradoxically, instead of being able to welcome it with open arms when it is finally achieved, Israelis find themselves threatened by this vast unknown.

The desire for peace is a sine qua non of Israeli life and society. But after 16 months of uncertainty, what Israelis want more even than peace itself is a respite from uncertainty, a return to known and settled conditions. Peace is unknown, and therefore very unsettling. We simply don't know what to do in times of peace, nor even how to celebrate it. We've never had it before, and can hardly conceive of what it means. We are aware that it will change the basic conditions of our everyday lives, yet we concentrate to a large extent on the negative changes like the worsening economy and increased terrorism. Since the treaty was signed, there has been a marked increase in terrorism, both in Israel and against Jews and Israelis abroad. Yet, this increase was anticipated, and therefore we can brace ourselves to meet it, however terrible it may be. But after more

than 30 years of enmity, our imaginations cannot grasp what peace might portend, and like human beings everywhere, we fear the unfamiliar and resist the changes we shall have to undergo. The enemy deeply entrenched in our psyches, in our dreams, is now the friend. The paper has been signed, yet the vast gap between law and psychology remains to be crossed.

No wonder the American reaction to the peace treaty was so much more enthusiastic than Israel's or Egypt's. Americans will have to make no adjustment to peace in their daily lives, in their very psychology, whereas peace will require a vast psychological adjustment on the part of both Israelis and Egyptians. The treaty has been signed; now peace must start, and for the individual Israelis and Egyptians who are the flesh and blood of the peace process, this is where the difficult work begins.

Meanwhile, our only definition of peace is what peace is not. Even President Carter, in his speech at the signing ceremony, concentrated on this aspect of peace. "At the end of this [peace] campaign," he said, "the soil of the two lands is not drenched with flowing blood. The countryside of both lands is free from the litter and carnage of wasteful war. Mothers in Egypt and Israel are not weeping today for their children fallen in senseless battle." But many Israelis, as they listened to the man who guided them to peace, must have looked deep in their hearts for an answer to the question: What then is peace? And as though to make the question more acute, there were the repeated assertions throughout the ceremonies, and throughout the commentary associated with them, of the difficulties to come, of the fact that, as Mr. Carter said, this was only "the first step of peace — a first step on a long and difficult road."

As Jews, Israelis know how difficult it is to adjust to a new reality that has as yet no definite shape. We see it in our recent history, in the Jews who stayed in Eastern Europe after Hitler came to power, who stayed even in Germany until they were taken away to the camps. Unable to grasp that this time the persecution would not pass, unable to adapt to a new reality and change their concept of what was happening, they perished for it. This is a terrible analogy, I know, for the difficulty of adjusting to peace. But this is the kind of analogy that we have to make in Israel. We

fear that the bad experiences of the past will be repeated, for the Jewish past is an integral part of the Israeli world view.

Despite the fact that I and my friends shrink in horror each time Mr. Begin hammers home the Holocaust, denigrating it by overemphasizing it, by playing it, as it were, for maximum emotional effect, despite this, the Holocaust and the Jewish past of persecution, isolation and the struggle for survival is part of each of us. Deep within us, we still carry on that struggle. Indeed it has been part of our daily lives for over 30 years, for Israel, which was our own solution to the problem of Jewish survival, has been one of the most insecure places in the world for a Jew to live. As the former director general of the Foreign Ministry, Prof. Shlomo Avineri, once put it: "We in Israel are bearing the Jewish cross more intensively than any other Jewish community."

To be able to accept even a modicum of security after so many years of chronic insecurity — this is the battle that we in Israel must now wage within our own minds. It is exacerbated by the fact that Israel will soon be smaller. Twelve years is barely a moment in historical time, but in human time it has proved more than sufficient for Israelis to adjust to the sense of a far larger country. After 12 years, Sinai seems a natural extension of Israel itself — indeed the Sinai and Negev deserts are basically one and the same desert, the division between them no more than a demarcation line drawn by the Turks and the British in 1906. It has been easy to forget that the Sinai is occupied territory, and that the original intention in 1967 was that it be the bargaining card with which to attain peace. So that although today the vast majority of Israelis support the return of Sinai to Egypt within the terms of the treaty, it hurts, as though an armless man had been given an artificial arm that functioned as well as a natural one, and then had to part with it. Israel is to withdraw to narrower borders, and the psychological similarity between smallness and vulnerability is felt as strongly in Israel as it is in the West.

Yet here too there is paradox. Though Israel will become smaller, its world of access will expand. For more than 30 years, Israel has been isolated in an arc of Arab enmity. Now Egypt has broken that arc, and raised the possibility that eventually Israel might be accepted into the Middle East, its borders open

on all sides. The prospect of this expansion of access provokes as much anxiety as giving back Sinai. For the binding power of Israel throughout the years has been the fact of siege and war. Hostility from without has drowned the chorus of social problems within. Now these will come to the forefront, and we do not know how to handle them.

If there were a comprehensive peace, Jews would travel to, and live in, Arab countries, and Arabs would come to Israel. In that case, a new question is already in the air: How then will we remain a Jewish country? How will we avoid being swamped culturally by the vast Arab world? Strict adherence to religion provides an answer for only about one-fifth of Israel's population. For the rest of us, another possibility is that being considered by such thinkers as President Yitzhak Navon, who appears to favor a Jewish ethic, a new cultural Judaism rather than a strictly religious one. But meanwhile, the fear of Israel's losing its identity through peace increases.

Despite the fears of a break in Israel's isolation, it is this very factor that may finally convince Israelis that peace is real. This will happen, I think, when the borders with Egypt are finally opened. When Israelis can go to Cairo, see the pyramids, struggle with the Cairo phone system, talk to Cairenes, float down the Nile in a felucca, venture into the Sahara, when Egypt becomes no longer merely a television image but a tangible reality for individual Israelis, then, finally, we may be able to say, "This is peace," and welcome the magic of it.

This may take a month, or three months, or maybe even longer. Meanwhile, we in Israel can prepare ourselves for the face-to-face experience of peace only by attempting that vast leap in concept that can persuade us, before we touch it, that peace is real, a process to be welcomed and sought out. Sadat, Begin and Carter have worked long and hard to achieve the treaty. Now Israelis and Egyptians, people like me and my friends, must work longer and harder on ourselves. We must examine our fears and doubts, and in doing so learn to exchange mistrust for hope and — eventually — trust. It seems to me that this is when peace will become a living reality, when we can grasp it not in terms of a treaty, or of trade and sports, or even travel, but in the most real and lasting terms of all: in our own minds. ■

April 29, 1979

YOUNG RESIGNS POST AT U.N. IN FUROR OVER P.L.O. TALKS; CARTER EXPRESSES REGRET

DELEGATE SPEAKS OUT

Says He Is Not a Bit Sorry, Promises to Continue Backing President

By BERNARD GWERTZMAN

Special to The New York Times

WASHINGTON, Aug. 15 — Andrew Young resigned today as United States delegate to the United Nations, asserting that he could not promise to muzzle himself and stay out of controversies that might prove politically embarrassing to President Carter.

One day after he was reprimanded for holding an unauthorized meeting with a representative of the Palestine Liberation Organization, Mr. Young submitted his resignation personally to Mr. Carter, who reluctantly accepted it. Jody Powell, the White House press secretary, announced the move in tears.

Mr. Young, as outspoken as ever, told a news conference at the State Department of his decision, which he said was not sought by Mr. Carter. He said he would work hard to help the President be re-elected.

"I really don't feel a bit sorry for a thing I have done," Mr. Young said today. "I have tried to interpret to our country some of the mood of the rest of the world. Unfortunately, but by birth, I come from the ranks of those who had known and identified with some level of oppression in the world. By choice I continued to identify with what would be called in biblical terms the least of these my brethren."

"I could not say to anybody that given the same situation, I wouldn't do it again, almost exactly the same way," Mr. Young said.

Mr. Young, who has produced controversy throughout his two and a half years at the United Nations, defended his meeting last month with the P.L.O. observer to the United Nations and his decision not to inform the State Department of what had happened and then to give only a partial and inaccurate version of events when he was asked.

As a result of that meeting, a mounting number of powerful voices, including that of Senator Robert C. Byrd, the majority leader, demanded his removal, even before Mr. Young conferred with Secretary of State Cyrus R. Vance and then with Mr. Carter today.

Mr. Young said that today there were "sharks" in the Washington waters "smelling blood."

"But I think I come before you not at all bloody, and in a way I come because I am unbowed," he said.

In his letter of resignation to Mr. Carter, which he said he composed last night, Mr. Young said he was grateful for the opportunity to serve the Administration. "I am afraid, however, that my conduct has created serious difficulties for the Administration on several occasions," he said.

"It has made me question my value as a continued part of your team. "I have always acted in behalf of what I felt was the best interest of our nation, though often it has been interpreted to the contrary."

As to the incident over the meeting with the P.L.O. representative, Mr. Young said it was "extremely embarrassing that my actions, however well-intentioned, may have hampered the peace process."

"In order to avoid any further complications, I would like to offer my resignation as the United States Permanent Representative to the United Nations," he said, offering to stay on the job into September. He is president of the Security Council this month.

Mr. Carter said he accepted with "deep regret" the resignation and said that Mr. Young had "earned the gratitude of all Americans with your superb performance in a most difficult assignment."

Mr. Young's meeting with the P.L.O. observer, in the face of a policy ruling out any substantive contact with that organization unless it accepted Israel's right to exist and also accepted Security Council Resolution 242, prompted a stream of condemnation from American Jewish groups, some of which called for his resignation. Some Republican leaders also joined in demanding his removal.

Envoy in Austria Not Rebuked

The State Department also acknowledged today that Milton A. Wolf, the United States Ambassador to Austria, had held an unauthorized meeting recently with a P.L.O. official in Vienna, Isa Sartawi.

But the department said that Mr. Wolf was not reprimanded, rather "reminded" of American policy. The implication of the remarks of a State Department spokesman, Thomas Reston, was that Mr. Wolf had not carried on any significant business with Mr. Sartawi, who had also met Mr. Wolf socially on two occasions.

Mr. Wolf, a Cleveland businessman and a prominent figure in the Jewish community in suburban Shaker Heights, has reportedly become close to Chancellor Bruno Kreisky of Austria, who met with the P.L.O. leader, Yasir Arafat, last month in Vienna. But Mr. Reston said that Mr. Wolf had played no role in arranging that visit.

Mr. Young said that after he had submitted his resignation to Mr. Carter, the President asked him to talk the matter over with Mr. Vance again before making a final decision. But Mr. Young said that he felt he had to be decisive and therefore chose to quit before seeing Mr. Vance once more.

At the news conference, Mr. Young stressed over and over his continued support for the Carter Administration and said he thought he could do more good for Mr. Carter as a private citizen than as a controversial public figure.

When asked if he thought the American policy of nonrecognition of the P.L.O. was a sound one, he said, "No, but I understand it." He said that the P.L.O. was gaining all the time in political and economic strength, even while decreasing in military power.

"It is in nobody's interest to ignore those forces," he said. He added that he could sympathize with Israel's refusal to deal with the P.L.O., "but I guess I think the United States is in a different position."

Mr. Young, the leading black foreign affairs official in the Administration, sought to leave the impression that he was relatively relaxed and unperturbed by the developments.

Mr. Vance was upset by Mr. Young's handling of the P.L.O. meeting and in particular by Mr. Young's failure at first to disclose that it was more than a social exchange.

In his news conference, Mr. Young said that when he spoke to Kuwaiti, Syrian and Lebanese diplomats last month seeking a postponement in a pending Security Council vote on Palestinians, he was told that only the P.L.O. could make that decision.

Mr. Young said he told them that he could not meet with the P.L.O. observer, but that if he and the P.L.O. official happened to be at the Kuwaiti Ambassador's house at the same time, that was something else. He said that he never informed Washington of the meeting, which did produce the postponement in the vote.

August 16, 1979

CONFRONTING THE P.L.O.

Israel refuses to recognize it. Sadat and Hussein may prefer to see it disappear. But the P.L.O. is a fact of Middle East life and a growing force on the international scene. It will have to be dealt with.

By Christopher Wren

Yasir Arafat, chairman of the executive committee of the Palestine Liberation Organization, likes to show visitors in Beirut a statuette of an Aztec, emphasizing that the Palestinians will not be swept into extinction as were those ancient Mexicans. And, he makes clear, the P.L.O. is the chief preventive force.

Created by Egypt and other Arab states 15 years ago, the P.L.O. embraces an assortment of rival political and guerrilla factions. The question of how seriously the P.L.O. deserves to be taken has been raised anew by the latest United Nations Security Council debate over the rights of nearly four million Palestinians — nearly half living under Israeli authority. The United States had promised Israel not to deal with the P.L.O. until it accepted the United Nations resolutions recognizing Israel's right to exist. When Andrew Young, the acting American delegate to the United Nations, resigned last month over the disclosure that he had met with a P.L.O. representative in New York, the issue of the P.L.O. was projected into American domestic politics.

Still, a variety of indirect contacts has already taken place between the P.L.O. and Washington, primarily through Saudi Arabia and, more recently, through other Western diplomats at the United Nations. After the American Ambassador to Lebanon,

Christopher Wren is the Cairo bureau chief of The New York Times.

Udo Schreiber/Gamma-Liaison

P.L.O. commandos patrolling in South Lebanon (left) lend muscle to the diplomatic efforts of Yasir Arafat (above, at a meeting in Vienna with former West German Chancellor Brandt, right, and Austrian Chancellor Kreisky in July.

Israeli soldiers carry a dead Palestinian commando who with three others terrorized the Israeli town of Nahariya last April.

The heart of the problem: Palestinian refugee camps, such as this one in Lebanon.

Francis Meloy Jr., was assassinated in June 1976, the security officers of the American Embassy in Beirut were authorized to keep in contact with their P.L.O. counterparts.

American diplomats stationed in the Middle East invariably find themselves rubbing shoulders with the P.L.O., though they are instructed to keep their distance. "You don't walk anywhere without meeting a Palestinian and you don't ask for his credentials," explained a senior diplomat.

But a dialogue with the P.L.O. implies an admission that it has something to offer toward a solution and this means ultimately conceding the P.L.O. its mantle of legitimacy. Whenever the United States has looked for an opening, it has not only been promptly chastised by Israel for breaking faith but also been rebuffed by P.L.O. officials themselves. Few issues in American foreign policy are as emotionally loaded as this one.

□

The case against recognizing the P.L.O., articulated by Israel and its partisans for almost two decades, holds that there is no point in talking to the P.L.O. because it is bent solely upon the destruction of Israel. The language of the P.L.O.'s charter speaks of creating a democratic Palestinian state on the territory of Palestine. But Israel's interpretation could be termed sound inasmuch as Palestinians themselves have viewed Article 27 as calling for the demise of Israel. Last January, the Palestine National Council which is the quasilegislature of the P.L.O., reaffirmed in Damascus that "armed, political and popular struggle in the occupied lands form the cornerstone of its policy" and demanded "the rejection of all capitulationist solutions as offered by the United States."

To the Israelis, the real intentions of the P.L.O. have been exemplified by its terrorist tactics — hijackings, market bombings, rocket attacks, the massacre of Israeli athletes at the 1972 Munich Olympics and the slaughter of schoolchildren at Maalot in 1974. Israel's Prime Minister, Menachem Begin, has been emphatic about the P.L.O.: "The most barbaric organization since the Nazis."

Some Palestinians, recalling Mr. Begin's own early years as a terrorist, have told me that they have only taken a cue from his Zionist underground group, the Irgun Zvei Leumi, which was responsible, among other acts, for the massacre of several hundred men, women and children in the Arab village of Deir Yassin in 1948. When the self-styled Eagles of the Palestinian revolution seized the Egyptian Embassy in Ankara last July, Israel cited it as proof that the P.L.O. was a "bunch of murderers." Yet it was a P.L.O. mission that talked the terrorists into giving up. While some Palestinians have become less sympathetic toward such terrorism abroad, popular support for operations against Israel itself seems undiminished.

The subject of terrorism and the P.L.O. came up repeatedly when Shafik al-Hout, a P.L.O. official in Beirut, made a three-week lecture tour in the United States over Israeli objections last spring. Mr. Hout replied that "if terrorism is to be condemned, let it be an unprejudiced condemnation. . . . To deny our people the right to repatriation and self-determination, and to condemn us to destitution forever, and at the same time ask us to remain silent, that would be impossible. It would be hypocrisy."

□

There have been indications in recent years that the P.L.O. might acknowledge the existence of Israel if the Palestinian right to self-determination were also accepted. In 1974, the Palestinian National Council expressed support for the establishment of a Palestinian state on any land "recovered" from Israel, suggesting for the first time that the P.L.O. might settle for the West Bank and Gaza Strip.

Yasir Arafat told the New York Times columnist Anthony Lewis last year that the "only possible solution" was a Soviet-American guarantee for both Israel and a Palestinian state. The P.L.O. leader has dropped similar hints to American congressmen. But, with his usual backtracking, Mr. Arafat told a recent rally in Beirut that "if the Palestinian state is to be a gift from Carter, we don't want it" and bragged that "our state will be established through our guns."

The ambiguity seems intended, for Mr. Arafat has been cautious about outdistancing his shaky consensus. According to one Cairo insider, Mr. Arafat had agreed in 1977 to accept the Security Council Resolution 242, which affirms Israel's right to exist, provided there was some understanding on Palestinian rights, but the P.L.O.'s executive council overruled him. Said an American official, "Arafat has come pretty close to making a statement that we'd call adequate."

The sticking point for the Palestinians on Resolution 242, which also calls for Israel's withdrawal from Arab land occupied in the 1967 war, is that it alludes to the Palestinians only in vague terms of "a just settlement of the refugee problem." In August 1977, Secretary of State Cyrus R. Vance recognized publicly that the P.L.O. had difficulties with Resolution 242 because it viewed them as nothing more than refugees.

"I think the policy makers in Washington have realized that Resolution 242 is outdated," said a European diplomat in the Middle East. "The Palestinians are a problem not in terms of refugees but of their national identity. The P.L.O. has said time and again that 242 is totally unacceptable if their rights are not included. The P.L.O. is being asked to recognize Israel when Israel has made it very clear that it will never recognize the P.L.O."

A senior American diplomat with whom I spoke doubted that the resolution, which took five months to negotiate, could be changed. "It is the only thing that all these Arab states and Israel have accepted. It is the basis for the peace treaty and for what we are

doing in the autonomy negotiations." In fact, Israel warned recently that if Resolution 242 were altered, it would consider the Camp David accords, which were based on the resolution and in turn form the groundwork for the Egyptian-Israeli peace treaty, null and void.

But the American official noted that "the pragmatists in the P.L.O. are sending out word that it might accept Resolution 242 with some additions recognizing Palestinian rights. I think we have to treat it very seriously." One possible compromise may be to pair the resolution with another existing United Nations resolution supporting Palestinian rights.

□

The P.L.O. is obviously unwilling to play its trump card, the acknowledgment of Israel's existence, if it gets nothing in return. And Israel has been consistent in its adamant refusal to have anything to do with the P.L.O. Moreover, with an American Presidential election campaign only months away, Mr. Carter would have to weigh any P.L.O. opening against the inevitable protests from Israel and its supporters.

The Israeli argument is that any peace overture from the P.L.O. would only be a stratagem to get a Palestinian state in the West Bank and Gaza Strip and that this entity, with ties to the Soviet bloc and to other radical regimes in the area, would pose a threat to Israel's existence.

The Carter Administration has already expressed doubts about letting a sovereign Palestinian state emerge and moderate Arab states, such as Saudi Arabia and Jordan, which profess support for the idea, would be unlikely to let the radicals take over for fear that could destabilize the entire region.

Mr. Arafat, however, told Anthony Lewis, "Assume that a Palestinian state has been founded. Would you believe that a state which is going to start from zero for the establishment of its institutions, its economy, culture, social problems — would such a state be able to form any serious threat against Israel?"

Even the Soviet Union, which is anxious to maintain a Middle East foothold, has been cautious about its links with the P.L.O., for all the professions of support. Though Mr. Arafat first visited Moscow in 1968, the P.L.O. was not invited to open an office there until 1974 and did not get

around to doing so until 1977.

To Israel's other objections must be added yet one more reservation — that inviting the P.L.O. into the peace discussions might undo all the painstaking compromises that paved the way to the Egyptian-Israeli peace treaty. The Camp David accords studiously ignored the P.L.O. "I don't see how the P.L.O. could be brought into the Camp David framework," said the European diplomat. "U. S. recognition of the P.L.O. would imply willingness to abandon the Camp David process in favor of a more comprehensive approach."

Predictably enough, Israel does not want the P.L.O. sitting in on its Palestinian autonomy talks with Egypt. Neither, it turns out, does President Anwar el-Sadat, who has virtually severed links with the organization over its hostility to his peace initiative. Mr. Sadat has derided the P.L.O. leadership as "nightclub militants," an epithet that took on fresh bite when Zuheir Mohsen, chief of the P.L.O.'s military department, was gunned down by assassins this summer after an evening at the roulette tables in Cannes on the French Riviera.

The problem is that the autonomy negotiations mired down even before they opened in late May, largely because of the inability to find any Palestinian alternative to the P.L.O. Prominent Palestinians in the West Bank and Gaza Strip are unwilling to get involved without the P.L.O.'s sanction. This is partly for fear of reprisal. Sheik Hasham Huzandair, one of the few Palestinian notables to back Mr. Sadat's peace policies, was stabbed to death outside his home in Gaza last June, and nervous colleagues stayed away from the funeral. But Palestinians living under Israeli occupation also feel that the talks are little more than a fig leaf for a bilateral peace between Egypt and Israel if the P.L.O. is not involved. "The only body that can represent us is the P.L.O. and nobody else," said Karim Khalaf, who was elected Mayor of the West Bank town of Ramallah on an openly pro-P.L.O. platform.

□

Against these considerations must be weighed the mounting argument in favor of dealing realistically with the P.L.O., which over the years has survived not only military retaliation from Israel but also the trauma of bloody defeats in 1970 and 1971 by the Jordanian Army and another in 1975 at the hands of the Syrian Army in the Lebanese civil war, emerging with a measure of widening international acceptance.

Abdel Mohsin abu Maizar, a member of the P.L.O.'s executive committee, boasted to me that "the P.L.O. is not only a political or military institution. What happened to us in Jordan in 1970 and Lebanon in 1975 would have been enough to demolish us. The P.L.O. means the national rights of the Palestinian people. The P.L.O. means the unity of the Palestinian people."

The overwhelming Palestinian backing for the P.L.O. has extended well beyond the refugee camps and guerrilla garrisons. Israeli policies like the es-

tablishment of Jewish settlements in the occupied West Bank and tough security controls have managed to push local Arab residents into the arms of the P.L.O. "As long as the Israelis offer us nothing, why should anyone consider a split with the P.L.O.?" Aziz Shehade, a Palestinian lawyer in Al Bireh, asked me. "The P.L.O. has lately gained more support on the West Bank because they have expressed more reasonableness. There's no longer any difference between moderates and the P.L.O."

There are signs that the P.L.O. may even be speaking for many Arabs within Israel itself. A survey released in June by Sami Samouh of the Jewish-Arab Center of Haifa University reported that half of the Israeli Arabs still did not recognize Israel's right to exist and that a clear majority favored the establishment of an independent Palestinian state.

The P.L.O.'s appeal is most evident among young Palestinians living under the Israelis. Nafez Nazzal, a Palestinian-American who taught at Bir Zeit University on the West Bank before the Israelis closed it, told me last summer that "99.9 percent" of his students supported the P.L.O. "To them, the P.L.O. is their dignity, their identity, their loyalty. The P.L.O. achieved international recognition of the Palestinian people."

Starting with the Arab summit at Rabat in 1974, the Arab world has formally recognized the P.L.O. as the only legitimate representative of the Palestinian people and accorded the organization full membership in the Arab League. Even Mr. Sadat, for all his differences with the P.L.O., has not renounced the Rabat decision.

A Jordanian official in Amman explained why his country had rejected an invitation from Camp David to help supervise a transition to self-government in the West Bank. "We agreed in Rabat not to represent the Palestinians and to accept the P.L.O. as their sole representative. Therefore our position is not to talk for the Palestinians but to talk in support of any position that they take. We canot consider ourselves a party to negotiations without the P.L.O. being represented," he said.

Even some Israeli politicians have conceded the significance of the P.L.O. Foreign Minister Moshe Dayan created a flap last February when he said of the P.L.O. that "we cannot deny their position or their value" in the Middle East. The next day, Mr. Dayan explained that he was not advocating negotiation but said that "it is impossible to ignore the position of the P.L.O. in the conflict and its influence on the Arab states."

The P.L.O. may be becoming as adept at diplomacy as it is at terrorism, for it scored some political coups this summer in Western Europe. To Israel's outrage, the Austrian Chancellor, Bruno Kreisky, and the former West German Chancellor Willy Brandt sat down and talked with Mr. Arafat in Vienna in July. At a press conference later, Mr. Brandt defended their meetings by contending that "Palestinian self-determination cannot be thrown into the same pot as seeking the de-

Zehdi Labib Terzi (left), the P.L.O.'s U.N. observer, gets support from black leaders, including the Rev. Joseph E. Lowery (center), president of the Southern Christian Leadership Conference.

struction of the state of Israel." Mr. Kreisky told reporters that "the situation in the Middle East is no longer a matter of Israel and Arabs only. It is a question of world peace."

That same month in Paris, the French Foreign Minister, Jean Francois-Poncet, received Farouk Kaddoumi, head of the P.L.O.'s political department. Jurgen Möllemann, a key aide to the West German Foreign Minister, Hans Dietrich Genscher, visited Mr. Arafat in Beirut. The impact of such meetings has been largely psychological, since none of the countries represented has formally recognized the P.L.O. Switzerland discouraged a similar call from Mr. Khaddoumi after protests from local Jews. But such events suggest that there is a declining tolerance in the world for the the American position. After Andrew Young resigned, he admitted that he found it "kind of ridiculous" not to talk to the P.L.O., which enjoys observer status at the United Nations.

□

Other developments in the Middle East have whetted the P.L.O.'s self-confidence. Recognition by the new regime in Iran, which welcomed Mr. Arafat as its first official guest, has given tha P.L.O. a maneuverability beyond the limits of traditional Arab patrons like Syria and Iraq. The P.L.O.'s financial status has been buttressed by an allocation of $150 million annually

over the next 10 years from the Baghdad conferences of Arab states opposing the Egyptian-Israeli peace treaty.

Ironically, Israel's military strikes also seem to have benefited the P.L.O., which showed that it could withstand massive search-and-destroy missions during the Israeli invasion of southern Lebanon in March 1978. Previously, the P.L.O.'s image had sagged because of the battering suffered in the Lebanese civil war and after being upstaged by Mr. Sadat's peace initiative. According to reports in Beirut, there are now a third more Palestinian guerrillas and weapons in Lebanon than before the Israeli invasion.

Probably the most alarming case made for dealing with the P.L.O. is that it will forestall any cutoff in Arab oil supplies to the United States. This summer, Saudi Arabia boosted production by one million barrels daily to 9.5 million barrels to offset cutbacks elsewhere in the Arab world. According to reports given credence in Cairo diplomatic quarters, the Saudis did so as part of a carrot-and-stick strategy worked out in consultations with the P.L.O. to make the Americans confront the Palestinian issue more directly.

Consequently, any cutback in oil from Saudi Arabia is bound to be interpreted as blackmail. But the Arab viewpoint is that the United States has been dodging the realities of the Middle East too long and must be shown that its real

self-interest lies beyond Israel's frontiers. "Other Arabs give us the benefit of the doubt but they don't think we're effective. They think we get cold feet and run up the white flag whenever the Israelis complain," said an American official.

It has been argued that failure to talk to the P.L.O. because of its philosophy perpetuates an inconsistency in American foreign policy. Thus, on the other hand, the United States has cultivated its ties with the Soviet Union to encourage a minimum level of civilized behavior from the Kremlin leadership. In a more analogous situation, the Carter Administration has tried to moderate betweeen the white-dominated regime in Zimbabwe Rhodesia and the pro-Marxist African guerrillas committed to toppling it. It has tried to defuse the confrontation between Morocco and the Polisario Front in the Western Sahara. "Almost every national liberation group is welcome in the United States except the P.L.O.," pointed out an American diplomat.

On the eve of the Egyptian-Israeli peace treaty, President Carter said that Washington wanted "direct relations" with the Palestinians but admitted that he had a "problem" with the P.L.O. However, Mr. Carter said, if the P.L.O. dropped its opposition to Security Council Resolution 242 and acknowledged Israel's existence, the United States would "immediately start

working directly with that organization."

Were the Carter Administration to soften this precondition and offer a dialogue, the prospect of American recognition would, diplomats expect, touch off a possibly violent power struggle within the P.L.O. over how moderate the response should be. "There would be splits, backbiting and assassinations, but what would emerge would be smaller and more cohesive," said a Western observer.

It remains to be seen whether Mr. Arafat could successfully reconcile the conflicting interests of four million Palestinians. The biggest opposition to a negotiated settlement would probably come from original refugees of 1948 who cling to the dream of return. The "rejection front" within the P.L.O., led by George Habash of the Popular Front for the Liberation of Palestine, would undoubtedly try to sabotage it. But, said a Palestinian official in the Jordanian Government, "I think the mainstream is prepared to accept the idea of recognizing Israel if they can have their own state."

Some Western experts contend that it is in the American interest to deal with the P.L.O. now, while the moderate faction led by Mr. Arafat is in charge. His group, Al Fatah, is believed to represent three-quarters of the guerrilla forces. The rising young generation of leaders is known to be more radical and militant.

In the end, the overriding argument for talking to the P.L.O. may be its studiously enigmatic nature itself. Too often, it has successfully obscured its intentions behind a smokescreen of ambiguous rhetoric. Would it ever be willing to let Israel live undisturbed in return for a viable Palestinian homeland? Is there a sufficient wellspring of moderation within the P.L.O. that it can be transformed into an honest negotiating partner and advance the Palestinian cause through a peaceful settlement? The questions should be intended not to bring the P.L.O. to heel but to seek a broader Arab acceptance of Israel's legitimacy together with overdue remedies for the sufferings of the Palestinian people. As Andrew Young put it, "Conversation does not mean recognition," and it need not begin as such. But such a dialogue seems the only way of finding out whether the United States and the Palestine Liberation Organization have anything in common to contribute toward a fair and durable peace. ■

September 9, 1979

INSIDE IRAN'S CULTURAL REVOLUTION

By Youssef M. Ibrahim

It was 4 o'clock in the morning, but hundreds of men and women crowded the stark, gray departure terminal of the Mehrabad Airport, waiting for the call to board their flight out of Iran. The young Iranian banker tried desperately not to show his nervousness. He studiously avoided the eyes of the fearsome, bearded Pasdars, the Islamic Revolutionary Guards, who roamed the terminal. In his khaki trousers and white shirt, a day-old stubble covering his face, he blended well with the crowd. His Pierre Cardin suit had been left at home.

The banker already had his passport in hand, with his exit visa stamped in it — no mean feat. Every day, at Mehrabad, Iranians seeking to leave the country have their passports confiscated "subject to further inquiries." As the banker put it, "You're never really sure until the plane takes off."

He had made it through the luggage search unscathed. Now he watched the

Youssef M. Ibrahim, a New York Times correspondent, was expelled from Iran last July because the regime was "unhappy" with his reporting.

Pasdars pulling carpets from the bags of other passengers, caught by a Government ban on carpet exports just announced that day. At last, the young banker was summoned for the final search in the curtained rooms leading to the exit door.

"How much money do you have?" one of the bearded men asked. "Three thousand dollars," the banker replied. "No," the Pasdar insisted, "try again." "Just the $3,000 we are allowed, I swear." The Pasdar contemplated him for a moment, then said, "O.K., you can go."

As the Pan Am Boeing 747 lifted off the tarmac in the dawn twilight, the banker glanced across the aisle at an acquaintance and raised a glass of orange juice in a toast. "Bye-bye, Ayatollah Khomeini," he whispered.

☐

For the thousands of technocrats, doctors, engineers, economists and teachers who have gone the same route as the young banker, before and since, the decision to leave their homeland has not been easy.

Most of them had been eager supporters of the revolution of last February that ended the dictatorial rule of Shah Mohammed Riza Pahlevi. Many, in fact, had returned to Iran just before or immediately after the revolution,

ready to serve in the Islamic Republic of Ayatollah Ruhallah Mussawi al-Khomeini, moved by the dream of a new age of freedom.

But within a few months, it became clear that Iran was on the verge of a new and different dictatorship. A puritanical, totalitarian regime presided over by the Muslim priests of the Shiite clergy, the mullahs, was tightening its grip on this strategic oil nation, sweeping away all forms of opposition, formulating new laws and rules designed to mold Iran into a theocratic state.

The young banker and those others who have taken flight leave behind them a nation where political opponents of the new regime are executed, where dissenting publications are banned, where music is forbidden, women's legal rights are disappearing and unemployment is soaring.

☐

The Parliament of Iran, a rubber-stamp debating society under the Shah, used to meet in a marble-faced building on Teheran's south side. At each corner, brightly garbed guards stood at attention. But on Aug. 19, when an assembly of experts — most of them elderly priests — gathered there to draft a new Constitution for Iran, the uniformed guards had been replaced by the ragtag militia, including a complement of teen-agers carelessly fiddling with Israeli-made Uzi machine guns.

Inside, beneath the mirrored, vaulted ceiling, the opening session began with the invocation of the name of God and his Prophet Mohammed and a long recitation of verses from the Koran by a young religious student. Then came a message from the Ayatollah that clearly pointed the way:

"You are here to create a Constitution that is 100 percent Islamic. Not a single clause, nor a single phrase, can be devoid of the Islamic spirit. . . . Only the religious leaders, some of whom, thank God, are among you, are competent to decide what is for Islam and what is against it."

Because of the secretive nature of the Iranian religious establishment and of the powerful Revolutionary Council that formulates all decisions in Iran, the precise workings of the future "Islamic Republic" are not public knowledge. But its general outlines can be pieced together from the moves already made by the Government and from interviews with members of Prime Minister Mehdi Bazargan's provisional Government, leading clergymen and Islamic ideologues.

The prospect is clear: The Koran will be the blueprint of the future. The mosque will become the center of national life. The mullahs will lead the country's sociological transformation.

In the matter of law, for example, the civil code that has been the basis of the judicial system since the Constitution of 1906 will be replaced. Justice will be based on the Sharia, the Islamic law of the Koran. The mullahs will serve as prosecutors, judges and jury. There will be no appeal system; all judgments will be final and all punishment will be rendered swiftly.

The impact has already been felt. The Ministry of Justice has not functioned since the February revolution; it will be scrapped and all civilian judges will be retired. Women may no longer serve as judges. Several religious leaders have declared that no women will be admitted to law schools and that those now in school will not be graduated. For that matter, no civilian lawyers — male or female — will play any significant role in the judicial system of the future.

For the new leadership of Iran, Islam is not simply a religion but a way of life, a complete and practical code of conduct. Thus, for example, the Ayatollah condemns the Western judicial system not only because it is in opposition to his religion but because it is inefficient and expensive: "The case on which the Sharia judge used to make a decision in two or three days," he wrote while in exile, "now takes 20 years to settle."

The Koran deals with every aspect of life, from taxation to defense to hygiene. And its wisdom is most available, the new leaders believe, to those who have devoted their lives to the study of the Koran, the mullahs.

Since the February revolution, Islamic adjudication of disputes has become a fact of everyday life in many towns and villages, the mullahs serving as arbiters between feuding merchants, between landlords and farmers, between husbands and wives. The decisions of the mullahs are final, and the sentences — which are clearly laid out in the Sharia — are not subject to qualifications or mitigating circumstances. A merchant found to be cheating his customers is given 50 to 80 lashes. One who drinks alcohol, which is against the tenets of the religion, receives 80 lashes. Unmarried adulterers receive 100 lashes; those who are married are condemned to death.

The most severe punishment is primarily reserved, however, for political crimes; opposition to the Islamic regime is considered to be opposition to Islam. A leading figure in the Islamic establishment, Seyyed Mohammed Beheshti, one of the founders of the Islamic Republican Party — increasingly referred to as the Party of God — explained the role of the revolutionary courts this way in a recent interview in a Teheran weekly:

"The reasons for holding each trial are so clear to both the public and the court that there is very little need for questions, and if some questioning and answering does take place, it is simply to further clarify the situation."

Among those who are executed are "counterrevolutionaries," officials of the previous regime, and Kurdish and Arab ethnic rebels who are demanding a degree of autonomy. They are charged with "sowing corruption on earth" or "warring with God and his emissaries."

The rule of Islam is being felt in the everyday life of the people. Iranians today may no longer purchase, possess or consume liquor or pork. In August, the consumption of caviar, Iran's most famous product other than crude oil, joined the proscribed list. The sturgeon that lays the eggs is considered *najas*, filthy.

Men and women may no longer swim on the same beaches. Soon boys and girls will be segregated in schools through all phases of education, including university. Religious training will be compulsory for all students. The voices of women singers have been banned from the airwaves, and a few weeks ago Ayatollah Khomeini declared all music a "corrupting influence" upon the nation's youth. For a land whose literature and art are steeped in the sensuality of wine, music and dance, such changes are deeply felt.

The new regime has been moving rapidly to alter the status of women. The Family Protection Act of 1975, a mild attempt to protect women against arbitrary divorce and polygamy, has been nullified. The clergy initially insisted that women use scarves to cover their heads, or the head-to-toe veils called chadors, to envelop themselves; protests led the leadership to back down, but this conciliatory attitude may be only temporary.

The basic view of the clerical establishment toward women is that their most important role is that of raising a family, and being "the pillar of the Moslem home." In fact,

Parivash Khajehnouri, an attorney and a women's rights activist, warns that Article XII of the new Constitution paves the way for confining women to child care.

□

Nahid Asghar — it is not her real name — is a 28-year-old advertising-agency executive, a dark, slim woman who finds the new Iran very different from what she had imagined. "We have waited for this revolution for so long," she said sadly, "like a child waiting for a precious gift."

It was March 10, and Asghar was sitting in her home in east Teheran, applying a bag of ice to her swollen forehead. She had been hit by a stone while marching with 10,000 women that morning in the protest against the Ayatollah's edict against Western clothing.

The marchers had gathered on a muddy soccer field at Teheran's university. Most of them were dressed in jeans and sweaters, though some even showed up in chadors. Sandbags still littered the area, mementos of the uprising against the Shah when the campus was patrolled by leftist students in support of the revolution; on this day, roving bands of young men, supporters of Ayatollah Khomeini, hurled insults at the protesting women. When the march began, insults became rocks. Asghar had been struck and suffered a surface wound.

Now, as she sat in her living room, watching a pot of tea brewing, she remembered another demonstration, last December, when she had marched in the streets of Teheran chanting "Independence! Freedom! Islamic Republic!" Along with literally millions of others, Asghar — a middle-class woman, married and divorced, who had spent six years studying in India — had come out in response to a call from Khomeini. She had known little about the exiled old man who had been preaching revolt for 15 years, but what she'd heard sounded good.

"They fooled us," she said now, only a few short months later, shaking her head as she poured the tea. "This is no revolution. It's a mullah's game. It's nothing more than a mullah's game."

□

During the years of the Shah's rule, organized opposition was effectively squelched, but the mullahs in their speeches to the faithful continued to speak out in protest. Seventy percent of the people of Iran are illiterate; to most of them, the leadership of the clergy is a matter of faith. After years of cultivating these masses — through an elaborate network of 200,000 mullahs spread across the nation, through thousands of mosques, and through the widespread distribution of taped messages reproduced on cassettes — the Ayatollah had brought them into the streets.

The departure of the Shah brought the Shiite clergy to power. They had waited a long time to root out what they saw as the corrupt way of life that had undermined all the noble values of their religion, and they felt they were prepared.

Young students of theology are carefully selected and groomed. In the Faizeyah Religious School in Qum, where Ayatollah Khomeini taught for years, they are encouraged to argue and debate with their teachers. Only those who stand out are promoted to higher ranks. The mullahs are taught the arts of speech, and most of them are engaging preachers. But they are also prepared to be community leaders. After leaving school, most lead simple lives devoted to their congregations, collecting money from the rich merchants of the bazaars, who are among their most fervent supporters, and distributing it to the poor.

In recent years, the mullahs have also recruited a corps of Islamic technocrats, intellectuals and ideologues who, in the Islamic Republic, are the only nonclerical persons allowed to hold any position of power. Many of them have been educated in the West, where they kept their distance from Western culture and nurtured their fervor in tightly knit Islamic societies.

One of the principal figures among the nonclerical Islamic ideologues is Foreign Minister Ibrahim Yazdi, a naturalized Iranian who studied and lived in the United States for 17 years, before joining Ayatollah Khomeini in Paris. Others include the chief economic adviser to the regime Abdul Hassan Banisadr and Sadegh Ghotbzadeh, the director of radio and television, both have lived in France.

For such leaders, the political doctrines of foreign lands are anathema. Capitalism and communism have both been condemned as the "creed of Satan" by Ayatollah Khomeini.

In an interview in late February, Foreign Minister Yazdi rejected both the secularism of the West and the atheism of the East: "Neither of these two is correct or applicable to Iran. There is only room for Islam. Islam is not an opiate of the people, and in Islam there is no room for the separation of church and state."

In a fiery speech on Aug. 17 in which he damned all the "corrupt intellectuals and the poisoned pens of conspiring writers and democrats," Ayatollah Khomeini declared that "there can only be one party in Iran. The Party of the Disinherited, the Party of God." He warned all intellectuals who have other thoughts in mind to "take the right way of Islam," or face the fate of all counterrevolutionaries. In revolutionary Iran, that usually means the death sentence.

There is little question about the anti-intellectual penchant of the Islamic regime. In his treatise on Islamic government, Ayatollah Khomeini stated his view on freedom of thought very simply: "There is no place for opinions or whims in the government of Islam." And that view has rapidly been put into effect. After some months of tolerating a freewheeling press and the airing of opinions by leftists and secularists, the Government last August shut down dozens of publications. All leftist parties, including the Tudeh, the Iranian Communist party, have been banned.

□

The rejection of all non-Islamic values has also been extended to the realm of economics. There is a profound belief among the theoreticians of the revolution that the industrial structure created by the Shah over the past 20 years, as well as the economic directions set by the old regime, are simply pretentious manifestations of the need to mimic the West, a "cardboard industrial structure" that should be allowed to collapse. These industries, they feel, most of them based on the assembly of consumer goods imported from the West, were meant only to make profits for the corrupt by encouraging undesirable habits of consumer consumption.

Last spring, the 37-year-old Deputy Prime Minister of Iran, Hossein Bani-Assadi, commented: "These policies made a little Shah out of every Iranian. Look at Iran today. The major concern of every Iranian is the pursuit of materialism. Those who do not have a car or a house think of nothing else but cars and houses. Those who have them, want bigger cars or bigger houses. Iranian Moslems have forgotten the values of their religion. Modesty, contentment, integrity, kindness and fairness to others have been replaced with greed. We have to change all this. This revolution has just begun."

A prime example of approved economic behavior is provided by Ayatollah Khomeini, himself. He lives in a bare house in the religious city of Qum, south of Teheran, sleeping on a mattress on the floor and eating simply.

Yet, the Islamic Republic is not a socialist venture. Wealth, as long as it is "rightfully earned," is protected. All forms of exploitation — interest on loans, for example — must be eliminated. The guiding principle, outlined by the regime's chief economic spokesman, Abdul Hassan Banisadr: "In Islam, only God is the real owner. Consequently, man is only the owner of his labor, his work, his effort. Islam regards every kind of position, advantage, wealth, acquired by force to be unlawful. In other words, it acknowledges only one lawful criterion and that is work."

It remains to be seen how the new Islamic economy will perform, but a general direction is now emerging from the few economic decisions already made. For one thing, all banks and insurance companies have been nationalized. The properties of the top 51 industrialist families in Iran have been confiscated as "unjustly accumulated wealth based on theft, bribery and the exploitation of others."

Iran has cut its oil production from 6.5 million barrels a day down to an average of three million barrels a day. The cutback was not only motivated by the need to preserve oil as a natural resource, but also because, as Ali Akbar Moinfar, Budget and Planning Minister at the time and last month named Minister of Petroleum, put it, Iran "does not need nor do we want to be an industrial power at any cost."

Gone are the grandiose plans to build 20 nuclear power generation stations, superhighways, a subway system for Teheran, a gas pipeline to Russia and Europe. The new motto, according to Moinfar, is: 'Smaller is better." In fact, the first budget submitted by the new regime for the Iranian calendar year that started last March 21 totaled $35 billion — a third less than the previous year's budget.

The new regime's decisions have already altered the profile of Iran. All around the country, and particularly in Teheran, hundreds of idle cranes stand out like scarecrows in the middle of abandoned construction sites.

Nearly a million construction workers have lost their jobs as the Government canceled the major projects that kept them employed.

At large factories, Islamic committees have either arrested or fired most members of management. Purges of employees deemed untrustworthy are going on every day in Government ministries, depriving them of many of the people who used to run the economy.

On the Kharaj Road outside Teheran, the big showrooms for American and other multinational companies stand empty, collecting dust. In Teheran, the shops are running out of imported furniture, foods and other goods. Essential foods are available, but once-crowded supermarket shelves have become bare.

With the ban on the import of cars, the few showrooms in town still exhibiting Mercedes and Cadillacs are torn between the need to sell their goods and the wisdom of keeping a low profile. They keep their shutters half-lowered and their showrooms dim. They are not likely to sell the cars anyway, because the sort of customers who can buy them have left Iran.

In a wide-ranging conversation at his modest home in northern Teheran last spring, Deputy Prime Minister Bani-Assadi explained that some of the plans to "reshape" Iranian society included cutbacks and rationing of gasoline "to stop this wave of car buying and the feverish consumption of electricity and gas by the rich." Thus, further cuts in oil exports and production seem likely.

The principal economic priority set by the new regime is the revitalization of agriculture. Officials of the new government point to the fact that 20 years ago, Iran exported food; today, food is Iran's largest import. In the new budget, the allocation for agriculture has been doubled from the previous year to almost $3 billion.

That figure includes close to $1 billion for the Fund for the Disinherited, which will be channeled to the mullahs in every village. They, in turn, with the help of village elders, will dole it out to the peasants. "In Islam," budget chief Moinfar explained, "we must trust the people. We cannot treat them like children. It is better that they assume responsibility for themselves."

A general decentralization of the economy seems to be the intent of the new Government. The nationalized banks will be reorganized on a regional basis with much greater provincial autonomy. The sorts of industries the new regime will be seeking to promote are local, independent entities based on available raw materials. Of all the multibillion-dollar industrial projects started by the Shah, the only ones now surviving are those which draw on Iran's oil and gas as base material, such as the $3 billion Iranian-Japanese petrochemical venture currently under construction. Car assembly plants, such as the General Motors and Iran National factories outside of Teheran, which are totally dependent on import kits from the United States, Britain, France and Germany, are likely to be restricted to the production of public-transport vehicles instead of private automobiles.

There seems to be little concern among Government leaders about the growth in unemployment that has accompanied these new economic policies. Plants that produced liquors, and employed thousands of workers, have been shut down. Advertising agencies, purveyors of the evil consumer habits the regime detests, have been starved for business. Beauty shops have been banned as "dens of moral corruption." Dozens of hotels, which catered to foreign businessmen in the cities and to rich Iranian tourists on the Caspian beaches, are being turned into orphanages and rehabilitation houses for heroin addicts. The regime does not mind handing out unemployment aid to idled workers if what they produced conflicts with the new values of Islam.

Significant changes have been made in the nation's commercial life. Under the Shah, the banking system had replaced the thriving bazaars as the money centers of Iran; now, the businessmen of the bazaars are making a comeback. In this, they have the blessings of the clergy, whom they have bankrolled throughout the revolution and continue to support. From the clerical point of view, the bazaar is the model of society, where life and business are organized around neighborliness, trust (a handshake seals a bargain) and devotion to the authority of the mosque.

However, this Islamic vision of the economic future leaves many questions unanswered: How can a huge banking system function without collecting interest? Is it feasible to scrap the entire taxation system in favor of an "honor" system whereby every person is trusted to pay 20 percent of his earnings to the poor, as decreed in the Sharia? How is Iran going to make up the loss of thousands of talented managers and entrepreneurs who have left the country? How long can Iran live with an unemployment rate of three million in a work force of 11 million?

The new regime, notwithstanding its problems and uncertainties, is supported by a powerful militia in the form of the Islamic Republican Guards, those dedicated, almost suicidally fanatic young men recruited by the local mullahs from households loyal to Ayatollah Khomeini. They would die to preserve his rule.

Moreover, the clergy's power in Iran is becoming more pronounced every day, enhanced by the actual exercise of government.

Yet the regime's immediate prospects are grim. This winter, the Kurdish minority, which fought the Ayatollah's forces all summer before retreating into the mountain hideaways in west Iran, will surely resume its guerrilla warfare. Such struggles could strain the Iranian Army's remaining resources to a breaking point. The Arab minority of Khuzistan is also likely to rise once more, with the attending risk of sabotage of the oilfields. Meanwhile, the Ayatollah's clergymen have been encouraging subversion by Shiites in Kuwait and in Iraq, and they have been renewing the claim to Bahrain that the Shah had made and then renounced. Yet these neighboring Arab nations are precisely those in a position to provide weapons to a rebellious Arab minority in Iran.

Within Iran, the disappointment with the regime seems bound to increase. Ayatollah Khomeini has promised the *muztazafin*, the destitute masses, free water, electricity and housing. But it seems unlikely that the Shiite leader will be able to keep those promises by turning off the oil tap, scaling down economic development, driving the educated elite into exile and generally shutting out the world. Man, the Ayatollah's enemies say, cannot live by faith alone. ∎

October 14, 1979

Standoff at Embassy Threatens U.S. Security, Regional Stability

By RICHARD BURT

WASHINGTON — It began on Nov. 4 as a test of nerves between the United States and an aging, Iranian holy man, Ayatollah Ruhollah Khomeini. But after three weeks of stalemate over the fate of Americans held hostage at the embassy in Teheran, Washington's travail was transformed into a full-blown regional crisis, threatening chaos throughout the Middle East and undermining Washington's strategic position in that vital area.

In that edgy atmosphere, the unrelated seizure of the Grand Mosque in Saudi Arabia's holy city of Mecca, which triggered a lethal mob attack against the United States Embassy in Islamabad, Pakistan, only substantiated the belief of many analysts that the world's most important oil-producing area was indeed caught up in an "arc of crisis."

While President Carter's primary concern still was to secure the release of the 49 hostages in Teheran, security planners were also deeply concerned about widening implications for stability in the Islamic world and for Washington's influence in the Middle East. "Lots of things begin to unravel, the longer the crisis goes on," a National Security Council aide said at the White House.

According to these experts, the Administration has been forced to focus more acutely on the repercussions of the 11-month-old revolution in Iran: the upsurge in religious fundamentalism in Moslem nations; Moscow's opportunities to expand its influence in the Middle East and South Asia; risks of political instability in the Persian Gulf, Pakistan and Turkey and the doubts raised — particularly in Saudi Arabia — about American commitments.

Officials saw an interesting contrast with the Jordanian crisis in 1970, when Palestinian terrorists hijacked several airliners and took 100 American and European hostages. While the Nixon Administration and other Western governments negotiated for the hostages, that incident widened as Palestinians, aided by the Syrian Army, sought to overthrow Jordan's ruler, King Hussein. Finally moving against the Palestinians, the King, with American military backing, obtained release of the hostages and restored order in Jordan. The net result, analysts such as Henry A. Kissinger contend, was to bolster moderate forces in the Middle East and to enhance American prestige.

Nobody expects moderate forces to emerge during or after the Iran events. Instead, many experts believe that Ayatollah Khomeini has exploited anti-Americanism as a tool for sustaining the fervor of the Iranian revolution. Since taking power 10 months ago, they note, the Ayatollah and his followers have fought off political challenges from well-organized leftist groups in Teheran and from separatist movements, principally the Kurds, around Iran's periphery. Some specialists believe that the Ayatollah, fearing the fragmentation of the nation, has seized the political initiative in Iran by turning an act of terrorism against the United States into a national crusade.

Officials are deeply concerned that the Ayatollah's efforts to revive his revolution have reduced Iran's freedom of action in negotiations for release of the hostages. They are distressed as well by potential repercussions of the Ayatollah's crusade on the stability of pro-Western Islamic countries such as Turkey and Pakistan. Nevertheless, they are confident, for the short run at least, that the Turkish and Pakistani Governments can control the situation.

Prompting even greater worries, some officials said, is the high probability that Iran eventually will slide into ever-increasing turmoil. A breakdown of central authority in Teheran would heighten chances for separatist violence, almost surely involving neighboring Iraq, Turkey and Pakistan. It could also jeopardize Western European and Japanese access to Iranian oil and, according to some Pentagon analysts, increase risks of Soviet intervention in Iran.

The Soviet reaction to the Iranian crisis has been ambiguous. Moscow has supported Washington's call for the release of the hostages in statements and at the United Nations. Recently a broadcast beamed into Iran from a radio station in the southern Soviet Union called for release of the hostages, but earlier the same station called the hostage-taking "understandable." White House aides note that East Germany, which often speaks for Moscow, expressed its full support earlier this week for students holding the hostages.

Most analysts believe that Moscow's interests are best served by staying in the background while the rhetoric exchanged by Washington and Teheran grows hotter. "Right now, the Soviets wouldn't dream of running afoul of Islamic nationalism by intervening," one Pentagon analyst said. "But if the situation deteriorated, their incentives to move in would grow." Soviet military power provides an instrument for quietly deterring American retaliatory strikes against Iran, during or after the crisis. But while the likelihood of Soviet military action remained small, one senior official suggested that if the turmoil worsened, Moscow might get "greedy," throwing its weight behind a drive to replace the Ayatollah's mullahs with a more disciplined, leftist government. A pro-Marxist takeover obviously would constitute a heavy strategic blow to the West, bolstering Soviet influence with radical Middle Eastern governments while putting new pressures on Pakistan, Turkey, Egypt and Israel.

For now, however, Saudi Arabia is the chief concern for American planners. Even short of a Soviet success in Iran, the new anxiety around the Persian Gulf has fed longstanding Saudi fears over the fragility of their monarchy. Their enormous, increasing wealth, ties to Washington and the presence of millions of foreign workers from neighboring countries have created staggering security problems. Saudi worries over the growing Soviet influence in Iran have been reinforced by Moscow's expanding presence in North and South Yemen, on the tip of the Arabian Peninsula. While maintaining close ties with the Marxist Government in South Yemen, the Soviet Union reportedly has just completed a large arms deal with the Government of President Mohammed Mottee'e in North Yemen. The deal, which includes Soviet MIG-21 aircraft, tanks and antiaircraft missiles, was said by a State Department aide to have caught Washington by surprise and to have deeply disturbed the Saudis. It was an effort to reassure the Saudis, as much as an attempt to put pressure on Iran, that led Mr. Carter last week to send a second aircraft carrier into the Arabian Sea and to hint that military force might be used.

If and when the hostages are freed, political pressure on President Carter to retaliate against Iran would be enormous. However, retaliation — strong trade sanctions or punitive military strikes — would probably risk still-worse turmoil, in Iran and outside, most experts contend. For a President seeking re-election, after an Administration in which no American had yet died in combat, it seemed clear that deciding whether to use military force against Iran would be especially difficult.

Pentagon officials said that the Iranian crisis had led the Administration to begin "serious contingency planning" for intervening in a Persian Gulf war. A senior official declined to discuss details, but he said, "we would be remiss if we didn't start thinking about these things, don't you think?"

November 25, 1979

CHAPTER **12**
The Cities

Charlotte Street, in New York's South Bronx, is a national symbol of urban blight.

United Press International

THE NEW ELITE AND AN URBAN RENAISSANCE

By Blake Fleetwood

People often snicker when they first hear of it. A renaissance in New York City? The rich moving in and the poor moving out? The mind boggles at the very notion. After all, what about the graffiti, the abandoned buildings, the

REDISCOVERING THE CITY

chronic fiscal crisis? Hard as it is to believe, however, New York and other cities in the American Northeast are beginning to enjoy a revival as they undergo a gradual process known by the curious name of "gentrification" — a term coined by the displaced English poor and subsequently adopted by urban experts to describe the move-

ments of social classes in and around London.

In most of the major cities in Europe, gentrification has been in full swing for some time. Affluent young professional people — the 20th-century gentry — have been moving into previously deteriorating city centers and driving out the working class and the poor. And

now it is becoming apparent that America's oldest cities, not only New York but Boston, Baltimore, Philadelphia and Washington, are also beginning to be resettled — an ironic twist to the blockbusting that turned many urban areas black and Hispanic in the 1950's and 60's. Indeed, the evidence of the late 70's suggests that the New York

of the 80's and 90's will no longer be a magnet for the poor and the homeless, but a city primarily for the ambitious and educated — an urban elite.

The signs are all around us. Young professional people are flocking into New York City. Rents are higher

Blake Fleetwood is a freelance writer.

271

REDISCOVERING THE CITY

than ever before, and vacant apartments are hard to come by. The price of a cooperative apartment has doubled in the last three years. Brownstoners have renovated large sections of the West Side and Chelsea in Manhattan and Park Slope in Brooklyn. It will take a long time for the effects of gentrification to make themselves felt in areas such as Harlem and the South Bronx — if indeed they ever will. At the moment the revival of New York City is by and large restricted to Manhattan below 96th Street. But some of the most ravaged areas in Washington, D.C. have already been reborn. And as the heart of New York revives, it may pump new life into old, outlying neighborhoods.

A disco; (above, right) pâté at Zabar's.

Lining up for theater tickets in Times Square.

Sampling new cosmetics.

Making the rounds of the boutiques.

West Side antique store.

A Rolls-Royce cruises Madison Avenue.

Steam-cleaning the Lyric Theater's facade.

A West Side bistro.

On line to get into Studio 54.

A salesperson behind a counter at Fiorucci's.

The bustling Museum Cafe on the West Side.

Sports car on Columbus.

In the atrium of the Citicorp Center.

Enjoying a night on the town.

For, as rents rise still higher and the prices of co-ops and brownstones continue to go up, the gentry will be forced to move into and upgrade marginal areas.

Merchants who cater to this trend-setting class are cashing in on the boom. In six years, Bloomingdale's has doubled its volume of sales. "We're the busiest we've ever been," says Jerry Berns, one of the owners of the "21" Club, "serving 8 to 10 percent more bodies than last year." European boutiques, gourmet shops, chic restaurants, sidewalk cafes and bookstores are popping up all over town — and thriving. "I was very unsure when we opened last August," confesses Burt Britton, co-owner of Books & Co., a new bookstore on Madison Avenue. "I didn't think anyone cared. But in two days, half of my stock was gone!"

New York may never woo back the middle class that fled to the suburbs after the Second World War, but it seems to be attracting a new professional upper class, an achievement-oriented gentry. As former New York City Housing Commissioner Roger Starr explains, "It really is gentrification rather than bourgeoisification. We are skipping a whole group of middle-class business people. The people who are moving in are professionals: young lawyers, architects, doctors, people in the investment community. Blue-collar

The new landmark: The World Trade Center.   Window-shopping at Gucci's on Fifth Avenue.   Living it up into the wee hours at a midtown discothèque.

A mural in SoHo's W.P.A. Restaurant.   At Elizabeth Arden's.   A renovated brownstone; (above, right) Fifth Avenue.

Crowd at the New York State Theater.   Passing a hair stylist's window display.   Promenaders; (above, right) Grand Central's new dazzle.

Eye-catchers on East 59th Street.   Dustin Hoffman shooting a film on the East Side; (above) a natural-food store.

workers are moving out because they have less social distance from the poor, whereas the professional person has eminent social distance. He's willing to put up with a reduced physical distance, as it were." And to these American-born professionals are being added increasing numbers of affluent foreigners who are accelerating the pace of gentrification.

The Exodus From the City

For years now, social scientists and urban specialists have been writing off the cities, predicting that soon our urban centers would be inhabited exclusively by poor minorities. Dr. E. A. Gutman, for example, a professor of urban studies, called today's city centers "the cancer of urban existence" and contended that we would be better off simply forgetting them and starting from scratch elsewhere.

Until recently, most of the evidence supported such a dim view. Since World War II, New York City has lost some two million white residents. This drop has been largely offset by the substantial immigration of blacks and Puerto Ricans. Unfortunately, these newcomers arrived when the city's manufacturing base, historically the greatest source of employment for new migrants with limited skills, was in serious decline. (Between 1950 and 1970, New York City lost nearly 300,000 jobs in goods-producing industries.) This massive influx of poor people unable to find work strained city services to the breaking point.

Faced with similar problems, many other cities around the world instituted policies that kept the newly arrived poor from strangling their central cores. Housing projects and factories were built on the urban peripheries. But the United States embarked on what some critics consider in retrospect to have been a deliberate policy to kill off its older cities. Federal money subsidized low-cost housing for the poor in urban centers, far from the higher-paying factory jobs that were seeping to the suburbs. Moreover, Government agencies concentrated most of their social programs — medical care, employment-opportunity projects, welfare and free legal services — in the cities. No wonder the poor flocked to them. As one politician put it, "During the 1960's, we seemed to be actively competing for the poor in order to qualify for Federal grants."

At the same time, Federal and local governments were encouraging in almost every way possible the flight to the suburbs of the middle and upper-middle classes. Federally subsidized low-interest housing loans, superhighways and cheap gasoline made the exodus possible, and no efforts were made to break down the artificial political boundaries between the wealthy "bedroom" towns and the cities — boundaries that in most other parts of the world were being knocked down. As John Kenneth Galbraith once said, "It's outrageous that a person can avoid income tax by moving to New

Jersey or Connecticut. Fiscal funkholes are what the suburbs are."

But just as the prospects for New York City looked bleakest, a dramatic statistical change occurred. From 1970 to 1976, the population of New York City, which had remained stable for two decades (while that of cities like St. Louis, Boston and Philadelphia was declining by 20 percent), plunged by nearly half a million people. The loss was not attributable to white flight. True, whites continued to leave the inner city, but at a slightly slower rate than in the 1960's. The primary reason for New York's plummeting population was that blacks, who had been streaming into the city at the rate of 32,000 per year, suddenly started leaving at an annual rate of 25,000. The rate of black flight in the 70's was greater even than that of white flight during the 60's. And during this same period, some 69,000 Puerto Ricans also left the city.

Much of this exodus could be traced directly to the disappearance of low-

dents who are too impoverished or enfeebled to pick up and leave. But what many people failed to realize during the agonizing period of the early 1970's — when New York seemed to be running out of money, jobs and even people — was that the bloodletting may well have been a natural adjustment to the problems that had been accumulating for the previous three decades.

A New Kind Of Industry: Ideas

Today, four years after the doomsayers delivered their eulogies during the worst moments of the fiscal crisis, New York City seems to be inching its way toward recovery. Its budgetary problems are still enormous, but the city's spirit has rallied and the business climate has taken a definite turn for the better. There seems to be a general perception that even though the fiscal crisis is far from over, at least the city

---

The young gentry gladly endure the urban indignities their parents ran away from. This new breed of professionals is willing to put up with smaller apartments, dirty streets and crime in order to live in chic neighborhoods.

---

skill jobs, causing black unemployment to rise at a much faster rate than white unemployment. Tightened city welfare requirements played a part, too. But perhaps most important was a feeling among poor blacks and Puerto Ricans that New York was no longer the mecca of opportunity it had once seemed. The booming economy of the Sun Belt held out the promise of more jobs, and welfare benefits improved considerably in Puerto Rico.

Census Bureau statistics indicate that black migration out of the cities is not limited to New York. In 1970, 59 percent of the nation's blacks lived in central cities. Seven years later, the figure was down to 55 percent. Fourteen of 15 Congressional districts in the United States represented by blacks have lost population since 1970, and the nation's only predominantly Puerto Rican district, in the South Bronx, has lost nearly a third of its population in eight years. Figures from Washington, D.C., in which blacks are a majority, back up this trend: Two-thirds of the immigrants to the central city in 1974 were white.

The mass exodus has had a devastating effect on many neighborhoods in New York. Vast tracts in Brooklyn and the Bronx have been abandoned to gangs that prey on the remaining resi-

has weathered the worst of it. The turnaround began in 1976 with the Democratic National Convention, Operation Sail and the Bicentennial celebrations, and tangible evidence of a revival came last year, during which private enterprise committed nearly a billion dollars to construction in midtown Manhattan. These projects include the A.T.&T., I.B.M. and Fisher Brothers buildings and the renovation of the Chrysler Building, as well as speculative new office towers and high-rise apartments.

In addition, the last two years have seen tremendous activity in the hotel industry. Occupancy rates in 1978 were 5 percent greater than in 1977; the number of major conventions was up 14 percent. Hilton, Hyatt and Holiday Inn are planning major new hotels in Manhattan. A long-stalled $172 million Times Square hotel project has been revived, and extensive renovations are under way at other hotels.

The hotel business is thriving in large part because New York is enjoying its biggest tourist boom ever. Only a couple of years ago, New York City was the butt of bankruptcy jokes throughout the country. Its image was tarnished further by the widespread looting during the blackout of 1977 and by the lurid publicity attending the "Son of Sam" slayings. But, thanks partly to an ag-

gressive advertising campaign, New York has suddenly become *the* place to visit, and it is attracting not only more American tourists than ever before, but also more foreigners, whose currency buys more because of the weakened dollar. In 1978, New York City played host to some 20 million visitors who contributed almost $2 billion to the city's economy. Tourism is now New York's second-largest job producer (the garment industry is still the biggest), and its growth follows the pattern for historic cities around the world — Paris, Amsterdam, London, Moscow, Rome — which have increasingly relied on tourism as their greatest source of revenue.

The job loss seems to have bottomed out in other sectors, too. In 1977, for the first time in 21 years, New York actually gained some manufacturing jobs. Moreover, employment has increased in banking, real estate, law and medicine. Last year was the first since 1969 during which the number of jobs in the city climbed.

In all, New York City seems to have become a much more desirable place in which to work and live — at least for people with money. Lewis Rudin, one of the city's largest commercial and residential landlords, says, "Our surveys have indicated a dramatic increase in foreigners, out-of-towners and young people coming back to the city. All of a sudden, from having a tremendous number of vacancies in both office space and apartments four years ago, we literally have no vacancies. Rents for some of our office space have more than doubled in three years."

A survey of recent graduates of the foremost law and business schools indicates that top students are eager and willing to come to New York. "Making it in Milwaukee just isn't the same," said a placement officer at Harvard Law School, pointing out that this represented a significant change in attitude. In 1977, Harvard Law School sent 117 young lawyers from a graduating class of 545 to New York — up from the 76 who came here in 1975, the worst year of the city's financial crisis. Now more than ever, New York is attracting the best and the brightest, not only in law but also in international finance, advertising, communications and publishing.

And for good reason. New York is the headquarters for 80 billion-dollar organizations. Of the 14 American companies with assets of $25 billion or more, 10 are based in New York. To be sure, in the last 10 years New York has lost a good number of corporate headquarters from the Fortune magazine list of the nation's 500 biggest industrial concerns. And American Airlines' recent announcement that it would move its headquarters — and 1,200 jobs — to Dallas is definitely a blow to the city. But New York has gained corporate headquarters in fields that are not included in Fortune's industrial list: retailing, diversified financial, banking and insurance companies. And the complex of ancillary services — banks, financial consultants, accounting and law firms, advertising agencies — upon which major corporations depend has actually grown, creating more white-collar jobs.

New York used to be a factory town full of blue-collar workers and immigrants, but it would be folly to try to re-create that bustling younger city. According to Robert F. Wagner Jr., chairman of the City Planning Commission, "If we try, we will be fighting against forces that are so profound that we will be doomed to fail." But New York can become — and indeed is already becoming — a different, more modern metropolis along lines the British urbanologist Peter Hall described in his book "The World Cities":

"At the very center of each world city there is found a small nucleus of highly skilled professionals who live, in one way or another, by creating, processing or exchanging ideas. The stockbroker, considering the fortunes of a hundred companies in a dozen countries; the company lawyer, pondering a difficult piece of patent law; the consultant, considering whether to recommend an operation; the university professor, arguing about urban growth in a seminar; the editor, looking for a specialist to write on the latest trouble spot in Asia; the television producer, discussing a script on the housing problem with a journalist; the advertising copywriter, talking about a campaign with an account executive and then a number of technical specialists; the free-lance photographer, taking varied assignments from a half-dozen editors.

"All these people live only on their ideas. The central business district therefore can be seen as a specialized machine for producing, processing and trading specialized *intelligence*. And the ideas industry is growing many times faster than industry as a whole.' "

What is happening is that the basis of New York City's economy is shifting from manufacturing goods to providing services and generating ideas, and as the nature of the jobs the city has to offer is changing, so is the population the city attracts. The Bureau of Labor Statistics predicts that a million jobs will open up in New York City during the next 10 years, and that three-quarters of these will be in skilled, white-collar occupations.

The Flight From The Suburbs

Many of the people who will fill these new jobs grew up in suburbia, which no longer seems as attractive as it did to their parents. Increased automobile traffic has made commuting a brutal chore, and higher gasoline prices have made it twice as costly. Many of the problems of the cities have moved out to the suburbs. Crime is rising faster in suburban areas than in urban centers. Even suburbia's public schools no longer seem to be living up to their promise. Suburban private schools are thriving because ambitious parents feel they are the best route to the most prestigious universities. In Manhasset, L.I., for example, which spends more than $4,000 per year on each public-school student and has what is regarded as one of the best school systems in the state, 38 percent of the children attend private schools.

At the same time that suburban life is becoming less inviting, it's also becoming much more expensive. One of the most important causes of the movement of the gentry into the central cities is the skyrocketing cost of single-family houses and ever scarcer land. When small suburban ranch houses sell for more than $100,000, it is no wonder that more and more young people are investing in co-ops, condominiums and brownstones in the city. Even the builders of Levittown, that archetypal suburb, recognize the change. "The demand for the house on a little piece of land is changing," says Levitt's vice president Edward Cortese. "Perhaps that is no longer the American dream."

The reasons young professionals are returning to the cities are not simply economic. Another important factor is the revolt of the so-called "baby-boom" generation against the values and life style of its parents. The suburbs are growing stodgy, dominated by old people as their children move to rural Vermont or back to the city that Mom and Dad fled. In 1955, for example, the average resident of Rye, N.Y., was 36 years old; today he is 47. Meanwhile, in many areas of New York — the Upper West Side, the East Side, Greenwich Village, Chelsea, Yorkville, Park Slope — the average age has been dropping dramatically as young professionals have moved into new quarters.

These quarters are adequate because families are smaller now — nationwide, the birthrate has dropped nearly 50 percent in the last 20 years.

Many young couples are opting not to have children — 54 percent of American households now consist of adults only. With fewer children to care for, wives are more likely to hold jobs (a situation that doubles the disadvantages of commuting from the suburbs). And with fewer mouths to feed, working couples have more money to spend on the amenities of urban life, travel and a vacation home. If you have a place in New Hampshire or the Hamptons, who needs Rye?

Moreover, the young gentry are willing to endure the urban indignities their parents could not abide. As John Kenneth Galbraith has observed, "The suburban movement was the response of the older city dwellers to the poverty and indiscipline of the new arrivals. As that shock effect loses its relevance, the superior quality of city life will naturally assert itself." The new generation of professionals is willing to put up with smaller apartments, dirty streets and crime in order to live in newly chic neighborhoods. SoHo still strikes many older suburbanites as grimy; to the new breed of urbanite, it's fashionable. There is an old saying, "The crowd on the boulevard never grows old." For a certain class in this country, Manhattan has become the boulevard, and nowhere else will do.

Housing consultant George Grier conducted a study of a number of these new urban settlers in Washington, D.C. Their average age was 35, and 90 percent of them had annual incomes of more than $20,000. "A lot of these people are ambitious, hard workers," says Mr. Grier. "They can't end their working day at 5 o'clock and commute two and a half hours a day to the far suburbs, because that's not where their interests are."

In a study of 678 suburban railroad commuters to New York City, Dr. George Sternlieb, director of the Center for Urban Policy Research at Rutgers, contrasted those who planned eventually to move to New York City with those who had no such intentions. The suburbanites who hoped to relocate had a median income of $48,000 a year, while those who planned to stay put earned a more modest median annual income of $34,000. Clearly, the city is more attractive to more affluent people.

Contrary to popular opinion, most of the people who migrated to the suburbs in the 1960's — and even those who are leaving the cities today — are not upper-middle-class professionals. Most of them are clerical or blue-collar workers, as Andrew Hacker pointed out in his book "The New Yorkers." According to census figures, even as early as 1970, more professional men were moving into than out of New York.

Reclaiming New York

Part of the reason pessimistic urban experts have failed to recognize the gradual revitalization of our cities is that they have underestimated the enormous influence of these trend-setting gentrifiers. Dr. George Sternlieb notes that "the development of a critical mass of such individuals may become a magnet for others like them." Consider, for instance, the magnetic pull of SoHo. Not long ago, SoHo was a dilapidated industrial area, but as the numbers of young professionals settling into renovated lofts have multiplied, rents have risen so steeply that people are spreading out and creating new neighborhoods like "NoHo" and "Tribeca" where neighborhoods never existed before. Thus gentrification marches on.

Among the most conspicuous signs of gentrification is the burgeoning brownstone movement. Urban homesteaders have moved into some of the city's most dismal districts and turned them into delightful neighborhoods. Columbus Avenue, which just a few years ago was one of the most depressed areas in Manhattan, is now lined with colorful shops and restaurants that are beginning to spill over onto Amsterdam Avenue. On 101st Street in East Harlem, a colony of young professionals lives in handsome renovated brownstones.

Conventional wisdom among urban specialists in the 1960's held that one of the biggest handicaps of the older industrial cities was their outmoded physical plants. Today that notion is out of date. The period from 1830 to 1917, when most of the homes in our older cities were built, was a unique era during which mass production meshed with individual craftsmanship. "It was an incredible time for building in this city," says Everett Ortner, a brownstone owner in Brooklyn's Park Slope and the president of Back to the Cities Inc., an urban homeowners' group.

"You had mills turning out mahogany doors which were being finished by hand. From this period came extraordinary houses that can last hundreds and hundreds of years. It would cost millions of dollars to re-create a brownstone today. New York has the best 19th-century quality housing of any city in the world. There are thousands of young people who bought these old houses because they were cheap and then grew to love them."

Though most brownstones in prime areas command from $150,000 to $350,000, tens of thousands of four-story houses with 5,000 square feet of floor space are still available for $25,000 — quite a bargain when you consider that the average suburban house has only 1,200 square feet and costs $55,000. According to one brownstone owner on the West Side, "The cities are as undervalued as the stock market was in 1974, when you could pick up fine stocks at modest prices. People who took a chance on buying a city house can make back four and five times their money in a few years if they want to sell." Urban co-op owners are also reaping windfall profits: A co-op on Central Park West purchased for $25,000 two and a half years ago was recently resold for $142,000.

Between 1973 and 1975, the number of urban homeowners in America increased by 1.4 million while the number of renters increased by only 500,000. In New York City, the rate of conversion from rental apartments to co-ops has more than doubled in the last three years. This increase in the percentage of owners as opposed to renters indicates that New York is attracting a more stable, committed population.

Perhaps the most positive proof that the gentry is revitalizing large tracts of New York City is that the banks, which only a few years ago had redlined most of the inner cities in the Northeast, have been opening their mortgage coffers to brownstone and co-op buyers in more and more neighborhoods even before the State Assembly's recent passage of anti-redlining legislation. Bankers have learned that these are among their safest loans, and they are now willing to invest in city real estate even if they are reluctant to invest in city government.

Foreigners are also investing and living in New York in ever greater numbers. A recent survey by the London Economist noted that "New

York City is becoming the capital of the world even as it is becoming noticeably less the all-dominating city of the United States." New York's pre-eminence as the world center of international finance has expanded tremendously in the last 10 years. Nearly one million of New York City's 7.5 million residents are foreign-born, and increasingly it is the affluent from overseas, not the poor, who are making New York their second home. To Europeans, Latin Americans and Arabs, New York offers political stability, a safe haven for themselves and their money, and unlimited cultural opportunities. A recently surveyed group of foreign businessmen overwhelmingly chose New York above London, Paris and all other major European cities as the place to live and work, and in the last three years, 60 branches of foreign banks have opened in New York, along with shops, restaurants and luxury apartment buildings.

What About the Poor?

Like any social phenomenon, gentrification has its negative as well as positive aspects. High rents and renovation are driving the poor and the working class out of their homes and neighborhoods. Ironically, the ethnic diversity that is drawing the gentry back to the city, the cultural heterogeneity that has always been the source of so much of New York's character and energy, may become lost in a forest of homogenized high-rises and rows of renovated brownstones. "This neighborhood is becoming as sterile as the East Side," a young lawyer complained about the Upper West Side, where he lives. "People don't hang out on the stoops anymore, and everybody is beginning to look the same. So what if 20 restaurants have opened on Columbus Avenue in the last three years? They all serve the same third-rate quiche."

The displaced working class can always move to the suburbs, where there are more jobs anyway. But what about the poor? In Washington, D.C., London and Paris — where gentrification has been proceeding at a faster clip than in New York — such dislocations have increased racial unrest and urban tension. Indeed, many observers fear that uprooted "urban nomads" will pose a major social problem in the coming decade. The Department of Housing and Urban Development has recently awarded more than 200 grants totaling $3 million for studies of the situation.

Still, as Dr. Peter Salins, an urban expert at Hunter College, reminds us, "there have been many more people pushed out by abandonment than have ever been moved out by so-called gentrification." Dispersing the poor will not reduce poverty, but it may reduce the pathology of welfare dependency, unemployment and crime that high-density slums foster. Statistics show, for instance, that some areas in Brooklyn have less crime than equally impoverished but more densely populated sections of Manhattan and the Bronx. And, as social critic Irving Kristol points out, "any policy which anchors poor people in a declining city — whether it be by generous welfare payments, subsidized housing or subsidized employment — is bound to be cruelly counterproductive."

Urban experts and politicians are beginning to understand that only the middle and upper classes — not the poor — can rebuild cities. The economic opportunities in manufacturing and construction that allowed previous generations of poor immigrants to work their way into the middle class no longer abound in New York City. Although some blacks and Hispanics have joined the professional class and contributed to the gentrification of New York, many others have risen only to the ranks of the lower-middle class. And most evidence indicates that minority blue-collar workers act just like their white counterparts: When they've saved enough money for a house in Patchogue, they turn their backs on New York.

The survival and recovery of New York City depend on an educated, integrated urban elite. As urban expert Charles Abrams said, "Cities inhabited [chiefly] by the poor are poor cities. And poor cities are poor for the poor as well as the rich." ■

January 14, 1979

Long Beach, Calif.: The Selling of a 2d City

By ROBERT LINDSEY
Special to The New York Times

LONG BEACH, Calif., Jan. 1 — The Queen Mary didn't do it, the Grand Prix didn't do it, and the Spruce Goose laid an egg. Now, promoters here have another plan to put Long Beach back on the map: They want to build the world's highest building.

It would stand 2,000 feet high and look, its architect says, like a spaceship rising from beside the liner Queen Mary.

Long Beach is a city where old-fashioned American civic boosterism still thrives. Although some cities in California are feverishly debating slow-growth and no-growth plans, Long Beach is trying to sell itself, as William Dawson, the city's principal salesman, puts it, "the way Procter & Gamble sells soap." There are indications that the efforts are producing suds.

Using the income from oil wells disguised as colorful offshore islands, Long Beach bought the Queen Mary in 1967 for $3.45 million. Since then it has spent close to $65 million to operate the dowager ocean liner as a tourist attraction. But the Queen Mary failed to live up to expectations.

Annual Grand Prix Race

The city thereupon obtained a franchise to stage an annual Formula One automobile race, part of the Grand Prix international racing circuit, through city streets. After three years, its promoters say the Long Beach Grand Prix made a profit in 1978. Skeptics doubt this and complain that tax money has been diverted to the promotion.

A plan to use the late Howard R. Hughes's giant flying boat, the Spruce Goose, which flew only once, as a tourist attraction has been less successful. A group of local citizens two years ago asked the Summa Corporation, the holding company for Hughes assets, to give the plane to Long Beach for a museum. Summa agreed. The organizers predicted they could raise several million dollars for a hangar, but they could not. Now there is talk that the plane may be donated to a museum in Florida.

The newest proposal to give Long Beach a special identity is Space Tower 2000, a spaceship-shaped structure of offices, shops, hotels and other facilities. Jean Claude Destievan, president of the Space Tower 2000 International Development Corporation, said it would be "the most unique tourist attraction in the world, of great prestige and international influence."

Those were the kind of words officials here, where the most prominent addition planned for the city's harbor horizon is an unglamorous if controversial terminal for transshipment of oil from Alaska, like to hear. Why the efforts to put Long Beach on the map?

Long Beach is a second city; it suffers from its proximity to Los Angeles, 15 freeway miles to the northeast; it is one of those cities, as Newark is to New York, and Oakland is to San Francisco, and Kansas City, Kan., is to Kansas City, Mo., that feel lost within a larger neighbor's shadow and endure a kind of

The New York Times / Jan. 2, 1979

municipal neurosis as a result.

Part of the problem, says Christopher Pook, the principal promoter of the Grand Prix race, is that practically no one knows where Long Beach is. Until the Grand Prix race, he said, "people would say, 'Well, Long Beach is a little south of Los Angeles; it's near Disneyland; north of San Diego, and it's somewhere south of Santa Barbara.'"

Three or four decades ago Long Beach was a fashionable beach resort for the gentry of Los Angeles. Fine summer homes were built overlooking the Pacific and the city's fine, white beaches.

The New York Times / David Strick

Oil wells on islands off Long Beach, Calif., disguised as brightly painted skyscrapers, have decorated skyline for more than a decade. Income from the wells has been used to buy the Queen Mary and finance other projects to promote city.

In World War II, Long Beach prospered as a Navy town, a homeport for warships and the site of a naval shipyard. The Long Beach Pike, an amusement park with a spectacular roller coaster, was a must-visit destination for teen-agers, and Long Beach became the first home of the Miss Universe contest.

Then things shifted into reverse: the Navy closed most of its operations; the Miss Universe contest moved out of town, and the descendants of the kids who rode the roller coaster went to Disneyland, which sprouted in once rural countryside in Anaheim, 10 miles southeast.

Meanwhile, Long Beach's downtown core, like others across the country, decayed into strips of empty stores, pornography theaters and shops with perpetual going-out-of-business sales.

The population, which was 358,000 in 1970, declined to 338,000.

Worse, the city's oil money was beginning to run out. Under an agreement with the state, the city has shared revenue from oil produced from its tidelands, receiving about $9 million a year. But the income is to decline progressively, starting in 1980, and by 1988 may be only $1 million a year.

It was partly concern that the oil money was running out that prompted

the city 18 months ago to establish a new promotion organization, and it hired Mr. Dawson, a 40-year-old marketing expert who had promoted amusement parks in Atlanta and New Jersey, as general manager. Emulating Madison Avenue, Mr. Dawson said the first thing he did was to make a market research study "to find out what people thought of Long Beach."

"We found out that, really, Long Beach didn't have a bad reputation; it just didn't have any reputation," he explained. "People were asked, 'What do you think of Long Beach?' and 60 percent said 'Nothing.'"

The city began a direct-mail advertising campaign, hoping to attract conventions, promoting itself jointly with San Diego, Palm Springs and Las Vegas as "Sun Country, U.S.A.," while continuing to emphasize the city's connection with the Grand Prix and the Queen Mary.

There are some indications that Long Beach has begun to make a comeback. Mr. Dawson cited statistics showing that more organizations were choosing Long Beach for conventions. Developers have announced plans for a large downtown shopping and a new hotel.

Arts Complex Opened

Earlier in 1978, the city opened a $51 million performing arts complex, an

architectural gem meant to help revitalize the downtown area and, incidentally, rival the Los Angeles Music Center with a program of quality theatrical productions.

So far, the city experiment as an angel financing such theatrical productions — with income from its oil wells — has been less than fully successful. A local newspaper summarized the city's first two productions in the impressive theater as "financial disaster."

Still, the promoters say, Long Beach has to spend some money to make some. A news release from Mr. Dawson's organization, to a large extent, summarized the city's attitude in attempting to promote itself. It reads:

"Dawson and the Long Beach Promotion and Service Corporation did not disguise the fact that they are utilizing time-tested attention getters to position Long Beach as a city with guts, away from the pack, in the fiercely competitive business of redevelopment and tourist-convention promotion. In the finest tradition of Barnum, Ziegfeld and Mike Todd, the one and only cardinal sin is to be ignored."

January 2, 1979

Chicago Bank Makes Money on Loans That Aid Deteriorating Neighborhood

By NATHANIEL SHEPPARD JR.
Special to The New York Times

CHICAGO, July 26 — Five years ago the new owners of the South Shore National Bank set out to disprove the notion of their fellow bankers that making development loans in deteriorating neighborhoods was like throwing money out of the window.

Today, buoyed by its success in helping to arrest blight in the South Shore district, a predominantly black community of 80,000 seven miles south of the Loop, the bank is preparing to embark on an even larger rehabilitation effort.

In a joint venture, it will finance its holding company's multimillion-dollar purchase and rehabilitation of about 500 units of deteriorating multifamily housing in the most blighted section of South Shore. It will be one of the largest rehabilitation efforts involving other than high-rise buildings.

National Model Is Hoped

Officers of the bank hope that the effort will help their neighborhood redevelopment project become a national model of how local banks and institutions can play a vital role in preserving communities while still turning a profit. Other banks in Chicago and in other cities have expressed interest in the bank's efforts, and it is setting up a seminar for bankers who would like to know more about the bank's methods and experience.

"While other banks are redlining neighborhoods we are greenlining them," said Susan Davis, vice president of the bank. "And we have been able to turn a profit while doing so."

Redlining is the practice by banks, insurers and other institutions of refusing to make loans, grant mortgages or write insurance in certain neighborhoods because of perceived risks. Redlined neighborhoods are usually poor or predominantly occupied by minorities, or they are close to such neighborhoods.

"In the new project we will take the biggest and the worst buildings in South Shore and make them as good as new and hope that smaller buildings and homes in the area will ride on our coattails," said Thomas Gallagher, president of the City Lands Corporation, an affiliate of the Illinois Neighborhood Development Corporation, the bank's holding company. "We believe the project will be sufficiently dramatic to get the community back on its feet."

99 Percent White in 1950's

In the 1950's South Shore was 99 percent white, a well-to-do community with a lakefront on its eastern boundary, an area of stately mansions and a mix of houses and three-story, multifamily buildings.

Ronald Grzywinski, left, chairman of the executive committee of South Shore National Bank in Chicago, working with Milton Davis, the bank's president.

Rapid growth of surrounding black communities increased the demand for housing in South Shore and prices began to rise rapidly. As blacks moved in, whites moved out, along with virtually all of the businesses and the exclusive shops that had thrived there. The area is now 85 percent black.

As the neighborhood changed, the South Shore Bank began to cut its services, granting only two mortgages totaling $59,000 in the community in 1972. As a result of its cutback in services, deposits declined from a peak of $80 million to $46 million and the bank's owners decided to bail out.

"We call this disinvestment," said Ronald Grzywinski, chairman of the bank's executive committee and the man who set up the redevelopment project. "As the former operators of the bank saw it, the new people in the community were not as credit-worthy, did not maintain accounts large enough to be profitable and presented the risk of increased fraud and higher delinquent loan payment rates."

"When a neighborhood changes racially," Mr. Grzywinski went on, "the people

in its institutions often throw up their hands and say there is nothing that can be done to save the community, and leave. Bankers have at least as much if not more prejudice as others and when a neighborhood changes they keep investment money out of that neighborhood."

By this time, Mr. Gryzwinski had developed his model for a bank's participation in community development through a holding company and was looking for the opportunity to try it out.

After obtaining about $4 million in financing from foundations and other groups, he and three partners took over the bank and began the neighborhood redevelopment effort

In the last five years the bank has granted development loans totaling $18 million in the South Shore community. In 1977 and 1978 the bank financed the rehabilitation of 625 multifamily housing units at a cost of $3.7 million. It also granted 114 mortgages for single-family homes, 37 mortgages for multifamily homes, 209 home improvement loans, 125 commercial loans and 293 student loans.

Despite the infusion of money into

South Shore housing, many storefronts remain vacant or in disrepair along 71st Street, the hub of the neighborhood.

Last year a group of South Shore residents joined forces with developers and formed the Phoenix Partnership. The partnership bought three and a half blocks of commercial buildings along 71st Street. With loans from the bank the group began a renovation effort that already has resulted in the restoration of a block of vacant storefronts.

And the owners of the Rosenblum Drug Store, the only business to remain in the neighborhood after its population shifted, have undertaken a $340,000 project, underwritten by the bank, to rehabilitate and expand their retail space.

"We knew that the community had to be revitalized at some point because everybody can't move to the suburbs or to the North Side of the city," said John Kelly, one of the drugstore's owners.

"And besides, the neighborhood wants us to stay."

Last year the bank extended $6.4 million in development loans to the South Shore neighborhood, raising its total outstanding loans to the area to $15.2 million. The 90-day delinquency rate on the loans, at 2.4 percent, was well below the industry average, according to bank officials.

The bank's net income last year, after subsidizing its redevelopment project for the $30,254 difference between development income and development expenses, was $208,000, representing a 5.2 percent return on its equity. The level was well below the industry average of 14 percent for banks of its size, "but it proved those critics wrong who argue that you can't make loans for commercial real estate ventures in black communities without losing your shirt," said Milton Davis, the bank's president.

"Conventional wisdom has it that the economics of rehabilitation don't work and that you have to bulldoze and build anew from the ground up," he said. "We've come to the opposite conclusion — that you can rehabilitate, get sound housing and put it back on the market at prices people can afford."

Two studies by a University of Chicago sociologist indicate that the rehabilitation effort has played a significant role in arresting blight, reducing crime and stabilizing the South Shore community.

In a follow-up to a 1974 survey of the community, the sociologist found a dramatic reduction in population turnover and the crime rate, a stable welfare rate and rising median family incomes and property values. He also found residents more optimistic about the neighborhood's future.

July 30, 1979

Salt Lake City Changing as Ties to Mormon Church Fade

By MOLLY IVINS
Special to The New York Times

SALT LAKE CITY, Utah — This city in a valley between the mountains, which has long served as the capital and as the fortress of the Mormon religion, is taking on a more secular character.

The Mormons fled to this area in 1847 to avoid religious persecution, and for years Salt Lake City has seemed to outsiders to be in a sort of Rip Van Winkle sleep, to be a place that remained improbably wholesome while other cities were wracked with problems.

The Mormon influence is still dominant in Salt Lake City, but now there is change as well as stability — physical change as the town grows, enormous economic change as it develops into a regional trading center, and emotional change as the church and its members adapt.

This weekend will mark something of a coming-out party for the city.

The occasion is the formal dedication Friday and Saturday of the $90 million Utah Symphony Hall, a handsome, angular, glass-and-buff-brick affair. The new $4.5 million Salt Lake Arts Center next door is already open.

There is to be a civic whoop-de-do for the grand opening, and the city's suspenders are already straining with pride, although there is some anxiety that all that gold leaf in the lobby may be too.

But Salt Lake City isn't a nouveau riche town that has gone out and bought itself some culture with a capital C. The Utah Symphony and the city's three dance companies, with their indefatigable touring, are known throughout the mountain West.

Salt Lake's best-known musical commodity is, of course, the Mormon Tabernacle Choir.

But the Utah Symphony has a long-established reputation for excellence. The recent retirement of its conductor of 32 years, Maurice Abravanel, has left it somewhat adrift, however, and a series of guest conductors will appear this season while the search for his replacement is under way. Stanislaw Skrowaczewski, formerly of the Minnesota Orchestra, will conduct on opening night at the new symphony hall.

The proof will be in the playing, but early predictions are that the hall is an acoustical gem. It was designed by Fowler, Ferguson, Kingston and Ruben of Salt Lake City.

The city's three dance companies, the Ballet West, the Ririe-Woodbury Company and the Repertory Dance Theater, use the recently restored Capitol Theater, which had gone into decline as a movie house. When the tacky neon-and-metal front was torn away, an old terracotta front, an endearingly ornate confection in the style of an Italian palazzo, was discovered.

Salt Lake City's entire downtown has been done over, in fact, to mixed reviews. The original renewal plan was to make Main Street a pedestrian mall, but there was a compromise and the street was simply narrowed and bedecked with trees and fountains.

While a church-owned shopping center at the head of Main Street is doing well, merchants further south are complaining that their customers were driven away by the long period when the sidewalks were torn up.

Elliott Wolfe of Wolfe's Sporting Goods Emporium said bitterly, "We have flowers, we have trees, we have benches, we have everything but customers."

Salt Lake City has always been a pretty and prosperous town. It is in the valley between the Wasatch and the Oquirrh (pronounced O-kwur) mountain ranges, near some of the loveliest mountain scenery anywhere in the West. Brigham Young laid out the town according to a plan drawn by Joseph Smith, founder of the Mormon Church, with streets wide enough for a full team of oxen and a covered wagon to turn around.

But, like all high-altitude cities, Salt Lake City is susceptible to air pollution, and the city's rapid growth, with its accompanying increase in automobiles, is causing more pollution than the Kennecott Copper Corporation's huge smelter.

Utah is the seventh fastest-growing state in the nation, and 70 percent of its population lives along the Wasatch front, from Ogden, to the north of Salt Lake, to Provo, on its south. The metropolitan area has grown from 557,000 in 1970 to 840,000 today. At least 85 percent of Salt Lake City's residents are Mormons.

The city was built on mining and agriculture, but World War II brought a huge investment in military bases and ordnance depots. It has most recently grown as a warehousing and distribution center. The state does not tax inventories, which makes the city especially attractive for warehousing, and it is a logical regional center since it is the only city of size between Denver and California.

A dozen ski resorts that attract tourists from all over the country are within an hour of the airport, including Alta and Snowbird, the twin meccas of deep powder skiers.

The growth in tourism has fostered a boom in hotels, including a Howard Johnson, a Sheraton, a Marriott and a Hilton, but the finest hotel in Salt Lake is still the grand old Hotel Utah, known for its superb service.

The question that fascinates outsiders most is the effect that all this growth is having on the Church of Jesus Christ of Latter-day Saints.

The Mormon Church has long had a

dual life: it is both a religion and a huge corporation. In addition to hotels, it owns the city's afternoon newspaper, The Deseret News; a commercial printing company; the city's largest bookstore; the shopping mall; a broadcasting holding company that owns six television stations, including the largest in Salt Lake City; about 30 blocks of prime downtown real estate, and much, much more. And it pays full corporate taxes on all its commercial holdings.

The church releases no figures, but five years ago, its income from industries was estimated by Nation's Business magazine at more than $125 million.

The church encourages tithing by its members and, worldwide, owns more than 5,000 buildings for religious purposes. In New York City, the church has a visitors' center at Broadway and 65th Street.

To an outsider the church would seem to have a strong grip still on the city and state. The Governor, the entire Congressional delegation, 90 percent of the Legislature and 70 percent of the people in Utah are Mormon.

But those who have known the city for a long time say the changes are striking.

The old antagonism between the "saints" and the "gentiles" — in Mormon parlance, all non-Mormons are gentiles — is gone.

From before the turn of the century to around 1920, the feud between the two groups was ferocious. The gentile-owned Salt Lake Tribune said in an editorial on the occasion of Brigham Young's death, "The most graceful act of his life has been his death."

There is now noticeable unanimity of purpose between the two groups when it comes to the betterment of Salt Lake City. The church, for example, contributed much of the land for both the Salt Palace sports and convention center and for the new arts complex.

Although the church is strongly opposed to alcohol and would not permit liquor by the drink, it gave way to tourist demand, in an unusual compromise: the itty-bitty bottle system. You cannot buy a drink as such in a public place, but you can buy an itty-bitty bottle from a tiny state liquor store tucked into the corner of many a fine restaurant and pour it into the set-up provided. However, signs warn that alcohol is harmful to your health.

The so-called private clubs downtown can be joined at the drop of a sawbuck.

The church is opposed not only to alcohol but also to tobacco, coffee, tea, abortion, the feminist movement, homosexuality, pornography and "permissiveness." Some gentiles find the climate oppressive, but others enjoy the good things it provides — a relatively low crime rate and emphasis on family life — and ignore the rest.

Non-Mormons now seek out Salt Lake City simply because it is such a pleasant place to live, with the mountains, the skiing and the cleanliness.

Jack Goodman, a reporter who has lived in Salt Lake for 25 years, said: "The only time I ever remember feeling different out here was sometimes I thought my kids felt left out because the Mormon kids have so many church-related activities. I am not what you would call an observant Jew — I go to temple on High Holy Days because it would please my parents — but I joined the temple here so my kids would have a group to feel part of."

So the town changes, builds, reconstructs, and the church bends, too. And more and more people are looking on Salt Lake City and saying, as Brigham Young is supposed to have said when he first viewed it, "This is the place."

September 12, 1979

THE GLORY THAT WAS CHARLOTTE STREET

Two years after the Presidential visit, this devastated block is nowhere near as alive as its past.

By Ira Rosen

It has been two years now since President Carter stepped out of his black limousine and, for 10 minutes, entered the world of Charlotte Street, making it a national symbol of urban blight. His visit was the final brushstroke of an 80-year tableau of construction and demolition, a process that has caused the lonesome South Bronx block to be renamed by area residents *yarda vacia*, "the empty lot."

The street is a mile from the Bronx Zoo, and two miles from Yankee Stadium. It extends for three blocks, two short and one long, from Crotona Park East to Jennings Street. No one lives on the street, and the only remaining building is Community School 61. Garbage is piled high; the Sanitation Department has joined the area dumpers in using Charlotte as a designated trash site for street sweepers.

There are few other areas in the South Bronx that are riddled with as much burned-out scenery and decay; it is a landscape that has been compared with Dresden after World War II. No other place seems as neglected and hopeless, in a terrain that has come to personify neglect and hopelessness: In the South Bronx, 35 percent of the population is on welfare, there are 1,500 vacant buildings on 500 acres of abandoned land, 6,900 buildings are in tax arrears and the per capita income of $2,340 is 60 percent below the national average.

For these reasons, Patricia Harris, then Secretary of Housing and Urban Development, selected Charlotte Street for President Carter's visit, intending it to be emblematic of his concern over urban decay. Reporters made the street a symbol of the problems inherent in a have- and have-not society, pointing out that Scarsdale, one of the richest communities in America, is only minutes away. Mr. Carter was also struck by the appearance of the street, and said he was "sobered" by it. His immediate response was to ask that plans be drawn for a recreation area, or for housing.

There was no need for a park; Crotona is at the top of the street. Instead, a $32 million, 732-unit low-income housing plan was moved from its scheduled site some blocks away, on Chisholm Street, to Charlotte Street as a "symbolic gesture." But, last February, the Board of Estimate voted against the plan

It would seem that the only legacy of President Carter's visit is that the street has been turned into a tourist attraction. The urban planning department of

Ira Rosen is a freelance writer.

Hunter College gives a tour of Charlotte Street every other Saturday for $8. Several bus lines in the city report that they have chartered trips for foreigners who come to see this American eyesore, and afterward take in the Bronx Zoo. "When I have judges come to visit from other cities, I ask them where they want to go," says Bronx District Attorney Mario Merola. "Radio City? Empire State Building? Statue of Liberty? 'No, they say. Take us to Charlotte Street.' " Unfortunately, the people in the area cannot make any money on the sales of souvenirs. Sightseers are too scared to leave the bus.

And with good reason. In five weeks of research for this story, during which I mingled daily with the residents, I was shot at while driving my car, robbed of $15 and once, while standing on the corner of Jennings Street and Southern Boulevard, punched in the face without any warning or provocation.

When I reported the shooting to a police officer at the 42d Precinct, I was met with a shrug and a one-line comment: "It happens every day." Such is the case now, but not very long ago, Charlotte Street wasn't a wasteland, but the home of thousands of new settlers in America. Among them were my grandparents, the Biglaisers, who lived on the street from 1932 to 1942.

The image is still clear, from the family tales, of my grandfather, Harry Biglaiser, kicking a quarter down Charlotte Street during the Depression. The day was Saturday, *Shabbat*, when Orthodox Jews are not permitted to work or to touch money. So, he didn't touch the coin. He kicked it. And he kicked it for five blocks until he reached the front of the house. He spent all Saturday afternoon by the window, to make sure that no one would touch the quarter that he had covered with leaves. Before he went to afternoon services, he told grandmother to watch the coin. She didn't; she thought he was crazy for not simply picking the coin up, since he had already touched it with his shoe. When he returned, after sundown, he went to the gutter where he had hidden the coin. It was gone. And he didn't sleep all that night.

On these pages, another family on the block, the Fiers, and their friends and neighbors, share their memories of the old block. Their stories span five decades of life on Charlotte Street.

The 1930's
A Fine View of the Street

The apartment was at 1514 Charlotte Street, only a half a block from his father's dairy store on Seabury Place. He remembers, and in fact will never forget, the way the street was in the spring of 1932, when he was 13: people sitting outside on folding chairs or milk boxes; kids eating jelly apples; the sounds of business coming from the block's two main markets, the sounds of a thousand people talking at once.

All these scenes were like a still life for Irving Fier as he rolled his homemade go-cart down the steep Charlotte Street hill. As he passed his house he was picking up speed, going faster and faster until the unexpected happened. The wheel fell off. He was thrown from the cart, and he tumbled on the pavement. He says now that the people didn't know first aid, and picked him up and carried him to the doctor. Moving him that way caused a permanent dislocation of his hip, a marking the street gave him as he entered manhood.

That year his sister Claire was born. The family moved to the front of the building to get a better view of the street. That was the only improvement they could have afforded. From the window, his father could also keep watch over the activity around his store. He didn't fear crime — he had had only one burglary, and that was by a hungry person stealing a loaf of bread — but he wanted to watch the crowd on the street to determine if he should open during the night.

Looking out the window at the rows of tenement houses, it was hard for Irving to imagine that Charlotte Street was once a huge farm. As late as the 1890's, the area was covered with bridle paths. The street was laid out across the estate of a pioneer Quaker, William Fox, the president of the first gas company of America. Fox sold the property to the American Real Estate Company, and as part of the agreement the street was named after his wife, Charlotte Leggett, while Fox had a street named for him three blocks away.

The purchase price for the 86 acres in the triangle that is now Boston Road, Wilkins Avenue and Southern Boulevard was $1 million, quite a sum in 1899. But within 30 days, the size of the figure was more than justified when a contract was signed with the city to bring a subway through the area.

For the next seven years, the section was in a constant state of chaos; rocks were being blasted, streets were being laid out, sewers were being constructed and millions of dollars more were being spent to transform the area into a part of the city. The growth rate throughout the Bronx was staggeringly high. In 1903, the population for the borough was 268,341. Ten years later it was 641,981, an increase of nearly 140 percent. Private companies took advantage of the population shift by opening new transportation facilities. Charlotte Street was close to the New York Central Railroad and the elevated trains on Southern Boulevard. Trolleys, at 5 cents a ride, crossed the neighborhood. The first bus line to operate in the Bronx ran between West Farms Square and City Island, via Boston Road and Pelham Parkway, in 1916. The charge was 20 cents a ride, or six for a dollar, but the line didn't get many passengers, and by 1922 the City Island Motor Bus Company went bankrupt.

It was this great network of transportation that attracted many here, including Irving's father, Abraham. The Fier family had lived on Cannon Street on the Lower East Side, with their fellow immigrants from Russia. But to operate a successful dairy the proprietor must get his fresh cheese daily. And the iceman had to have clear roads for his deliveries, for one day without ice during a hot summer might be disastrous. A popular item like Brinzer cheese, for example, was kept in a tub, and had to be covered with ice or the warming smell would drive customers from the store. Transportation also brought customers from as far away as the Lower East Side, people who would spend two hours traveling to save a few pennies on goods.

The presence of the bargain hunters made for a rather lusty competition among the store owners. Ralph Romisher owned a house-furnishing store at the Jennings Street market, at the end of Charlotte Street. A similar store opened, and tried to undercut the prices of Romisher. Romisher noticed that if he marked in his display window a certain item at 49 cents, the competitor would mark it at 45 cents. The profit margin amounted to pennies, so a few cents made the difference between paying the bills and being broke. Romisher, the story goes, thought of a plan. He knew the competitor walked past his store each night to check the prices. So every night before he went home, Romisher changed all the prices. A 45-cent item became 25 cents. The other man continued to keep pace. In the morning, Romisher took *(Continued)* the night signs away and put up the regular prices. Within five months, the competition was forced to close.

No matter how hard the store owners worked, they always managed to make an end to their day. After work, a great relaxation was to go up to Crotona Park, at the top of Charlotte Street. When Irving and his brother Reuben en-

1946
In the block's heyday, Alma Katz, left, and Shirley Bransky pose in front of 1517 Charlotte Street.

1979
At right, the street-scape as it appeared last month; the area has been compared to Dresden after World War II.

Leslie Wong

tered the park from the gray and brown world of Charlotte, he felt like he was traveling to a foreign country. In the middle of the park was a big lake where he could rent rowboats or fish for carp or sunfish. And there were the handball courts, where on weekends grunting men would battle and sweat and curse. Near the handball courts was a bocci rink, a home for the old Italians who talked and played as dozens of spectators watched. One of the competitors was a one-eyed old man who at first scared the Fier boys, but they were told later that he was the finest bocci player in the Bronx. Past the handball courts were two dozen golden clay tennis courts, always crowded. And there was a big field usually filled with picnickers. This, too, was part of the Charlotte Street way of life: a big lawn that thousands shared.

It was here that Irving took his first date. "If you wanted to make out, you walked the long way through the park to Tremont Avenue, got a soda and then walked back." After a few years, he had the route down pat. But so did many others, and if there wasn't enough privacy, he usually ended up kissing in the lobby of the date's apartment house. He met his future wife, Shirly, who lived on the block, and he courted her with the help of Crotona Park and a relatively empty lobby.

In the 1930's the Fier family and other first-generation Europeans felt fortunate to live in the area. For now Irving and his father, Abraham, looked out the window of the apartment and talked about this wonderful neighborhood; though not rich, it was fraternal and stimulating.

The 1940's
Carmen and the High
Holy Days

In 1941, a few weeks before the Japanese bombed Pearl Harbor, the Fernandez family became the first Puerto

Ricans to move to Charlotte Street. If the former event took America by surprise, the latter was also something new for the people of Charlotte Street. But the family was accepted, if not royally welcomed, by the residents. Almost immediately, the Fernandezes' daughter Carmen became friendly with Claire Fier.

"Who ever heard of Puerto Ricans? I thought she was from Spain," Claire Fier said. So did many of the people on the block. In 1940, there were fewer than 60,000 Puerto Ricans in all of New York City. By 1960, there would be nearly 700,000. At the time, only Rhode Island had more people per square mile than Puerto Rico, a land whose large population and lack of many resources made for hard living conditions. It wasn't until 1952, when Puerto Rico became a Commonwealth of the United States, that the great transfer of humanity began. It was understandable why the Fernandez family was considered to be Spanish.

Their immigration had occurred out of necessity; the parents couldn't find work, and were forced to go elsewhere. But the search for a job was no easier in the United States. Mr. Fernandez took odd jobs for many years before getting lucky: he landed a night-shift job mixing flour at the Sunshine Biscuit Company in Staten Island.

Claire and Carmen were a team. They went to the Freeman theater on double dates, they were in the same classes and talked on the stoops till late at night, till 11 o'clock. "Then the old ladies got the cold water and threw it down at you," Carmen remembers. "They would wet the steps and force us to change to another stoop, or go home."

There were other things they couldn't share. At times, Carmen wished she were Jewish so that she, too, could get dressed up on the high holy days with the others in the neighborhood. Instead, she and her brother functioned as

shayner malach hamoves, the "cute mischievous ones" who were called to turn the lights off on Saturdays for the observant Jews. For that, they would be left quarters.

In school, too, Carmen would be treated differently. At Herman Ridder Junior High School, the teachers were noticeably surprised if Carmen, a Puerto Rican, did as well as the others. She hated that. "They had the idea that Puerto Ricans were dumb," she says. "I just wanted to be treated equally, without discrimination."

Raymond, Carmen's older brother — in the 1940's it seemed that brothers were always older, sisters younger — found happiness on Charlotte Street through playing in the stickball games, or pitching cards or shooting marbles. The kids were loyal to one another, and that loyalty was tested in the first youth gang confrontation remembered by the early Charlotte Street residents.

Gratosky, a boy from a nearby block, was causing trouble. He stuck a fountain pen belonging to the teacher down Raymond Fernandez's pants, and when the pen was found, Raymond was blamed for its disappearance. Later, Gratosky beat up a Charlotte Street boy. The pen prank was a minor matter, but picking on a fellow stickball player was too much, recalls Raymond. So the Charlotte Street kids, "Roamer," Leftie, Artie's uncle and others banded together and beat Gratosky and his gang up. "No big deal. I think one of the Gratoskys had his nose broken in it."

The event passed, provoking little gossip. If there had been any talk, it would have been at one of the five candy stores on Charlotte Street. It was at the candy stores that people gathered to discuss politics and unionism. If a poll had been taken to determine the political leanings of these sweetshop talkers, the finding might have been "extremely liberal." From these shops, and others like them in the Bronx, the views of a Congressman

named Fiorello H. La Guardia, who railed against crime and poor administration, became known, and helped to elect him to three terms as the Mayor of New York. The neighborhood kids, especially Raymond Fernandez and Reuben Fier, Irving's brother, made pocket change by hanging out at Leboff's candy store, and calling neighborhood people to the phone. Not many residents had phones in the early 1940's, and when calls came in at the store for people on the block, the boys would fetch them and collect tips.

Reuben was especially keen, and listened to the parlor banter attentively. Though the area was 71 percent Jewish, Reuben felt that life on Charlotte Street would not always continue as it had. One day he went to his father and told him he should look for another location for his business. "Dad, this area will change. One day the Jewish people will leave," he remembers saying. His father looked at Reuben in disbelief. "No, no," his father said, as he shook his head. "Who would give up the Crotona Park?"

The 1950's
Flickering Television and
Empty Streets

It was in 1951 that the Fier family first noticed the crime increase in their neighborhood. Claire's friend Lottie Rosenbaum was walking home one night, and a man followed her up the stairs to her Charlotte Street apartment. He had a knife. Scared, she fought back, and the man fled. It was an isolated attack, but it spread, like fire in a pile of dry wood, throughout the neighborhood, and throughout the candy store network.

"Hear what happened to Lottie?" someone would ask.

"No. What?"

"She was mugged by someone with a knife."

And then the story would be retold, becoming worse with each retelling. The 41st Precinct, later to be known as Fort Apache, patrolled the area, and reported that the level of crime did not increase measurably from the time of the late 1940's to the early 1950's. But, to Claire and her

friends, the situation seemed to be getting dangerous. Conversations that were once dominated by little worrisome problems now took on a larger scope. "We were brought up in a safe world. We never did anything where a lot of money was required. Suddenly, we were frightened of the street. And that made me very mad," says Claire.

The first of her friends to leave was Gail Alken. The pattern became repetitive after a while: Someone would get married, live on the block for a year or two, and then depart — usually to Westchester, Queens or Long Island. First Gail, then Lottie and finally Carmen. "It didn't pay to stay anymore. Our friends were leaving in panic," Carmen said.

Social theorists have developed various hypotheses to explain the flight from the city to the suburbs. Edward C. Banfield, in "The Unheavenly City," writes that "much of what has happened — as well as much of what is happening — in the typical city or metropolitan area can be understood in terms of three imperatives. The first is demographic: if the population of a city increases, the city must expand in one direction or another. . . . The second is technological: if it is feasible to transport large numbers of people outward (by train, bus and automobile) but not upward or downward (by elevator), the city must expand outward. The third is economic: if the distribution of wealth and income is such that some can afford new housing and can spend the time and money to commute considerable distances to work while others cannot, the expanding periphery of the city must be occupied by the first group (the 'well-off') while the older, inner parts of the city, where most of the jobs are, must be occupied by the second group (the 'not well-off')."

These elements contributed to Charlotte Street's demise, but other factors also played their part. For example, when more centralized heating was brought into the tenements, the residents welcomed it. Coal was dirty and expensive, and fouled the air, and each building required its own furnace. But central oil heating

President Carter's brief tour of Charlotte Street in October 1977 made it a symbol of urban blight in America. To his right, gesturing, is Patricia Harris, now the Secretary of Health, Education and Welfare, who had suggested the visit.

could warm four or five apartment buildings at once. And the landlords looked upon this as a godsend: the system could open up as many as 40 new apartments on Charlotte Street that were previously taken up with the storage of coal. But the landlord soon discovered that the Jewish, Italian and Irish families who lived on the street did not want the basement apartments. Like the Fiers, who improved their lives by moving to the front of the building for the view, many saw the prospect of living in the basement apartments as a degradation. So the vacancies were filled by Puerto Rican and black families who had been previously discriminated against in housing. The families that moved in found the low rent ($65 for a four-room basement apart-

ment in 1953), the established neighborhood and the convenient transportation system better than the conditions in overcrowded Harlem, which was located 15 minutes away on the el train.

An increase in the number of television sets and telephones also greatly affected the spirit on the block. When trouble came on the street, the boys or men there would band together to meet the challenge. But in 1950, when the Fiers had their first phone installed, and when the TV aerials began appearing above the street's tenement row, it meant that fewer and fewer people were on the street at night. Each TV light flickering in an apartment meant that at least one less person was outside. The candy stores began to disappear. And the streets became

more empty, and then, more sinister.

"When the TV's and the phones were installed, you hardly saw anyone in the streets. Instead of sitting outside after work, people would watch TV and then fall asleep," Claire Fier remembers. "I was longing for the type of life I saw slipping away."

Though their friends were leaving, the Fiers were determined to remain. Irving was married, working as a car mechanic, and he was living across the street from his parents' apartment. Reuben, an accountant, was also married, and decided to move closer to his job; he took an apartment on 149th Street and the F.D.R. Drive. Claire got married when she was 18 years old, and continued to live on the block

with her husband. Throughout the 1950's, she says, leaving the block never was a serious consideration.

The 1960's
Hitting the Rats
With a Shoe

In 1960, new stores with Spanish signs opened in the area. The *Se Hacen Mudanzas*, "We Move," the Spanish gypsy moving company, appeared more and more on the block. So the Fiers often sought out the familiar, the holdouts from the old days. One of those places was the Jennings Street market. It was a bargain-as you-will marketplace, with a whole spectrum of merchandise being sold on tables, in the stores or on the sidewalk. The store owners stood behind the tables and motioned for the customers to come to them. Sometimes they would even drag people off the street to show them the "specials" they had. But Irving Feuer, who owned a deli in the heart of the market, never had to call customers over. His shop was always packed.

Feuer was said to have been the best lox slicer in the Bronx. When his knife touched the fish, cutting the pieces off thinly and cleanly, nothing ever remained on the knife. That was the sign of a good lox cutter: neatness.

"People came in and spent money," remembers Al Morgenstern, Feuer's assistant for 20 years, who took over the store after Feuer's death. "In this neighborhood, they mainly had babies and ate. People found enjoyment within their means."

They came to buy lox, selling for 25 cents a quarter pound, or they bought herring at a nickel apiece. Good pickled herring took seven to 10 days to make. First it had to be soaked in water. Then the spices were added in a combination that Morgenstern still protects as though it were the Colonel's secret fried-chicken recipe. Morgenstern will only say that the secret is in the cream. "To be successful in the market all the food had to be prepared daily. The only people who sold old food were the banana sellers, and the customers bought it since it was cheap."

The symbolic king of the marketplace was Jacob Shertzer, known to many as Jake the pickle man. He had a stand at Jennings Street and Charlotte, where he stood with several barrels brimming over with pickles. Customers remember the long lines of those waiting to buy his pickles.

"But he was crazy, selling his pickles," recalls Fanny Biglaiser, who lived next door to the Fiers at 1512 Charlotte. "You bargained for the pickles that he sold, one by one. He used to inspect each one. He'd say, *"gey a heim, gey a heim, clean arouse te haus, gey a heim in mach en kugel, ih hob-nicht a pickle for de,"* ("Go home, go home, clean the house, make a pudding, I don't have a pickle for you!") The lines were a block long and, if you gave him a jar, no matter what size, he charged you the same price. He used to have barrels four feet high, and he'd have the best pickles. Lots of times he would take a jar from a woman's hand and throw the top in the gutter. He was crazy, really. His wife wore diamonds on all her fingers, and every time she would put her hands in the pickle brine, she would clean them, so her rings were always shiny."

The memory of Jake brought a smile to most of those who remember him. "They used to say that you know the neighborhood would go bad, if Jake left. He was there for 45 years and, sure enough, in 1961 he retired," says Morgenstern. "That's when we knew we were finished."

Somehow, Jake knew that his end would not be good. After his retirement, Jacob Sklar, his neighbor at 885 Jennings, remembers Jake saying Kaddish, the mourner's prayer, for himself, because he didn't think his children would. "Each night he would call up the stairs because he didn't want to be surprised by muggers in the hall. People thought he was nuts."

On June 10, 1963, Jake's intuitions proved to be correct. He was found dead in his apartment with a rag stuffed in his mouth; his hands were tied up behind him, and the gas was on. Police said he'd been the victim of a robbery.

At the time of Jake's death, a number of the street's long-standing stores began to leave the block. Among them were

THE POLITICS OF CHARLOTTE STREET

It was on Feb. 8 last winter that the Charlotte Street housing project, a 732-unit dream that the Koch administration had pledged to make a reality, was wiped off the map of the South Bronx amid a welter of bad communications, ill will and deep misgivings. The $32 million project was envisioned as the keystone of an ambitious $1.5 billion redevelopment plan. But 16 months after President Carter had paused amid the rubble to affirm his committment to rebuilding the area, the Charlotte Street development was defeated 7 to 4 by the city's Board of Estimate.

The reasons were varied. Several of the five borough presidents objected to such an allocation going to one borough. "If President Carter's chauffeur had taken a right turn instead of a left turn on his trip here, we'd have all the money going to Bushwick now," the Borough President of Brooklyn said.

Other members of the board believed that the Federal funding commitment was vague, and that the location of the project was ill-conceived. More telling than those objections, however, was the fact that the Koch administration had not seemed to know they existed: apparently, when the Mayor entered the majestic blue-and-white board chamber to vote on the Charlotte Street question, he was sure that he had the votes to carry the project; and so, the defeat was to emerge not only as the bitterest rift until then between the Mayor and the board, but also as a monument to poor politicking. The Mayor's administration had not lobbied for the plan and paid for its self-confidence.

After the vote, that self-confidence gave way to hard feelings. Herman Badillo, the Deputy Mayor who had been Borough President of the Bronx and is one of the nation's best-known His-panic leaders, immediately resigned his leadership position in the project and bitterly denounced those who had defeated it as "demagogues." Mayor Koch — who, only 48 hours before the defeat, had received a letter from Presidential assistant Jack H. Watson Jr. reaffirming the President's commitment to aid the South Bronx — announced, to the astonishment of many members, that the defeat of the Charlotte Street project meant an end to the entire plan for the South Bronx. Three days later, Mayor Koch admitted that he was "eating crow" and going ahead with the plan minus the housing project, but Mr. Badillo remained both angry and unmoved. He did not participate in plans for the projected South Bronx revival, and when he resigned in August, it was widely believed that the defeat of the Charlotte Street project had been a large measure of his disaffection.

In fact, even Mayor Koch incurred Mr. Badillo's wrath. When the new — and considerably more conservative — $375 million preliminary proposal for the South Bronx project was presented to the Mayor recently by urban planner Edward J. Logue, Mr. Badillo called it "the most cowardly kind of retreat a Mayor could indulge in." The plan — which stressed the rehabilitation of existing housing rather than construction of new buildings, and which asked for far less initial Federal assistance than its predecessor had — was clearly designed to allay some Board of Estimate objections, but it did not allay Mr. Badillo's. The Charlotte Street site is mentioned in the plan; it is consigned to the "increased demolition, greening and cleaning" the city favors for unsightly vacant lots that are not slated for construction.

— ANNA QUINDLEN

'We were brought up in a safe world,' said one Charlotte Street resident. 'But suddenly, we were frightened of the street. And that made me very mad.'

Stern's bakery, Rosenblatt's dry goods, Mazel's drugstore, Wald's shoe store, Weintraub's ladies' wear, and Ralph's grocery. "When these stores were leaving, it was like the Dodgers departing Brooklyn. Somehow, you knew life would never be the same again," says Raymond Fernandez.

In the first two years of the 60's, the street became dirtier and dirtier. Landlords, stymied by rent-control laws that prevented them from passing on their expenses to the tenants, began to neglect the buildings. At first they did this in little ways — leaks weren't fixed, bronze doorknobs or mailboxes weren't shined, garbage was left to clutter up halls; soon, the garbage got so bad that tenants had to side-step the loose chicken bones.

Nourham Kechejian was a landlord at 1134 Intervale Avenue, and owned a number of other buildings near Charlotte Street. "They were good buildings. Beautiful ones. But the welfare people came in, and they didn't pay the rent. I couldn't pay the taxes, one year. The tenants would come and leave. I had no money to repair the buildings."

In Claire Fier's building, the situation was very bad. One day she found that her husband's shirts had been partly eaten by rats. "Roaches you can get rid of, but not rats. I remember my husband hitting at the rats with his shoe. It was then that we decided to leave. When all my friends and neighbors were leaving I wondered, 'why are they deserting this area? How could they do it?' I think back now that I should have moved out then, too."

By the end of 1962, Claire, Irving and their parents found a new home in the Rockaways. It was a hard thing to do, but they finally moved out of the neighborhood. The father's dairy was no longer successful, since farmer cheese, pot cheese and other types of cheeses were not selling well in the neighborhood. The store was taken over by a Spanish grocery.

The 1970's
The Javelins, Turbans
And Other Survivors

Peggy Long, 35, sat on a park bench outside Crotona Park and watched as a city-hired demolition team tore down the last remaining apartment building on Charlotte Street, across from Community School 61. One man ran the forklift, and two yellow-helmeted workers separated the old Bennett bricks of red clay into piles. The workers got 3 cents a brick for them, and the company sold them for 20 cents each to builders in Westchester County. There, the bricks were used to build expensive homes: builders like to use the bricks because they give the houses an antique look.

As she watched, Peggy Long began to cry. "This reminds me of when I got on the bus one day, and saw that my elementary school was gone. *I mean gone.* I went to P.S. 54, at Freeman and Intervale. 'What did you do with my school?' I asked. There was just an empty lot there."

Since 1954, Peggy Long had lived on Crotona Park East and Wilkins Avenue, and knew the Fiers and their store. But, being black, she had her own set of friends — people who remained during the years of terror in the late 1960's and the early 1970's.

"What happened here is simple to understand. People stopped paying rents, the landlords neglected the buildings and didn't pay taxes, and then either the city took over, or the landlords started fires to collect the insurance," Long says.

For some, a house fire in the Bronx became an expected part of a weekly schedule. The Rev. John Luce of St. Ann's Church in the South Bronx recalled the time when a young boy came up to him one Friday. "Father, I can't serve in mass this Sunday," he said. "Why?" "Because my house will burn down."

Peggy Long has never got used to looking at the burned-out buildings. They scare her. "They resemble empty, black eyes," she says.

These are the buildings that attract the curious, armed with Polaroids, who come in the buses and the cars. "These people come by, and stop, and take pictures. We are not animals. We are suffering. There is nothing to look at here," Peggy Long says. "This is serious, this is people's lives."

Another survivor of the 1970's on Charlotte Street was 19-year-old Rodney Green. He had a unique perspective on life on the changing block. His youth gang, the Javelins, ruled the street.

He was a student at Herman Ridder Junior High School when he joined. "I needed some backing," Green explains. "Because I was getting beaten up every day." He was 11 years old.

After he joined, the beatings stopped. There was loyalty among the Javelins, but it was the kind of respect that soldiers had for each other during war. During the late 1960's and the early 1970's, the Javelins were battling two other street gangs that ruled adjacent areas, the Turbans and the Outlawmakers.

"People would be getting killed over nothing," Green recalls. "People were getting shot over little things. I saw one guy getting killed with a golf club. He was in a car, and two guys dragged him out and began to beat on him, and they clubbed him to death." Green was 14 at the time.

The gang dues were $2 a week, quite a lot for a junior high school student. "We used the money to buy guns. They were easy to purchase. A guy would come around in his car. Everyone knew who he was. And you would select."

"I didn't do any killing. Wounding, I did a lot of that."

Green's life in the gang began to change when he was 16. He was playing with a .38 in front of the mirror. He put what he thought was an unloaded gun to his head, and pulled the trigger. Empty. He pointed the gun to the floor and pulled the trigger again. This time it fired. After that, he sold his gun to a friend for $60.

All this happened three years ago. The gangs have for the most part died out since then, and so has their kingdom. But Green still lives on the turf. Green, a likable and ambitious worker who has directed his energy at finding a good-paying job, is living independently from his mother. He never knew who his father was. When he is not looking for work, he might be found walking in Crotona Park.

One recent weekday afternoon, we both walked through the park. I told him stories about what the park was like 20 years ago when my grandfather brought me there, and he showed me what it was like today. He said the lake has been renamed the "ghetto pool," since it is covered with marsh, and is layered with broken glass and bottles. We walked where the tennis courts are, and they were unused and desertlike. The handball courts were blue with graffiti. The bocci rink was now a big box filled with weeds.

"Do you think this area will ever improve?" I asked him.

"The only way the Bronx will be saved is for more jobs to come in. We need jobs. Everything else will work out. Didn't those people 40 years ago have jobs, and create a meaning for this place?" he said.

As we walked out of the park, Green turned toward it again and stopped. "You know, with everything else coming down, one thing is the same," he said. "No one can take the breeze out of that park." ■

Los Angeles Community Redevelopment Agency

A proposed "people mover" system is drawn over a photograph of Los Angeles, above, with the Bonaventure Hotel prominent in background. Inset, left, a detail.

'People Mover' for Los Angeles Gains

Special to The New York Times

LOS ANGELES, Oct. 15 — After making dozens of studies on proposed mass transit systems over the last 25 years and then shelving them, Los Angeles appears finally to be edging to-

ward construction of a transit line of at least moderate proportions.

Local officials say that they expect to solicit bids this fall for an elevated "people mover" system in the city's central business district, which has

been undergoing brisk revitalization in recent years. More than a dozen office buildings and hotels have gone up in the city center in the last decade.

The $44 million-a-mile transit system is aimed at unclogging traffic in the central city. City engineers planning the system say that, if their schedule holds, it will begin hauling passengers in the summer of 1983, a year before the city plays host to the Summer Olympic Games.

First Test in Major City

If the project is completed, Los Angeles will become a laboratory for testing a type of public transportation that has long been espoused by some transit experts but has never had a test in a major city, largely because of technical problems.

Operating much like an elevator but moving horizontally instead of vertically, the system is designed to carry passengers in automated cabs directed by a central computer on a 2.9-mile guideway above the streets and sidewalks.

For more than a decade, people movers, or "personal rapid transit" systems, have been called a potentially important way to deal with urban congestion. According to their proponents, they offer an alternative to the automobile, thereby reducing traffic, and because labor costs are low. The passenger cabs require no motormen. But except for relatively simple systems at amusement parks and airports, few people movers to date have been very successful.

Largely because of much-publicized

problems at certain facilities, including one in Morgantown, W.Va., and a troubled system at the Dallas-Fort Worth International Airport, several cities including Denver suspended plans to build people movers, and much of the early enthusiasm cooled. Other critics have called the elevated systems unsightly.

However, proponents say that flaws in the early systems have been worked out, and that innovations in computer technology and other fields now make them reliable and attractive. And officials in several cities, including Detroit and Miami, have said that they favored development of people movers to help relieve traffic congestion.

As the first operational system in a major city, the Los Angeles project could show not only evidence that the technical problems have been solved but also whether people movers can do much to help unclog downtown traffic.

The project here has been on the drawing board for almost a decade, and there has been skepticism about whether the latest proposal would get under way. There have also been threats by critics of the project to attempt to block it in court.

But Daniel T. Townsend of the city's Community Redevelopment Agency, who is director of the Downtown People Mover Project, said in an interview that he did not expect any serious prob-

lems. He said that $175 million, the total that he said would be needed for the project, had been pledged by Federal, state, city, and county governments; public hearings have been concluded and, he said, "I don't see any obstacles ahead now."

The sum includes $131.7 million for the transit system itself, plus $43.3 million for new parking facilities for 3,750 vehicles and a three-level bus station that will be built to accommodate commuters who use the people mover. The Federal Government is providing $143 million of the total cost.

The transit line will run through the heart of the business district between Union Station, the city's train depot, and the city's convention center, passing many of the city's newest office buildings and hotels.

Mr. Townsend said that up to 60 automated cars would run between 13 stations, with one vehicle leaving a station every 1.5 minutes at peak travel times, up to every 4.5 minutes in the evening and on weekends. It would take about 15 minutes to travel the full route for a fare now expected to be 25 cents. "During peak hours, we expect to carry about 9,000 an hour," Mr. Townsend said.

October 16, 1979

IDEAS & TRENDS Continued

Urbanism

It's Growing, Growing, Gone For the Skyscraper Syndrome

By GLADWIN HILL

"Bigger-Better-Busier" was once the ethic of the nation, reflecting the frontier conviction that growth was intrinsically good. Today that cliché is seldom heard, and then only sotto voce and with qualifications. Communities by the hundreds from coast to coast have adopted growth controls, and more are constantly doing so.

San Franciscans will vote Tuesday on a proposal that would lower permissible building heights from 500 feet (about 50 stories) to 260, virtually banning skyscrapers in the downtown area. Similar proposals have been defeated twice in the last decade in San Francisco, which is one of several big cities that are beginning to think that high-rise buildings may represent a net deficit in municipal accounting. The city has to extend water, sewer and power lines, enlarge fire protection, and cope with traffic congestion, while much of the revenue generated by a skyscraper flows out of town every night. Skyscrapers are only one of many types of

development that are giving municipalities second thoughts.

For generations most communities had room for comfortable, orderly expansion. But in the post-World War II population boom, developers bought cheap outlying land, threw up subdivisions, and left contiguous communities to pick up the tab for roads, waterlines, sewers, schools, parks, mass transit, and other amenities.

Much of the development was opportunistic and unsightly, and established residents soon began to clamor for brakes on growth. Such devices have been many: numerical population ceilings; restrictions on building permits and the water supply and sewage hookups; large-lot zoning; bans on multiple dwellings; floor-space and setback minimums; and special imposts on new developments.

Such restrictions were quickly challenged in the courts — both by developers and by civil rights campaigners, who pointed out that growth controls tended to discriminate against the less affluent home-seekers, who often were members of minorities.

A new era of growth-control litigation began in the 1960's. Its main point of departure was a 1926 ruling, in which the United States Supreme Court upheld zoning as a legitimate exercise of a community's police power in the general welfare, even if it limited some people's freedom of action and potential profits.

But how far could that power reasonably be extended? The Pennsylvania courts early on pinpointed a key growth control issue when they said that a community could not simply bar newcomers because growth is then deflected to other communities, and that one community didn't have the right to shape other communities' futures.

A score of Federal and state court decisions have left it that, while arbitrary population ceilings are unconstitutional, reasonable programs for phased community growth over a period of years toward a plausible maximum are valid.

Plaintiffs alleging discrimination have both won and lost. Many courts have held that control programs cannot favor elites but must encompass economic cross sections. However, the Supreme Court has ruled that discriminatory impacts that are not the manifest objective of a control scheme are not unconstitutional, since countless public and private actions can have incidental discriminatory effects.

Such litigation has impelled communities by the thousands to buckle down and formulate reasonable growth programs. But citizens have not stopped taking direct action either. Residents of Santa Barbara, Calif., last March turned down a bond issue to augment the city's water supply, and San Jose, Calif. is considering a plan to subject the admission of new industries to community vote. Both Hawaii and Alaska have essayed statewide population controls with laws discriminating against newcomers in employment, unemployment benefits and welfare.

Growth controls, by limiting supply, increase the cost of land and housing. There are exceptions. Where there has been no planning, regulation may produce more economical use of land and provision of public services; "sprawl" is inherently costly. But generally developers now have to pay special levies, not only to finance streets, sewers and schools, but to provide other community benefits, such as parks. In one San Diego suburb, the cost of fees and permits for an average house is said to have jumped from $43 to $1,283 in a couple of years. Bernard Frieden, a professor of urban studies at the Massachusetts Institute of Technology, asserts that "environmentalist" growth-control measures have helped push the cost of housing in the San Francisco Bay area up to one and a half times the national average.

In many communities growth controls have become the reigning political issue, with advocates of "managed growth" tending to prevail, while the other side denounces procontrol people as advocates of "no growth" and stagnation. And neither Federal nor state government, whose allocations of money and choices of projects can profoundly influence concentrations of population, have yet approached any comprehensive planning of population distribution. The question implicitly raised by the Pennsylvania courts — "Who decides whether community A or community B should absorb more people?" — remains unsettled.

November 4, 1979

The Frontier Style Clings To the Police In Houston

By JOHN M. CREWDSON

HOUSTON — For nearly three years, no single issue has more bedeviled and divided Houston than the conduct of its police department. Once again, the force is the most controversial topic in town, and, with municipal elections on Tuesday that will give the city's minorities a new influence in city government, the hottest political issue as well.

The latest debate was touched off last week by the sentencing of three former policemen to a year and a day in jail for their roles in the death two years ago of Joe Campos Torres, a Mexican-American arrested in a barroom brawl and forced to jump into a swift-moving bayou where he drowned. Federal District Court Judge Ross N. Sterling initially placed the officers on probation. When ordered by an appeals court to send them to prison, he did, but set the sentences to run concurrently with those meted out last year on a separate count; they have not yet served any time. Leonel Castillo, the former Federal commissioner of immigration and a leading mayoral candidate, called the decision "most incredible." Other Hispanic leaders, who have made the Torres case and the Houston police focal points for their campaign against police brutality in the Southwest, called for Judge Sterling's impeachment.

The death of Mr. Torres is the most notable, but by no means the only, instance of unjustifiable or suspicious deaths. Barely a month ago, Reggie Lee Jackson, a black, was killed by officers who claimed he had pointed a pistol in their direction after being ordered out of a car. Witnesses said Mr. Jackson had no gun and pleaded for his life before shots were fired. The community's questions about the fate of Mr. Jackson seem only natural. In separate Federal trials on two other cases in which young white men were killed, officers admitted having tossed "throwdown guns" beside the bodies of their victims.

The Trouble is More than Racism

Justice Department officials who have tried to bring civil rights prosecutions in such cases rank Houston's among the nation's worst big-city police departments. But while police excesses in Philadelphia and Memphis seem chiefly against black citizens by white officers, in Houston the difficulty is not simply institutionalized racism. In a city that its boosters like to describe as having entered the 21st Century, the police force is undermanned, underpaid, and struggling to shed its links to frontier law and its brand of horseback justice.

It has only been within the past two years, for example, that Houston's officers have been forbidden to wear cowboy boots or to carry high-powered deer rifles in their cruisers — the kind of thing that exemplified the force's reputation as a private army set up to keep minorities in their place as a pliant labor force on whose backs progress was being built.

There have been other, more substantive reforms instituted by Harry G. Caldwell, the scholarly, contemplative police chief, who worked his way up the ranks to be appointed head of the force in 1977 by

Gordan Kansas

Fred Hofheinz, the city's only liberal mayor. His reputation as the department's resident intellectual — he is working for a doctorate in police administration — does not sit well with some of the officers he commands. But despite internal resistance, the Chief is slowly modernizing and professionalizing his force, issuing new regulations governing the use of deadly force and highspeed chases. A sure sign that the bad old days of Houston's "Blue Meanies" are on the wane is that nearly a decade has passed since one of Mr. Caldwell's predecessors referred at a news conference to a local reporter as a "spic lover" because he was married to a Mexican-American woman. Back then, a favorite saying among officers was, "You may beat the rap, but you can't beat the ride [to the stationhouse]."

Mr. Caldwell is an ardent public defender of his department. But he is quick to acknowledge that there are some "bad apples" in the ranks, and quick to discharge those involved in untoward incidents; 19 officers have been dismissed, for brutality, bribery and suspicious deaths. He would like to weed out all of them, aides say, but is hampered by civil service

regulations. Instead, he has embarked on a counterstrategy: hire more professional personnel and, among them, more minorities. He has not come close to achieving his goals. Houston, the fastest-growing large city in the country, has outstripped many of its municipal services. With two million residents and 3,000 police officers, it has a police-to-citizen ratio less than half that of most Eastern cities.

One reason is money. The current starting salary for a Houston police officer is $16,000 a year, $2,000 less than in Los Angeles. Mr. Caldwell feels that to compete for talent in Houston's booming job market he must offer more, as much as $24,000 to start. The funds have simply not been forthcoming. He is also far short of his goal of 40 percent minority officers (roughly the percentage of the city's black and Hispanic population). Some, like Jim McConn, the incumbent Mayor who is seeking a second term, believe that Justice Department intervention in the cases of Mr. Torres and others like him and the resulting adverse publicity has handcuffed officers on the street and made it hard if not impossible to attract the kind of officers the force wants.

The Justice Department's position is that it has stepped in, reluctantly, where local grand juries and prosecutors seem to fear to tread. When the local District Attorney suggested recently that Justice might have waited to convene a Federal grand jury in the Jackson case until it had conducted its own inquiry, Federal prosecutors pointed out that no such inquiry had been begun, much less completed. Between 1966 and 1978, local grand juries declined to charge officers involved in the deaths of 155 citizens.

Except on the salary issue, Mayor McConn has been a steadfast supporter of Mr. Caldwell. But Mr. Castillo, who is running second to Mr. McConn in the polls, and who has vowed that if elected he will "see that the police are controlled," set off a minor shock wave last week by refusing to say that he would retain the chief. Whether he is elected or not, Mr. Caldwell, who has been both praised and criticized by the minority community, would have to pay even closer attention to their concerns. Because of a Justice Department order, Houston is choosing its City Council members by district rather than by the at-large system, that had placed one black and no Spanish-Americans on the council. Now, political observers say, there are likely to be two of each.

November 4, 1979

BOSTON: THE PROBLEM THAT WON'T GO AWAY

By Howard Husock

More than ever, Boston is a city of postcard beauty and charm. Tourists are impressed by its architecture and history. To those who troop its Freedom Trail, to the brigades of parents who park their station wagons at its dormitories, Boston looks like a city that blight has somehow bypassed. Even late on mid-week nights, downtown is alive, the subways relatively safe. The renovation of the Faneuil Hall area — three once-dilapidated 19th-century wholesale food markets — has created an urban buffet of open-air restaurants and shops on a revitalized waterfront.

Mayor Kevin H. White is fond of reminding voters they are lucky not to live in bankrupt Cleveland. He counted heavily on Boston's much-publicized

Howard Husock is a staff reporter for WGBH-TV, the public television station in Boston.

renaissance to propel him to a record fourth four-year term Nov. 6.

For many Bostonians, theirs *is* an exceptionally comfortable city: compact, cosmopolitan and — the word its boosters never tire of — livable. Such qualities are proving to be magnets to affluent young professionals, and after years of decline, the city's population has stabilized. "Boston," says a transplanted Detroiter, "is New York without all those people."

But if there are urban problems with which Boston is having greater success than other cities, there is one with which it is having significantly less. race. The tourist who makes a wrong turn can find himself in another Boston — a city of "turf," of neighborhoods to avoid, of hostile graffiti. This Boston, in contrast with most American cities today, is a place where blacks fear whites, where race affects the places **blacks can drive, work and play. This Boston is also a city in which many**

angry and bitter whites feel that they, too, are the victims of race, not its exploiters.

This is the Boston that surprised the nation with the violence of its reaction to school desegregation in September 1974, when the city became known for jeering mobs surrounding busloads of black students in white neighborhoods, and later for a 1976 Pulitzer Prize-winning photograph of a black businessman being beaten by white teen-agers wielding the pole end of an American flag on City Hall Plaza, and where a Federal court has continued to oversee day-to-day operations of the city's public schools to this day. And this Boston has not gone away.

□

The city's sixth year of school desegregation has proved every bit as tense as the first. In mid-September, busloads of black students bound for South Boston High were pelted with rocks and nails by youths in ski masks. On Sept. 28, three days before the visit of Pope John Paul II, a black high-school football player was shot in the neck as he huddled with his coach on the field of Charlestown High, in a neighborhood which, like South Boston, is poor, white and Irish. Fifteen-year-old Darryl Williams remains paralyzed from the neck down. Three Charlestown teen-agers charged with the crime are alleged to have sniped at the field from the roof of an adjacent housing project. And, although there is no hard evidence that the incident was racially motivated, Mayor White confided to reporters that his "gut" told him so.

The shooting put race back at the top of the municipal agenda, but conclaves of political and religious leaders did not head off a new round of incidents. In mid-October, violence occurred on four consecutive days. Black and white students battled on the steps of South Boston High as they waited to pass through metal detectors before entering school. At East Boston High, in a blue-collar

Italian neighborhood, there were no such devices and a white student was stabbed by a black with a hunting knife. Citywide, students walked out of schools. And on Oct. 19, white students pursued a black couple eating lunch on the Boston Common, while on the same day blacks pummeled a white speech therapist leaving work at a Roxbury high school.

Far from being the fallout from a single shooting, these incidents were merely the latest symptoms of an infection which, since the first major outbreak in 1974, has never fully abated. During the past two years — when the press and city officials generally referred to racial problems in the past tense — numerous incidents have occurred. For example:

■ In the Charlestown section, a group of black Bible students from Pennsylvania were beaten as they visited the Bunker Hill monument.

■ In Irish South Boston, the neighborhood that has most staunchly opposed "forced busing," a car belonging to a friend of a Jamaican woman, who was the first black to take part in a voluntary public-housing desegregation plan, was firebombed, despite 24-hour police surveillance of the area. She has since moved out.

■ In Italian East Boston, the public-housing apartment of a black family, which had lived there for 10 years, was firebombed. In the two previous years, some 20 black families had moved out of the neighborhood.

■ In the racially changing section of Hyde Park, a black homeowner had to wave a shotgun to fend off white youths coming through his front door at 2 A.M.

■ At a June hearing called by a city-wide parents organization, public school teachers told of regular harassment of racially mixed groups of students on field trips.

□

Racial violence in Boston is common enough for the police to have created a special unit — and euphemism — to handle "community disorders," involving black and white victims. The unit has been reluctant to release hard data on the number of crimes it has classified as racial. But earlier this year, in an attempt to demonstrate that the city's racial climate was improving, police figures were released showing 300 racial incidents in the first six months of 1978 as compared to roughly 200 in the first six months of this year — before the current round of high-profile problems.

Martin Walsh, the New England regional director of the United States Justice Department's Community Relations Service, says that conversations with his counterparts around the nation have convinced him that his office gets more reports of racial violence in Bos-

ton than in any of its other offices, including the South. (Whether there are more such incidents in the city than in other large cities is difficult to say with certainty. New York police, for instance, do not keep racial data per se and, according to the Washington-based Police Foundation, neither do most police departments.)

Boston's racial incidents do include black attacks on whites. The city's Roxbury-North Dorchester black ghetto is spoken of in the same fearful way many white Americans view such areas elsewhere. White motorists traveling on Blue Hill Avenue, the community's main drag, have frequently been stoned at red lights by gangs of black teenagers and, on occasion, have been pulled from their cars and beaten. Such beatings usually occur in the wake of highly publicized incidents. After the 1975 American flag incident at City Hall Plaza, for instance, a white motorist was pulled from his car near a black public housing project and fatally beaten on the head with bricks.

But in hard numbers, police data show that Boston blacks are far more often the victims of racial violence than whites. Although they make up only a fifth of Boston's 640,000 population — one of the smallest percentages of major cities — blacks accounted for 71 percent of the victims of racially motivated crimes in the first half of 1979.

However, statistics do not fully reveal the deep effect of racial tension on Boston. For whites, this tension is the catalyst for further deep, self-feeding resentment; for black Bostonians, it produces an all-embracing fear.

□

Says Melvin King, a black Massachusetts legislator who ran unsuccessfully for Mayor this year, "There is anger and frustration that is very, very pervasive, a feeling that people of color simply don't have access to large portions of this city."

As a rule, blacks avoid the city's beaches, located in South Boston, the scene of confrontations two summers ago when some blacks did try to use them. They also avoid Red Sox games — the Fenway Park bleachers have a reputation for problems. (With Jim Rice its only black player, the Red Sox organization has been rebuked by the Massachusetts Commission Against Discrimination for not hiring enough minority members in other capacities as well.) While whites have been attacked in Roxbury and Dorchester, blacks have been set upon in more than half of the city's neighborhoods. As a result, they (Continued)

Boston's racial violence is vividly captured in this 1976 Pulitzer Prize-winning photograph.

Stanley Forman/Boston Herald American

tend to spend much of their time in their own neighborhoods.

"That's the first thing people mention," says a Boston Redevelopment Authority employee who shows off the city's revitalized downtown area to visitors. "They all ask, 'Where are the blacks here?' "

Isaac Graves, a black aide to Massachusetts Senator Paul E. Tsongas, noted that his presence even in white neighborhoods not known for anti-black hostility — the affluent Beacon Hill or Back Bay sections — prompts comment. "People always ask, 'What are you doing here?' " he says. "I ask them if this is Johannesburg, do I need a pass?"

The Rev. Gilbert Thompson, a Boston minister who grew up in Philadelphia and attended school in Chicago and Boston, says, "Before I came to Boston, no one ever told me, 'Don't go here, don't drive there.' But I wasn't here long before the people in the church made me understand there are places black people don't go. 'You don't drive through South Boston, you don't use their beaches, you're taking your life in your hands if you do.' So I simply don't."

For the average black Bostonian, the effects of race on daily life are generally subtle, a series of small frustrations and changes of plan. Mary Crawford, in her mid-30's, came to Boston from North Carolina with her family as a girl. She lived first in public-housing projects, now in a two-family frame home like most others in racially changing Dorchester. Her block, however, has become predominantly black and she has not been harassed there. When her three children were younger, she used to take them to the city's beaches, but she has not done so for 10 years. "When I want to swim," she says, "I drive down to the Cape."

She does most of her shopping at grocery and discount stores outside of Dorchester, even though there are a number of stores at the opposite end of her own street.

"A girlfriend of mine got stones thrown at her on the other end of my street," she

Last month an unruly band of demonstrating white students flaunted their hostility from atop

a sculpture depicting white officers and black soldiers who served the Union during the Civil War.

Peter Southwick/Boston Herald American

Black students being bused from South Boston High in October after a flareup of fighting caused the school to cancel classes.

explains. That other end remains predominantly white. She also adjusted her route to her job when she worked at a Defense Department office on the fringe of the South Boston peninsula; she drove 20 minutes out of her way to and from work to avoid South Boston streets.

Once, when her car was being repaired, a white woman in her office offered her a ride home — and took the South Boston route. "A bunch of kids started jumping all over the car," she recalls. "That woman never did offer me a ride again."

Many white Bostonians no doubt adjust their travel habits to avoid black neighborhoods, but the problems are not really comparable. Whites can be seen daily driving and

walking on Roxbury streets. They own stores in black neighborhoods, they eat in black restaurants.

Blacks avoid much more than South Boston. Charlestown is considered flatly off-limits, East Boston and the North End risky. In addition, even if they avoid white neighborhoods, blacks have no guarantee of insulation from racial harassment. Although the most extreme examples of racial violence have occurred in the city's all-white enclaves, the most common occur in changing neighborhoods. In the first six months of 1978, there were 164 racial incidents in the changing sections of Dorchester and Hyde Park alone. Most of those, say police, were stonings of homes of black families by white youth

gangs. As a result of such incidents, middle-class black professionals have been turning down jobs in Boston.

Juan Evereteze was, until September, the top minority recruiter for the Boston public-school system, now under Federal court order to increase its percentages of black teachers and administrators. Evereteze traveled around the country, particularly the Deep South, looking for candidates.

"Inevitably," he says, "the question about the racial climate came up and I had to be candid. I told them there are parts of the city we must actively discourage them from living in. Many are from the heart of the South and this upsets them." Evereteze says that fully half of the job offers he made to blacks from other

states were rejected — 25 to 50 a year over the past four years.

Private employers, too, report problems attracting blacks. Edward Blanchard recruits middle- and upper-management personnel for the Gillette Company, one of the city's largest firms.

"People don't come right out and say they're not taking the job because of the racial situation," he says. "But employment agencies we deal with will say, after we've dealt with them for awhile, 'If you're talking about minorities, you have to realize you have a problem in Boston.' "

Isaac Graves, the aide to Senator Tsongas, has seen many black friends leave the city. "They might go to school here," he says, "but they don't want to settle here. They want to go somewhere they can be comfortable, where they can participate at all levels of the city, at all levels of the society.''

□

By and large, Boston proper — the city accounts for less than 25 percent of the population of the metropolitan area — is white, blue-collar, mainly Irish and Italian. The breadwinners, generally, are small tradesmen, policemen, firemen and utility-company employees; or, as is natural in a city with large local, state and Federal bureaucracies, they may be low-level government workers. These residents identify their neighborhoods by parish, outer boroughs which are psychologically, if not physically, distant from ''in town'' — even more distant from the universities and consulting firms of Cambridge.

There is widespread unemployment in many white neighborhoods, the result of a long-term decline in the number of unskilled and semiskilled jobs in the Boston area. The naval shipyards of Charlestown and South Boston are long-closed, their waterfronts nearly deserted. New jobs, such as those in New England's electronics industry, have not provided employment for those displaced by the closing of shipyards and shoe factories.

"There is a mismatch between the educational attainments of the population and the requirements of new jobs around here," says Sara Wermiel, an economic analyst for the Boston Redevelopment Authority. In a city poll in 1977 to determine neighborhood problems, some 20 percent of the residents of Charlestown questioned identified themselves as unemployed. It is a city with a white underclass.

It is from blue-collar Boston that the bulk of the city's racial violence has stemmed. In part, it is violence that can be traced to economics. Blacks and whites have been competing here for a dwindling number of blue-collar jobs, as well as for jobs with the police and fire departments.

Robert Coles, the Harvard

psychiatrist and sociologist, has looked at the city as closely as anyone, having written about its poor, black and white, and lived among its wealthy in suburban Concord.

"What people have to understand," says Coles, "is that before blacks even came to the city in any numbers, there were a hundred years of racism of another kind. Blacks are taking it on the chin because of a rage that was here before they were. And it stems from an arrogant upper class."

What Coles refers to is a history that dates back to the arrival of the first waves of Irish immigrants in a mid-19th-century Boston dominated by a Yankee gentry that appeared little different from the English manor lords whom the immigrants had left behind in Ireland. The middle and upper strata of society — those who work at jobs outside the docks and the mills — were hard for immigrants to crack. "No Irish Need Apply" was to Boston what the back of the bus

was to the South. The side effects of class tension permeate the city's history, no better symbolized than by the famous police strike of 1919, when the underpaid Irish police force sought to unionize and was replaced, in part, by volunteers from Harvard.

Mayor White likes to put current racial problems in the context of previous battles among Irish, Italians and Jews. "We are an ethnic city," White said in early September. "It was worth your life to be an Italian and go into Charlestown 20 years ago and it's worth your life, if you're Hispanic, to float into some black areas. We're an ethnic city with that type of consequence."

Paradoxically, blacks in Boston — unlike those in the South, who shared with poor whites roughly the same rung on the social ladder, as well as much of the same culture — have had an historic link with the city's upper class, dating to the abolition movement. The city was the scene of Civil

War draft riots when the Irish servants of upper-class abolitionists balked at fighting to free black slaves. Following the war, a small intellectual black community developed. Until the late 1950's, it remained relatively small and middle-class, somewhat aristocratic in style, as reflected by the black summer colony on Martha's Vineyard.

In the late 50's and early 60's, there came a large influx of blacks from the Southeast and Caribbean. Many believed they were coming to a liberal and affluent city, a center of education for their children. Their demands relating to education ultimately came down to one thing.

"Busing," says South Boston City Councilor Raymond L. Flynn, a populist-style racial moderate, "created a deep feeling of exploitation."

For a long time, elected officials declared that the buses would never roll. Many blue-collar whites, however, felt that blacks and the city's aristocracy — who are now mostly suburban liberals — were in league against them. And, says state legislator Mel King, "You cannot have leaders running around saying 'never' for years and years and not have it fester."

It was fully 11 years from the time a group of black parents approached the city's five-member School Committee in May 1963 — asking first for improved schools in black neighborhoods, then for recognition that the school system had been unintentionally segregated — until busing arrived in September 1974. In the interim, the issue came to dominate the city's politics. A generation of politicians was able to use antibusing pledges as a ticket to power — most notably Louise Day Hicks, who went from the School Committee to Congress. Indirectly, the issue put Mayor White in the office he still holds — he was first elected in 1967 as an alternative to Mrs. Hicks, whom he labeled an extremist. White has been about as liberal on the race issue as a mainstream white politician can be in Boston. He has opposed busing, but not aggressively; he has

Louise Day Hicks, a staunch opponent of busing, leads a 1975 demonstration. Her position on race carried her into Congress.

Eugene Richards/Magnum

In a racially changing area, neighbors argue over whose children can use the basketball court.

denounced racial violence, but only when it has reached dramatic and publicized levels.

White's racial strategy has so far kept him in office. Blacks, under the city's "at-large" voting system, possess sharply limited political influence. Since all public office-holders run citywide, minority candidates are virtually never elected. (The system was originally adopted by Yankee city fathers in part to dilute the power of a growing immigrant Irish vote.) With the advent of the busing issue, blacks became a made-to-order political punching bag: criticizing their demands guaranteed a white vote that was all most candidates needed. The situation was aggravated by the continuing absence of a strong black political leader able to build coalitions.

Mrs. Hicks and her allies were not, of course, able to deliver on their no-busing pledges. In fact, their own ac-

tions made busing inevitable: the Federal court found that, in order to avoid mixing the races in a city where white and black neighborhoods are often contiguous, the School Committee had adopted a complex assignment pattern which was designed to segregate.

Most white Bostonians, however, were acutely aware not of a wrong righted, but that those who did not live within the city limits were insulated from the busing order. Suburban liberals — many of whom have been outspoken busing proponents — came to be seen as contemporary abolitionists asking the city to bear the burden of integration. And suburban liberals have, indeed, made it clear that their sympathies are more with blacks than with blue-collar whites.

Suburban communities, for instance, have been the driving force behind the Metropolitan Council for Educational

Opportunity (METCO), by means of which more than 4,000 black students are bused to superior suburban schools. Begun in 1965, the program aimed to expose affluent whites to integrated education and to aid blacks then attending what was an unconstitutionally segregated Boston school system. From the outset, however, Boston whites viewed the program as one which simply favors blacks, a feeling even more pronounced today when METCO continues despite the Boston schools' desegregation. Under METCO, no white suburbanites are bused to black neighborhoods and no blue-collar whites are bused to the suburbs, although any black who arrives in the city today is eligible. Says a white from Charlestown, a policeman who attended Boston University: "I would see those METCO kids getting the chance to go to Wellesley or somewhere and

there I was in Charlestown." He and six other "townies" he knew managed to go to college only on hockey scholarships.

The liberals' feeling of antipathy toward the city's poor whites goes deeper than any one program. A Boston Public Library historian once declared: "There's a widespread feeling among the wealthy here that if blacks are poor, it's society's fault, but if whites are poor it's their fault." Among the affluent, it is acceptable at parties to scoff at the accents of working-class Bostonians. Ethnic stereotypes, unthinkable if applied to blacks, have not gone out of style in relation to whites.

Ultimately, it was a judge from suburban Wellesley — a friend of the Kennedy family named W. Arthur Garrity — who gave the hated "forced busing" order. Within a week after busing initiated the exchange of students between South Boston and Roxbury, a black Haitian bakery worker unwittingly cut through South Boston and came upon a mob that had gathered near South Boston High, perhaps to surround a school bus there. The bus had taken an alternate route, so the crowd, instead, attacked the Haitian, some wielding hockey sticks. French-speaking Yvon Jean-Louis died.

☐

Busing coincided with the beginning of racial change in large, lower-middle-class white sections of Hyde Park and Dorchester, touching off a white exodus and racial violence.

Robert Whitley, a Boston police sergeant who had moved from South Boston to Hyde Park, moved again in 1975 to a middle-class white suburb on the South Shore. "When you see two or three black families on a street, you get panicky," he says. "Maybe you stay. But then, with busing, your kids start going to parts of the city you worked to get away from. You say, 'The neighborhood is not as I remembered it.' You leave."

Those who leave sell their homes at rock-bottom prices.

Today, well-kept homes in "changing" sections of the city can be had for $30,000 or less. The market is depressed by a lack of white buyers willing to move in. Whites who sell, moreover, face high home prices and mortgage rates in the suburbs. Often they go deeply into debt, a condition that heightens their resentment of blacks. Police have regularly arrested teen-age whites who return to their old neighborhoods and, with white teen-agers still living there, throw rocks and bottles at homes bought by blacks.

Representative Brian Donnelly, who has always lived in Dorchester, represents most of Boston's racially changing neighborhoods and many of its white working-class suburbs. "The kids throwing the rocks hear it from their parents, all the racial stuff," he says. "Maybe they want to be a cop or a firefighter and they're told they can't get those jobs anymore or there's a quota. Among a lot of the people I represent, there is a very deep feeling that blacks are getting a better shake."

Robert Coles confirms this observation: "There is a great unfocused resentment among many whites in the city and blacks become the scapegoat."

There are, however, some optimistic notes in Boston's changing neighborhoods and the city at large. Block organizations of blacks and whites often come to the aid of black families whose homes are "rocked," in the parlance of white youth gangs. Such groups see violence as a threat to *all* residents — a physical threat for blacks, an economic threat for both whites and blacks.

However, such positive signs remain minuscule. Tolerant whites who stay in changing neighborhoods fear to aid blacks because they themselves could become targets for youth gangs. Neighborhood stabilization groups often feel their efforts are not supported by city government.

Indeed, in recent years, Boston officials have worked less at solving racial problems than at assuring themselves and the city that the troubles had passed. The tone of a pamphlet called "Living in Boston," published by the Mayor's office and aimed at newcomers to the city, is typical: "South Boston's image has been negatively affected in the past few years as opposition to court-mandated busing made headlines. Although there were some difficult times and feelings still run high on occasion, 'Southie' has much that makes it a desirable place to live."

Until recently, the city's social and religious leaders have skirted the race issue. Mainstream white politicians in the city have also preferred to avoid the race issue when possible. It no longer pays political dividends: It is hard to run for office on an antibusing platform when busing is a fact of life that has been upheld by the Supreme Court. But neither is it considered prudent to do anything that might appear overly pro-black. This fall, when Mayor White was endorsed by former Senator Edward Brooke, White distributed notices of the endorsement primarily in black neighborhoods. His opponent, however, made sure that leaflets drawing attention to it were distributed in South Boston.

☐

It would not be fair, however, to say that the city is doing nothing to deal with its racial dilemma. It has made a series of important beginnings. The police community-disorders unit has stationed plainclothesmen in front of homes plagued by attacks. Sgt. Francis Roache, the unit's head and himself a South Boston native, believes the city will eventually adjust to racial change. "It's a process of some very different groups getting accustomed to each other," he says. The unit has also been involved in a number of prosecutions, including one in which two East Boston men were given prison terms for a racially motivated firebombing.

Police Commissioner Joseph Jordan has announced that coping with racial attacks is his department's "highest priority." The police as well as State Attorney General Francis X. Bellotti have lobbied hard for passage of special state civil-rights legislation that would allow those suspected of crimes, such as the stoning of homes, to be charged with more than just vandalism. "We think more serious charges could have a deterrent effect," says Assistant Police Commissioner Sherwin Wexler.

The city's political response to racial tension has also been stepped up because of this fall's incidents. Mayor White, the day after his re-election, told reporters that assuring access for all citizens to all sections of the city would be the the first goal of his fourth term. Perhaps most significantly, the city's clergy, led by the Catholic Church, announced a campaign against racial violence, with priests and ministers working to cool things in their neighborhoods. It was, in fact, front-page news when Humberto Cardinal Medeiros, head of the Boston archdiocese, pledged that parish priests would speak out against racial violence on Sundays.

Other programs, not explicitly aimed at the race problem, could produce even greater benefits over the long run. For example, the Boston School Department, led by Supt. Robert Wood, former president of the University of Massachusetts, is set to open a major new Occupational Resource Center, in which students will be trained specifically for those jobs available in the Boston area.

But even when official Boston does address itself to the race problem, its efforts often seem counterproductive. In the wake of this fall's Charlestown shooting, there were a series of well-publicized meetings of black ministers, politicians and neighborhood leaders to discuss ways to improve the racial climate. The sessions produced little beyond statements denouncing racism and again left many white Bostonians with the feeling that they were being singled out for castigation by the liberals. The South Boston Information Center, the city's oldest anti-busing group, denounced the press for playing down black attacks on whites. Observed James M. Kelly, the group's president, "White people from South Boston and other neighborhoods in Boston are again being told they're to blame for all the problems of black people."

Boston awaits a leader or movement to rally a cross section of blacks and whites around issues of mutual self-interest — a leader who will say explicitly that tolerance can be economically beneficial for both whites and blacks in changing neighborhoods, that black use of South Boston beaches could mean that whites would feel more comfortable visiting the city's largest park and zoo in Roxbury. And so on. If Boston's racial problems are more obvious than elsewhere, so are the common economic problems of its whites and blacks. Public housing, white and black, is equally rundown; development of new industry, a common need.

But, as of today, Boston is still waiting and hating. ■

November 25, 1979

Suggested Reading

DRUGS

Etons, Ursula. *Angel Dusted: A Family's Nightmare*. New York: Macmillan, 1979.
Kaufman, Edward and Pauline Kaufman. *Family Therapy of Drug & Alcohol Abusers*. New York: Halsted Press, 1979.
Silver, Gary T. *The Dope Chronicles: Eighreen Fifty to Nineteen Fifty*. New York: Harper & Row, 1979.

THE MASS MEDIA & POLITICS

Saldich, Anne. *Electronic Democracy: Television's Impact on the American Political Process*. New York: Praeger, 1979.
Smith, Anthony. *Television & Political Life: Studies of Six European Countries*. New York: St. Martin's Press, 1979.
Twentieth Century Fund. *With the Nation Watching: Report of the Twentieth Century Task Force on Televised Presidential Debates*. Lexington, Mass: Lexington Books, 1979.

CHINA

Domes, Jurgen, ed. *Chinese Politics After Mao*. Short Hills, N.J.: Enslow Publishers, 1979.
Louis, Victor. *The Coming Decline of the Chinese Empire*. New York: Times Books, 1979.
Schaller, Michael. *The United States & China in the Twentieth Century*. New York: Oxford University Press, 1979.

WOMEN: THEIR CHANGING ROLES

Ardener, Shirley, ed. *Defining Females: The Nature of Women in Society*. New York, Halsted Press, 1979
Ehrenreich, Barbara and Deirdre English: *For Her Own Good: 150 Years of Expert's Advice to Women*. New York: Doubleday, 1979.
Kennedy, Susan E. *If All We Did Was to Weep at Home: A History of White Working-Class Women in America*. Bloomington, Ind.: Indiana University Press, 1979.

BLACK AFRICA

Adamu, M. *The Hausa Factor in West African History*. New York: Oxford University Press, 1979.
Carter, Gwendolen. *Southern Africa: The Continuing Crisis*. Bloomington, Ind.: University of Indiana Press, 1979.
Hargreaves, John D. *The End of Colonial Rule in West Africa: Essays in Contemporary History*. New York: Barnes & Noble, 1979.

VALUES AMERICANS LIVE BY

Fitzgerald, Francis. *America Revised*. Boston: Little, Brown, 1979.
Horowitz, Irving L. *Ideology & Utopia in the United States, 1956-1976*. New York: Oxford University Press, 1977.
Miller, Douglas T. and Marion Nowak. *The Fifties: The Way We Really Were*. New York: Doubleday, 1977.

CRIME AND JUSTICE

Allen, John. *Assault with a Deadly Weapon: Autobiography of a Street Criminal*. Diane H. Kelly & Phillip Heymann, ed. New York: Pantheon, 1977.
Congressional Quarterly Staff. *Crime & Justice: Trends & Directions*. Washington, D.C.: Congressional Quarterly, 1978.
Silbermann, Charles. *Criminal Violence-Criminal Justice: Criminals, Police, Courts & Prisons in America*. New York: Random House, 1978.

THE PRESIDENCY

Di Clerico, Robert. *The American President*. Englewood Cliffs, N.J.: Prentice-Hall, 1979.
George, Alexander. *Presidential Decisionmaking in Foreign Policy*. Boulder, Colo.: Westview Press, 1979.
Pious, Richard M. *The American Presidency*. New York: Basic Books, 1979.

POPULAR CULTURE

Logan, Joshua. *Movie Stars, Real People & Me*. New York: Delacorte, 1978.
Pichaske, David. *A Generation in Motion: Popular Music & Culture in the 1960's*. New York: Schirmer Books, 1979.
Skolnik, Peter L. et al. *Fads: America's Crazes, Fevers & Fancies From the 1890's to the 1970's*. New York: T. Y. Crowell, 1978.

SCIENCE IN THE TWENTIETH CENTURY

Keylin, Arleen, ed. *Science of the Times, 2: A New York Times Survey*. New York: Times Books, 1979.
Sagan, Carl. *Broca's Brain: Reflections on the Romance of Science*. New York: Random House, 1979
Sullivan, Walter. *Black Holes: The Edge of Space, the End of Time*. New York: Doubleday, 1979.

THE MIDDLE EAST

Bernstein, Burton. *Sinai: The Great & Terrible Wilderness*. New York: Viking Press, 1979.
Edens, David G. *Oil & Development in the Middle East*. New York: Praeger, 1979.
Kazziha, Walid W. *Palestine in the Arab Dilemma*. New York: Barnes & Noble, 1979.

THE CITIES

Geller, Evelyn, ed. *Saving America's Cities*. New York: H. W. Wilson, 1979.
Goldfield, David R. & Blaine A. Brownell. *Urban America: From Downtown to No Town*. Boston: Houghton Mifflin, 1979.
Rifkind, Carole. *Main Street: The Face of Urban America*. New York: Harper & Row, 1979.

Index